RUSSIAN TRANSLATIONS SERIES

(continued)

(continued on title facing page)

Historical Geotectonics
PRECAMBRIAN

Historical Geotectonics
PRECAMBRIAN

V.E. Khain
N.A. Bozhko

RUSSIAN TRANSLATIONS SERIES
116

A.A. BALKEMA/ROTTERDAM/BROOKFIELD/1996

Translation of : *Istoricheskaya geotektonika - Dokembrii.* Nedra Publishers, Moscow, Russia. Revised and updated by for the authors the English edition in 1994.

Translation Editor : Mr. P.M. Rao

Technical Editor : Dr. V.E. Khain

General Editor : Ms. Margaret Majithia

ISBN (Precambrian Volume) 90 5410 225 X
ISBN (Set) 90 5410 224 1

Distributed in USA and Canada by: A.A. Balkema Publishers, Old Post Road, Brookfield, VT 05036, USA.

Preface

This book is the translation of the first volume of the series "Historical Geotectonics" conceived by V.E. Khain and realised with collaboration of his former graduate students, (later Doctors) N.A. Bozhko, K.B. Seslavinsky and A.N. Balukhovsky. The second volume "Historical Geotectonics-Paleozoic" (authors V.E. Khain and K.B. Seslavinsky) has already been published in English in 1995. The third volume "Historical Geotectonics-Mesozoic and Cenozoic (authors V.E. Khain and A.N. Balukhovsky) is in press.

The present volume was originally published in Russian by Nedra Publishers in Moscow in 1988. Since then new important data has appeared in the literature. This necessitated a complete revision of the original text and therefore the English edition represents a fully reworked and updated version of the Russian edition.

We preserved, however, the original plan common to all the volumes of the series—the subdivision of the text into regional overviews with descriptions of factual data and global synthesis with interpretation of these data. The text of the first part of the book on Early Precambrian was written by V.E. Khain and the text of the second part on Late Precambrian by N.A. Bozhko. The conclusions belong to both authors.

May 1996

V.E. Khain
N.A. Bozhko

Contents

PART II
LATE PRECAMBRIAN

PART I
EARLY PRECAMBRIAN

1

Origin of Solar System and Formation of Planet Earth. Hadean (4.5–4.0 b.y.) Stage of Development

In order to better understand the course of structural evolution of the Earth's crust, a knowledge of the primordial state of our planet, i.e., the degree of its stratification, temperature distribution in its interior etc., is essential. Since this state is directly related to the method of evolution of the Earth, an acquaintance with modern concepts on the origin of the Earth and Solar System is necessary. These concepts are yet far from clear and largely contradictory but nevertheless impose certain restrictions in selecting different models of the subsequent development of the Earth.

1.1 PREHISTORY OF THE SOLAR SYSTEM

It has presently been established that the universe is twice older than the Solar System (in any case, than its planets!). Its age is 10–11 (up to 15 b.y.). A hot gaseous nebula, originally pure gas (mainly hydrogen), should have served as the primary material for the evolution of the Solar system. The fact that the material composition of this nebula during formation of the planet already contained all the chemical elements of Mendeleev's periodic table, including now extinct radioactive elements and isotopes, compels us to seek a mechanism which could have resulted in the generation of this diversity of atoms, i.e., a mechanism of the process of nucleogenesis. Such a mechanism, more accurately different mechanisms for different groups of elements, has been worked out by space chemists and briefly reviewed by Voitkevich (1983) but the question arises about the conditions under which these mechanisms took place. Let us examine two hypotheses in this context.

According to the first hypothesis (after Sobotovich, in: *Early History of the Earth*, 1976), the near-explosion of a Supernova star acted on the primordial interstellar gaseous nebula, which led to its emission of photons and plasma. As a result, a heterogeneous mixture of elements and their isotopes was formed, initiating the evolution of the Solar System. During this process there occurred intense condensation of matter, formation of dust particles and chondrules, i.e., congealed drops of silicates, separation of metals of the ferrous group and silicates and accretion of matter. In the next stage condensation led to central thickening, i.e., the Proto-Sun, and to formation of the planets of the Solar System. Accretion of the planet proceeded quite rapidly, in the course of 10–100 m.y. A far greater rate of accretion assumed by some researchers, is contradicted by the fact that this would have resulted in very intense heating of the planet (gravitational energy is converted into thermal energy) and its complete fusion with no chance of cooling until now; the contemporary thermal regime of the Earth contradicts such an assumption.

The second hypothesis of nucleogenesis (after Voitkevich, 1983) is based on the early formation of a hot and massive star of the type Wolf-Rae in the interior of which hydrogen-helium combustion occurred, followed by the conversion of helium into light elements. Later, as the core of the star became heated to 10^9 K, fast nuclei of helium (α-particles) were set free from the light nuclei and reacted with the remaining light nuclei and formed a group of elements from Mg and Si to Ca and Ti. Later, the ferrous group of elements was formed. This was followed by the formation of free neutrons in the interior of the stars and their 'seizure' by the already existing nuclei gave rise to heavy elements, including transuranium and numerous radioactive isotopes which rapidly became partially extinct. This process could have proceeded only in the interior of a major star, far greater than the present-day Sun's mass, probably its ancestor. Soon after the synthesis of heavy elements this Proto-Sun should have discarded its excess mass (as against the mass of the present-day Sun) as a result of explosion or gradual flow of matter into its equatorial zone. Thus was formed the Sun and the proto-planetary nebula around it in the form of a gaseous disc in the plane of the solar equator. The cooling of this gaseous nebula gave birth to molecules, chemical compounds and metallic phases. The gaseous nebula transformed into gas and dust.

In the next stage material differentiation began within the cooling hot gaseous nebula; gases were forced out to the periphery where the outer giant planets (Jupiter, Saturn etc.) arose subsequently. Differentiation was stimulated by shock waves originating from the Supernova star. There is also a hypothesis that the reason for such differentiation of the protoplanetary shell, apart from the lowering of temperature towards its periphery, was the pressure exerted by the rays of the primordial Sun and the effect of fast particles discharged by it. In either case condensation of the hot gaseous nebula and its differentiation either immediately preceded or even proceeded parallel to commencement of formation of the planet.

1.2 ACCRETION—HOMOGENEOUS OR HETEROGENEOUS?

Thus condensation gave way to the accumulation of proto-planetary matter, initially in relatively small, of the size of present-day asteroids, *planetesimals*, and later in the nuclei of present-day planets. Internal planets, i.e., planets of the Earth group whose nuclei arose close to the Sun, were formed by thickening of the high-temperature fraction. In the intermediate, asteroid belt, the low-temperature fraction had already contributed to accretion (accumulation), which ensured the presence of carbonaceous chondrites in the composition of asteroids. These chondrites are characterised by the presence of water-bearing silicates and complex organic matter of abiogenic origin. The outer planets, formed far away from the Sun, consisted of nearly unfractionated solar matter, mainly gases (predominantly hydrogen).

It was originally assumed that the accretion of the planets began immediately after differentiation of the hot gaseous nebula was complete. Correspondingly, planets should have formed from a relatively homogeneous, well-mixed matter (its accumulation accompanied by fractionation) of chondrite composition for planets of the Earth group. Differentiation commenced in the interior of protoplanets with separation into metallic, mainly ferrous, core and silicate mantle only as a result of heating caused by the impact of planetesimals, compaction of protoplanetary matter and heat liberation by radioactive elements and isotopes.[1] This represents the so-called model of homogeneous accretion. Some important geological aspects of the evolution of planets emanate from this model. Firstly, formation of the core should have been accompanied by very significant additional heating, which could have stimulated differentiation of the mantle with early liberation of the Earth's crust. Secondly, fusion with formation of a molten metallic layer should have commenced not at the centre of the Earth where extremely high pressure would have precluded such, but at relatively shallow depths, followed by the flow of heavy melt into the central zone. This flow, according to Elsasser (1963), did not proceed uniformly throughout the area of the layer but rather in a definite region under the influence of a swell caused by the tidal effect of the Moon, which was closer then. Thus arose the prerequisite for formation of global asymmetry of the Earth with its separation into oceanic (zone of subsidence of the 'iron core' towards the centre) and continental hemispheres. Thirdly, and most importantly, it is quite easy to assume that the process of differentiation of the primary homogeneous matter of the Earth did not cease in the early stage but has continued right to the present. Since

---

[1] This process is easy to represent if it is assumed that the Earth passed through a phase of total fusion. Such an assumption was indeed made at the beginning of this century by V. Goldschmidt who drew an analogy between formation of the Earth's shell and material separation into metal and slag in a blast furnace. It was subsequently shown, however, that the viscosity of silicate at high pressures prevailing in the interior of the Earth is so high that no large-scale mixing of matter and its separation into core and mantle of different composition could have occurred. A no less serious objection is the fact that the volume of the crust is too small to be regarded as a direct product of the fusion of the mantle as a whole.

this process should be accompanied by considerable heat liberation, gravitational differentiation of the matter of the Earth can be regarded as the most important factor of tectonics. This view has been adopted in various modern geotectonic concepts.

However, in recent decades the model of homogeneous accretion has come under serious criticism on the basis of new data on the structure and composition of planets of the Earth group and meteorites. This model cannot explain such facts as the systematic variation in density of planets of the Earth group with increase in distance from the Sun, difference in density of the Moon and Mercury which are nearly equal in size, presence of an iron core in Mercury which constitutes about two-thirds of its mass (such a core cannot have been the product of differentiation of a planet of chondrite composition[2]), division of meteorites and asteroids based on composition into a few significantly differing groups etc. The chondrite (average) composition of the Earth has also proved unacceptable; a mixture of 15% iron meteorites, 45% common and 40% carbonaceous chondrites is more likely (Barsukov, 1981).

All these contradictions can be overcome by adopting a model of *heterogeneous accretion*, according to which the accumulation of protoplanetary matter into a planet occurred fairly simultaneous with condensation and fractionation of this matter as its temperature fell. Correspondingly, after formation of the core of the Earth the highest temperature fraction, i.e., the metallic phase corresponding to iron meteorites, should have condensed and accumulated in the first instance. Fractionation could have continued even in the period of mantle formation and been reflected in a change of its composition along the vertical. As a result of this, the lower mantle could still contain a significant admixture of Fe-Ni particles (iron and ironstone meteorites) and Al-Ca silicates. The main portion of the mantle probably corresponds in composition to chondrites and is represented predominantly by Fe-Mg silicates and the upper mantle by carbonaceous chondrites enriched with volatiles, including complex hydrocarbons and other organic compounds (Barsukov, 1981; Galimov et al., 1982). Degassing of the upper mantle could have been responsible for formation of the atmosphere and hydrosphere and could have played an important role in the evolution of organic life on the Earth (Galimov et al., 1982).

The most detailed model of formation of the Earth based on the concept of heterogeneous accretion was put forward by Sobotovich (Rudnik and Sobotovich, 1984; Sobotovich, 1984). In this model (Fig. 1) the origin of crust (protocrust) is also 'inscribed' in the accretionary process as its concluding phase. The model further assumes that the matter of the Supernova star

---

[2] According to Vinogradov (1962), this is also relevant to the core of the Earth to some extent.

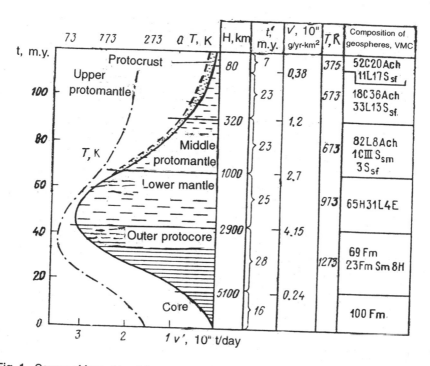

Fig. 1. Course of formation of the protogeospheres of the Earth (after Sobotovich, 1984).

Abscissa, bottom—rate of accrotion (v); top—surface temperature of the Earth in Kelvin degrees (T, K). Ordinate—duration of accretion (t). H—depth of level of corresponding protogeosphere; t'—approximate period of its formation; v'—specific rate of accretion.

The composition of each protogeosphere has been calculated in terms of virtual meteorite constituents (VMC); Ach—achondrites; C, CIII, L, H and E—chondrites (carbonaceous in general, carbonaceous of the third type, low-iron, high-iron and enstatitic respectively); Fm—ferruginous; Fm-Sm—ironstone; S_{sf} and S_{sm}—felsic and mafic constituents of the matter of the Supernova star.

enriched with radioactive elements played a significant part in forming the original protocrust, as evidently explained by the high content of such elements in the crust. The origin of the molten outer core is associated with the heat-insulating effect of the lower mantle under conditions of overall heating of the Earth.

At present, most specialists, geochemists (space chemists) in particular, support the model of heterogeneous accretion. Nevertheless, some (Ringwood et al., 1979) do respect the model of homogeneous accretion. According to Ringwood (1979), the hypothesis of heterogeneous accretion does not explain the absence of a distinct vertical zoning in the chemical composition of the mantle nor the presence of large quantities of light elements in the core or siderophile and volatile elements in the mantle. In the model of homogeneous accretion as modified by him, planetesimals varied considerably in their composition (from iron meteorites to carbonaceous chondrites) due to fractionation during condensation and were mixed up in the process of accretion as a result of gravitational disturbance of their orbits.

Besides these two extreme viewpoints on accretion, there are also intermediate models. Particular mention should be made of the concept of *partial heterogeneous accretion* described by Voitkevich (1983) and Kuskov and Khitarov (1982). It is assumed in this model that only the internal core of the Earth was formed directly by accumulation of metallic and sulphide-metallic phases of protoplanetary matter. The outer core and mantle were primordially characterised by gradual reduction of the content of these phases upwards and were distinctly separated only in the course of further differentiation of the matter of the Earth over a prolonged period. This view, in our opinion, is most plausible. It should be emphasised that, according to the latest data, differentiation of the matter of the Earth commenced quite rapidly following its accretion. This is supported in particular by the isotope structure of noble gases (Tolstikhin, 1991). Furthermore, from these concepts even the inner core of the Earth cannot be regarded as constant in volume and composition. The inner core could have enlarged and been supplemented with Fe and Ni due to differentiation of the outer core, this process more importantly bearing an exothermal character.

Thus the following premises can be regarded as relatively well established:

1) The primordial matter for the build-up of the Solar System was a hot gaseous nebula whose matter was either ejected by a massive star, i.e., the Proto-Sun or was the result of a mixture of interstellar matter at the early stage of development of our galaxy and plasma, i.e., a product of the explosion of a Supernova star.

2) Cooling of the nebula plus the effect of solar radiation or shock wave generated by the Supernova explosion led to fractionation and condensation of matter contained in the nebula with the generation of diverse size sand particles and planetesimals (right up to $n \times 100$ km, according to Safronov, 1969).

3) Accumulation of this matter occurring partly simultaneous with condensation and possibly partly after its completion, led to the formation of planets differing in ratio of gases and solid matter (outer and inner planets) and in ratio of metallic (and sulphide-metallic) and silicate phases.

4) Stratification of the Earth into its main shells, i.e., inner and outer core and mantle, arose in part even in the stage of accretion and probably partly after its completion during differentiation of the newly formed planet. But even according to the model of homogeneous accretion, the core could not have formed later than 0.4 to 0.5 b.y. after the conclusion of accretion (Ringwood, 1979), perhaps far earlier.

5) Accretion of the Earth was accompanied by significant heating of its interior and surface, especially in the case of late formation of the core. The temperature of the interior could have reached 1800 to 2000 K or more and of the surface a few hundred to 1000 K.

1.3 HADEAN (4.5–4.0 b.y.) STAGE OF DEVELOPMENT OF THE EARTH—LUNAR OR VENUSIAN STAGE?

So far, rocks older than 4 b.y. are not known with certitude in the composition of the Earth's crust; there is only information about detrital zircons aged 4.3–4.2 b.y. detected in Western Australia. This denotes that the fairly prolonged period between completion of formation of our planet (4.6 b.y.) and the beginning of its documented history (4 b.y.) is a matter of conjecture based on general concepts and planetological analogies. These conjectures are restricted to some extent by the fact that even at the beginning of the subsequent stage, the Earth possessed a sialic crust, atmosphere and hydrosphere, while the nature of geological processes (endogenous and exogenous) did not differ essentially from those observed in the Phanerozoic aeon.

Pavlov (1922) was the first to suggest that the Earth's landscape in this period of its evolution resembled the present-day lunar landscape since it was still devoid of atmosphere and hydrosphere, underwent intense meteoritic bombardment, was covered by innumerable craters and its surface was a vast lifeless desert. This concept, later developed by Muratov (1975) and especially by Pavlovsky and Glukhovsky (1982), has been confirmed time and again. Firstly, it was established that the Moon actually experienced intense meteoritic bombardment in the period 4.2–3.8 b.y. ago, resulting in the formation of numerous large craters. Secondly, space photographs have enabled detection on the surface of Precambrian shields, especially on the Aldan and Canadian shields, of a large number of ring-shaped structures, which supporters of the lunar stage do not fail to explain as resulting from meteoritic bombardment. In fact, there is no wholly reliable proof of the antiquity of these formations since the age of the rocks constituting them has not been established with certitude and, in any case, ages older than 2.8 b.y. are not known for them.

In recent years, with the commencement of studies on Venus, which by all accounts is most proximate to the Earth in size and developmental features of the planets of the Solar System, an alternative view has been proposed regarding the nature of the earliest stage of the Earth's evolution, based on an analogy not with the Moon, but with Venus. This implies that the Earth should have been enveloped in a dense and massive oxygen-free atmosphere (carbon dioxide prevails exclusively in the composition of the Venusian atmosphere), its surface should have attained a very high temperature (of the order of 500°C) and pressure on this surface should have been significantly greater than at present. Such a view was expressed in particular by Shuldiner (1982), who recognised the need to explain the evolution of rocks of granulite facies of metamorphism requiring for their formation high temperatures and pressures even in the Early Archaean under conditions when the crust was still thin. According to him, this was possible from the existence of high temperatures of a few hundred degrees (300 ± 100) and pressures of 15 MPa on the surface of the Earth. Although this presumption is quite debatable, as shall be seen below, the Venusian model nevertheless is no less, perhaps even more probable than the lunar model for the earliest stage of development of the Earth since the degassing of our planet at this stage, especially under the influence of meteoritic bombardment, should have proceeded very intensely while the dimension of the Earth, contrary to that of the Moon, helped it to hold the products of this degassing, including water vapour.

Irrespective of the model adopted, intense meteoritic bombardment in the interval 4.2–3.8 b.y. ago remains a valid conclusion as its remnants are traceable even on Venus. The geological consequences of this bombardment, as pointed out by Goodwin (in: *Early History of the Earth*, 1976), Barsukov (1981) and others, could be highly significant. Under conditions of a markedly warmed upper mantle enriched with volatiles, it could have promoted formation of major centres of fusion, i.e., formation of whole magmatic basins. The products of this fusion could be basalts or andesite-basalts filling basins of the type of lunar seas while anorthosites and norites, like rocks on lunar continents, could have crystallised at the base of basalt covers. The primordial crust of the Earth could have formed in this manner but its relics have yet to be detected and possibly have not survived, barring the xenoliths in the oldest gneisses, in particular in Minnesota (USA) (see Sec. 2.1.1).

2

Early Archaean (4.0–3.5 b.y.): Oldest Sialic Crust of the Earth and Problems of Its Formation

In the 1950s, when radiogeochronometry recorded its first successes, a search began for the oldest rocks on the Earth. Attention was primarily devoted to the Archaean greenstone belts since it was quite naturally assumed that the oldest crust of the Earth is of basaltic composition. Greenstone belts were regarded as nuclei of the crystallisation of the primordial Earth's crust, whose chemical weathering, erosion and metamorphism of erosion products gave rise to a sialic crust (M. Wilson, C.R. Anhaeusser, E.V. Pavlovsky and M.S. Markov). This assumption was apparently confirmed by data for Canada, South Africa and India where granites and gneisses forming together with greenstone belts of the region, subsequently called granite-greenstone regions, have yielded much younger age values than rocks of greenstone belts. However, doubts about the universality of such a conclusion, especially in relation to India as well as Canada, were expressed even in the 1960s (Salop, 1966; Laz'ko, 1966). In the early 1970s, Canadian data gave rise to the concept of 'grey gneisses' as rocks constituting the basement of greenstone belts everywhere. Subsequently, analogues of grey gneisses of the Canadian Shield were detected in many ancient continental shields and it is these that have recorded the oldest radiometric dates of 3.5–4.0 b.y. Thus the view arose that the oldest crust of the Earth represented a complex of 'grey gneisses', i.e., that this crust has a sialic although not wholly granitic (keeping in view the normal, potash-soda granites) composition. In fact, some scientists, especially in Russia and the Ukraine, regard granulites, which in some cases give proximate radiometric age values, and not grey gneisses usually metamorphosed not beyond amphibolite facies, as the oldest rocks. Early Archaean age has also been established for some greenstone belts (for example, the Isua belt of Greenland) which complicates the problem of identifying the oldest crust. We have also to reckon with the fact that in most regions grey gneisses represent the oldest formations. Hence in our review of Early Archaean rocks, we have mainly concentrated on the distribution of complexes of this type. However, we shall also deal with their possible analogues of highly metamorphosed and melanocratic type.

But first, what are these grey gneisses? Judging from the published data, they represent an extremely generalised concept. Apart from purely external characteristics (grey colour and gneissic appearance), the characteristic features of formations of grey gneisses are a composition corresponding on average to andesite, dacite, tonalite, trondhjemite, plagiogranite, distinct predominance of sodium over potassium, low content of Rb, U, Th, Ti, F, Zr, Nb, Ba and B, enrichment with Ni, Cr and V compared to normal granitoids and very low $^{87}Sr/^{86}Sr$ ratio (0.699–0.701). According to their origin, the rocks of the complex are intrusive, extrusive, pyroclastic and sometimes also sedimentary. Quite often, the main criterion for classifying the rocks in the complex of grey gneisses is the fact that they constitute the basement of greenstone belts or that their age exceeds 3.2–3.3 b.y. Of late, the terms tonalite-trondhjemite-granodiorite (TTG) or tonalite-trondhjemite-dacite (TTD) complexes are more often used to describe complexes of this type.

2.1 REGIONAL REVIEW

Our regional review will commence with data on the northern row of cratons, from the North American to the Sino-Korean, followed by the southern row, from the South American to the Australian and the Antarctic.

2.1.1 North American Craton

Within this craton, complexes of grey gneisses have been established in four regions: Great Slave Lake and Lake Superior provinces (Canadian Shield), eastern Labrador peninsula and south-western Greenland.

2.1.1.1 *Great Slave Lake province.* The oldest rocks known on the Earth, aged 3.96 ± 0.003 b.y., were recently established for the Akasta gneisses emerging in the westernmost part of this Epi-Archaean block of the Canadian Shield by the ion microprobe U/Pb method on zircon. Protoliths of these gneisses are of tonalite-granite composition. Their model neodymium age of 4.1 b.y. points to the possibility of the initial rocks of these gneisses having been basalts enriched with neodymium (Bowring et al., 1989).

2.1.1.2 *Lake Superior province.* The main region of development of reliably dated oldest rocks lies in the extreme south-western part of the province, predominantly in Minnesota (USA). Here these rocks are represented by orthogneisses of tonalite-trondhjemite composition with inclusions of amphibolites and pyroxene-granulites, which may be regarded as relics of much older greenstone belts. The age of these gneisses determined by the Rb-Sr method on whole rock and the U-Pb method on zircon is 3.7–3.5 b.y.

The rocks have been metamorphosed to amphibolite and partly granulite facies.

Apart from this region, two others of granulite development are known in this province, which might be older than the main portion of its green-stone belts. One lies north-west of the province near the Nelson-Thompson River fault zone, separating it from the adjoining Early Proterozoic Churchill province. The rocks are older than 3 b.y. Another, more extensive region is situated north-east, in Ungava peninsula, where several large outcrops of granulites have been detected. Such outcrops are also present in the central part of the province.

2.1.1.3 *Eastern Labrador*[1]. Much of the Labrador peninsula falls in the Early Proterozoic Churchill province (Canadian Shield) but an elongated narrow (up to 100 km) band of Archaean gneisses was long ago identi-fied along the eastern coast of the peninsula and delineated as the Nain province. Radiometric age determination of these rocks revealed them to be an analogue of the oldest formations on the opposite coast of south-western Greenland. Here Uivak orthogneisses play the main role in the composition of the Archaean complex. These orthogneisses represent fairly typical grey gneisses of quartz-feldspathic tonalite composition with inclu-sions of metasedimentary and metamafic volcanic rocks. The isochron age of these gneisses determined by the Rb-Sr method on whole rock is 3.622 ± 0.072 b.y. with primary $^{87}Sr/^{86}Sr = 0.7014$. Later, the ion microprobe U-Pb method on zircon yielded a slightly older age for these gneisses, i.e., 3.863 ± 0.012 b.y. which by the Sm-Nd method was 3.732 ± 0.006 b.y. Thus the age of 3.62 b.y. was recognised as the period of migmatisa-tion. Moreover, volcanics of mafic, tholeiite and komatiite-basalt composi-tion transformed into amphibolites as well as marbles, calc-silicate rocks and banded ferruginous quartzites have been detected in the form of inclu-sions in the Uivak gneisses; these were combined into the Nulliak complex aged 3.8 b.y. which, as shall be seen later, is wholly similar to the Isua complex of south-western Greenland. It has been suggested that the base-ment of these supracrustal formations could be 3.9 to 3.85 b.y. old (Nut-man et al., 1989). Besides these gneisses (Uivak I), porphyric granodiorite gneisses (Uivak II) are also known here; both groups are intruded by Saglek amphibolite dykes. The supracrustal rocks of Upernavik, metasedimentary, from lutites to quartzites and marbles and also metaigneous, i.e., amphibo-lites, lenses of ultramafics and stratified bodies of mafic composition, are much younger. Quartz-feldspathic gneisses aged 3.133 ± 0.156 b.y. with $^{87}Sr/^{86}Sr = 0.7063$ are younger still and are regarded as a product of remo-bilisation of the Uivak gneisses. The entire complex is intensely deformed

---

[1] After Collerson, Jesseau and Bridgwater, in: *Early History of the Earth,* 1976.

right up to the formation of low-angle thrusts, possibly even nappes. Even after deformation and metamorphism to amphibolite (at places to granulite) facies, intrusion of anorthosite and later reactivation of the Uivak gneisses and repeated granulite metamorphism occurred. In the southern part of the province Archaean gneisses reveal the presence of synforms of amphibolites (metavolcanics), lenses of ultramafics and metasedimentary rocks; these are evidently analogues of greenstone belts. Intrusions of granite and pegmatite along the surface of overthrusts and intrusions of diabasic dykes represent the oldest pre-Proterozoic events.

The geological section of eastern Labrador and the course of evolution of this region in the Early Archaean correspond in general features to those established for south-western Greenland, thus confirming the structural entity of these parts of the North American Craton in the past before formation of the Labrador Sea basin at the end of the Cretaceous period.

2.1.1.4 *South-western Greenland* [2]. Archaean rocks are exposed on the western coast of Greenland in a belt about 600 km long and about 200 km wide (Fig. 2). Eastward they are concealed under ice but reappear on the eastern coast. These rocks have been best studied in the west, in the region of Godthåb and Godthåb Fjord, where they have attracted particular attention from the time the oldest supracrustal rocks of the Isua complex were established in their composition. These rocks form the mantle of a gneiss dome made up of Amitsoq gneisses on which they were evidently tectonically superposed and, in turn, surrounded by outcrops of the same gneisses. The following rocks constitute the Isua complex: 1) amphibolites, judging from the relics of pillow structure, formed after mafic volcanics (most predominant variety); 2) siliceous and carbonate-containing biotite-muscovite shales with bands consisting of fragments (?) of metarhyolites; 3) banded ferruginous quartzites; 4) intraformational conglomerates with quartzite pebbles; 5) carbonate-siliceous and carbonate rocks; and 6) metalutite rocks. Metamorphism of the complex is mainly to low-pressure amphibolite facies. The Rb-Sr isochron age of the ferruginous quartzites was determined as 3.76 ± 0.07 b.y. Amitsoq gneisses tectonically in contact with Isua rocks are quartz-feldspathic calc-alkaline in composition and the Rb-Sr isochron age on whole rock determined as 3.65–3.55 b.y., i.e., actually very close to the age of the Isua complex. Nevertheless, they are usually regarded as somewhat younger although it is assumed that a part of the gneisses may be even older. Determinations by Pb-Pb, U-Pb and Sm-Nd methods gave age values of 3.8–3.7 b.y. for the Isua complex and Amitsoq gneisses, thus confirming their antiquity (Yan Zhin-shin, 1983). Magmatic zircons aged

---

[2] After Myers and Allaart, in: *Early History of the* Earth, 1976.

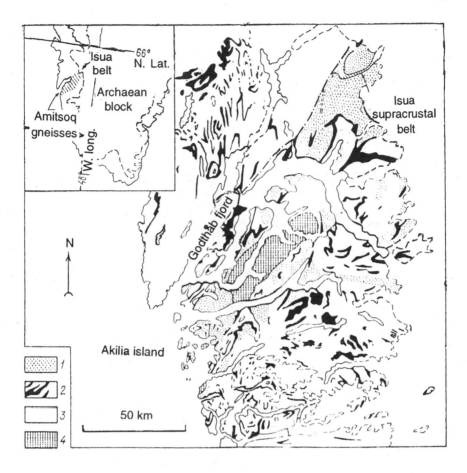

Fig. 2. Schematic geological map of the region from the Isua supracrustal belt to Akilia Island in south-western Greenland (after Baadsgaard et al., 1984).

1—Amitsoq gneisses; 2—Malene supracrustal rocks, their equivalents and anorthosites; 3—Late Archaean Nuk gneisses; 4—Korkut granites.

3.822 ± 0.005 b.y. (or even older) were detected in the gneisses by the ion microprobe method.

The Isua complex and Amitsoq gneisses are intersected by Ameralik metadolerite dykes. Gneisses, petrographically and geochemically similar to the Amitsoq gneisses are also widespread in the region but differ from the latter in containing dolerite dykes; they have been named the Nuk gneisses. Their Rb-Sr isochron age is actually very much less at 3.04 ± 0.005 b.y. The supracrustal amphibolite Malene formation with partings of quartz para-gneisses (whose relationship to the dyke series is debatable—the former could be even older) and bedded, stratimorphic anorthosite intrusives fall between the Ameralik dykes and Nuk gneisses.

The structure of the Archaean complex of south-western Greenland is highly complex up to the development of nappes. This is a consequence of the prolonged and complex history of deformation and metamorphism that commenced right in the period of origin of the Amitsoq gneisses (3.75 b.y.) and ended 2.85 b.y. ago before the intrusion of Korkut granites (2.53–2.59 b.y.).

Thus we see in Greenland a complex combination of two types of Early Archaean formations: on the one hand typical grey gneisses (Amitsoq) and on the other, rocks characteristic of greenstone belts (Isua, Akilia and Malene). Their age relationship is still not very clear but the latter are positively older than the former.

2.1.2 East European Craton

Formations of the grey gneiss type have been identified in both the shields of this craton—the Baltic and the Ukrainian; attempts are being made to delineate them on the basis of drilling data even in the buried salients of the basement of the Russian Platform. It must be acknowledged, however, that the Early Archaean age of these rocks, especially in the Baltic Shield, has not been established with certitude (Semikhatov et al., 1991).

2.1.2.1 *Baltic Shield*[3]. Outcrops of the complex, comparable to grey gneisses, are known mainly in the eastern, much older part of the shield, i.e., in the Kola peninsula (Murmansk block, central part of Kola block), in Karelia, including southern Belomorie (White Sea) and in eastern Finland. Grey gneisses evidently form the basement of numerous greenstone belts here but stratigraphic contacts with the latter have nowhere been detected. The rocks of the complex form a typical volcanoplutonic association. In the Kola peninsula this association is represented by metagranodiorite porphyries, metaporphyrites and gneisses after them, as well as granitoids which have intrusive contacts with all of them. Among the volcanics, andesite-basalts and andesites predominate; basalts and dacites are encountered less often. Intrusives correspond to quartz diorites, tonalites, plagiogranites and granodiorites belonging to the sodium series of normal alkalinity. In Karelia this complex is made up of gneissic plagiogranites, in the west bipyroxene-crystalline schists and hypersthene-biotite plagiogneisses and in Belomorie tonalites with relict granulite associations. In Finland, the oldest rocks are 'cantagranites', i.e., plagiogranites, trondhjemites and more mafic rocks of the typical sodium series. All these rocks appear in the Baltic Shield as domal structures with a predominantly north-eastern or meridional strike.

---

[3] After Bel'kov, Batieva and Vetrin, in: *Oldest Granitoids of the USSR* (1981).

The radiometric age of the oldest granitoids and volcanics of the Kola peninsula determined by the U-Pb isochron method is $3.13 \pm 0.1(0.25)$ b.y. and by the Sm-Nd method on apatite roughly 3.2 b.y. Quite recently, gneisses aged 3.7–3.5 b.y. (U-Pb method on zircon; Lobach-Zhuchenko et al., 1989) were detected in eastern Karelia in the Vodlozero region. The oldest greenstone complex was also identified in this region, dated 3.391 ± 0.076 b.y. (after komatiites; Lobach-Zhuchenko et al., 1989).

2.1.2.2 *Ukrainian Shield*[4]. The oldest rocks of this shield belong to the Aulian series of the Middle Dnepr block whose minimum isochron age determined by the U-Pb method on zircon from ultramafics is 3.7 ± 0.2 b.y. The composition of this rock series consists of two formations: lower (crystalline schists and gneisses) consisting of biotite-hornblende plagiogneisses, hornblende schists and amphibolites, and the upper (crystalline schists and amphibolites) consisting of biotite-hornblende and hornblende schists and amphibolites. The volcanosedimentary genesis of these rocks is quite evident. The series additionally comprises bands of ferruginous and aluminous rocks, metamafic and metaultramafic bodies and intersecting tonalite intrusions. The incomplete thickness of the series (basement not known) has been put at 5 km. Metamorphism corresponds to high-temperature amphibolite and low-temperature granulite facies.

It should be pointed out that rocks belonging to the Aulian series are encountered only as relics among fields of younger Archaean granitoids and hence a stratigraphic relationship between the rocks constituting the series remains indistinct; thus it would be more correct to term the Aulian series a complex. Ultramafic and mafic rocks of this complex, aged 3.91 ± 0.22 b.y. as determined by the Sm-Nd method, are intruded by plagiogneisses aged 3.19 ± 0.14 b.y. (same method; Zhuravlev et al., 1987). Shcherbak and colleagues (1986) subsequently reported somewhat different values: 3.45 ± 0.008, 3.441 ± 0.01, 3.633 ± 0.016 and 3.64 ± 0.011 b.y. for inclusions of pyroxenites and 3.609 ± 0.005 b.y. for their host gneisses (ion microprobe U-Pb method on zircon).

Somewhat more eastward, in Orekhovo-Pavlograd zone, between the Middle Dnepr and Azov blocks of the same shield, Early Archaean or proximate age values have been obtained (by U-Pb method on zircon) for tonalites and intrusive enderbites of the Novopavlovka complex, viz., 3.65 ± 0.05 and 3.4 ± 0.1 b.y. respectively (Semikhatov et al., 1991).

In the Bug block (south-western part of the Ukrainian Shield) a similar old age has been suggested for the Dnestr-Bug series metamorphosed to granulite facies of moderate pressures and formed by bipyroxene-crystalline schists and biotite-hypersthene plagiogneisses with complex-bedded bodies

---

[4] After G.I. Kalyayev (pers. comm.).

of enderbites. Some researchers (Laz'ko et al., 1982) regard this granulite complex as the oldest in the shield, older than the Aulian series. But radiometric data (U-Pb dates on heterogeneous zircons from enderbites) puts the age of metamorphism as Late Archaean (2.75 ± 0.05 b.y.) although the interpretation of analogous ages for premetamorphic zircons suggests these rocks to be older than 3 b.y. and places this series parallel to the Konka-Verkhovtsevo (Shcherbak et al., 1986). Evidently its comparison with the Aulian cannot be ruled out.

2.1.3 Siberian Craton

The basement complex of the Aldan Shield, recently distinguished (Kitsul et al., 1979), can be tentatively placed among analogues of the Early Archaean grey gneiss complex in this craton. This complex underlies the well-known Yengra series and emerges in the central part of the shield, notably in the Aldan-Timpton megadome (Drugova et al., 1984). According to researchers, the composition of this complex contains quite diverse formations: granite-gneisses formed after metavolcanics or orthogneisses, enderbites and charnockites constituting the core of domes, amphibolites, mafic and ultramafic crystalline schists and pyroxene-biotite plagiogneisses. The Pb-Pb thermoion emission method on first-generation zircon gave the age of these rocks as more than 3.3 b.y. (Drugova et al., 1984). More reliable data was obtained for the Kurul'ta series of the same Aldan complex, i.e., 3.54 ± 0.37 b.y. by the Sm-Nd method on mafic crystalline schists and 3.45 ± 0.05 b.y. by the U-Pb method on metamorphogenic zircon (Semikhatov et al., 1991). Analogous ages were recorded for similar formations in the Omolon median massif of the Verkhoyansk-Kolyma Late Mesozoic fold zone, probably representing a detached mass of the Siberian Craton, i.e., a microcontinent. The corresponding rocks are represented by hypersthene, amphibole-biotite, biotite plagiogneisses, interstratified with bipyroxene-crystalline schists and aluminous gneisses. Bibikova (in: *Precambrian Geology,* 1984) determined the age of first-generation zircon in a sample of plagiogneiss by the U-Pb method as 3.4 ± 0.15 b.y. and of second-generation zircon as about 2.8 b.y. The second epoch is regarded as one of granulite metamorphism. The presence of rocks of the same oldest age is also suggested in the lower sections of another massif of the Verkhoyansk-Kolyma province, i.e., the Okhotsk.

Still another exposed region on the periphery of the Siberian Craton where there are serious grounds for assuming the presence of a grey gneiss complex is K a n b l o c k (southern Yenisei range). Hypersthene-garnet gneisses of tonalite composition and charnockites yielded the age of 4.1 ± 0.17 b.y. (Pb-Pb method) and 4.05 b.y. (U-Pb method) on zircon (Musatov et al., 1983). These rocks appear in the cores of domes but such an antiquity has not been confirmed so far by more modern methods. The Sharyzhalgai series of the south-western frame of the Siberian Craton can also be regarded as a member of Early Archaean formations but such an age has not yet been confirmed by radiometry.

And lastly, it has been opined (*Oldest Granitoids of the USSR,* 1981) that the deeply metamorphosed rocks of the Anabar massif reveal an affinity with this complex. These rocks are hypersthene plagiogneisses (enderbitoids according to Rosen, 1989), mafic bipyroxene-crystalline schists, garnetiferous with bands of marbles, calciphyres, magnetite-quartzites and hypersthene granites (charnockites); lenses of metaultramafics are also encountered. Volcanic island arc origin has been suggested for the bulk of the rocks constituting the Anabar complex. Their premetamorphic age, as concluded by E.V. Bibikova, is not less than 3.32 b.y. The age of granulite metamorphism and of basic deformation is 2.7 ± 0.1 b.y. and of diaphthoresis of amphibolite facies 1.97 ± 0.02 b.y.

2.1.4 Sino-Korean Craton

The oldest rocks found in northern China occur in the Qianxi (Zhinin) complex in the eastern part of Hebei province. This complex, metamorphosed to granulite or amphibolite facies, was quite recently subjected to geochronological study by the U-Pb method on zircon. The oldest zircons (3.67–3.65 b.y.) were encountered in quartzites in Qianang region associated with amphibolites aged 3.5 b.y. by the Sm-Nd isochron method. These rocks form inclusions in grey gneisses aged 2.9 b.y. where leptynites of dacite composition with a Pb-Pb age of 3.3 b.y. for zircon were also encountered. All these formations constitute the western frame of the granite-gneiss dome, the core of which contains much younger Archaean granitoids. In addition to amphibolites and felsic gneisses, more likely metavolcanics, the oldest supracrustal complex also comprises marbles, calc-silicate rocks and banded ferruginous quartzites (Liu et al., 1990; Kaiyi et al., 1990).

2.1.5 South American Craton

This craton is divided into two major Precambrian shields—Guyana and Brazilian—by the latitudinal depression of the Amazon River valley, which is filled with a Phanerozoic cover. The latter shield is often differentiated into the Central and East Brazilian shields. Originally the two must have formed a common Amazon Craton. Early Archaean formations are evidently present within both shields but as yet have not been well studied; thus reliability of their identification requires confirmation.

2.1.5.1 *Guyana Shield.* The most reliable analogue of the grey gneiss complex within this shield is the Imataca complex emerging in the extreme north in the eastern part of Venezuela along the right bank of the Orinoco

River. This complex is well known for its rich iron ore deposits. In the region of Caroni River the radiometric age for this complex has been determined at over 3.4 b.y. with recurrent metamorphism at around 2.7 b.y. (Late Archaean) and about 2 b.y. (Early Proterozoic). It is usually assumed that granulite metamorphism of the complex is confined to the latter age; in other words, the Imataca complex in this respect belongs to the Early Proterozoic granulite belts of the Stanovoy type (see Sec. 4.1.3). The Imataca complex at present includes felsic, intermediate and subordinate amounts of ultramafic granulites, orthogneisses, paragneisses, amphibolites, ferruginous rocks, dolomite-marbles and anorthosites. The felsic granulites and gneisses of amphibolite facies encountered were derived from felsic, predominantly calc-alkaline magmatic rocks of the continental type.

2.1.5.2 *Brazilian Shield.* This shield, like the Guyana, has not yet been thoroughly studied. Presumably, the grey gneiss complex here may be quite extensively developed but at present only formations outcropping on the surface of the Goyas median massif between the two branches of the Central Brazilian Baikalides within the same-named state and in the western part of Minas Gerais state are regarded as most positively belonging to this complex (Danni et al., 1982). These formations serve as the basement of the Archaean greenstone belts (see Sec. 3.1.5.2) with ages up to 3.2 b.y. (Rb-Sr method) for gneisses and 2.9–2.8 b.y. (by the same method) for granites at $^{87}Sr/^{86}Sr = 0.701$.

In petrographic composition these rocks represent quite monotonous biotite, less frequently hornblende-quartz-feldspathic orthogneisses of tonalite or granodiorite origin. Metamorphism is to upper amphibolite and at places granulite grade. Banded paragneisses, quartzites, jaspilites, aluminous schists and gneisses, amphibolites, and also lenses and bands of altered pyroxenites and peridotites are found in subordinate quantities. The complex contains granitoid plutons (see age values given above) which are partly older and partly younger than the greenstone belts. The Early Archaean granite-gneiss complex is overthrust from the east by a complex of ultramafics and mafics metamorphosed to granulite facies. The latter consist (from the bottom upwards) of peridotites and banded alternation of peridotites and norites. The age of these rocks has not been established but can be regarded as Archaean from indirect evidence.

The presence of Early Archaean formations has also been proven in the north-eastern part of the Sao Francisco eocraton in which a series of outcrops of rocks has been established in a band of meridional strike at 40°W long. The age of these rocks has been determined as 3.4–3.5 b.y. These in particular are charnockites of the Mutuipe region (3.5 b.y.; $^{87}Sr/^{86}Sr = 0.7015$) and tonalite granodiorites of the Bona Vista region (3.433 ± 0.061 b.y.; $^{87}Sr/^{86}Sr = 0.7006$). The age of granodiorites at four more points in Bahia state has been determined as about 3.1–3.16 b.y. In petrochemical characteristics these are proximate to the rocks of the Iron Ore Quadrangle in Minas Gerais state in the extreme south of the Sao Francisco eocraton.

Rocks aged 3.2–3.4 b.y. are also present among the oldest Xingu complex of the Central Brazilian (Guapore) eocraton, which underwent intense reworking in the Transamazonian epoch (about 2 b.y.).

2.1.6 African Craton

Early Archaean formations have been most reliably studied in the southern part of this craton—in Swaziland, Zimbabwe and on Madagascar island.

2.1.6.1 *Swaziland (Kaapvaal) eocraton.* Here an ancient gneiss complex, probably representing the basement of greenstone belts, can be regarded as Early Archaean. The ancient gneiss complex of Swaziland covers a bimodal association of tonalite and trondhjemite, biotite-hornblende, quartz-feldspathic gneisses and amphibolites, quartz-feldspathic paragneisses, metabasalts from tholeiite to komatiite composition and metasedimentary rocks (quartzites, jaspilites and marbles) in extremely subordinate amounts.

This complex was recently subjected to extremely detailed geochronological study (ion microprobe U-Pb method on zircons) and the following ages were recorded: for the main part of the bimodal gneiss suite 3.644 ± 0.004 b.y., migmatite-gneisses 3.59–3.50 b.y., metamorphic suite, i.e., the oldest greenstone belt of Dvalile, consisting of remnants of komatiites, basalts, basic tuffs and quartzites including ferruginous 3.565 ± 0.01 and 3.423 ± 0.03 b.y., tonalite-gneisses of Tsavela 3.458 b.y. and for the metamorphic Mhondo Valley suite (amphibolites, gneisses, ferruginous quartzites etc.) 3.48 ± 0.01 b.y. The age of much younger intrusive anorthosites and gabbroids has not been determined but quartz-monzonites as also the intruding gneiss complex have been dated 3.3–3.0 b.y. (Kröner et al., 1989).

2.1.6.2 *Zimbabwe eocraton.* Ancient gneisses constituting the base of greenstone belts and up to 3.5–3.6 b.y. in age have been delineated in this eocraton. The superposition of the oldest (Sebakwian) generation of greenstone belts (see Sec. 3.1.6.2) on granite-gneisses and migmatites containing inclusions of ultramafic rocks which have undergone granulite facies of metamorphism has been established in the region of Shurugwi (Selukwe). The granite-gneiss complex has been intruded by pegmatite of K-Ar age 3.48 ± 0.07 b.y. for muscovite. Unconformable superposition of Sebakwe formations on tonalite-gneisses of Rb-Sr age 3.57 ± 0.12 b.y. has been noted in the region of Zvislavne (Shabani) and the age of gneisses in the region of Mashaba has been determined as about 3.6 b.y. by the Rb-Sr method.

In the Limpopo mobile belt extending between Zimbabwe and Kaapvaal eocratons, the oldest rocks are the Sand River gneisses, 3.79 b.y., as determined by the Rb-Sr and U-Th-Pb methods on whole rock and on zircons. According to Barton (1983), these gneisses could have arisen after greywackes formed due to erosion of an island arc or volcanoplutonic belt.

2.1.6.3 *Madagascar*[5]. The oldest complex on Madagascar is most reliably delineated in the north in the Mazura and Antongil massifs for which the age has been determined as over 3.28 b.y. An age of 3.190 ± 0.244 b.y. has also been determined by the Rb-Sr method (Vachette, 1979). The complex is represented by plagiogranite-granodiorite-tonalite-gneisses and plagiomigmatites unconformably overlain by melanocratic and high-alumina paragneisses. Analogues of a lower complex outcrop also in central and southern Madagascar where its sequence is overlain by the Androgen 'system' with a predominance of leptynites with which calc-magnesian and high-alumina paragneisses, orthoamphibolites and quartzites are associated at places. This system houses potash (!) granite plutons 3.02 b.y. old.

Besides South Africa, the presence of a grey gneiss basement is probable in several regions of central and western Africa. These are: K a s a i   m a s s i f with .Luani and Kanda-Kanda granite-gneisses of tonalite-granodiorite composition in amphibolite facies of metamorphism with an Rb-Sr age of 3.49 ± 0.17 and 3.98 ± 0.165 b.y. for pegmatites (Delhal and Ledent, 1973); W e s t e r n   N i l e   m a s s i f with charnockites 3.605 b.y. old, Gangu galenas 3.38 b.y., Bomu gneisses 3.31 b.y. and Nzangi tonalite-orthogneisses 3.417 ± 0.019 b.y. (Lavreau, 1982); and K i b a l i   m a s s i f with Upper Ituri paragneisses 3.349 b.y. old (Pb-Pb; Lavreau, 1982). The ancient basement emerges on the surface probably also in the L e o n-L i b e r i a n   m a s s i f where, however, ages older than 3.1 b.y. have not been recorded. More northward, in the In-Ouzzal spur of Ahaggar massif, charnockites with an Rb-Sr age of 3.5–3.1 b.y. and in Reguibat massif Gallaman gneisses 3.27 ± 0.347 b.y. are known (Cahen and Snelling, 1984). All these age values need to be defined more precisely, however, as does the data on the composition and distribution of corresponding rocks.

2.1.7 Indian Craton

The oldest rocks of India according to radiometric data (Sm-Nd isochron on whole rock) are the tonalite-gneisses exposed in the core of the S i n g h b h u m   d o m e in the north-eastern part of the South Indian

---

[5] After Zabrodin (Explanatory Note to *International Tectonic Map of the World,* 1988).

Shield (East Indian Shield) whose ages were placed at 3.775 ± 0.089 b.y. These ages were subsequently revised downwards as follows: 3.378 ± 0.098 b.y. (Pb-Pb method), 3.43 b.y. ($^{40}Ar/^{39}Ar$ method) for the same gneisses, 3.369 ± 0.057 b.y. (Pb-Pb method) for trondhjemites and 3.3 b.y. (the same method) and 3.163 ± 0.126 b.y. (Pb-Pb method) for granites in which they are contained as inclusions (Sengupta et al., 1991). Their Rb-Sr ages are 3.2 ± 0.085 b.y. ($^{87}Sr/^{86}Sr = 0.7018$) and 3.18 ± 0.054 b.y. ($^{87}Sr/^{86}Sr = 0.703$). Singhbhum gneisses contain inclusions of mafic, calc-magnesian and metalutite rocks forming an 'old metamorphic group'.

Archaean rocks are widely distributed in the southern region of the shield (Karnataka). Here Gorur-Hassan tonalite-trondhjemite gneisses of the following age are known: 3.315 ± 0.054 b.y. by the Rb-Sr isochron method on whole rock ($^{87}Sr/^{86}Sr = 0.7006$), 3.305 ± 0.013 b.y. by the Pb-Pb isochron method on zircon. Gorur-Hassan gneisses contain inclusions of metavolcanosedimentary rocks of various size, right up to greenstone belts that could be mapped. In the latter, delineated into the Sargur group, komatiites, basalts, jaspilites, chert, quartzites, micaschists, marbles and also dunite and anorthosite bodies are developed. Gneisses show with these bodies distinct intrusive contacts. At the same time, some authors assume that these intrusive relations are secondary, being the outcome of remobilisation of the Presargur gneiss basement 3.0–2.9 b.y. ago. All these gneisses (according to this interpretation) belong to the complex of 'Peninsular Gneisses' extensively developed in southern India.

Peninsular gneisses in the southernmost part of India have been transformed from amphibolite to granulite facies. The age of granulite (and high-temperature amphibolite) metamorphism in southernmost India has been determined as 2.6 b.y., i.e., end of the Archaean. Before this, at the verge of 3.0–2.9 b.y., these rocks were metamorphosed into amphibolite facies (Raith et al., 1982).

The presence of an Early Archaean grey gneiss type formation can be assumed also in north-western India in Rajasthan. Here a sequence of banded gneissic complex has formed, containing, as in many other regions, inclusions of metasedimentary rocks including quartzites and marbles. These gneisses (called Mewar gneisses here) in Udaipur are 3.5 b.y. old (Sm-Nd and Pb isochron methods; McDougall et al., 1983; Deb and Sarkar, 1990). In a more recent work (Gopalan et al., 1990) this age value was more accurately determined as 3.31 ± 0.07 b.y. (Sm-Nd isochron).

2.1.8 Australian Craton

The Precambrian of Australia is distinguished by a fairly high degree of radiogeochronological coverage (see review of Page et al., in: *Precambrian Geology*, 1984). According to available data, the presence of an Early

Archaean complex, an analogue of grey gneiss, can be assumed in Western Australia in the Pilbara and Yilgarn blocks.

2.1.8.1 *Pilbara block* represents an immense dome consisting of a family of granite-gneiss domes surrounded and separated by greenstone belts. Gneisses with metavolcanic inclusions (reference to a very old basaltoid crust) housing intrusive bodies of tonalites, granodiorites and granites emerge in the core of these domes. The most recent ages are 3.429 ± 0.013, 3.443 ± 0.01 and 3.448 ± 0.008 b.y. for banded tonalite-trondhjemite gneisses in the Mount Edgar batholith by the U-Pb method on zircon (Williams and Collins, 1990). The Sr model age of the same rocks was determined as 3.65 b.y., which permits their placement in the Early Archaean. Granites of the same batholith are aged 3.31 b.y. The gneisses were probably formed from bimodal volcanics and their age is proximate to that of similar bimodal volcanics of the lower section of the Warrawoona greenstone complex, i.e., 3.452 ± 0.016 b.y. (see Sec. 3.1.8.1). The period of intrusion of granites (tonalites and granodiorites) from 3.5–2.85 b.y. and culminating between 3.3 and 2.95 b.y. coincides with the age of metamorphism (amphibolite facies) and deformation, which led to the evolution of the contemporary internal structure of the Pilbara block.

2.1.8.2 *Yilgarn block.* This is the largest block in the basement of the Australian Craton. One of the oldest gneiss complexes on the Earth occurs in its western part. It is made up of biotite-adamellite gneisses with Sm-Nd age 3.63 b.y. interlayered quartzites with beds of conglomerates. Clastic zircons with concordant U-Pb ages of 4.1 to 4.2 (!) b.y. have been detected in the quartzites of Mount Narryer. One grain recorded an age of 4.276 ± 0.006 b.y. (!!). Later, zircons of the same age were detected in metaconglomerates of Jack Hills 60 km to the north-east; the age values were determined by the method of successive evaporation of lead (Kober et al., 1989). Gneisses and quartzites have been intruded by granite-gneisses aged 3.51 b.y. The age of deformation and granulite metamorphism of this complex has been evaluated by the Rb-Sr method on whole rock as 3.348 ± 0.043 b.y. and of superposed tectonothermal processes as 2.6 b.y. (Myers and Williams, 1985). The Western Gneiss complex is regarded as a basement on which greenstone belts formed in the Late Archaean (see Sec. 3.1.8.2).

2.1.9 Antarctic Craton

The oldest rocks of the Antartic Craton were identified and studied by Soviet geologists (Grikurov et al., 1976) in Enderby Land and Prince Charles Hills. Ravich and Kamenev (1975) distinguished in these regions two complexes—Napier and Rayner. The latter probably represents products of reworking of the former. The N a p i e r c o m p l e x is made up of enderbites and mesoperthitic granite-gneisses belonging to the highest temperature subfacies of granulite facies of relatively low pressure. This complex recorded a Pb isochron age of 4.0 ± 0.2 b.y., which has generally been confirmed by the age of tonalite-orthogneiss determined as 3.93 b.y. (Black et al., 1986).

The subsequent history of the complex, as revealed by isotope data of various methods, proved to be extremely complex (Kamenev, 1991). Two peaks of granulite metamorphism aged about 3.5 and 3.0 b.y. (charnockite-gneisses and paragneisses) have been recorded. The stage of retrogressive metamorphism to granulite and amphibolite facies and intrusions of porphyroblastic granitoids and their pegmatites appeared at the level of 2.5–2.4 b.y. while dykes and sills of mafic rocks including anorthosites were manifest at the level of 2.4–2.35 and 1.1–1.0 b.y. Pegmatites along the boundaries with the much younger Rayner complex (see Sec. 3.1.9) are much younger, about 0.58 b.y. (K-Ar method).

2.2 TECTONIC REGIME OF EARLY ARCHAEAN AND FORMATION OF PROTOCONTINENTAL CRUST

The foregoing review reveals that, firstly, formations older than 3.2–3.5 b.y. were established by now in all the continents and, secondly, their composition is quite variegated. Rocks of the grey gneiss type are present in this complex almost everywhere and predominate quantitatively, thus justifying the collective name for this oldest complex. In most cases they are intrusive and their composition corresponds to tonalites, trondhjemites, granodiorites and quartz-diorites. In this context the term grey gneiss has begun to be substituted of late by the more accurate term tonalite-trondhjemite-granodiorite (TTC) complex, sometimes termed the tonalite-trondhjemite-dacite complex. Amphibolites formed after volcanics of mafic composition evidently fall in the second place. Metavolcanics of intermediate and felsic composition are probably less developed and the association bears for the most part a bimodal character, but not everywhere. According to prevalence, paragneisses along terrigenous (particularly lutitic) rocks and also jaspilites are inferior to magmatites; quartzites, clastic as well as chemogenic (after cherts) and also carbonate rocks are less frequent. This list points to the diversity of magmatic and sedimentary rocks right in that remote geological period. The isotope age of grey gneiss complexes has been determined as 3.9–3.2 b.y. but some of these dates were obtained by the Rb-Sr method. A comparison of these results with those of Sm-Nd and U-Pb methods points to the possibility of significant rejuvenation. Therefore, it is wholly possible (Kratz and Lobach-Zhuchenko came to a similar conclusion, in: *Oldest Granitoids of the USSR,* 1981) that the true age of the grey gneiss complexes is not less than 3.5 b.y., if not older.

The relations of grey gneisses with the earliest generations of greenstone belts are highly variable although they are usually regarded as the basement of the latter. The isotope dates are very proximate for grey gneisses and the oldest rocks of greenstone belts in south-western Greenland (Akilia, Isua and Amitsoq), Swaziland, Zimbabwe, in northern Zaire and in Western Australia (Pilbara). Stratigraphic relations in Greenland, the eastern Labrador peninsula (Uivak gneisses and rocks of Upernavik group) and the southern Indian peninsula are not clear. Sometimes a close interlacing of rocks characteristic of grey gneiss complexes and greenstone belts is noticed or melanocratic apovolcanic variety predominates among grey gneiss complexes (Aulian complex of central Dnepr region). Inclusions of mafic and ultramafic and also sedimentary rocks, characteristic of greenstone belts, are encountered in grey gneisses. A definite impression is created that soda-granitoid magma intruded into a much earlier crust, consisting of mafic and subordinate amounts of intermediate and felsic volcanics and also the products of their weathering, erosion and redeposition (presence of aluminous schists and quartzites). Since, however, isotope dates reflect predominantly the age of the grey gneisses per se, this crust may be older than 3.9 b.y. In any case, true grey gneisses should not be regarded as the oldest crust of the Earth and the present state of our knowledge justifies more the old conclusion of J. Hutton (1795) that the history of the Earth reveals traces of neither its beginning nor any sign of its ending.

If, however, it is recognised that the accumulation of mafic (and possibly ultramafic) volcanics and sediments of the type of greywackes and lutites preceded the formation of grey gneisses, the question arises: How did this accumulation become widespread? In spite of the proximity of composition and nature of these formations to Archaean greenstone belts, there is no basis for assuming that they were also confined to definite quasi-linear zones. Taking into consideration the discussion at the end of the preceding chapter, it can more readily be said that the accumulation of this actually oldest (simatic and not sialic) crust and products of its redeposition occurred in basins arising at the site of large craters formed as a result of impact. These very sequences subsequently underwent Na-metasomatism and granitisation, resulting in the formation of grey gneiss protocontinental crust. The magma giving rise to grey gneisses (i.e., orthogneisses), judging from Sr and Nd isotope ratios, was of mantle origin. As pointed out by Bogatikov and colleagues (1980), the high content of water and volatiles in the Early Archaean mantle and also the high values of heat flow may have promoted the direct melting of magma of such composition from the mantle.

Moorbath (1977) and Cloud and Glaessner (1982) expressed a similar view. However, it has been demonstrated in many works that such a process might have led to the formation of rocks only perceptibly more basic than dacites. Also, the fractional crystallisation of basalt magma cannot serve as an effective mechanism for the formation of a fairly large volume of rocks of the type TTG. Contemporary researchers therefore conclude that the TTG association represents a product of partial fusion of basalt crust under the condition of its subduction (Drummond and Defant, 1990; Rapp et al., 1991), obduction (DeWit et al., 1992) or simply subsidence (Kröner and Layer, 1992). A high geothermal gradient and/or heating of the young crust would have promoted the foregoing process.

Thus by the end of the Early Archaean, 'islands' of sialic crust had already arisen at the site of craters of the Hadean stage. Judging from the already identified distribution of the grey gneiss complex and the highly probable even more extensive development of it on the surface and especially deep inside the Early Precambrian crust, these sialic islands (protocontinents according to Goodwin, 1985; 'nuclears' according to Glukhovsky and Pavlovsky, 1984) were quite extensive and abundant, at least within the contemporary platforms and their miogeosynclinal margins. As suggested by Goodwin (1985), Glikson (1983), Naqvi (1982), Gintov (1978), Moralev (1986) and others, it is wholly possible that the major dome structures noticed in the contemporary structural plan of ancient shields, well expressed in relief in aerial photographs, may represent relics or, in any case, successors of these sialic islands of the end of the Early Archaean. The most probable examples of such dome structures are: Aldan-Timpton dome (Aldan Shield), Ungava dome (Canadian Shield), Singhbhum dome (India) and Pilbara dome (Western Australia). It can further be assumed that the protosialic islands formed the nuclei of Early Archaean lithospheric plates beneath which oceanic crust began to subduct, giving rise to the next generation of grey gneisses (TTG), subsequently generating these nuclei. This process should have expanded further in the Middle Archaean, as demonstrated by the example of Slave province in the Canadian Shield (Kusky, 1990).

A question was posed at the beginning of this chapter about the relationship between grey gneisses and granulite complexes. The above discussion points out that gradual lateral transitions are noticed from the former complexes to the latter in several regions; this is true of Lake Superior province (Canadian Shield), Labrador peninsula, Greenland, Indian peninsula, Western Australia and Brazil. In some regions the probable analogues of grey gneisses have been wholly metamorphosed to granulite facies (for example, within the Siberian or Antarctic cratons). This suggests that with increasing depth, grey gneiss complexes have usually metamorphosed to amphibolite facies and may have transformed into granulite facies. It is important to know when the granulite metamorphism occurred. In many cases it has been seen to have occurred already in the subsequent stage of development of the crust, for example in the southern part of India only at the end of the

Archaean (but this could have been repeated metamorphism; Friend, 1984). It is highly possible that Rb-Sr ages of the oldest complexes may correspond to the epochs of granulite metamorphism. However, on the example of the south-western part of Greenland, Kan salient (Yenisei range), Limpopo belt (South Africa), western belt of Yilgarn block and Napier complex of Enderby Land, we come to the conclusion of an extremely early manifestation of granulite metamorphism, within the Early and Middle Archaean, i.e., from 3.7 (Greenland) to 3.1 (Antarctica) b.y. Evidently, the protocontinental crust had already by this time attained a thickness of 25 to 30 km at places, providing a pressure of up to 800 MPa (the high level of heat flow generated the required temperature).

Thus already in the Early Archaean, lateral differentiation of the Earth's crust had begun with its differentiation into regions of thickened crust of the subcontinental type, and others of thinned oceanic type, as also vertical differentiation with separation into granulite (basic granulite), diorite (grey gneiss) and volcanosedimentary (in the intervals between domes) layers. In the Early Archaean the hydrosphere (as judged from the deposition of sediments and pillow basalts of the Isua series in an aqueous medium) and atmosphere already existed. However, water reservoirs on the surface of the Early Archaean Earth should have been flat and small, somewhat deepening in the basins between the growing protocontinental nuclei. Water was devoid of dissolved oxygen and sulphates but contained chlorides. A very low concentration of hydrogen ions compared to the present level (pH = 7 against 7.5 to 8.5) prevented precipitation of carbonates from the solution (Monin, 1983). The temperature was probably higher than at present, which promoted the evolution of primordial life, as propounded by Shidlowski (1984) and Cloud (1983) based on carbon isotope analysis of the sediments of the Isua series.

In so far as the Early Archaean atmosphere is concerned, it differed significantly from the present atmosphere and resembled that of Venus, i.e., consisted mainly of CO_2 and water vapour with admixtures of CO, CH_4, NH_3, N_2, H_2S and H_2. The Earth revolved far more rapidly (duration of day was about 5 h) while the greater proximity of the Moon to the Earth caused powerful tides. Solar radiation was extremely intense (Monin, 1983).

The role of the mechanism of plate tectonics in the Early Archaean continues to be an open question. It has been supported by Borukaev (1985) and Monin (1983) but opposed by others (Kröner and Layer, 1992). The division of the upper solid shell of the Earth into lithosphere and asthenosphere probably occurred already in the preceding stage since fusion centres arose in the asthenosphere. But the viscosity of the lithosphere should have differed significantly from the present viscosity and approached that of the material of the present-day asthenosphere in stable regions. Correspondingly, the lithosphere could not have separated into rigid, internally undeformed plates, which is the main feature of plate tectonics. This, however, does not exclude far more powerful convection in the mantle than known today. The effect of meteorite and asteroid impacts in the interval between 4.2 and 3.8 b.y. probably determined the localisation of ascending mantle currents and subsequently even the growth of sialic nuclei above them; in the intermediate depressions, however, compensatory descending convective currents, which later became ascending, should have existed. Thus arose the conditions for manifestation of plate tectonics, i.e., spreading and subduction, and ultimately for the transition from plume tectonics to plate tectonics.

3

Middle and Late Archaean (3.5–2.5 b.y.): Granite-Greenstone Terranes and Granulite Belts. Formation of Mature Continental Crust

As seen in the preceding chapter, already by the commencement of the Middle Archaean, a protocontinental, grey gneiss crust had formed over significant expanses of future continents. It should have corresponded in physical properties to a diorite layer, which has been identified by some geophysicists and geologists, but there are paramount grounds for assuming that intense deformation, probably caused by convective mixing during and at the end of the Archaean, may have led to a complex tectonic interleaving of this primordial sialic crust with the crust formed in the Archaean. This mixing and interleaving could not have been a general phenomenon, however, because many sections of ancient shields have preserved large regions within which g r e e n s t o n e b e l t s prevail. These belts have preserved a relatively simple, although highly compressed internal structure and were subjected only to a relatively weak metamorphism. These belts represent the most characteristic type of structural members of the Archaean (Ar$_{2-3}$) crust although they are encountered all through the Early Precambrian—from Early Archaean (Isua!) to Early Proterozoic. Greenstone belts are divided and fringed by more extensive fields of granites and gneisses, i.e., a second type of Archaean structure among which salients of the grey gneiss basement are present along with much younger granitoids. Granite-gneiss fields together with greenstone belts form g r a n i t e-g r e e n s t o n e terranes occupying much of the area of the shields, probably not less than two-thirds. The remaining area of the shields (and massifs) was covered by a third type of Archaean structure, g r a n u l i t e (granulite-gneiss) b e l t s, arising perhaps already in the Early Archaean and continuing to grow in the Proterozoic. A regional review of these types of structures follows.

3.1 REGIONAL REVIEW

3.1.1 North American Craton

3.1.1.1 *Superior (Lake Superior) and Slave (Great Slave Lake) provinces.* Within the Canadian Shield lies one of the largest granite-greenstone provinces of the world, delineated as the Lake Superior province. It is overlain in the south-west by the Phanerozoic cover of the Great Plains.

Greenstone belts extend through Lake Superior province in a near-latitudinal, east-north-easterly direction. They alternate with belts consisting

Fig. 3. Greenstone belts of Lake Superior province of the Canadian Shield (after Condie, 1981, modified).

1—Post-Archaean formations; 2—highly metamorphosed supracrustal rocks; 3—greenstone belts; 4—paragneisses; 5—gneissic complexes and granites; 6—suggested boundary of province; 7—dykes.

mainly of paragneisses and granites (Fig. 3). Several such belts have been encountered in a cross-section of the province; belts of the western and central parts of the province have been set off from those of the eastern part by the oblique-transverse Kapuskasing zone of transgression faults. Some belts run several hundred kilometres in length and tens or even several hundred kilometres in width. The largest greenstone belt is Abitibi, which is 800 km long and 200 km wide; its extent reaches 1300 km with its western continuation into the Wawa (Michipikoten) belt.

Greenstone belts of the Superior province were formed in the Late Archaean in the interval from 3000–2950 to 2760–2660 m.y. The presence of a much older, probably sialic (tonalitic gneisses) basement, older than 3.0 b.y., may be assumed only in the northern and southern parts of the province. The major part of the greenstone belts arose on oceanic crust, however. The oldest of them, situated in the northern part, were formed in the interval 3.0–2.9 to 2.85–2.80 b.y. but most of the belts between 2.76 and 2.70 b.y.; furthermore, the duration of the period of intense volcanic activity has been estimated to be 50 m.y. The concluding deformation, metamorphism and intrusion of granitoid plutons occurred between 2760 and 2660 m.y. and has been called the Kenoran orogeny. It has been established that volcanic activity concluded and deformation, metamorphism and granite formation commenced 20 to 60 m.y. earlier in the north than in the south.

Besides the Lake Superior province, greenstone belts were formed in another Archaean province of the Canadian Shield, i.e., Slave province. Five such belts, of smaller size, have been encountered here. Their volcanosedimentary filling is assigned to the Yellowknife supergroup aged 2.7–2.65 b.y. intruded by Kenoran granitoids aged 2.6–2.5 b.y. The Archaean rocks in the south-western part of Slave province are concealed under the Phanerozoic cover of the northern part of the Great Plains but re-emerge on the surface in the eastern Rocky Mountains of Montana and Wyoming (USA). This region is called the W y o m i n g   p r o v i n c e. Originally, the extensive area between the Slave and Superior eocratons was regarded as that part of the Churchill province which was cratonised at the end of the Early Proterozoic. Recent geochronological investigations have revealed the presence of at least two more major blocks of Archaean crust in this province, at places overlain by an Early Proterozoic cover. Hoffman (1989) distinguished these blocks as the Rae and Hearne provinces. Another Archaean block, encompassing the north-eastern coast of Labrador and south-western Greenland, has been named the Nain province.

Fig. 4 shows generalised sections of some Canadian greenstone belts (Condie, 1981). A preliminary study of the greenstone belts of Canada enabled A.M. Goodwin to distinguish three stages in their development: 1) extensive submarine outpourings of predominantly tholeiite-basalts forming a vast mafic platform; 2) explosive eruptions of magma of increasing siliceous felsic content, forming a superstructure rising high above the mafic platform; and 3) partial erosion of volcanic structures and formation of volcanoclastic aprons.

Dimroth et al. (1982–1983) worked out this scheme in greater detail on the example of the Abitibi belt and the adjoining Belcombe (Pontiac) gneiss belt. The sequence of the Abitibi belt comprises products of two cycles of volcanism, each commencing with extrusions of primitive komatiites and tholeiite-basalts and ending with outpourings of highly differentiated lavas ranging from basalts to rhyolites of the tholeiitic and calc-alkaline type. The succession of events which led to the formation of this belt are detailed below:

1) Submarine ultramafic-mafic lava plains are formed as a result of fissure eruptions from deep subcrustal pockets with lava thickness of about 5 km and extend for more than 200 km in meridional and latitudinal directions.
2) Volcanic activity ceases in the southern part of the region and fissure eruptions continue in the north with the formation of a less extended second lava plain with tholeiite-basalts reaching 5 to 7 km in thickness. Central volcanic complexes exceeding 30 km in diameter and up to 7 km in thickness are formed in the east.
3) Central volcanic complexes are formed in the south-west and north-east. Lavas vary in composition from basalts to rhyolites, belong to the low-potash tholeiitic and calc-alkaline series and originate from shallow magmatic pockets. In addition to lavas, hyaloclastic pillow breccia and volcanic tuffs are present at places. The lava at the centre of volcanic structures exceeds 10 km in thickness. These structures are intruded by stocks, sills and small batholiths of tonalites and trondhjemites.
4) Volcanic structures rise above sea level, forming permanent islands and bringing in material for volcanoclastic formations. Rocks belong to the calc-alkaline series and are intercalated with larger turbidites occurring from a belt of volcanoplutonic land elevated at the rear.
5) The volcanic zone experiences uplift along the latitudinal flexural fault. Planar, fan-like fluvial sediments formed on the land abruptly transit into deepwater debris cones towards the south. Land volcanic activity is represented by lavas and tuffs, intercalated with sediments of the Timiskaming group. The K_2O/Na_2O ratio increases in the volcanics; the rocks are highly differentiated and belong to the calc-alkaline and locally, the alkaline series.
6) Folding commences with compression of the Abitibi belt in a meridional direction. In the Belcombe belt folds inverted southwards are formed and these together with thrusts at the boundary of the belts suggest, according to Dimroth and colleagues (1982–1983), thrusting of the Belcombe belt under the Abitibi. The growth of pre- and synkinematic granite batholiths commenced simultaneously and was accompanied by their diapiric uplift

34

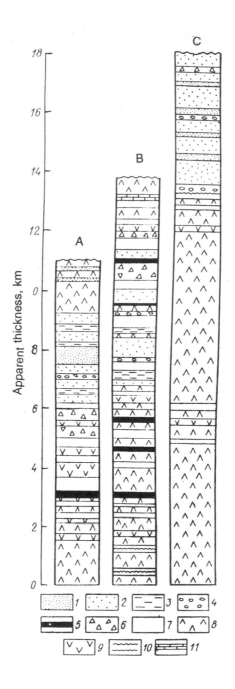

Apparent thickness, km

in the already solidified state during folding. In this process the granites were transformed into gneisses. Late and post-kinematic batholiths rose in the form of a melt, hardening in the course of uplift and elevating the host rocks. The granitoids reveal a calc-alkaline differentiation trend and the K_2O/Na_2O ratio in them increases as age decreases.

7) Folding concludes with displacement along faults and formation of kink-bands associated with the same meridional compression.

The last to form are batholiths of granitoids. In Section 3.2 we shall examine the genetic model proposed by Dimroth (Dimroth et al., 1982–1983) to explain this succession of events. Let us only mention here that most modern researchers interpret the development of greenstone belts of the Canadian Shield from the viewpoint of plate tectonics by explaining the formation of intermediate and felsic volcanics and granitoids as due to the subduction of oceanic crust represented by the lower sections of a belt sequence. This view would also explain the formation of metamorphosed sedimentary complexes, mainly greywackes and metalutites synchronous with volcanics and constituent belts that fall between greenstone belts (see Fig. 3). On the example of one such belt, Quetico, a model has been proposed according to which these belts represent ancient forearc accretionary wedges. The appearance of alkaline lavas, i.e., shoshonites, trachytes and leucite-trachytes, in the concluding stage of formation of the Abitibi belt is noteworthy. These alkaline lavas form a part of the Timiskaming detrital group concluding the sequence of this belt and have been dated 2702 m.y. Alkaline plutons of syenites and dykes of nordmarkites corresponding to the end of the Kenoran orogeny in this part of the Superior province, also arose in this same age interval.

Latest research has shown that the greenstone belts of Superior and Slave provinces have an extremely complex deformation history and structure. They experienced several phases of tangential compression and were subjected to the action of diapiric plutons of granitoids which complicated their structural orientation; longitudinal zones of plastic displacement and steep faults bearing gold mineralisation arose in the concluding stage of deformation.

Seismic profiles of reflected waves through Kapuskasing in the southern part of Abitibi belt have revealed a three-layered crustal structure. The upper

Fig. 4. Generalised stratigraphic sequences of Archaean greenstone belts of North America (after Condie, 1981):.

A—Michipikoten group, Ontario province; B—Vermilion group, Minnesota (USA); C—Yellowknife supergroup, Northwest Territories. 1—greywacke-argillite complexes; 2—quartzite-arkosites; 3—shales; 4—conglomerates; 5—siliceous and iron-ore rocks; 6 to 8—felsic and intermediate volcanics (6—breccia and agglomerates, 7—tuffs, 8—lava streams and sills); 9—mafic volcanics and sills; 10—ultramafic volcanics and sills; 11—carbonate rocks.

structural layer is made up of volcanosedimentary rocks and constitutes a complex fold-overthrust structure; furthermore, overthrusts flattening with depth possibly inherited the listric faults of the early stage of formation of the belts. This layer is bound at the bottom at a depth of about 15 km by a subhorizontal plane of decollement separating it from the middle layer made up of felsic intrusives and gneisses. The latter is underlain by a lower layer, presumably consisting of mafic gneisses and intrusive rocks.

3.1.1.2 *Greenland and Labrador.* Commencement of Archaean history proper of the south-western part of Greenland was marked by uplift and erosion of the Early Archaean Isua-Amitsoq massif described in the preceding chapter. Later, some 3.1 b.y. ago, Malene mafic metavolcanics and paragneisses (garnet-sillimanite, quartz-cordierite-anthophyllite-staurolite and mica-sillimanite-gneisses) began to accumulate. The amphibolites and metalutites of Upernavik in Labrador probably correspond to them. This was followed by the intrusion of layered plutons of anorthosites, tectonic deformation manifest in thrusting with the formation of tectonic intercalation of Malene rocks and Amitsoq gneisses and later, around 3 b.y. ago, formation of inverted folds and profuse injection of Nuk tonalite-granite-gneisses (Rb-Sr isochron age 3040 ± 50 m.y.). Their analogues are known even in the eastern Labrador peninsula where their Rb-Sr age for rocks is 3133 ± 156 m.y. at $^{87}Sr/^{86}Sr$ = 0.7063 ± 0.0012. These granite-gneisses evidently represent a product of remobilisation of Amitsoq gneisses. Later, around 2.8 b.y. ago, in both regions granulite metamorphism and formation initially of recumbent folds and tectonic nappes and later upright folds (in Greenland) occurred. Archaean development of Greenland and the Labrador peninsula concluded with formation of the Korkut (Greenland) and Igukshuak (Labrador) plutons of potash granites aged 2.6–2.4 b.y.

Recent data shows that the extent of Archaean rocks is not restricted to the block in which they were almost unaffected by later transformation. Archaean dates have been recorded at several points on the western coast of Greenland north of the Archaean block (Kalsbeek, 1982). The Archaean basement also extends some distance southwards of this block but subsequently signs of its existence disappear (Kalsbeek et al., 1984).

3.1.1.3 *Hebrides massif*[1]. Early Precambrian formations emerging on the north-western coast of Scotland and the Hebrides islands are regarded as a former direct continuation of very similar formations of Greenland and eastern Labrador, constituting together with them a single North Atlantic Craton (in: *Early History of the Earth,* 1976) separated during formation

---

[1] After Bowes (in: *Early History of the Earth,* 1976; Bowes, 1980).

of the Labrador Sea and the North Atlantic. Archaean-Lower Proterozoic rocks of the Hebrides massif were known together under the name L e w i s i a n   c o m p l e x; radiometric data has enabled delineation from it of the G a l e   s u p e r g r o u p pertaining to the Late Archaean and experiencing polyphase deformation and metamorphism in the Scourian epoch aged 2.7–2.6 b.y. The Gale supergroup is mainly made up of calc-alkaline volcanics, partly intrusive rocks, with relics of metasedimentary rocks and layered peridotite-gabbro-anorthosite plutons. Formations of a different degree of metamorphism from greenschist to granulite facies are present in it; gneisses of amphibolite facies (evidently diaphthoresis) predominate. In some quartzite specimens round grains of zircon aged 3.2 b.y. (U-Pb method) have been encountered, pointing to erosion of the much older continental crust. But the low strontium isotope ratios recorded for orthogneisses point to their mantle origin (in: *Early History of the Earth,* 1976; Bowes, 1980).

3.1.2 East European Craton

Greenstone belts are known within both the main salients of the basement of this craton, namely, the Baltic and Ukrainian shields, and were also identified from drilling data in the Voronezh massif.

3.1.2.1 *Baltic Shield.* The Karelian granite-greenstone province, which also encompasses eastern Finland, is the main and fairly typical zone for this shield. Moreover, Archaean greenstone belts are known also in the Kola peninsula to which the following zones belong: Kolmozero-Voronya zone at the boundary of Murmansk and Central Kola blocks; Varzuga zone delineated by Fedorov (1985) from Imandra-Varzuga Early Proterozoic protoaulacogen (see Sec. 4.1.2.1); a band of development of ferruginous-siliceous ore formation in the Central Kola block described by Goryainov and Fedorov (Terskei-Allarechensk zone) (1986); and finally, the Korva-Kolvitsa zone in the southern part of the peninsula (Grudinin, 1992).

According to the data of Vrevsky and Kolychev (1987), these belts are of two generations: 3.1–3.0 b.y. (marginal zones of Kola megablock) and 2.9–2.8 b.y. (central Terskei-Allarechensk zone).

In the most typical Karelian granite-greenstone province[2], no less than six greenstone belts extending in the west in a meridional and in the east in a south-easterly direction are known (Fig. 5). The period of formation of belts in central Karelia has been dated by the U-Pb method on zircon from felsic volcanics of upper sections of the Semchen-Koikarsk belt at 2935 ± 20 m.y. and in adjoining structures at 3020 m.y. The age of gabbro-diorite and leuco-gabbro intrusions is 2890 and 2840 m.y., that of granite porphyries

---

[2] After Krylov et al. (in: *Precambrian Geology,* 1984).

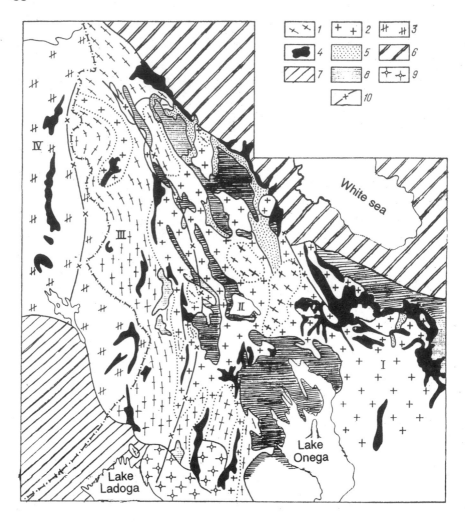

Fig. 5. Scheme showing structure of granite-greenstone province of Karelia and eastern Finland (after Krylov et al., in: *Precambrian Geology,* 1984).

Archaean: 1—gneiss-granite fields; 2—Lopiya greenstone belts. Early Proterozoic: 3—Sumian-Sariolian complex; 4—Jatulian complex; 5—Svecofennian (Svecokarelian) complex; 6—Belomorides; 7—intrusions of Proterozoic mafic rocks of age (?); 8—rapakivi granites; 9—granodiorites; 10—faults; I to IV—zones: I—eastern Karelia; II—central Karelia; III—western Karelia; IV—eastern Finland.

2830 m.y., and post-folding and post-metamorphic granites 2740 ± 20 m.y. Duration of development of the belts has been assessed at 150–200 m.y. (in: *Precambrian Geology,* 1984). The age of the Gimola series was recently determined as 3.2 b.y. (Kapusta et al., 1985).

In the eastern Karelian belt volcanics constitute a bimodal series with a predominance of basalts and tuffs of dacite and rhyolite composition in the upper sections. Quartzites, including those containing magnetite and black schists are present to a small extent. In the central Karelian zone basalts predominate in the composition of belts and are intercalated with komatiites in the lower sections; they are overlain by andesites and dacites, the thickness of the former sometimes commensurate with that of basalts. The upper part of the section is made up of polymictic conglomerates, greywackes and sandstones. Thus a classic threefold sequence with successively differentiated volcanics is present here. Metamorphism is zonal—from greenschist to amphibolite facies. In the western Karelian zone we again encounter bimodal volcanism: basalts, dacites and rhyolites are developed in the lower sections and ferruginous-siliceous and sandy-clay in the upper. The age of the felsic volcanics of Kostomuksha varies from 2.8–2.7 b.y. Finally, in the eastern Finland zone, as in central Karelia, we again find a complete suite of volcanics ranging from komatiites and basalts through andesites to dacites and rhyolites. The age of intruding granites is 2.65 b.y. On the whole, basalts in the sequences of greenstone belts of this province constitute 40 to 75% and correspond to olivine, less often quartz-tholeiites with low contents of alkalis, titanium and iron.

Greenstone belts are separated by much wider fields of granitoids and gneisses. These formations are extremely heterogeneous, as suggested by the variable nature of the magnetic field from low intensity with almost total absence of local anomalies to high intensity with linear anomalies of meridional orientation. A part of the gneisses is older than 3 b.y., namely the grey gneisses (see Sec. 2.1.2.1). Granitoids sometimes form dome or dome-block structures. Radiometric ages vary from 2.92–2.65 b.y. The latter value similarly corresponds to the age of late granulite metamorphism.

In eastern Finland the age of tonalite, trondhjemite and granodiorite gneisses forming the basement of the Kuhlo-Suomossalmi greenstone belt has been determined as 2.86–2.62 b.y., of the calc-alkaline volcanics of the upper section of the belt sequence as 2.50 and of the post- kinematic granites as 2.41 b.y.

3.1.2.2 *Ukrainian Shield and Voronezh massif.* The main granite-greenstone province of the Ukrainian Shield is the central Dnepr block bound from the west by Krivoy Rog-Kremenchug and from the east by Orekhovo-Pavlograd fault zones. This is the same block in which the Aulian grey gneiss complex was delineated (see Sec. 2.1.2.2) but granite-gneiss domes of the Saksagan type and greenstone formations of the Konka-Verkhovtsevo series overlying the Aulian series and filling the gaps between the domes play a major role in the block structure. Amphibolites of the lower sections of the Konka-Verkhovtsevo series rest on the metamorphosed weathered crust of gneisses of the Aulian complex (Esipchuk et al., in: *Stratigraphy of Precambrian Formations of the Ukrainian Shield*, 1983). The Konka-Verkhovtsevo series reveals the following succession of formations (after Laz'ko et al., in: *Stratigraphy of Precambrian Formations of the Ukrainian Shield*, 1983): metabasite-andesite-tholeiite, komatiite-tholeiite, jaspilite-tholeiite, metaandesite-dacite-tholeiite and komatiite; this succession is not generally acknowledged, however. According to Shcherbak and Bibikova (in: *Precambrian Geology*, 1984), the metamorphosed sedimentary and volcanic rocks constituting this series do not persist in the facies composition, are distributed very unevenly in both vertical and lateral directions and their quantitative proportions vary widely. On the whole, however, the following general tendency is observed: 'Upward along the stratigraphic sequence the thickness of metasedimentary rocks increases.' A general idea of the sequence of the Konka-Verkhovtsevo series can be drawn from column VI in Fig. 6. The thickness of the series has been evaluated at 6–7 km; it is metamorphosed mainly to greenschist facies. The age of the volcanics of the lower part of the series has been assessed as 3250 ± 120 m.y. by the U-Pb method on zircon and at 3.3–3.24 b.y. by the Sm-Nd method on rock-forming minerals. The oldest granites of the Dnepr complex of tonalite-granodiorite composition intruding the Konka-Verkhovtsevo series, together with the rocks of the Aulian complex constituting the granite-gneiss domes and arches, are aged 2970 ± 20 m.y. (U-Pb method on zircon). Thus the period of formation of greenstone belts of the central part of the Ukrainian Shield may go up to 250–300 m.y.; they belong to a much older generation than the belts of the Baltic Shield.

Within the V o r o n e z h m a s s i f, Krestin (1980) delineated Archaean greenstone belts made up of the Mikhailovka series. This series overlies with a sharp unconformity the gneisses of the Oboyan series—an analogue of grey gneiss complexes (see Sec. 2.1.2.2). The lower section of the Mikhailovka series is represented by metamorphosed peridotite and pyroxenite komatiites and magnesian basalts, on top of which lies a suite of amphibolites formed after tholeiite-basalts. The thickness of this part of the sequence is 1.5–3.0 km. It is overlain at places by a suite of lavas and tuffs of intermediate and felsic composition constituting volcanic structures of the central type. The thickness of this suite is 400–500 m. Rocks of the Mikhailovka series are metamorphosed to greenschist, partly epidote-amphibolite facies. Prefold intrusions of peridotites, pyroxenites and gabbro-plagiogranitoids of the Saltykovka complex are associated with the Mikhailovka series. The youngest are the microcline-albite granites of the Veretenin complex with radiometric age of 2960 ± 40 m.y. (Rb-Sr method). Thus the Mikhailovka series represents a possible analogue of the Konka-Verkhovtsevo series of the Ukrainian Shield. It is unconformably overlain

41

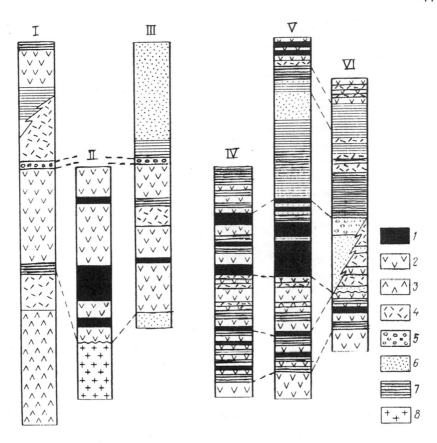

Fig. 6. Comparison of stratigraphic sequences of greenstone belts of the Baltic and Ukrainian Shields (after Shcherbak et al., 1986).

1—picrites, basalt and peridotite-komatiites; 2—metabasalts; 3—metaandesites; 4—metadacites and metarhyolites (tuffs and lavas); 5—conglomerates; 6—metasandstones; 7—ferruginous-siliceous rocks and cherts; 8—basement granites. I to VI—structures of the Baltic (I—Khautovara, II—Palaya Lamba, III—Kostomuksha) and the Ukrainian (IV—Samotkanskaya, V—Granovskaya, VI—Belozero) shields.

by locally developed formations of ancient weathering crust and detrital products of its redeposition.

Further, Archaean formations have been discovered by drilling into the basement of the Volga-Urals province in the eastern part of the Russian Platform. These formations in the Tatar arch, drilled to a depth of up to 3.5 km, revealed two petrographically different series: Otradnaya and Bolshoy-Cheremshan. The former is made up of 'enderbitoids', metagabbro and bipyroxene crystalline schists presumably formed after bimodal volcanics (gabbro of the calc-alkaline type). This series is intruded by granites with U-Pb age of 2.7 b.y. The latter series is formed by high-alumina gneisses-metalutites. Both series are metamorphosed to granulite facies and on this basis the second series may also be regarded as Archaean (Bogdanova, 1986).

3.1.3 Mediterranean Belt

This belt includes a large number of blocks and slices of Precambrian continental crust but Archaean complexes among them were established or assumed only in its Afghan-Pamirs segment. This is true primarily of the Badahshan massif (south-western Pamirs and adjoining part of Afghan-Badahshan) where radiometric studies proved the Archaean age of the Goran and Vakhan series made up of gneisses (biotitic, amphibolitic, garnetiferous and kyanitic) and also amphibolites, quartzites, crystalline schists, marbles and calciphyres with the participation of plutonic rocks (granite-gneisses, granosyenites, plagiogranites, charnockites, gabbro and ultramafics). The oldest radiometric age recorded is 2.7-2.4 b.y. The complex was metamorphosed to amphibolite and granulite facies and migmatisation is manifest regionally.

In the Kabul block, according to Karapetov (1979), Archaean age is probable for the Khairkhana series formed of gneisses, granite-gneisses, amphibolites and marbles and metamorphosed to amphibolite facies with relics of granulite facies to which charnockitoids-mangerites in particular belong.

On analogy with these formations of Badahshan and Kabul blocks, S.S. Karapetov places the gneiss and granite-gneiss (migmatised) complex widely distributed throughout the segment in the Archaean. However, in the Greater Himalaya this complex has been dated 1.8 b.y. by radiometry and hence an Early Proterozoic age is more probable (see Sec. 4.1.3).

At the same time, there is interesting evidence of the presence of an Archaean substratum in other parts of the Mediterranean belt. In the extreme western part of the European Hercynides, on both sides of the Bay of Biscay but mainly on its southern side, specimens of mafic and felsic granulites of metasedimentary and metamagmatic origin aged 2.7 b.y. recorded by the U-Pb method on zircon and confirmed by the Sm-Nd method were brought up by dredging and submersibles. Signs of reworking of these rocks at around 1.9 b.y. were simultaneously detected; according to the authors, this represents the age of granulite metamorphism (Guerrot et al., 1989).

Farther away in the east, in the Alpine belt of southern Turkey (Menderes-Taurus block), grains aged 2.4-2.5 and 3.1 b.y. (Pb-Pb method) have been found among clastic zircons in greywackes of a metamorphic complex and granites of pre-Middle Cambrian age. These grains suggest the erosion of an Archaean substratum (Kröner and Şengör, 1990). In the first case the authors suggest affinity with rocks in Africa and in the second a Siberian origin.

3.1.4 Siberian Craton

Within the Siberian Craton, the north-western Olekma block (Aldan Shield) may be placed among the granite-greenstone provinces. In this block 'trough' structures with a meridional strike are delineated among granites and gneisses of granulite facies (Fig. 7).

The basins of the Chara and Olekma rivers represent the main region of development of greenstone belts. It was first thought that corresponding formations pertaining to the Late Archaean (only their upper boundary has been established with certitude) are confined here to just narrow (4–5 km) grabens. The more recent detailed investigations of Akhmetov and colleagues (1983), Kudryavtsev and Nuzhnov (1981) and other researchers suggest a much wider distribution (Fig. 8). These formations fill, as in other shields, complex synclines with isoclinal folding (Fig. 9), their structure reflecting the superposition of several phases of deformation. In so far as grabens ('troughs') are concerned, they represent the most recent Archaean structures cross-cutting the internal facies zoning of greenstone belts.

Rocks of the Chara series metamorphosed to granulite facies and pertaining already to analogues of the grey gneiss complex (see Sec. 2.1.3)

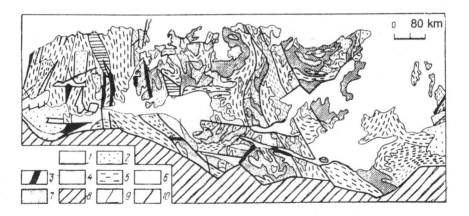

Fig. 7. Scheme showing disposition of Early Precambrian structural complexes of the Aldan Shield (after Kitsul, Petrov and Zedintsov, 1979).

1—formations of the platform cover; 2—Karelian megacomplex, Udokan and Maimakan complexes of the Lower Proterozoic; 3 and 4—Subgan megacomplex of the Upper Archaean (3—Olondo complex and 4—Borsala complex); 5 to 7—Aldan megacomplex of Lower Archaean age (5—Dzheltula complex; 6—Yengra complex, 7—basement complex); 8—Dzhugdzhur-Stanovoy region; 9—boundaries of complexes; 10—faults.

44

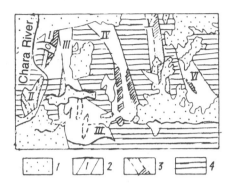

Fig. 8. Scheme showing locations of Late Archaean structural zones of the western Aldan Shield (after Akhmetov et al., 1983).

1—Post-Archaean formations; 2—Late Archaean structural zones (I—Imalyk, II—Chara, III—Tokmov-Khana, IV—Temulyakit-Tungurchik, V—Olekma-Ashinsk, VI—Amedichin); 3—'troughs' and near-fault depressions (after V.S. Fedorov; A.F. Petrov; and others); 4—zones of development of Late Archaean granite-gneiss domes with relics of metamorphic belts of Middle, less frequently Upper Archaean.

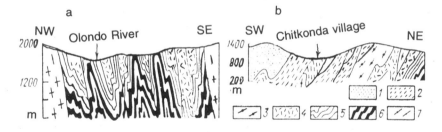

Fig. 9. Geological sections through Khana-Olondo graben (after Kudryavtsev and Nuzhnov, 1981).

a—northern and b—southern parts of the graben; 1—metalutites and metasiltstones of Ikabii suite of the Lower Proterozoic Udokan complex; 2—apopsammitic and apoaleuritic shales and gneisses of the Tungurchik series of Upper Archaean age; 3—high-alumina gneisses; 4—meta-andesites and orthoschists of intermediate, less often mafic composition; 5—metabasaltoids and basic orthoschists; 6—metabasaltic and metakomatiitic orthoschists; 7—Upper Archaean granitoids.

serve as the substratum for the greenstone belts of the Chara-Olekma region. An unconformity exists between these two complexes which is concealed either by fractures or intrusions of granitoids. The filling of greenstone belts is described in this region as the S u b g a n   (O l o n d o) c o m p l e x, its thickness at places exceeding 7 km. This complex has a three-layered structure, so characteristic of most greenstone belts. In the lower part (Borsala series) komatiites, picrites and diabases are developed and above them lie ferruginous quartzites and lutites; in the middle part (Tungurchik series) andesites, psammites, siltstones and lenses of carbonates; in the upper part (Tasmyely series) andesites, dacites, rhyolites, quartz-sandstones and conglomerates. This complex bears a molassoid character. The three series are separated by unconformities.

Shcherbak and colleagues (1986) recorded an age of 2.96 b.y. for the felsic volcanics of the upper Olondo series by the U-Pb method on zircon.

Bedded bodies of serpentinites, peridotites, orthoamphibolites, plagiogranites and granodiorites are associated with the lower series; intrusions of gabbro, diorites, granodiorites, granite-gneisses and granites with the middle series; and microcline granites with the upper series. The age of the latter is 2600 ± 100 m.y. The lower series of the Subgan complex was metamorphosed to epidote-amphibolite and amphibolite (at contact with granites) facies and the upper series to greenschist facies.

Greenstone belts (there are six here) are separated by several broader fields of granite-gneisses. The predominant structures of the latter are domes of Late Archaean age. Relics of much older Presubgan granulites can be seen at places in these fields.

Based on a study of the facies zoning of Late Archaean deposits, Kudryavtsev and Nuzhnov (1981) concluded that they were formed in a common basin whose structure, however, experienced significant reorganisation on transition from the period of accumulation of one series to the period of accumulation of the next.

Analogues of the greenstone belts of the Olekma province are distinguished with some degree of confidence into peripheral salients of the basement of the Siberian Craton: in the west in the Yenisei range and the Onot graben of the Sharyzhalgai block and in the south the Stanovoy belt. These formations are intersected by granitoids aged 2.8-2.5 b.y. They are in complex tectonic relations with gneisses and charnockites, more intensely deformed and metamorphosed to granulite and less often amphibolite facies of the Kan, Sharyzhalgai and Zverev series unreliably dated in the range 2.8-3.4 b.y., pointing more to their Middle Archaean age although Early Archaean (see Sec. 2.1.4) and partly Late Archaean ages cannot be ruled out.

Based on geophysical data, it may be suggested that much of the basement of the Siberian Craton concealed under a Late Precambrian-Phanerozoic sedimentary cover is also made up of Archaean formations.

3.1.5 Ural-Okhotsk Belt

Salients of Archaean crystalline basement are known in the western, south-western and eastern peripheral zones of this belt tending to its cratonic frame.

On the western slope of the Urals, at the northern tip of the Bashkirian anticlinorium, Early Archaean age (3.2–3.0 b.y. by α-Pb and K-Ar methods) was established for rocks of the T a r a t a s h   c o m p l e x: bipyroxene crystalline schists, gneisses with garnet, sillimanite, hypersthene, sometimes graphite, magnetitic quartzites, magnetite-hypersthene rocks, hypersthene gabbro-diorite-gneisses etc. These rocks experienced granulite metamorphism not later than 2.7 b.y. ago with the formation of high-alumina granitoids, enderbites and charnockites. At 2.2–2.1 b.y. the complex underwent diaphthoresis to amphibolite facies with additional granitisation. The Taratash complex evidently represents a direct continuation of the basement of the Volga-Kama anteclise of the Russian Platform.

Archaean formations are present not only on the western slope of the Urals, but also on its eastern slope where they are known in the Sysert'-Ilmenogorsk salient of the basement, representing either a microcontinent or further continuation of the basement of the craton. Here granulites aged over 2.3 b.y. are exposed in the Selyankin block. They have experienced diaphthoresis to amphibolite facies at the level of 1.82–1.78 b.y. (U-Pb method on zircon).

Archaean age has also been suggested, based on the composition of rocks and the nature of their metamorphic transformation, for the Mugodzhary metamorphic complex made up of amphibolites, gneisses with garnet, kyanite, sillimanite, quartzites and leptynites, forming the southern Mugodzhary and Taldyk series. Metamorphism is of almandine-amphibolite facies with regional plagiomigmatisation and development of autochthonous plagiogneiss-granites, gneiss-granites and pegmatites. A similarity is seen with the Belomorie and Stanovoy complexes of the Baltic and Aldan shields (in: *Precambrian of Central Asia,* 1984).

Farther south-east, Archaean rocks are known or assumed to take part in the composition of the basement of the Kokchetav-Muyunkum massif of central Kazakhstan and northern Kirghizia.

The A k t y u z   c o m p l e x, developed in the western part of Transili Alatau, is made up of garnet-potash feldspar binary mica and muscovite gneisses, amphibolites and amphibole-gneisses with rare lenses of marbles. Migmatites are extensively developed and bodies of eclogites and garnetiferous amphibolites are present. The rocks were initially metamorphosed under the condition of low temperature and high pressure and later changed under higher temperature to kyanite-sillimanite subfacies of amphibolite facies. The age of metamorphism of the amphibolite facies is 2240±50 m.y. (Kiselev et al., 1992). At the end of the Early Proterozoic (1820±180 m.y.), they underwent diaphthoresis to epidote-amphibolite and greenschist facies. In primary composition these rocks are more probably metasedimentary, sandy-clay formations with mafic volcanics although another view supports their wholly volcanic (andesite-dacite-rhyolite) nature. Probable analogues of the Aktyuz complex are also known in other regions of Northern and Central Tien Shan right up to the Talas-Ferghana fault (Dol'nik and Vorontsova, 1974). In Sary-dzhaz range in the eastern part of Central Tien Shan, they are represented by the K u y l u   c o m p l e x consisting of crystalline schists, primarily metamorphosed to granulite facies and gneisses dated 2616 ± 50 m.y. by the U-Pb method on zircon and later to amphibolite facies 1922 ± 30 m.y. ago (same method).

More northward, within the Kokchetav-Muyunkum massif, in Ulytau, the Bekturgan series of gneiss-schist-amphibolite composition is dated as Archaean by some researchers. It is metamorphosed only to epidote-amphibolite facies. A much older age has been suggested on the basis of a much higher grade of metamorphism (amphibolite or granulite) for the Zerenda series of the Kokchetav block represented by gneisses, eclogites, amphibolites and granulites.

The K a r a t e g i n   s u i t e in the extreme south of Tien Shan, made up of gneisses (biotite, garnet-biotite, cordierite, sillimanite and amphibole), biotite, garnet-staurolite-biotite and amphibole schists with bands of marbles, quartzites, granulites and amphibolites, has been tentatively placed in the Archaean. Its age was determined by the Pb-isochron method as 3.0–2.6 b.y.

On the other side of the axial zone of the Ural-Okhotsk belt, Archaean formations have been delineated within the frame of the Siberian Craton commencing with the Yenisei range and beyond to the south and south-east. The K a n   c o m p l e x of hypersthene gneisses, granulites and charnock-ites with metamorphic age of 2.7–2.55 b.y. is exposed in the southern part of the Yenisei range, south of Angara; in the composition of this complex much older rocks of Early Archaean age were detected (see Sec. 2.1.3). Farther south the Biryusa horst situated between the Eastern Sayan and Siberian craton is made up of the same-named series of Archaean gneisses and crystalline schists of amphibolite facies.

Southward, between Biryusa and Sharyzhalgai horsts and within the latter, two greenstone belts extend in the form of graben-synclinoria with a north-north-easterly strike. These are Urik-Iya and Onot. The first of these graben-synclinoria extends for 500 km between the basins of the left tributaries of the Angara, i.e., Biryusa and Oka, with a width of 10–15 to 60 km. It is made up of several kilometre-thick terrigenous, primary sandy-clay sediments with a thick member of dolomites in the lower part (but not at the base) of the sequence and volcanics of predominantly felsic composition in the upper part of the sequence. In the uppermost section intercalation of psammites with lutites and carbonates bears a flyschoid character. Metamorphism ranges from amphibolite in the lower to greenschist in the upper parts. The deposits are quite intensely dislocated and intruded by granitoids ranging from gabbro-diorites to normal granites. The 'age of the granitoids has been determined as 2.65–2.6 b.y. by the Rb-Sr method.

The O n o t g r a b e n-s y n c l i n o r i u m is slightly smaller compared to the Urik-Iya but its continuation is concealed in the north under the Phanerozoic cover of the central Siberian Platform. The visible lower section of the graben sequence is made up of marbles, dolomites, amphibolites and biotite-amphibole schists; they are intruded (?) by small ultramafic bodies and very large plutons of a gabbro-plagiogranite formation. The latter is unconformably overlain by a suite of amphibolites, garnet-mica, quartz-biotite schists, ferruginous quartzites and gneisses, and stratigraphically even higher by intermediate and felsic volcanics, biotite and binary mica schists. As in the case of the Urik-Iya graben, the sequence terminates in terrigenous flyschoid formations. Granite plutons intrude this entire complex of formations. Shcherbak and colleagues (1986) determined the age of the plagiogranites as 3250 ± 100 m.y. by the U-Pb method on zircon.

Thus the Onot greenstone belt belongs to the oldest generation of these structures and the Urik-Iya to a more recent one.

In the T u v a-M o n g o l i a m e d i a n m a s s i f (microcontinent) in the Sangilen upland, which contains the most complete and best-dated sequence of Precambrian formations, Volobuyev and colleagues (1980) established a Middle Archaean age for the E r z a c o m p l e x metamorphosed to granulite facies and containing biotite-garnetiferous, biotite-garnet-cordierite gneisses, biotite-hypersthene, biotite-hypersthene-garnetiferous plagiogneisses, bipyroxene, bipyroxene-amphibole crystalline schists and amphibolites, as well as marble and calciphyre in the form of thin veins. The suite houses charnockites and enderbites. The age determined by the U-Pb method on zircons from gneisses and charnockites is 3100 ± 200 m.y.

The M o r e n c o m p l e x, formed of gneisses and ferruginous quartzites, marbles and calcareous-silicate rocks falls in the Upper Archaean in the same Sangilen upland. Gneisses are represented by metavolcanics of the bimodal series including tholeiite-basalts and, among felsic rocks, by rhyolites and rhyodacites of the calc-alkaline series. The age of this complex has been established by analogy with the Bumbuger complex of Central Mongolia (see below).

In the western part of Central Mongolia, Early Precambrian rocks are exposed in the Baidaryk (Baidaragyn) salient (microcontinent) between two much younger ophiolite zones: Lake and Bayan-Khongor. The Middle Archaean is represented by the B a i d a r a g y n c o m p l e x of differentiated volcanics similar to island-arc volcanics, metamorphosed to granulite and later amphibolite facies (*Evolution of Geological Processes ...*, 1990). According to another interpretation, these are metaultramafics and mafics that experienced intense migmatisation. The age of metamorphism of amphibolite facies is 1850 ± 20 m.y. The B u m b u g e r c o m p l e x, reliably dated by the U-Pb method on zircon as 2650 ± 30 and 2437 ± 35 m.y., corresponds here to the Upper Archaean.

Smirnov (1976) assumed a fairly extensive development of Archaean in the Mongolia-Okhotsk segment and its southern frame. In the Hingan-Bureya, Khanka and adjoining massifs of north-eastern China, he delineated a Manchurian complex of gneisses, migmatites, crystalline schists, amphibolites and marbles metamorphosed to granulite facies and intruded by alaskites. Shuldiner (1982) indicated the development of analogues of this complex even in eastern Transbaikalia, especially in the Argun massif. In Jiamuxi-Bureya massif, stretching from Russia to China and traversed by the Amur valley, Mamyn granulite and Bureya schist-gneiss complexes delineated in the Russian part belong to the Middle and Upper Archaean. The latter was metamorphosed to amphibolite and epidote-amphibolite facies; evidently the age recorded on the Chinese side pertains to the Bureya complex. In the Khanka massif (in: *Geological Structure of the USSR ...*, 1984) the Iman series, consisting mainly of graphite and clinopyroxene marbles (in the lower part), biotite, sillimanite, biotite-garnet, biotite-sillimanite, garnet-cordierite, clinopyroxene, clinopyroxene-amphibole, hornblende gneisses and schists, has been dated as Late Archaean.

3.1.6 Sino-Korea Craton

Formations of the Jiangxi group, typically developed in eastern Hebei, are placed in the Middle Archaean. Two nearly contemporaneous complexes are distinguished in its composition: Jiannan granulite and Zunhua greenstone. The former consists predominantly of mafic, pyroxene granulites, biotite-hypersthene-plagioclase gneisses and ferruginous quartzites; volcanics of the calc-alkaline series (andesites) and also shallow-water sediments served as protoliths of these rocks. The basalts of the complex are proximate to continental tholeiites. The Zunhua greenstone belt is characterised by a predominance of high magnesia (poor in K_2O) tholeiites and also grey gneisses and development of layered mafic-ultramafic intrusives. Radiometric ages by the U-Pb method on zircon as well as by the Rb-Sr method on whole rock lie in the range 2.59–2.48 b.y. but some determinations by the U-Pb method (concordant values) on zircon and the K-Ar method on hornblende have given values of 2.8–2.9 b.y. Granulites of the Jiangxi group have partially experienced retrogressive metamorphism to amphibolite facies.

Upper Archaean formations within China have been grouped into the F u p i n g or B a d a o h e group with thickness assessed at 7–10 km. Among them, separation into granite-greenstone provinces and granulite belts bordering them could be distinctly established. The largest of such belts extends in a latitudinal direction from Inner Mongolia into northern Hebei and eastern Liaoning. Another belt of the same general strike has been traced within central Henan south of the lower course of the Huanghe River to east-south-east of Xian city. The rocks constituting it have been identified as the T a i h u a   c o m p l e x and are represented by garnet-bipyroxene-granulite, plagioclase-amphibole, quartz-feldspar and graphite-gneiss, marble and magnetite-quartzite. This belt is distinctly overthrust northwards onto the adjoining granite-greenstone province. A third granulite belt lies in eastern Shanxi along its border with Henan south-west of Beijing. Like the preceding one, it is composed of mafic granulites, paragneisses, quartzites and amphibolites. The granite-intruding Fuping group has been dated as 2560 ± 9 m.y. by the U-Pb method on zircon and as 2.83–2.80 b.y. (same method) on clastic zircon from the lower part of the group.

These three granulite belts constitute the frame of the granite-greenstone province of the central regions of Henan and Liaoning. Greenstone belts have been established in this province in central Henan and northern Liaoning, in the regions of Fushun and Anshan. The Fushun sequence of the Qinshan greenstone belt, formed on a basement of granulites (from mafic to felsic) and charnockites, has been rather fully described (Zhang et al., 1985). This sequence, similar to classic belts, consists of three subdivisions (bottom upwards): mafic-ultramafic volcanics (tholeiites and komatiites); calc-alkaline volcanics (tholeiites, andesites and dacites) with sulphide ores and ferruginous quartzites, quartzite-sandstones, dolomite-marbles, greywackes and mafic lavas. The total thickness exceeds 9 km. These formations have been intruded by tonalites, monzonites and granites aged 2.8–2.6 b.y. A similar sequence is encountered in the Anshan belt except that ferruginous quartzites are contained in all three sections but mainly in the third. In so far as the central Henan Denfeng belt is concerned, the distinctive feature here is the bimodal character of the volcanics of the middle section (tholeiites and felsic volcanics). The rocks of all the belts described are metamorphosed to amphibolite facies and form complex synforms occupying the area between granite-gneiss domes.

The end of the Archaean here, as everywhere, is marked by a general cratonisation consequent to fold and fault (overthrust) deformation, regional metamorphism predominantly of amphibolite facies, migmatisation and granitisation. Metamorphism of the granulite facies in the mobile Inner Mongolia-Hebei-Liaoning belt may belong to the same epoch of diastrophism, named the F u p i n g   e p o c h after the complex of formations affected by it. Folding proceeded in several phases. The predominant strike of metamorphic lineation and of later broad folds is sublatitudinal; the strike of much younger folds is meridional. Granitoids are mainly of the soda type (tonalites) but normal granites are also present. The formation of swarms of mafic and pegmatite dykes at the beginning of the

Proterozoic in Shandong and Shanxi provinces confirms the formation of a fairly consolidated crust.

3.1.7 South China (Yangtze) Platform and Adjoining Massifs

Rocks of Archaean age, proven or assumed, are exposed in four marginal salients of the basement of the South China Platform[3]: Shonglo massif in the south (Vietnam), Sikan-Yunnan range in the south-west, Jiannan uplift in the south-east and Dabei massif in the south-eastern extremity of the Qinling range (China). In the core of the Shonglo salient augen migmatites of the Shongtai complex are found with scialites of crystalline schists containing zircons of 2652, 2452 and 2050 m.y. K-Ar age. In the Sikan-Yunnan range (Kannging axis) the lower part of the Kannging complex made up of bipyroxene-gneisses, granulites, sillimanite-cordierite gneisses with rare bands of marbles and ferruginous quartzites can be placed in the Archaean. The similarity of this suite of rocks with the Kannak complex in Vietnam and Mogok in Myanmar (see below) is evident.

Archaean rocks are least reliable in the Jiannan uplift in the south-eastern part of the craton where they may be represented by gneisses, marbles and migmatites, intruded by granites of Archaean age. On the contrary, ages determined by the U-Pb method on zircons at 3120 and 2500 m.y. are known for the Dabei massif. These age values point to the probable presence not only of Upper but also Middle Archaean in this massif. The rocks themselves are represented by gneiss, marble, ferruginous quartzite, amphibolite and migmatite. Metamorphic zoning from granulite facies in the north-west to greenschist in the south-east is noticed. The presence of eclogites, possibly the oldest in the world containing pseudomorphs after coesite, is interesting and points to the uplift of rocks from a depth of about 90 km.

3.1.7.1 *Tarim Craton.* The presence of Archaean rocks has been assumed in the southern part of the basement of the small Tarim Craton situated south of Tien Shan.

3.1.7.2 *Indosinian massif.* Development of Archaean formations in the Kontum salient of this massif, represented by the Kannak complex, has been suggested on the basis of the presence here of a young Ngoklin metamorphic complex dated 2.3 b.y. and the lower degree of metamorphism of the latter. The Kannak complex is wholly metamorphosed to granulite facies

---

[3] According to the data assembled by Fan Chyong Thi (1981), Le Zui Bath (1985) and *Metamorphic Complexes of Asia* (1977).

and probably belongs to the Upper Archaean. Its lower part (300 m) consists of bipyroxene crystalline schists, plagiogneisses, garnet-hypersthene gneisses, with lenses of garnet-cordierite-sillimanite and biotite-hornblende gneisses. The upper part of the complex (1 km) is made up of garnet-sillimanite-cordierite gneisses, marbles with graphite, wollastonite, diopside and calciphyre with graphite. In primary composition the rocks of the lower part of the Kannak complex represent most probably basaltic volcanics and those of the upper part clay and carbonate sediments. Among the metamorphics of the complex are found small bodies of serpentinised ultramafics and metagabbroids and also autochthonous granitoids—enderbites, charnockites, granite-gneisses and gneiss-migmatites.

3.1.7.3 *Sinoburman massif.* Within the massif, Archaean formations can be distinguished in the lower part of the Mogok complex in the Shan plateau region of Myanmar only by petrological analogy with the Kannak complex of Vietnam. They are the same bipyroxene crystalline schists and garnet-cordierite-sillimanite gneisses and marbles with bodies of enderbites, granite-gneisses and syenite-gneisses. The upper part of the Mogok complex is most likely an analogue of the Ngoklin complex of the Kontum salient and correspondingly can be placed in the Lower Proterozoic.

3.1.8 South American Craton

3.1.8.1 *Guyana Shield.* Archaean rocks have been reliably identified in this shield only in Venezuela in the north (Gibbs and Barron, 1983). Here El Cedrel gneisses aged 2.7 b.y. lie unconformably on the Early Archaean Imataca complex or in tectonic contact with it. Amphibolites and accompanying rocks of the Santa Barbara and Esperanca groups (2.7–2.5 b.y.) succeed these gneisses conformably. In so far as the rest of the Guyana Shield is concerned, some researchers place the rocks of granulite facies of metamorphism, especially charnockites of the Falovatra group in Surinam (Bosma et al., 1983) in the Archaean, considering them older than the Early Proterozoic greenstone belts (see Sec. 4.1.8.1).

3.1.8.2 *Central and East Brazilian Shield.* Three Archaean granite-greenstone terranes have been reliably delineated within this shield to date. One falls in the eastern part of the Central Brazilian (Guapore) eocraton, another corresponds in general to the Sao Francisco eocraton (Wernick et al., 1978) and the third of small dimension to the Goyas median massif (Danni et al., 1982) situated among the central Brazilian Baikalides (Brazilides in local literature).

In the Central Brazilian Shield most reliable outcrops of Archaean rocks are known in the interfluve zone of Xingu and Araguaia in Serra do Carajas massif. These structures extend in a sublatitudinal direction. The rocks are metamorphosed to amphibolite, partly granulite facies and are included in the extensive field of development of the Lower Proterozoic formations common with the Guyana Shield (Maroni-Itacayunas belt; see below in Sec. 4.1.8.2) and intruded by Early proterozoic granitoids. The Gran-Para group and its analogues comprising mafic and ultramafic (komatiites), partly felsic metavolcanics and ferruginous quartzites, fall among the Archaean formations of greenstone belts here. Rich deposits of iron ore are formed in the weathering crust of these formations, which also contain conglomerates and phyllites. A U-Pb age of 2759 ± 3 m.y. on zircons is known for them.

The greenstone belts are fringed by fields of gneisses, amphibolites, crystalline schists and migmatites including also metavolcanics and ferruginous quartzites. Rb-Sr ages of the same order, i.e., 2700 ± 150 m.y., have been recorded for them.

Small greenstone belts associated with granite-gneisses and migmatites are also known on the south-eastern margin of the Central Brazilian Shield in Mato Grosso state. Their Archaean age is supported by K-Ar dating of gabbro-granite intrusions at 2.8 b.y.

Greenstone belts (nine of them) have been established in the north-eastern and extreme southern parts of S a o   F r a n c i s c o   e o c r a t o n (Fig. 10) in Bahia and Minas Gerais states respectively. In the north-east the belts are oriented meridionally and superposed on a granite-gneiss substratum aged up to 3.2 b.y. The volcanic part of their sequence includes lavas, volcanic tuffs and hypabyssal bodies, predominantly of mafic and also ultramafic, intermediate and felsic composition while the sedimentary part consists of lutites (predominate in the lower part), psammites and psephites. Ferruginous and manganiferous varieties are usually associated with volcanics. The Rb-Sr age of the substratum of the greenstone belts has been determined as 3.4 b.y. and of the greenschist-amphibolite metamorphism of their filling as 2.7 b.y. East of the eocraton the granite-greenstone province adjoins the A t l a n t i c   g r a n u l i t e   b e l t of the polymetamorphic type. Radiometric datings vary from 3.1 b.y. (through 2.7 and 2.0 b.y.) to 550 m.y. and suggest a prolonged and complex history. This belt is made up of various granulites, enderbites and khondalites and formed after various sedimentary (sandstones, arkoses, greywackes, lutites, marls and dolomites) and mafic-ultramafic magmatites. Evidently it lies, at least in part, on the eastern continuation of the granite-greenstone terrane.

The R i o   d a s   V e l h a s-L a f a y e t t e belt located south of the Sao Francisco eocraton is also situated among much older (2.7-3.2 b.y.) but at places remobilised and intrusive gneisses and magmatites. Its age exceeds 2.7-2.9 b.y. and the sequence contains the usual suite of rocks, subdivided into two groups: Nova Lima and Maquine. The first group is made

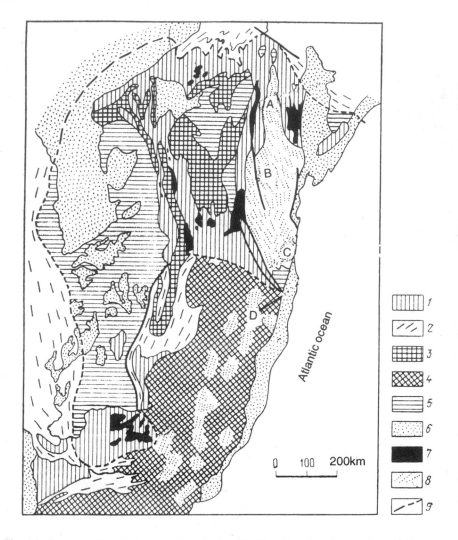

Fig. 10. Proved and probable greenstone belts of the Sao Francisco Craton (from *O Precambriano de Brasil*, 1984).

1—Transamazonian or much older basement; 2—Late Precambrian mobile belts; 3—Espinhaco belt and Chapada-Diamantina cover associated with this belt; 4—much older basement, deeply remobilised in the Brasiliano cycle; 5—sedimentary cover of the Brasiliano cycle; 6—Phanerozoic sedimentary and sedimentary-volcanic cover; 7—greenstone belts; 8—Atlantic granulite belt: A—Caraibas complex, B—Jequie complex, C—Salvador complex, D—Paraiba complex; 9—approximate boundaries of Sao Francisco Craton.

up of shales, phyllites, ferruginous quartzites, metagreywackes, sericite-quartzites and conglomerates, and also mafic and ultramafic magmatites transformed into serpentinites, talc, chlorite, tremolite, actinolite schists and amphibolites; its thickness varies from 1600 to 1400 m. This group contains rich deposits not only of ferruginous, but also of manganese ores and is also distinguished by rich gold mineralisation. The second group is represented by quartz-sericite schists in the lower part and conglomerates, chlorite and sericite-quartzite in the upper part. Its thickness varies from 250–400 m.

The greenstone belts of the Sao Francisco eocraton are characterised by complex isoclinal folding; folds are often overturned and their strike is often complicated by granitic diapirs; zones of tectonic faults are traced. But the modern structure of the entire part of this eocraton was formed not only as a result of deformation at the end of the Archaean (Jequie orogeny), but also of much later deformation manifest 2.0 and 0.6 b.y. ago. Granite-gneiss fields separating greenstone belts are predominantly made up of biotite gneisses, amphibolites, migmatites and auto- and allochthonous granitoids ranging from quartz-diorites to normal granites. Among these rocks ferruginous quartzites, calc-silicate rocks, dolomite-marbles, shales and cordierite, sillimanite and kyanite gneisses are encountered in subordinate amounts.

C e n t r a l   B r a z i l i a n   (G o y a s)   g r e e n s t o n e   b e l t s (Fig. 11) are considered Archaean based on indirect data, i.e., presence of xenoliths of mafic and ultramafic volcanics in granites aged 2.9 b.y. As seen elsewhere, these belts also form complex synforms with isoclinal synclines separated by faults. In the lower part of their sequence ultramafic metavolcanics with typical spinifex structure and fine interlayers of cherts and graphitic phyllites are found. The thickness of the komatiite beds reaches 1.2–1.5 km. The central part of the sequence consists of metabasalts, partly pillow lavas, with beds of ferruginous and carbonate rocks, cherts and graphite schists. Basalts are replaced laterally and vertically by intermediate and felsic volcanics (from andesites to rhyolites) of the calc-alkaline type. Their thickness varies from 800–2400 m. The upper part of the sequence is mainly sedimentary, consisting of graphitic phyllites, iron-bearing cherts, sericite-quartzite and chlorite-muscovite schists, and quartzites with bands of intermediate and felsic volcanic tuffs. Metamorphism is manifest predominantly in the lower stage of the greenschist facies rising to amphibolite facies along the periphery of the belt.

All the Brazilian greenstone belts are characterised by chromite, gold, copper and partly uranium mineralisation. But the distribution of these belts is evidently not restricted to the above two provinces. Another granite-greenstone terrane, not yet well studied, occurs in Uruguay and Argentina and the southernmost part of Brazil. It is known as the Rio de la Plata Craton because of its disposition on both sides of this estuary. Yet another small Archaean nucleus, Luis Alves, has been identified south-west of the Ribeira Late Proterozoic fold belt (see Sec. 7.1.14.2). This province is made up of felsic gneisses, migmatites, ultramafics, quartzites including

56

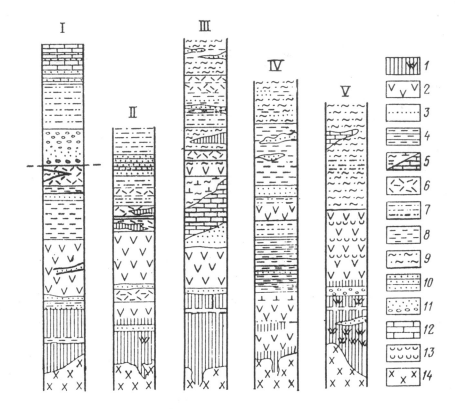

Fig. 11. Scheme showing stratigraphy of the Central Brazilian greenstone belts (after Danni et al., 1982).

1—serpentinites, talc schists, tremolite-talc-chlorite schists; 2—orthoamphibolites, pillow lavas and greenschists; 3—metasilica (Fe, Mn, CO_3) and graphitic phyllites; 4—magnesite-chlorite-muscovite schists (metatuffs); 5—carbonate rocks, marbles, biotite-calcite schists; 6—quartz-sericite schists (metarhyolites); 7—muscovite-quartz schists; 8—graphite-sericite phyllites; 9—chlorite and sericite-chlorite schists; 10—banded iron ores; 11—conglomerates and quartzites; 12—marbles with banded iron ores; 13—tuffs; 14—granites. I to V—stratigraphic columns.

ferruginous ones and calc-silicates. U-Pb and Rb-Sr isochron datings point to an Archaean age of 2.8–2.5 b.y. and other Rb-Sr and K-Ar datings to Early Proterozoic, 2.1–1.9 b.y., reworking.

3.1.9 African Craton

In South Africa two major granite-greenstone provinces, corresponding to Kaapvaal and Zimbabwe eocratons, represent classic zones; they are separated by the mobile Limpopo granulite belt. Greenstone belts are also known on Madagascar. In western Africa the Leon-Liberian eocraton represents a granite-greenstone province of considerable size. The Central African province has been poorly studied compared to the other regions and its contours have not been accurately established.

3.1.9.1 *Kaapvaal Craton* (Fig. 12). Archaean formations emerge on the surface mainly in Swaziland, in the easternmost part of the craton. Here,

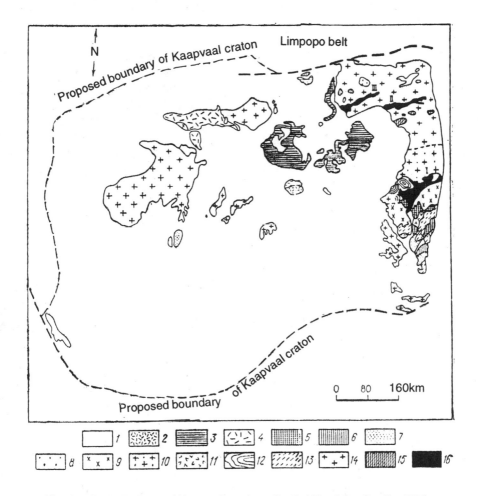

Fig. 12. Geological map of Kaapvaal province, South Africa (after Condie, 1981).

Main greenstone belts: I—Barberton, II—Murchison, III—Petersburg, 1—non-granitic rocks; 2—post-Waterberg granites; 3—Bushveld granites (1.95 b.y.); 4—Gaborone and Palala granites (2.3 b.y.); 5—Mpageni granites (2.65 b.y.); 6—Kwekwe granites; 7—Pongola granites and granites aged 2.7 b.y.; 8—Dalmein granites; 9—Lochiel granites (3.0 b.y.); 10—Nelspruit migmatites; 11—granodiorite series and granites aged 3.2 b.y.; 12—diapirs of tonalites; 13—tonalite-gneisses; 14—unidentified granites; 15—granitic plutons; 16—greenstone belts.

among diverse granites and gneisses, six greenstone belts extend latitudinally; the best known belts are Barberton in the south and Murchison in the north. Detailed stratigraphy has been worked out for the B a r b e r t o n b e l t (Fig. 13) and numerous radiometric age determinations are available. Three groups have been differentiated here: Onverwacht, Fig Tree and Moodis. The thickest (15–16 km !?) is the Onverwacht group, divided into two subgroups, each of which comprises three formations. The lower subgroup is characterised by an abundance of ultramafic and mafic rocks; its upper formation, Komati, was the source of identification of a special type of highly magnesian mafic-ultramafic volcanic rocks, namely, komatiites, differing from peridotites and picrites. Basalt volcanics of the upper subgroup are represented by normal tholeiites; here komatiites occupy a very subordinate position but felsic lavas and tuffs as well as cherts are present in significant amounts. Volcanics of intermediate composition occur to a limited extent. Sills and dykes of peridotites, less frequently pyroxenites and norites and also felsic rocks, i.e., quartz and quartz-free porphyries, are noticed in the sequence. Relicts of primitive micro-organisms, whose organic character was confirmed geochemically, have been detected in the cherts of the Onverwacht group.

The Fig Tree group (thickness 2 km) lies predominantly conformably on the Onverwacht group and consists of sedimentary rocks, coarse clastic in the south and greywackes and shales with subordinate cherts and jaspilites in the north. These formations are regarded as submarine debris cones. The content of sialic material increases upward along the sequence.

The Moodis group (thickness 3.5 km) is connected with the preceding group generally by gradual transition but at places it lies directly on the Onverwacht group and commences with basal conglomerates. Quartzites, subgreywackes and arkoses predominate in its composition; shales, cherts and jaspilites are present in small quantities. The sequence ends with sandstones and conglomerates. These formations have been identified as sediments of alluvial plains and deltas. The Fig Tree and Moodis groups are usually regarded as formations of margins of a continent situated in the south and made up of sialic, volcanic and siliceous rocks.

The metamorphism of the Swaziland supergroup does not extend beyond the greenschist facies with the exception of contact aureoles of granites. In structure the formations of this supergroup form an extensive synclinorium consisting of intensely compressed (up to isoclinal) synforms, usually overturned north-westward and separated not by anchiforms, but by faults (slides). Recent works have reported the complex internal structure of the Barberton belt and the sequence established earlier as well as the immense thickness of the Onverwacht group (contradicted by its low degree of metamorphism) have justifiably been questioned. In particular, the Komati formation is regarded not as an extrusion, but as a complex of parallel dykes

59

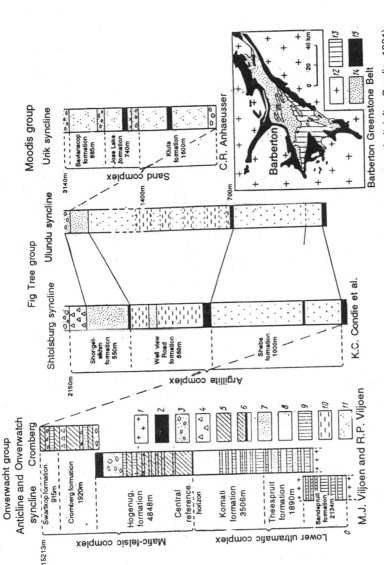

Fig. 13. Stratigraphic columns of the Swaziland supergroup in Barberton greenstone belt (after Condie, 1981).

1—intrusive tonalite gneisses; 2—siliceous rocks with subordinate amounts of shales and limestones; 3—felsic lavas, tuffs and porphyries; 4—mafic pyroclastic rocks, agglomerates, pillow breccia etc.; 5—mafic metalavas (tholeiites); 6—central reference horizon: cherts, limestones and shales; 7—felsic tuffs (often siliceous and aluminous); 8—gaps in sequence; 9—ultramafic lavas (metaperidotites); 10—shales; 11—sandstones, subgreywackes, gritstones, shales and quartzites; 12—intrusive granites; 13—sedimentary subdivisions of Moodis and Fig Tree groups; 14 and 15—Onverwacht group (14—mafic-felsic complex, 15—lower ultramafic complex).

feeding the volcanics of adjoining formations. The deformation history of the Barberton belt, comprising no less than three phases, and the history of granitoid plutonism associated with it are also complex.

Quartzites of the assumed basement of the Onverwacht group have been dated by the U-Pb method on zircon at 3456 ± 4 m.y. and felsic volcanics upwards along the sequence at 3438 ± 2 m.y. Clastic zircons in the Fig Tree group have given an age of 3453 ± 9 m.y. and clasts of granitoids and gneisses in the Moodis group 3570 ± 6 to 3518 ± 11 m.y. These values fall close to the age of the 'Ancient Gneiss Complex' of Swaziland determined as 3644–3450 m.y. and show that this complex could have been the basement and source of supply for the clastic rocks of the Barberton belt. The lower age limit of the belt itself has been determined by diapir intrusions of Kaap Valley and Stentor granites, still corresponding in composition to tonalite-trondhjemite-granodiorite, and dated 3229 ± 5 and 3347 ± 67/60 m.y. respectively. From this the duration of accumulation of volcanics and sediments of the Barberton belt should work out to about 100 m.y. (3450–3350 m.y.).

The southern part of the Kaapvaal eocraton had finally stabilised about 3.0 b.y. ago, i.e., by commencement of the Late Archaean. This is demonstrated by the development of platform cover-type sediments of the Pongola supergroup in the south-eastern part of the eocraton. These sediments lie unconformably on Lochiel granites aged 3.07 b.y. (Rb-Sr method) containing in the lower part lava aged 2940 ± 22 m.y. (U-Pb method on zircon), 2934 ± 114 m.y. (Sm-Nd method) and 2883 ± 69 m.y. (Rb-Sr method on whole rock; Henger et al., 1984) and are intruded by mafics-ultramafics of the Usushwana complex aged 2.87 b.y. (Rb-Sr and Sm-Nd methods). The Pongola supergroup is divided into two groups: lower, mainly volcanic (basalts, andesite-basalts, dacites and rhyolites) and upper, almost wholly sedimentary (conglomerates, sandstones, siltstones and shales). Based on some petrochemical features, lavas of the lower group can be regarded as tholeiites but according to other features as calc-alkaline (nevertheless, the former predominate). They were metamorphosed to greenschist facies (low level) and their thickness reaches 7 km. The sediments of the upper group (thickness 1.8 km) were formed in an environment of alluvial plain, tidal zone and shelf; land should have been in the north and sea in the south. In the lower group stromatolites have been detected, the oldest known in the world.

The intrusive layered Usushwana complex consists of major dykes and sills. Following its intrusion at the end of the Archaean, small hypabyssal granitoid plutons aged 2.8–2.5 b.y. began to form. They are distinguished by high K_2O content and variable $^{87}Sr/^{86}Sr$ ratio.

Latest geochronological data shows that, apart from the Pongola supergroup, two more thick sedimentary complexes of the Kaapvaal eocraton,

Witwatersrand and Ventersdorp, should also be regarded as Late Archaean since their ages are well known to be older than 2.5 and even 2.7 b.y. (they were regarded earlier as Lower Proterozoic). In the period of accumulation of these complexes the zone of sedimentation enlarged significantly northwards compared to the Pongola basin. Commencement of accumulation of the W i t w a t e r s r a n d   s u p e r g r o u p corresponds to a phase of rifting; deposition of the Dominion group, consisting of clastics including coarse clastic sediments and bimodal volcanics, prove this rifting phase. The Dominion group was earlier commonly considered a stratigraphic subdivision equivalent to the Witwatersrand and Ventersdorp supergroups and the three together were known in the literature as the 'Witwatersrand triad'. The Witwatersrand supergroup per se includes clastic rocks of two cycles of sedimentation commencing with coarse and terminating in fine clastic rocks; volcanic sheets play a particularly subordinate role. Sediments accumulated in debris cones of rivers that entered the lake basin. They contain the world's largest concentrations of gold and uranium; the highest concentration is confined to unconformities between the main and subordinate cycles of sedimentation. The total thickness of the Witwatersrand supergroup (including the Dominion) goes up to 11 km; the period of its formation fell in the interval 3096–2714 m.y. (U-Pb method on zircon; Robb et al., 1991).

It must be pointed out that diamictites, regarded as glacial formations (oldest in the world!) or as depositions of mud flows, occur among the Witwatersrand deposits.

According to some researchers, the asymmetry of the Witwatersrand basin distinctly manifest in the structure and sedimentation (erosion of coarse material from the north-west) reflects its nature as a foredeep associated with the Limpopo mobile belt (orogen) (see below).

The V e n t e r s d o r p   s u p e r g r o u p deposited with an interruption and weak unconformity on Witwatersrand formations or much older granite substratum is significantly volcanic and its lower part accumulated under conditions of rifting. Volcanics are bimodal, ranging from basalts to dacites; basalts are tholeiitic but perceptibly alkaline; sediments are clastic right up to coarse clastic. Their thickness is 3–5 km and age about 2.7 b.y. Metamorphism is of greenschist facies, the intensity of fold-fault deformation is generally weak and, together with metamorphism, diminishes upwards along the sequence of the 'Witwatersrand triad'.

3.1.9.2 *Zimbabwe eocraton and Limpopo belt.* The granite-greenstone terrane of Zimbabwe is remarkable for the development of three generations of greenstone belts within its limits. These belts are aged about 3.5, 2.9–2.7 and 2.7–2.5 b.y. Their formation concluded with tonalite plutonism. The first generation is represented by the Sebakwe group, the second by the Lower Bulawayo and the third by the Upper Bulawayo group

(Fig. 14). The rocks of the first group are distributed mainly in the Selukwe belt, contain stromatolites and are intruded by Mushandike granodiorite and Mont d'Or granite aged 3369 ± 72 and 3340 ± 60 m.y. (Rb-Sr method). The conglomerates in the Selukwe belt contain granite boulders, revealing its formation on a sialic basement. On the whole, the Sebakwe group is formed of basalt and komatiite volcanics, with sills of ultramafics, jaspilites and coarse and fine metasedimentary rocks. The Lower Bulawayo group commences with dacites which are overlain by komatiites and coarse clastic metasediments. At places this group lies on gneisses (Chingezi gneisses 2.8 b.y.) and elsewhere on the Sebakwe group. The Upper Bulawayo group, the most widely spread out, lies on Gwenoro-Rodsdale gneisses aged 2.7 b.y.

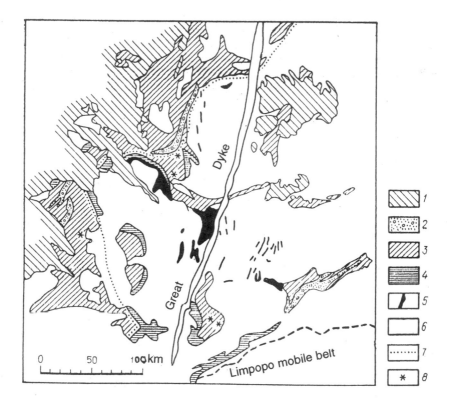

Fig. 14. Subdivision of the greenstone belts of the central part of Zimbabwe (after Condie, 1981).

1—post-Archaean cover; 2—Shamvayan group; 3 and 4—greenstone belts of the Bulawayo group (3—Upper and 4—Lower); 5—Sebakwe group; 6—granites and gneisses of different ages; 7—tentative boundary between western and eastern sequences of the upper greenstone belts; 8—stromatolites.

and consists of komatiites, tholeiites, andesites and dacites, as well as sedimentary rocks, including stromatolitic limestones. Its sequence is capped by the Shamvayan group of poorly graded clastic rocks: conglomerates, greywackes, quartz-mica schists, phyllites, with bands of limestones and jaspilites. The structure of the greenstone belts of Zimbabwe becomes more complicated from north to south in the direction of the Limpopo mobile belt (see below)—from narrow upright folds to nappes in the Selukwe belt with overthrusting of basement gneisses on greenstone rocks. The rocks of greenstone belts are metamorphosed from prehnite-pumpellhite to amphibolite facies and, together with the gneisses underlying them, are intruded by granitic plutons. The youngest complex of granites commences with soda tonalite pretectonic varieties of Sesombi aged 2.63 b.y., extends with syntectonic Chilimanzi adamellite and ends in post-tectonic potash granite (2.57 b.y.).

The intrusion of diapiric tonalite-granites is basically, although not completely responsible for the characteristic reticulate structure of the Zimbabwe eocraton, which came into prominence after the classic work of MacGregor (1983).

The final cratonisation of the Zimbabwe massif set in 2.46 b.y. ago when the Great Dyke of mafic-ultramafic composition intruded into it. The Great Dyke, up to 11 km in width, extends for more than 500 km from north to south, from Zambezi to Limpopo. Thus cratonisation of the Zimbabwe massif was delayed by almost 0.5 b.y. compared to cratonisation of the Kaapvaal massif, in any case its southern part.

The L i m p o p o b e l t separating the aforesaid massifs maintained its mobility for a long time. The belt is about 320 km wide at the centre and extends for a distance of over 700 km in an east-north-easterly direction, roughly along the boundaries of the Republic of South Africa and Zimbabwe, and enters Botswana in the west and Mozambique in the east. The belt is delineated relative to the adjoining eocratons primarily as a zone of very high metamorphism attaining granulite and higher stages of amphibolite facies. Transition to the restraining eocratons is gradual. Marginal zones, northern and southern, with the central zone in the middle, are delineated on both sides of the belt. The oldest basement of the belt emerging in the central zone is represented by Sand River paragneisses aged about 3.8 b.y. and is comparable to the oldest gneiss basement of Zimbabwe and Kaapvaal. About 3.55 b.y. ago, Sand River gneisses were intruded by dykes of tholeiite composition, differing from oceanic tholeiites in higher K_2O and $^{87}Sr/^{86}Sr$ ratio explained by the assimilation of a much older sialic basement. This manifestation of mafic magmatism is roughly synchronous with commencement of formation of the greenstone belts in the adjoining eocratons. Sand River gneisses are overlain by Batebridge supracrustals metamorphosed to granulite and amphibolite facies. They consist of sandstones, clays, carbonates, jaspilites and volcanics that accumulated under shallow marine conditions. In the northern marginal zone volcanics metamorphosed to granulite facies, ferruginous, carbonate and siliceous rocks with pools of mafic and ultramafic intrusives probably forming a continuation of the Sebakwe greenstone belts and occupying a similar stratigraphic position.

Intense compressive deformation commenced at the end of 3.35 b.y.; nappes with northern vergence arose on the southern periphery of the Zimbabwe eocraton. Manifestation of early granulite metamorphism, which may have continued up to around 2.9 b.y., coincides with the first phase of deformation. The Messina ultramafic and anorthosite complex arose almost immediately thereafter (3.3–3.2 b.y. ago) in the central zone. Charnockites and enderbites formed in the northern marginal zone about 2.9 b.y. ago. Later (2.7–2.6 b.y. ago), further phases of deformation occurred with the formation of nappes in the central zone, and folds parallel to the boundaries of the belt and intersecting the much older folds in the marginal zones. Deformation was accompanied by retrogressive metamorphism of amphibolite facies and intrusion of porphyry granites similar to those in the adjoining eocratons. The deformation history of the belt ended in the interval 2.6–2.5 b.y. with the formation of marginal displacements with overthrusts and folds. Soon thereafter the southern continuation of the Great Zimbabwe Dyke and its satellites intruded into the northern marginal and partly central zones leaving an imprint on all the much younger structures.

Contemporary scientists, commencing with Van Viljoen (1977), hold that the Limpopo belt originated as a result of the collision of the Zimbabwe and Kaapvaal eocratons with the former underthrusting the latter. This collision, commencing 3.35 and ending 2.5 b.y. ago, was preceded according to Light (in: *International Symposium on Archaean and Early Proterozoic Geologic Evolution and Metallogenesis,* 1982), by separation of the two eocratons due to the formation between them of an oceanic area more than 1000 km wide in the epoch of formation of the Batebridge group of sediments, which are regarded as sediments of a passive continental margin. According to Light (1982), spreading soon gave way to subduction along the suture of the central zone, particularly evidenced by the presence of aplitic riebeckite-gneisses, which are regarded as the older analogues of 'blueschists'. The direction of relative movement of the eocratons changed over time.

The collision model of the origin of the Limpopo belt well explains the characteristics of its structure, metamorphism and magmatism but there is no confirmation for admitting complete destruction of the continental crust and formation of a new oceanic crust on a significant scale in this region. No traces whatsoever of such a crust have been detected here and the Messina complex, presumably associated with its subduction, represents a typical layered complex. Furthermore, palaeomagnetic studies show that 2.3 b.y. ago the Kaapvaal and Zimbabwe eocratons represented one single entity and became a constituent of the Kalahari protocraton. But according to new information, deformations lasted in this belt until 2.0 b.y.

3.1.9.3 *Madagascar*[4]. On Madagascar typical greenstone belts (Andriamena and others) are of Late Archaean age (2756–2712 m.y.). They extend in a submeridional direction and are predominantly made up of amphibole gneisses, ortho- and para-amphibolites, with subordinate diorites and high-alumina gneisses, leptynites, quartzites and marbles. Intrusions of layered gabbroids and ultramafics are associated with them. Intrusions of tonalites, diorites, granodiorites and granites arose in the concluding stage of development of the belts. Metamorphism of the rocks of the belts attains epidote-amphibolite facies.

Rocks of the greenstone belts are conformably underlain by Middle Archaean formations of variegated composition ranging from amphibolite-gneissic with lenses of ultramafics to quartzite-high-alumina and leptynitic rocks.

3.1.9.4 *Central Africa*. The granite-greenstone province of Central Africa extends in a south-easterly direction from north-eastern Zaire through Uganda into Kenya and Tanzania (Lake Victoria region). The K i b a l i g r o u p covers the greenstone belts in Zaire, Nyanza and Kavirondo in Kenya and Tanzania. The N y a n z a g r o u p is distinguished by a comparatively high content of andesites and rhyolites as also turbidites and jaspilites along with mafic volcanics. The overlying Kavirondo group is predominantly clastic-sedimentary (greywackes, shales, tuffs, arkoses and conglomerates) and molassoid, resembling in this respect the Shamvayan group of Zimbabwe, to which the Kavirondo is proximate in age (2.7–2.6 b.y.). The Nyanza and Kibali groups are older than 2.74 b.y.

Minor relics of greenstone belts are also known in central Tanzania where they form the D o d o m a g r o u p. These relics are made up of metaquartzites (partly ferruginous), marbles, talc, chlorite and corundum schists, amphibolites and hornblende-gneisses. The age of the evidently much younger orthogneisses is 2573 ± 34 m.y.

Greenstone belts have also been recognised in the Kasai eocraton in western Equatorial Africa, south-western Angola (de Carvalho, 1983). Gneisses and migmatites aged about 2.7 b.y. serve as their basement. These rocks comprise three volcanosedimentary complexes (groups). The oldest (Y a m b a) consists of schists, greywackes, volcanics including pillow

---

[4] After Zabrodin (1988).

lavas, phtanites, jaspilites and itabirites. The S h i v a n d a volcanosedimentary suite, consisting of itabirites and conglomerates, at places gold-bearing, lies unconformably on the Yamba. A molassoid formation of clastic rocks, O o n d o l o n g o, lies on top of the Shivanda, again with unconformity. The Yamba group has been intruded by gabbroids and together with the Shivanda group by microgranites as well; all three groups are intruded by granites with Rb-Sr age 2.2–1.8 b.y. Consequently the upper part of this sequence may be of Early Proterozoic age.

In so far as gneisses and granites surrounding and separating the greenstone belts of Central Africa are concerned, extremely old rocks are present among them in the Tanzania eocraton, probably Early Archaean (Bomu granites in Zaire, aged 3.5 b.y., gneisses of upper Luani and Kanda-Kanda in Kasai province, also in Zaire, aged 3.4 b.y.), Middle Archaean gabbronorites and charnockites in Luisa region, also in Kasai aged 2820 (age of granulite metamorphism) and up to 2680 m.y. (retrogressive metamorphism), as well as the much younger Prekavirondo aged 2.74 b.y. and post-Kavirondo aged about 2.54 b.y. (Rb-Sr isochron method) at $^{87}Sr/^{86}Sr$ ratio 0.701–0.702. The age of the youngest Archaean granitoids of the Kasai eocraton is 2593 ± 20 m.y.

Relics of the Middle and Late Archaean granite-greenstone terrane are still preserved in the western part of Central Africa, in the Congo, Gabon and Cameroon; they reveal greater Pan-African reworking in a northerly direction. In Chailu massif (Congo-Gabon) their upper age limit has been put at 2.6–2.9 b.y.; rocks are metamorphosed to amphibolite facies. More northward, in Gabon and Cameroon, metamorphism rises to granulite facies and ages go up to 2.8–3.0 b.y. for granitoids. From here, this granite-greenstone terrane extends eastwards into the Central African Republic, northern Zaire and western Uganda where it forms the K i b a l i e o c r a t o n. Here two generations of greenstone belts, Middle and Late Archaean, separated by granites aged 3.0–2.9 b.y., are known.

3.1.9.5 *Leon-Liberian eocraton.* This Archaean granite-greenstone terrane covers the territory of the eastern part of Guinea, Sierra Leone and Liberia but has been best studied in Sierra Leone (Umeji, 1983). Four greenstone belts with a submeridional strike are known in it. These are formed in a background of granites, granite-gneisses and metamorphics which most probably represent their basement; in 1956, A. Holmes and L. Cahen recorded an age of 3.1 b.y. for them by the lead isotope method. Ultramafics predominate in the lower part of their sequence, mafic and felsic volcanics in the middle and sedimentary rocks in the upper, i.e., the usual sequence is observed. The quantitative ratio of ultramafics, mafics and sedimentary rocks has been determined as 2:5:3. Deposits of chromites and asbestos are associated with the ultramafics. Magnetitic quartzites are encountered on transition from the mafic to the sedimentary part of the sequence. Greywacke turbidites, sometimes transiting into quartzites, predominate in the sedimentary suite; conglomerates are rare. Clastic material is predominantly of quartz composition. Rocks are fairly intensely dislocated (with steep dips predominating) but fold hinges are rarely noticed; schistosity and cleavage are evident.

According to recent data, two generations of greenstone belts are present here, as in the Kibali eocraton. These generations are Middle Archaean (Loco and Cambui) and Late Archaean separated by the Leonean epoch of diastrophism and granitisation about 3.0 b.y. ago. The concluding epoch of diastrophism, Liberian, is dated 2.75 b.y.

3.1.9.6 *Reguibat and Ahaggar massifs.* Archaean rocks form the western part of the Reguibat massif in the north-western part of the African Craton. Amsaga and Gallaman are among the two better known of these complexes. They are intensely migmatised and granitised, especially the latter. Relics of supracrustal rocks are metamorphosed to granulite or amphibolite facies and include pyroxene-amphibolites, granulites, sillimanite and biotite-gneisses, amphibolites, marbles and ferruginous quartzites. In addition to granitoids, the aforesaid are intruded by gabbro-anorthosites and serpentinites. An analysis of radiometric ages by the Rb-Sr method, mainly on gneisses, granites and pegmatites, gave a value of about 2.71 b.y. (Cahen and Snelling, 1984), corresponding to the epoch of amphibolite metamorphism and granite formation. Some age values suggest the possibility of a much earlier (3.2–3.0 b.y.) granulite metamorphism. The youngest Gallaman granites are aged 2539 ± 54 m.y.

More eastward, in the south-western spur of In-Ouzzal in the Ahaggar massif, rocks of magmatic and sedimentary origin emerge, including marbles and ferruginous quartzites, metamorphosed to granulite facies. These rocks have been dated by the Rb-Sr method at 3.2–3.0 and by the Sm-Nd method at 3.4–3.1 b.y. They experienced the later Eburnean and Pan-African tectonomagmatic reworking. Archaean gneisses and granulites are known in central Ahaggar too; the age of granulites goes up to 3.48 b.y.

3.1.10 Indian Craton

As already pointed out in Chapter 2, Karnataka in southern India represents a typical granite-greenstone terrane. Here several greenstone belts with a submeridional strike are delineated in a field of 'peninsular' gneisses (Fig. 15). These belts belong to two generations; the older is known as the S a r g u r   g r o u p and the much younger as the Dharwar supergroup. Rocks of the Sargur group have been preserved, compared to Dharwar, in a very small quantity since they have been replaced by granite-gneisses aged

68

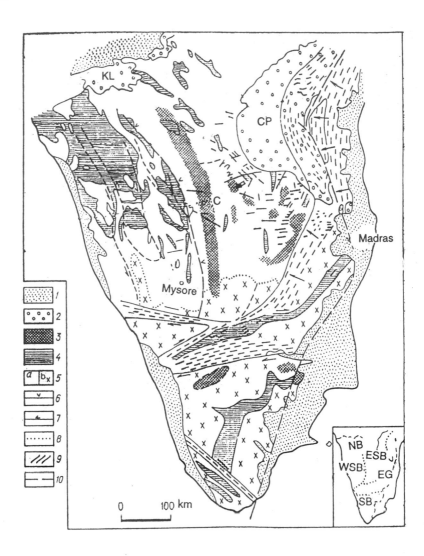

Fig. 15. Tectonic map of southern India (after Drury et al., 1984).

1—Phanerozoic cover; 2—Proterozoic supracrustal formations; 3—migmatites and granitoid plutons; 4—Archaean supracrustal formations; 5—Archaean gneisses (a) and granulites (b); 6—Archaean overthrusts; 7—boundaries of high Archaean stress in western Karnataka; 8—northern boundary of granulites; 9—major Proterozoic dykes; 10—boundary of high Late Proterozoic (?) lineation. Thick broken line depicts the western boundary of positive Bouguer anomaly. C—Closepet batholith; CP and KL—Middle Proterozoic (Lower-Middle Riphean—V.E. Kh.) Cuddapah and Kaladgi basins. Inset shows the main Archaean blocks in India; EG—Eastern Ghats; ESB—Eastern subblock; NB—Northern block; SB—Southern block; WSB—Western subblock.
Arrows—direction of fold vergence. Dotted line—fold zones.

3.0–2.9 b.y. over large areas. At the same time, these formations are evidently younger than tonalite-trondhjemite gneisses dated 3358 m.y., which places them in the Middle Archaean. Quasi-platformal metasedimentary formations were deposited in the basement of the Sargur group: quartzites, quartz-mica and mica schists, paragneisses and limestones; ultramafic and mafic metavolcanics (amphibolites) and ferruginous rocks overlie the former. The metamorphism of the Sargur group of belts is quite high ranging from amphibolite to granulite facies.

The Dharwar complex itself is usually subdivided into two groups: Bababudan and Chitradurga. The sequence of the B a b a b u d a n g r o u p commences with quartz-quartzitic conglomerates unconformably lying on peninsular gneisses and granites aged 3080 ± 80 m.y. (Rb-Sr isochron method) and 3185 ± 60 m.y. (Pb-Pb method) and the Sargur group of rocks. On top is a suite of subaerial metabasalts, quartzites, often magnetitic, with cross-lamination, and metalutites. The upper part of the sequence is more typical of greenstone belts; it consists of metavolcanics, from mafic to felsic, blackschists and ferruginous rocks. The C h i t r a d u r g a   g r o u p (Fig. 16) is over 7 km thick at places and is underlain in the west by gneisses and granitoids aged 2970 ± 100 m.y. (Rb-Sr method) and 3044 ± 150 m.y. (Pb-Pb isochron method; Myers and Williams, 1985). Its sequence is similar to that of the Bababudan group but its basal conglomerates are polymictic and not quartzose; laterally they are replaced by volcanics. On top of them were formed cherts, phyllites, volcanics and later ferruginous and manganiferous cherts and felsic volcanics. The upper part of the sequence is formed of greywackes, phyllites and metavolcanics ranging from mafic to felsic (pillow lavas are present). Some unconformity is noticed between the two groups. The Bababudan group metamorphosed to greenschist or lower stage amphibolite facies and the Chitradurga group to greenschist facies. The age range of the Bababudan group has been determined as 3100 ± 40 m.y. and of the Chitradurga group as 3030–2065 m.y. (Drury et al., 1983). Fig. 16, compiled from aerial photographs, provides an idea of the nature of deformation of the Chitradurga belt.

The Archaean rocks of Karnataka are intersected by the anatectic batholith of Closepet granites aged 2.6–2.4 b.y. extending for 500 km in a meridional direction at a width of up to 50 km. By the time of intrusion of this post-tectonic batholith and its satellites, cratonisation of the South Indian block (including metamorphism which attained granulite facies in the south) was essentially complete, as in the case of the Zimbabwe eocraton by the time of its intersection by the Great Dyke. But the southern tip of the Closepet batholith has been affected by repeated metamorphism, leading to the formation of charnockites (Friend, 1984).

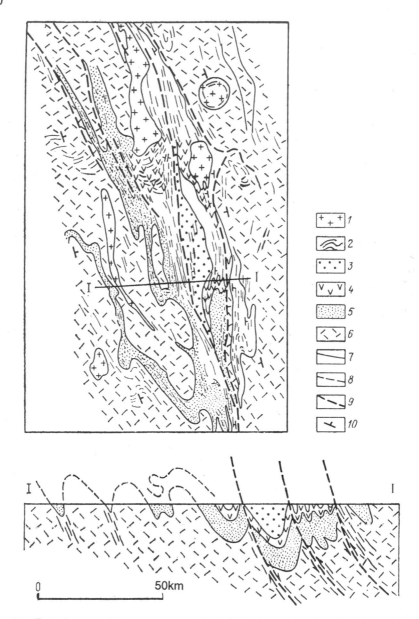

Fig. 16. Tectonic map of the greenstone belt of Chitradurga based on the interpretation of Landsat aerial photographs on scale 1:500,000 (after Drury, 1983).

1—post-kinematic granitoids; 2—unidentified supracrustal formations; 3—greywackes; 4—metavolcanics; 5—possible basal formations; 6—basement gneisses; 7—continuous boundaries of complexes; 8—broken lineation corresponding to bedding or schistosity; 9—main faults; 10—direction of regional dip of beds based on earlier works; I—I—geological profile. Thin lines in gneisses—conventional representation of shears. Faults shown as listric forms.

Thus a special feature of the South Indian greenstone belts is the predominance of sedimentary material in the lower part of all three generations of belts (no wonder they are called 'schist belts' in India), which is proximate in type to sediments of platform covers. This is evidently associated with the presence of a thick sialic basement, a two-phase (at least) complex of 'Peninsular Gneisses'.

The distribution of greenstone belts in India is not restricted solely to South India (Karnataka). They are developed in at least four regions in central India but are poorly exposed and have not been adequately studied. Rajasthan, between the Aravalli range corresponding to an Early Proterozoic fold system (see Sec. 4.1.7) and the field of young traps of the Deccan plateau, represents one such region in the north-west. Here a suite of Prearavalli metasediments consisting of quartzites, dolomites, biotite schists, phyllites, shales etc. was recently identified (in: *Archaean of Central India*, 1980). Metamorphism and migmatisation of this suite increase eastwards where it is known as the Banded Gneiss Complex; in the west it was earlier erroneously placed among the analogues of the Aravalli complex. The Archaean age of these formations has been proved by the intrusion of Berach granites aged 2585 m.y. (analogues of Berach granites—Bundelkand granites).

Westward, in the valley of the Son River, formations of the B i j a w a r g r o u p are known. These formations are represented by metamorphosed sedimentary rocks: quartzites, subgreywackes, jaspers (including haematitic), phyllites and marbles (stromatolites are known); metavolcanics of mafic and ultramafic composition are present in the lower parts. In the upper parts psammites and lutites are found in flysch interstratification. The rocks of the group are intruded by granites transiting into gneisses and are partly migmatised.

The third region of Archaean development in central India lies in the states of Madhya Pradesh and Maharashtra, east of the eastern extremity of the Deccan trap field and south of the latitudinal Satpura range. Here the corresponding formations are known as the Sausar group and are notable for their manganese mineralisation. The Sausar group lies unconformably on Tirodi biotite gneisses (evidently analogues of the 'Peninsular Gneisses' of South India) and, like the Bijawar group, is wholly metasedimentary. It consists of mica schists, transiting either into gneisses or into phyllites, quartzites with bands of gondites, dolomites and calcite-marbles. Eastward it transits into a greywacke-shale flyschoid formation. Metamorphism is predominantly of amphibolite facies; garnets, sillimanite, staurolite and kyanite are also present.

Southward, within the northern frame of the Singhbhum dome, formations of the 'I r o n   O r e   G r o u p', i.e., mafic metavolcanics, jaspilites, shales and sandstones, lie unconformably on an 'Ancient Metamorphic Group', an analogue of the Sargur, and granites aged

3.4–3.3 b.y. These formations are intruded by granites aged 3.1–2.9 b.y., outcropping in the central part of the dome. They represent close analogues of the lower Dharwar.

Yet another region of probable presence of Archaean formations in India occurs in the extreme north-eastern part of the country, in the state of Assam. Here, in the western part of the Shillong block, the so-called Gneiss Group outcrops. It consists of biotite and hornblende gneisses, biotite-chlorite schists with subordinate ferruginous quartzites, felsic and mafic granulites, pyroxenites, charnockites and leptynites. Metamorphism varies from epidote-amphibolite (superposed ?) to granulite facies. The rocks have experienced many phases of deformation and are intensely dislocated with a north-easterly strike. They are intruded by granitoids whose age, unfortunately, has not yet been established.

The above examples allow us to conclude that the possible analogues of Dharwar in the northern part of central India are almost wholly sedimentary and, in primary composition, are closer to a platform cover than their South India counterparts. Thus the northern half of the Indian Shield may have cratonised mainly by the Late Archaean, like the Kaapvaal Craton in South Africa.

3.1.11 Australian Craton

The Pilbara and Yilgarn blocks in Western Australia represent typical granite-greenstone terranes. The Yilgarn block constitutes one of the classic regions for greenstone belts.

3.1.11.1 *Pilbara block.* The greenstone belts of this block fringing the dome of Early Archaean granite-gneisses are amongst the oldest, having formed between 3560 ± 30 (age of mafic volcanics) and 3300–2950 m.y. ago. They comprise two complexes. The lower is known as the W a r r a w o o n a g r o u p and consists predominantly of mafic and ultramafic volcanics with subordinate metasedimentary rocks. In the east it is unconformably overlain by the G e o r g e C r e e k g r o u p consisting of clastic, siliceous and ferruginous formations. Metamorphism decreases from amphibolite in contact with granitoids to greenschist and even prehnite-pumpellyite at the centre of synclines. Complete cratonisation of the Pilbara block set in after 2.95 b.y. with post- tectonic intrusion of granites and eruptions of felsic volcanics aged 2.85 b.y. (Plumb, 1979). Thus it constitutes yet another example of pre-Late Archaean cratonisation.

3.1.11.2 *Hamersley protosyneclise.* The Hamersley basin is made up of clastic sediments, ferruginous quartzites, basalts and tuffs and lies uncon-formably on the southern part of the Pilbara block. Much of the filling is of

Late Archaean age, ranging from 2.8–2.5 b.y. The basin is evidently of rift origin; its formation was accompanied by effusions of basalts and the lower part of the sequence, the Forescue group, is made up of alluvial-lacustrine coarse clastic sediments. The latter are replaced in the Hamersley group by shallow marine formations, among which ferruginous quartzites are extensively developed. These quartzites form one of the largest iron ore deposits in the world. The uppermost retrogressive part of the sequence of the basin can be placed already in the Lower Proterozoic.

The structure of much of the basin is determined by major gentle brachyfolds; it becomes complex southwards where the basin adjoins the Early Proterozoic Ashburton fold system (Capricorn orogen). The degree of metamorphism rises in the same direction, from prehnite-pumpellyite to greenschist and local amphibolite facies.

3.1.11.3 *Yilgarn block.* Greenstone belts are developed in the northern and south-eastern parts of this block while its western part is covered by a granulite belt (Fig. 17), discussed in the preceding chapter. The belts become younger east of the Murchison belt (more than 3050 m.y.) and Southern Cross (more than 2.98 b.y.) to Eastern Goldfields belt (more than 2.79–2.76 b.y.). Recent data of Sm-Nd age determinations of the mafic and ultramafic lavas in Kambalda region in the southern part of Norsman-Willona belt in the eastern part of the Yilgarn block point to the existence there of a much older generation of greenstone belts, comparable in age to that observed in the Pilbara block. Mafics and ultramafics of the Kambalda region carrying sulphide nickel mineralisation have been dated 3262 ± 44 m.y. In another region of the same block, Dimals-Marda, the greenstone rocks are reportedly 3050 ± 100 m.y. in age, confirming the existence of Early Archaean belts (Claoue-Long et al., 1984). These age values have been questioned, however.

Granitoid magmatism began in the Yilgarn block 2.8–2.7 b.y. ago with the formation of 'banded gneisses'; deformation, metamorphism of greenschist facies and formation of granite-gneiss domes occurred in the period 2.67–2.5 b.y. ago. The last events were formation of post-kinematic granites 2.5 b.y. ago and metamorphism and metasomatism 2.4 b.y. ago, all of which points in general to very late cratonisation of the Yilgarn block compared to the Pilbara block.

The sequences of the greenstone belts of the Yilgarn block are characterised by the usual succession of rocks (bottom upwards): ultramafic and mafic metavolcanics including komatiites, with bands of argillites, and also layered ultramafic and mafic sills enriched with sulphide nickel and quartz porphyries; metabasalts, meta-andeśites, meta-argillites, banded ferruginous quartzites and clastic rocks (greywackes and conglomerates). The

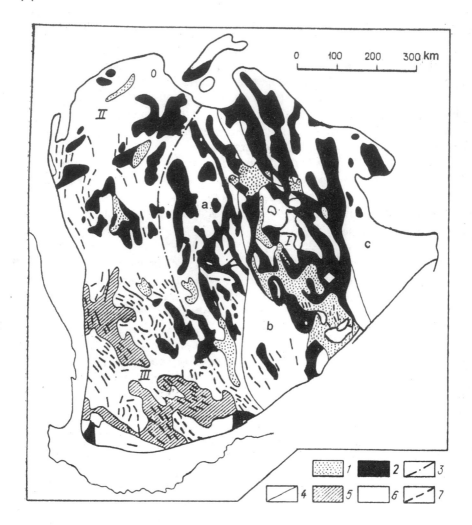

Fig. 17. Geological map of the Yilgarn province, south-western Australia (after Condie, 1981).

Subprovinces: I—Eastern Goldfields; II—Murchison; III—South-western (Wheat belt). Region of subprovinces of the Eastern Goldfields: a—Southern Cross, b—Calgoorlie, c—Laverton. 1—weakly metamorphosed sedimentary rocks; 2—metasedimentary rocks of the greenstone belts; 3—boundaries of subprovinces; 4—boundaries of regions; 5—highly metamorphosed sedimentary rocks; 6—unidentified rocks, gneisses and granites; 7—structural directions.

thickest (about 19.5 km) sequence of Coolgardie-Coorawank in the Calgoorlie belt (Eastern Goldfields), recorded by Condie (1981) based on the data of A.Y. Glikson, is characterised by a very complex and cyclic structure (Fig. 18). Here a greenstone bed of mafic and ultramafic volcanics with pillow texture or thick and intrusive quartz porphyries of bimodal character is repeated twice. On top lies another volcanic suite with volcanics from mafic through intermediate to felsic with a calc-alkaline trend. These are separated by bands of sedimentary rocks. The upper part of the sequence, as in the rest of the belts, is coarse clastic, with horizons of mafic and felsic volcanics. On the whole, volcanics comprise 75% of the sequence of the greenstone belts of the block. Hence the foregoing interpretation of the sequence of the Calgoorlie belt began to be questioned—the repetition of volcanic beds may be tectonic in character and thus its large overall thickness appears grossly exaggerated (as in the case of the Barberton belt in South Africa).

Greenstone belts of the Yilgarn block form synforms with a north-north-westerly strike and a fairly complex internal structure. Metamorphism of the rocks of the belts ranges from prehnite-pumpellyite to amphibolite but green-schist facies predominates. In the west it increases to granulite facies; the corresponding formations here are included in the Western Gneiss Belt (see Sec. 2.1.8.2).

The extent of Archaean on the surface of the Australian Craton is not restricted to the Pilbara and Yilgarn blocks. It emerges also in a small area in the core of Pine Creek uplift in Northern Australia and constitutes the base-ment of the Gawler block in the south-eastern part of the craton in South Aus-tralia. The rocks in both regions are of Late Archaean age. In the first region these are granite-gneisses aged 2.5–2.4 b.y. (U-Pb and Rb-Sr methods). In the second region Archaean ages (2.6–2.3 b.y.) were obtained for gneisses (augen and paragneisses), granite-gneisses and granites. These concomi-tantly represent the age of granulite metamorphism of the metasedimentary rocks.

3.1.12 Antarctic Craton

In the East Antarctic Craton formations of Middle and Late Archaean are difficult to identify among the deeply metamorphosed and very complexly dislocated formations of the Early Precambrian (Kamenev, 1991).

In the west, in Dronning Maud Land, the H u m b o l d t   c o m p l e x made up of supracrustal rocks, predominantly mafic metavolcanics but also terrigenous-carbonate metasediments, transformed into crystalline schists of amphibolite and granulite facies, falls in this age interval.

In Enderby Land, within the frame of the Early Archaean Napier block, the Rayner charnockite-granulite belt extends with similar composition and relics have been dated 3.0–2.5–2.3 b.y. and repeated manifestation of meta-morphism 1.7–1.5, 1.1–0.8 and 0.57–0.52 b.y.

More eastward, in the montane frame of Lambert and Amery glaciers, the Rooker granite-greenstone belt was established. This belt is made up of volcanic rocks ranging from mafic to felsic and sedimentary rocks including coarse clastic and ferruginous quartzites, metamorphosed to amphibolite

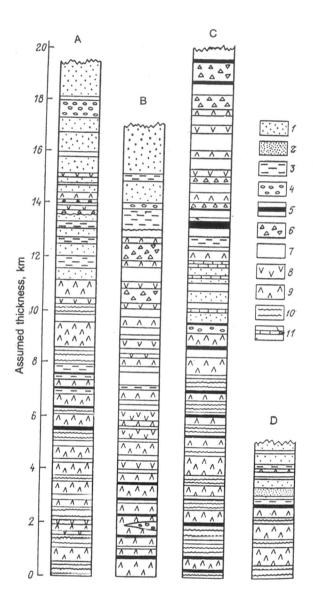

Fig. 18. Generalised stratigraphic sequences of Archaean greenstone belts of Africa and Australia (after Condie, 1981).

A—Coolgardie-Coorawank, Western Australia; B—Bulawayo and Shamvayan groups in the vicinity of Kwekwe, Zimbabwe; C—Tati belt, northern Botswana; D—Nimini Hills belt, Sierra Leone. 1—greywacke-argillite complex; 2—quartzites-arkosites; 3—shales; 4—conglomerates; 5—siliceous and ferruginous rocks; 6 to 8—felsic and intermediate rocks (6—breccia and agglomerates, 7—tuffs, 8—flows and sills); 9—basic volcanics and sills; 10—ultramafic volcanics and sills; 11—carbonate rocks.

and greenschist facies and very complexly deformed. Isotope datings are concentrated in the range 3100–2750 m.y. but much older (3.3–3.7 b.y.) values for granite-gneisses and much younger values right up to 500 m.y. have also been recorded.

Towards the end of the Archaean, possibly even earlier (3.0 b.y. ago), the area of the Antarctic Craton (judging from its exposed part!) experienced general and fairly complete cratonisation. This is suggested by the unconformable bedding of Lower Proterozoic formations, absence in them of signs of migmatisation and granitisation and their affinity to protoaulacogens (see Sec. 4.1.9).

3.2 MIDDLE AND LATE ARCHAEAN TECTONIC REGIME. FORMATION OF MATURE CONTINENTAL CRUST AND FIRST PANGAEA

As pointed out in the preceding chapter, protocontinental crust with bimodal composition and a predominance of orthogneisses with tonalite composition in the upper part and granulites in the lower part already existed at the beginning of the Middle Archaean over considerable expanses within contemporary ancient cratons. It has been suggested that this crust attained maximum thickness within large domal structures and became thinner and possibly wedged out in intermediate depressions, which also served as a preferred site for the location of greenstone belts now known in all the cratons. In most cases it is not clear whether greenstone belt formations accumulated within the same narrow belts in which they are now seen (probably in a somewhat broader frame) or whether these belts become isolated only in the concluding stage of development of granite-greenstone terranes as a result of the buoying up of intermediate granite-gneiss uplifts and represent only relics of more extensive basins. The second assumption is more probable since there are no signs whatsoever (neither sedimentary nor tectonic) of antiquity of the present-day boundaries of greenstone belts. It is wholly possible, however, that contemporary greenstone belts correspond to sites of most intense accumulation of sediments, especially volcanics. This is confirmed on the example of belts in Lake Superior province with their alternation of greenstone and paragneiss belts.

The granite-greenstone terranes and in particular the greenstone belts of all continents, cratons and shields reveal marked similarity in general structural plan, internal structure, composition of magmatic and sedimentary rocks, sequence and degree of metamorphism, i.e., in almost all characteristics, thus pointing to the commonality of their origin. Along with common features, however, certain differences are also apparent in the size of the belts, their shape, thickness of beds, measuring in some cases 15–20 km[5], ratio

---

[5] Such high figures sometimes give rise to justifiable doubts (for example, in the case of the Barberton belt in South Africa and Calgoorlie in Western Australia, discussed above).

of volcanic and sedimentary constituents, structural complexity and depth of metamorphic transformation (from the lower stages of greenschist facies to amphibolite and even granulite). The main difference is the chemical composition of the volcanics, however; this characteristic helps distinguish two main types among greenstone belts: bimodal and tholeiite-calc-alkaline, i.e., contrasting and successively differentiated. A combination of these two types is sometimes noticed with the former overlain by the latter (for example, in Calgoorlie region of Western Australia). Differentiation between these two types, as shown below, is of great genetic importance.

In the Archaean at least three generations of greenstone belts can be distinguished: Middle Archaean (over 3.0–2.9 b.y.) and two Late Archaean (2.8–2.6 b.y.). Their formation wherever they are encountered together is usually divided by an epoch of granite formation. The belts of the Ukrainian Shield and Voronezh massif, Guyana Shield, Pilbara massif (Australia), Swaziland, partly Zimbabwe, north-eastern Zaire, the Central African Republic and India belong to the Middle Archaean generation. In some places this generation is preceded by an even earlier one since, as pointed out in the preceding chapter, many cratons (Greenland, South Africa and India) contain inclusions of rocks which could belong to Early Archaean greenstone belts. Late Archaean greenstone belts are developed in the Canadian, Baltic and Aldan shields, in Brazil, in the Leon-Liberian and Zimbabwe eocratons in Africa, in north-eastern Zaire and in the Yilgarn block of Western Australia, i.e., are widely distributed (a much younger age would have made for excellent preservation!). The formation period of the belts ranges from a few tens to a few hundreds of millions of years, the earliest figures pertaining to the radiometrically well-studied regions of Canada, South Africa and Australia. In any case, these values are quite proximate in duration to, or less than geosynclinal Phanerozoic cycles.

The conditions of formation, nature and geodynamics of the development of greenstone belts continue to evoke lively discussion (Condie, 1981; Windley, 1977). There are two main models: rifting and plate tectonics. The Russian supporters of the former are Grachev and Fedorov (1980) and of the latter Borukaev (1985) and others. According to the rifting model, the belts formed on an older sialic crust during rifting and sagging above mantle plumes or diapirs. Calc-alkaline volcanism is explained either by partial melting of early tholeiites during their subsidence or melting of much older sialic material. The clastic material should have come initially from uplifts surrounding volcanic troughs and later from the growing granite diapirs. Folding must be associated with the gravitational instability of a volcanosedimentary prism above a less dense layer, giving rise to granite diapirism. According to the plate-tectonic model, commencement of development of the greenstone belts is signified by an extension and break-up of the sialic crust and formation of a new oceanic crust under the influence of rising convective currents in the mantle. During subsequent subduction the 'oceanic' (or quasi-oceanic) crust led to island-arc calc-alkaline magmatism. Deformation of the belts occurred due to lateral compression caused by subduction and collision of island arcs; granite diapirism only introduced an additional complication in their structure. The first model can be illustrated by the theoretical scheme of Kröner (Fig. 19) and the second by the concrete example of its application by Dimroth and colleagues (1982–1983) to the Abitibi and Belcombe belts in Canada (Fig. 20). The latter model was meticulously developed theoretically by Windley (in: *Early History of the Earth,* 1980; Windley, 1977) and is currently endorsed by most researchers.

The following phenomena are critical for evaluating the probability of either model: 1) nature of interaction between filling of the greenstone belt and adjoining granite-gneiss (tonalite-trondhjemite-granodiorite) substratum; 2) nature of magmatites constituting the lower sections of the greenstone belt sequences and the degree of their proximity to typical ophiolites; 3) bimodal or successively differentiated nature of volcanics; correlation, or lack of it, of intermediate and felsic volcanics to island-arc calc-alkaline type; 4) type of sediments present in the filling of the greenstone belts.

A study of the greenstone belts in relation to the above criteria revealed a diversity of conditions in their origin and development. For some groups formation on continental (more accurately protocontinental, tonalite-trondhjemite-granodiorite) crust is quite positively established, in particular the Dharwar group of India and the belts of Zimbabwe. In other cases a much older age has been firmly established at least for part of the granite-gneisses fringing the belt relative to the oldest members of the sequence of the belts themselves. Such is the situation in Western Australia (Yilgarn block), in Canada, in the Aldan Shield of Siberia, in northern China, the Leon-Liberian Shield in Africa and, recently proven, the classic belt of Barberton in Swaziland. But data on the nature of contact of ancient granite-gneisses and filling of greenstone belts is contradictory. For example, for the Lake Superior province of the Canadian Shield, Hoffman (1989) has opposed the stratigraphic contact concept and justified its tectonic nature. Finally, for the third group of greenstone belts we are not aware of the existence of much older protocontinental rocks in their framing, probably because the erosion level is not sufficiently deep. Or, as in the case of the Pilbara block in Western Australia, the age of the volcanics of the lower section of filling of these belts appears proximate, if not identical to the age of the granite-gneisses cropping out between them.

From the foregoing brief review, it may be concluded that in most cases direct or indirect proof is available of the formation of Middle and Late Archaean greenstone belts on much older protocontinental crust or between nuclei of such crust arising towards the commencement of the Middle Archaean, i.e., 3.5 b.y. ago.

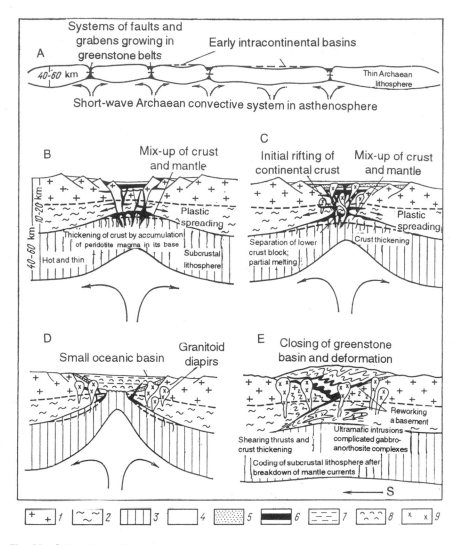

Fig. 19. Schematic profiles showing the suggested evolution of Early Archaean lithospheres in the context of small-scale convection in the mantle (A) and different stages of development of the greenstone belt (B to E) (after Kröner, 1991).

1—upper continental crust; 2—lower continental crust; 3—subcrustal lithosphere; 4—asthenosphere; 5—primitive tholeiitic and komatiitic (B) and bimodal (C) volcanics; 6—mafic and ultramafic rocks of oceanic type; 7—terrigenous sediments; 8—biogenic carbonates; 9—early granitoids (C); granitoid diapirs (D) and late K—granitoids formed mainly by crustal melting (E).

The next point concerns the nature, i.e., ophiolite or non-ophiolite and hence oceanic or non-oceanic nature of the lower, komatiite-basalt part of

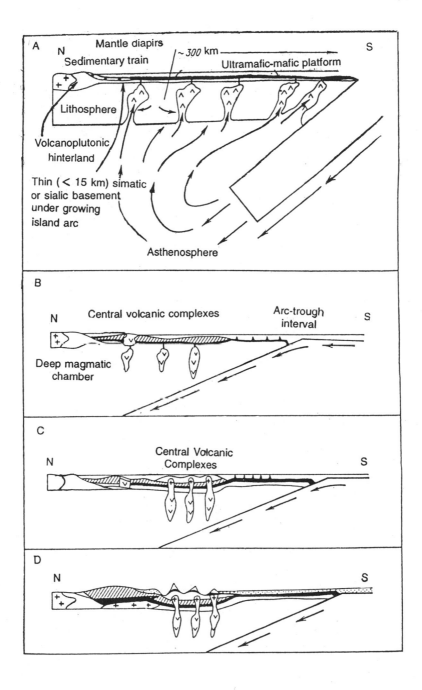

Fig. 20

82

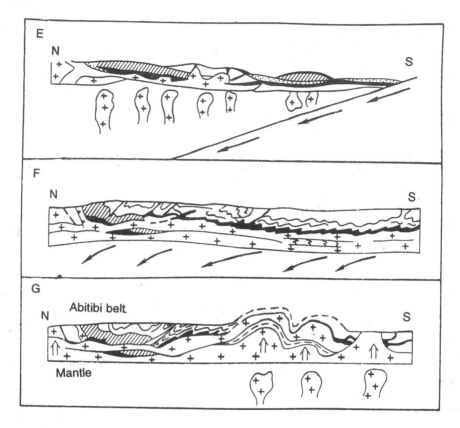

Fig. 20. Model of evolution of the Abitibi greenstone belt, Canadian Shield (after Dimroth et al., 1982–1983).

A—initial phase; B—zone of subduction shifts southwards (tholeiitic magma generated); C—subduction zone continues to shift southwards (generation of tholeiitic and primitive calc-alkaline magma); D—accretion of granites and uplift (calc-alkaline magmatism continues); E—abundant granite formation commences, subduction zones again migrate southwards (generation of calc-alkaline magma replaced by generation of tholeiitic magma); F—folding under the influence of stress associated with subduction (calc-alkaline magma rises and flows into the crust); G—granitic diapirism (last addition of calc-alkaline magma to the crust).

greenstone belt sequences. There are undoubtedly elements of similarity of corresponding complexes with ophiolites and contemporary oceanic crust. Therefore, several scientists in Russia (E.B. Nalivkina, G.I. Makarychev, for example) and abroad (A. Glikson, among others) show no hesitation in labelling these complexes ophiolites and the literature provides actual descriptions of some as ophiolites, e.g. in the Slave eocraton (Kusky, 1990) or in the Barberton belt (Cloud and Glaessner, 1982; DeWit et al., 1992). In fact, extensive development of tholeiite-basalts with pillow structure, presence of ultramafic and mafic sills and, in some sequences, even sheeted dykes (Slave), though not forming a continuous layer, as well as cherts—all these features are strongly reminiscent of ophiolites and oceanic crust. But significant differences also exist, primarily difference in structure of the sequence, different succession and a different set of rocks, i.e., absence of komatiites in the ophiolites per se and systematic replacement of 'layered' cumulative complex by isotropic gabbroids; furthermore, the 'dispersed' nature of distribution of dykes as described for the Slave province suggests a diffuse rather than normal spreading as more likely. Therefore, at best Archaean homologues may be regarded as p r o t o-o p h i o l i t e s to avoid confusion with ophiolites *sensu stricto* falling under the definition of the Penrose Conference. A possible model of the structure of an Archaean proto-ophiolite quasi-oceanic crust is shown in Fig. 21. The average composition of the crust is komatiitic. The initial melt of komatiite differentiates into double diffusion chambers with the formation of komatiitic cumulates in the upper crust and olivine cumulates in the lower. Although this model shows a crust 15-km thick, the average density contrast between the lithosphere and astenosphere does not depend much on thickness of the crust since the lithosphere became denser more rapidly than the hot forsterite asthenosphere, if the temperature of the latter is taken as about 1700°C.

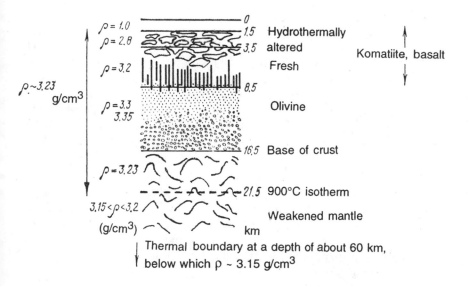

Fig. 21. Petrological model of the ancient Archaean oceanic lithosphere of 20-km thickness with assumed asthenosphere temperature of 1700°C (after Nisbet and Fowler, 1983).

In any case, a certain and even significant similarity of the lower part of the sequences of the greenstone belts (not all, a typical exception being the Dharwar group of India) with ophiolites shows that the crust in the early stages of development was proximate to an oceanic crust. Together with the conclusion from the preceding point that most greenstone belts were formed on continental crust, this denotes that their formation occurred under conditions of extension, sliding above a rising hot convective current into the mantle; it is not important how it is designated—as a mantle plume or mantle diapir[6], or anything else, e.g., advection. Under the condition of a brittle upper lithosphere characteristic of the Proterozoic-Phanerozoic stage of the evolution of the Earth, rifting should have been a consequence of it but the Archaean lithosphere was still intensely heated: the geothermal gradient in the Archaean is assessed at 54°C/km (Grambling, 1981) while the heat flow exceeded the present level by 2 or 3 times. It should thus be assumed that the lithosphere possessed the quality of a wholly viscous material of plastic consistency, sufficiently plastic so that its extension and thinning could not have resulted in the formation of long-living fractures of large amplitude of extension, but only in the formation of cracks which served as courses for the uplift of magmatic melts and were soon closed by them. In fact, faults which are fixed at present in the structure of greenstone belts and in their framing usually arose in the concluding stage of development of these belts, as was demonstrated on the examples of the Aldan and Canadian shields. Therefore, the stage of formation of greenstone belts can be termed rifting only very tentatively; in fact, it would be better called p l a s t i c r i f t i n g as this could ultimately lead to a total break-up of the sialic layer. Whatever the case may be, thinning of the protocontinental crust, the origin of fissures and crustal break-up at places increased its penetrability and provoked commencement of the tholeiite-basalt and komatiite volcanism so characteristic of the early stages of development of greenstone belts. Products of this volcanism are usually intimately intercalated with mafic and ultramafic sills, peridotite etc. The formation of komatiites representing rocks, extremely rare in the subsequent evolution of the Earth, is usually associated with very intense (up to 70%) melting of mantle material ensured by high heat flow. The large thickness (up to 7–8 km) of tholeiite-komatiite suites and their saturation with sills suggest very slow spreading and conditions resembling those of the marginal zones of the contemporary Gulf of California and the Red Sea. In some cases spreading could have been manifested on an even more significant scale; this could have been the case perhaps with the Barberton belt for example, if it is confirmed that the Komati formation represents a complex of sheeted dykes.

---

[6] Some researchers assume that the reason for mantle diapirism could be the low density of an impoverished (depleted) mantle.

The nature of the volcanism of the greenstone belts is an extremely important aspect.

In the middle stage of their development two main types of volcanism are distinguished: bimodal and calc-alkaline. This difference is evidently determined by the rate of extension of the protocontinental basement of the belts. Bimodal volcanism should suggest the preservation of a sialic basement under the entire belt even though thinned and penetrated by intrusions of mafic-ultramafic magma. Calc-alkaline volcanism should be characteristic, on the other hand, of belts in which rifting in their initial stage of development was replaced by spreading with the new formation of 'proto-ophiolite' crust and most probably with its subduction. Since this crust and the entire lithosphere were intensely heated and the scale of spreading was also most probably restricted (the relatively brief period of development of most greenstone belts should be taken into consideration), subduction could have been possible only with a high content of komatiites and peridotite sills in the proto-ophiolite crust (Arndt, 1983) or low density of Archaean asthenosphere (Nisbet and Fowler, 1983) or jointly by both factors. This subduction should have been responsible for the calc-alkaline volcanism that so strikingly resembles contemporary island-arc volcanism (Luts, 1978).

Thus the first, or bimodal type of greenstone belts can be regarded as b e l t s   o f   i n c o m p l e t e   d e v e l o p m e n t and the second type, with calc-alkaline volcanism, as f u l l y   d e v e l o p e d   b e l t s. The rifting model is applicable to the first (with considerable limitations as detailed below) and the plate-tectonic model to the second, thus eliminating the contradiction between them. In this context let us mention that attempts to resolve this apparent contradiction through petrochemical processing of the analytical results of as large a mass of volcanics of greenstone belts as possible without subdividing them according to type and age, including those of Early Proterozoic belts, are totally invalid.

A. Kröner recently expressed the view that bimodal volcanism is exclusively characteristic of Early and Middle Archaean generations of greenstone belts, suggesting a purely rift origin, while much later belts are characterised by successively differentiated volcanism. He based his view essentially on two well-studied examples of proved pre-Late Archaean age, i.e., Barberton and Pilbara. In our view it would be more correct to refer only to a certain tendency associated with progressive destruction of the protocontinental crust since there are examples of ancient belts with successively differentiated volcanism, e.g. the Central Dnepr region, and young belts with bimodal volcanism, e.g. Dharwar and partly Yilgarn.

The fourth criterion of the nature of greenstone belts, i.e., nature of sediments, especially basal sediments, is also very important. The development of quartzites, more rarely arkoses in the lowest sections of several greenstone belts, in particular the Dharwar and Barberton, in part the Yilgarn, once again confirms the conclusion that greenstone belts formed on a protocontinental crust or close to its outcrops.

The end of the middle and commencement of the concluding stage of development of greenstone belts is marked by the relative dormancy of volcanism, its partial substitution by Na-granitoid plutonism and substitution of volcanosedimentary by predominantly or even exclusively sedimentary deposition. Further, the lower part of the sedimentary section is characterised by a predominance of sandy-clay formations; development of turbidites and flysch type of alternation of psammites and lutites is often noticed. This suggests relatively deep basins whose slopes should have been fairly steep in order to ensure the prevalence of turbidity currents. The depth of basins could have gone up to 1–2 km or possibly more; on the whole (according to their parameters and configuration but not location), they may have resembled the modern interior or marginal seas. If the large number of such basins in all the contemporary continents is taken into consideration as well as the possibility that the space between basins may have been covered by a shallow sea, one could concede the conclusion regarding the significant size of the Archaean hydrosphere compared to its present volume. Its pH and salt composition have undergone very little change compared to the Early Archaean. The same may apply to the composition of the atmosphere.

Greenstone belts in the concluding stage of their development underwent extremely intense deformation. Isoclinal-imbricate or imbricate-overthrust structure is quite common for them and tectonic nappes are revealed in an ever-increasing number. Such a complicated structure cannot be explained by granite diapirism as this set in later and only deformed the already existing structural plan. It undoubtedly arose due to tangential compression. In fully developed greenstone belts this compression is the natural result of subduction of proto-ophiolite crust and collision of island arcs with protocontinental margins of the greenstone basins. The conditions of its manifestation in incompletely developed belts are less clear. One precondition might be the thinning and reworking of the sialic crust in the early stage of evolution of these belts; these zones later became sites of volcanosedimentary material piling up under the influence of compressive impulses from fully developed greenstone belts and especially granulite belts. Thus the Zimbabwe belts experienced compression under the influence of impulses from the south, from the Limpopo granulite collision belt.

The conclusive phase of evolution of the greenstone belts is the extensively manifest granite formation phase which, to an even greater extent, encompasses the granite-gneiss terranes separating them (secondary ?). Most researchers regard granitic magma as a product of remobilisation, of partial remelting of the more ancient sialic substratum and tonalite basement of the belts. Probably, such melting occurred primarily under the thick accumulations of volcanics and sediments but the granite melt or partially molten granite-gneiss material was squeezed from here in the direction of the periphery of the greenstone basins, where the largest bodies of Late Archaean granitoids have localised. The composition of these granitoids became increasingly felsic over time and enriched with potassium. New areas came up above the water surface, initially composed of island-arc volcanics and later non-volcanic material due to the piling up of volcanosedimentary material of the belts and subsequently plutons and diapirs of granitoids. The basins correspondingly became shallow and were filled increasingly with coarse clastic material.

The development of greenstone belts should have led at the end of the Archaean aeon to a significant growth in crust, up to 30–40 km, judging from the pressure required for the metamorphism of granulite facies; the metamorphism of their filling and granitisation must have resulted in the formation of a new granitic layer covering the tonalite, 'grey gneiss'. Hence three layers can be distinguished today in the consolidated crust: granulite, tonalite and metasedimentary-volcanic. This process of cratonisation proceeded in two phases: 1) at the verge of the Early and Late Archaean (about 3 b.y. ago) and 2) at the end of the Archaean (about 2.6 b.y. ago). Large areas within the future Gondwana supercontinent had already experienced cratonisation as a result of the first phase, namely: the Kaapvaal eocraton, northern part of the Indian Shield, Pilbara eocraton and probably huge areas of the Antarctic Craton. Cratonisation at the end of the Archaean encompassed large expanses of Gondwana and Laurasia. It has been suggested that this may have led to an increase in thickness not only of continental crust, but also of the entire lithosphere due to depletion of mantle as a result of magmatism.

It may be recalled that in the Archaean, apart from granite-greenstone, one more type of structural unit of the Earth's crust appeared, i.e., belts of tectonothermal reworking or granulite-gneiss belts, which were still few at that time. The earliest of them is evidently the Limpopo belt, which began developing by 3.35 b.y. and the first granulite metamorphism by 3.1–2.9 b.y. In the western granulite belt of Yilgarn block (Australia) this metamorphism may have been older, about 3.3 b.y. The same is true of the Antarctic Craton. But in the Atlantic belt of Brazil and in the belt encompassing South India and the Eastern Ghats, this metamorphism pertains only to the end of the Archaean but possibly was already a secondary one (Friend, 1984). The origin of granulite belts on the example of the Limpopo belt can be associated with the collision of the earliest cratonised microplates along sutures formed by rifts or faults under conditions of converging, descending, convective currents.

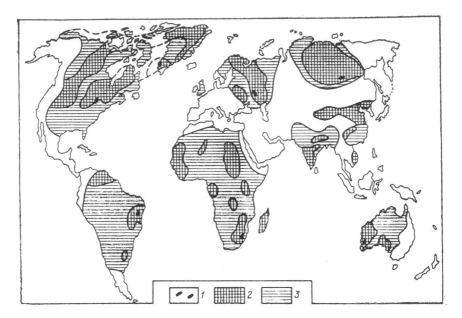

Fig. 22. Archaean eocratons (protocratons) and Early Proterozoic mobile belts in the composition of the basement of ancient cratons (compiled by Khain).

1—areas of outcrops of oldest (over 3.5 b.y.) crust ('grey gneisses'); 2—Archaean eocratons; 3—continental crust formed in the Early Proterozoic (partly regenerated Archaean).

The present distribution of Archaean rocks (Fig. 22) and their exposure at the base of Proterozoic mobile belts (e.g. in the Canadian Shield and in Greenland) suggest their presence almost everywhere in the consolidated crust of ancient cratons, their margins and many microcontinents of Phanerozoic mobile belts. It follows from this that cratonisation at the end of the Archaean was largely responsible for development of the continental crust of the Earth in general. According to the calculations of various researchers, 60 to 80% of the present volume of continental crust was formed by the end of the Archaean, i.e., by 2.5 b.y., against 10 to 15% at the end of the Early Archaean. Possibly, the overall piling up of continental crust at this time, manifest in closing and deformation of the basins of greenstone belts and formation of granulite belts, led to contraction of the sialic layer of the crust in one hemisphere which became continental and liberation of the other hemisphere from sial rendered it oceanic. This process could have continued in the Early Proterozoic.

In conclusion, let us revert once again to the subject of the manifestation of plate tectonics in the Archaean.

While for the Early and probably Middle (according to Kröner, 1991) Archaean, remelting of primary mafic-ultramafic crust can be assumed as responsible for the formation of protocontinental (tonalite-trondhjemite-granodiorite) crust by its simple subsidence into the mantle (this process is also called 'sagduction' as distinct from 'subduction'), for the Late Archaean in general it was no longer relevant. Formation of volcanic arcs with the magmatism typical of them is best explained by subduction.

In all the granite-greenstone regions where granite diapirism did not significantly distort the primary plan of disposition of greenstone belts, the latter are disposed in roughly parallel groups but not wholly identical in age. In the Slave and Superior provinces of the Canadian Shield, the Karelian block of the Baltic Shield and the Yilgarn block of Western Australia, the belts become younger in a definite direction across their strike. In the Slave province this tendency is clearly directed towards the ancient protosialic nucleus; such a tendency may be assumed in other cases too. This suggests that the successive accretion of volcanic arcs to the ancient nuclei occurred with the closure of intermediate small quasi-oceanic basins and the opening up of such new basins at the rear of these arcs. The continental crust of granite-greenstone terranes around some ancient nuclei could have formed in this manner. And all this against the background of migration of uprising convective cells

The granite-greenstone terranes thus formed represented nuclei for future cratons and continental lithospheric plates. Their number should have been significant and their size did not generally exceed a thousand kilometres. This permits defining Archaean plate tectonics as multiplate or miniplate tectonics. It could also be called embryonal plate tectonics if one considers that the processes of spreading and subduction were rather different then from those in evidence today.

4

Early Proterozoic: Partial Destruction and Restoration of Pangaea

The fact that the verge of the Archaean and Proterozoic is marked in many regions of the world by major unconformities and manifestations of granitic plutonism and regional metamorphism, as reflected by radiometric ages, gave rise to the concept that this verge represents one of the major benchmarks in the history of the Earth. However, factual data collected subsequently showed that the importance of this verge has been somewhat exaggerated. Many features of the tectonic regime characteristic of the Early Proterozoic, i.e., appearance of protoplatforms, origin of deep faults, formation of dyke swarms and some others, are noticed already in the Late Archaean and, on the other hand, such typical Archaean structures as, for example, greenstone belts, continued to exist in the Early Proterozoic (Condie, 1982; Windley, 1984). Nevertheless, the structural plan of the Early Proterozoic as a whole differs quite significantly and vitally from the Archaean, as is evident from a regional review.

4.1 REGIONAL REVIEW

4.1.1 North American Craton

As pointed out in the preceding chapter, there is every justification to assume that a sufficiently mature and consolidated continental crust was already formed over much of this craton by the end of the Archaean (Fig. 23). Doubts have been expressed only with regard to the more southern part of the craton, namely south of Lake Superior and in southern Wyoming where Archaean rocks do not outcrop, but isotope data for magmatites of Lower Proterozoic age does not suggest the presence of a very old sialic crust and points to mantle origin. Ignoring this disputed region, it may be stated that commencement of the Proterozoic over much of the North American Craton was characterised by partial destruction of the Epi-Archaean continent. As a result, the area of development of Lower Proterozoic formations within the exposed (Canadian-Greenland Shield and Rocky Mountains) as well as sediment-covered parts of the North American Craton exceeds the

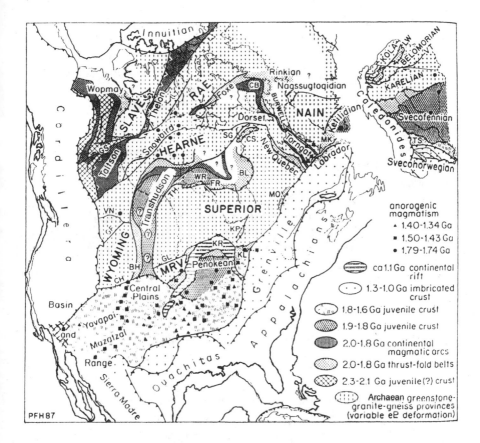

Fig. 23. Precambrian tectonic elements of the North American Craton (platform cover removed) and Baltic Shield (after Hoffman, 1989).

Upper case names are Archaean provinces; lower case names are Proterozoic and Phanerozoic orogens. BH—Black Hills inlier; BL—Belcher fold belt; CB—Cumberland batholith; CH—Cheyenne belt; CS—Cape Smith belt; FR—Fox River belt; GF—Great Falls tectonic zone; GL—Great Lakes tectonic zone; GS—Great Slave Lake shear zone; KL—Killarney magmatic zone; KP—Kapuskasing uplift; KR—Keweenawan rift; LW—Lapland-White Sea tectonic zone; MK—Makkovik orogen; MO—Mistassini and Otish basins; MRV—Minnesota foreland; SG—Sugluk terrane; TH—Thompson belt; TS—Trans-Scandinavian magmatic zone; VN—Vulcan tectonic zone; VT—Vetrenny tectonic zone; WR—Winisk River fault.

area of the Epi-Archaean eocratons (protoplatforms) which have remained stable. The quiet tectonic regime of the latter is demonstrated by the gentle homoclinal attitude of the weakly metamorphosed Lower Proterozoic shelf formations on the western and eastern slopes of the Slave eocraton and on the southern slopes of the Wyoming and Superior eocratons. Further, structures of aulacogen type were established within the Slave and Superior eocratons, this phenomenon being very characteristic for cratons.

The Superior (Lake Superior) eocraton represents the largest Archaean eocraton of North America, the rest being of minor proportions: Wyoming, Slave and Nain in eastern Labrador and southern Greenland. There are three more such minor eocratons—Hearne, Rae and Burwell—which were recently singled out by Hoffman (1989) from Churchill province, formerly regarded as wholly Early Proterozoic. The following Early Proterozoic mobile belts have been delineated along the periphery of eocratons and between them: folded protogeosynclinal Wopmay system west of Slave eocraton, Transhudsonian between Hearne eocraton in the north-west, Wyoming in the south and Superior in the south-east; Penokean system south of Superior eocraton with continuation in the east into the Ketilides of extreme southern Greenland. The Wyoming, Hearne, Superior and Nain eocratons divide the branches of the Transhudsonian system (eastern part known as the Labrador system) (Fig. 24).

Let us commence the review of these structural zones with the well-studied Wopmay system[1] (orogen) in the extreme north-western part of the Canadian Shield (Hoffman and Bowring, 1984).

4.1.1.1 *Wopmay protogeosynclinal fold system.* This system extends for 600 km in a submeridional direction with a width of up to 220 km and

---

Fig. 24A. Major tectonic elements of western Canadian Shield and basement of western Canadian Platform (modified and extended by Hoffman, 1989).

Archaean provinces (1–6): 1—Superior, 2—Wyoming, 3—Medicine Hat, 4—Hearne, 5—Rae, 6—Slave, including Simpson Islands terrane (SN). Zones of Archaean crustal reactivation (7–10): 7—Thompson, 8—Queen Maud, 9—Peter Lake, 10—thrust-fold belts involving Early Proterozoic cover; 11—mylonitic mafic-felsic granulites of uncertain age occurring within Snowbird orogen; 12—metasedimentary-volcanic rocks of unknown age in Lacombe zone of Hearne province; 13—volcanic (1.93 Ga) and sedimentary rocks of Wilson Island terrane (WN) within and adjacent to Great Slave Lake shear zone (GSLsz). Early Proterozoic terranes (14–16). 14—Buffalo Head, Chincaga and Thorsby (all 2.3 to 2.1 Ga), 15—Hottah (1.95 to 1.90 Ga arc built on 2.3 to 2.1 Ga crust), 16—Nahanni (age unknown, possibly equivalent to 2.3 to 2.1 Ga, Yukon-Tanana terrane of northern Cordillera). Early Proterozoic continental magmatic arcs (17–20): 17—Taltson-Thelon and Ksituan (both 2.0 to 1.90 Ga), 18—Great Bear (1.88 to 1.84 Ga) and Fort Simpson (ca. 1.85 Ga), 19—Wathaman-Chipewyan (1.86 to 1.84 Ga), 20—Rimbey (1.82 to 1.78 Ga). Internal zones of Transhudsonian orogen (21–22): 21—subsurface extension underlain by North American central plains conductivity anomaly (NACP) and Archaean subcrops, 22—juvenile crust exposing 1.91 to 1.85 Ga island arcs and interarc metasediments; 23—edge of platform (including Middle-Late Proterozoic) cover; 24—edge of Cordilleran deformation.

---

[1] Formerly known as the Coronation geosyncline because of its proximity to Coronation Gulf in the Arctic Ocean.

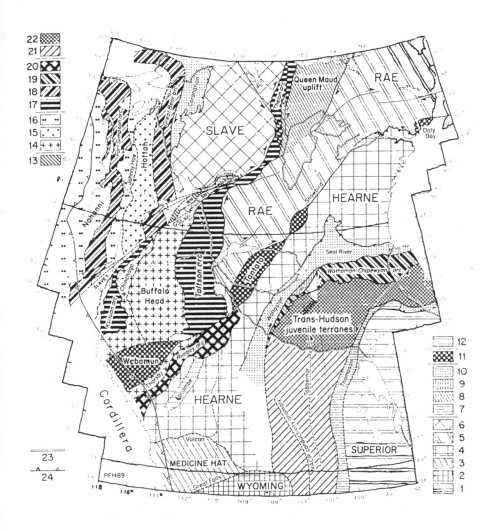

94

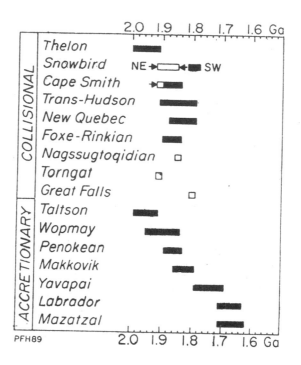

Fig. 24B. Age spans for Early Proterozoic orogens as indicated by U-Pb ages for orogenic igneous and/or metamorphic suites (from Hoffman, 1989). Igneous suites represent foredeep, island arc, continental arc and collisional anatectic magmatism; metamorphic ages are based on zircon and monazite, but not titanite, dating. Open boxes represent single ages and arrows indicate minimum or maximum limits based on pre- or post-orogenic rocks. Age spans are shown for the Snowbird orogen for the shield (NE) and platform (SW).

is concealed in the north in the region of Coronation Gulf under a Riphean-Palaeozoic mantle of the Arctic slope of the shield and in the south under the Phanerozoic cover of the Great Plains. Three zones are distinguished in its structure: 1) the ancient passive continental margin intensely deformed subsequently; 2) plutonometamorphic belt of Hottah; and 3) volcanoplutonic belt of Great Bear Lake situated on the suture between the first two zones. Formations of the passive margin, i.e., Coronation supergroup, include three sedimentary complexes: synrift, passive margin proper and orogenic (Fig. 25). The synrift Akaitcho group fills the grabens unconformably lying on Archaean granites aged 2.5–2.51 b.y. (here and subsequently the age determinations were done by the U-Pb method on zircon). Mature clastic and chemogenic sediments with flows of continental tholeiites occur at the base of the group. They are overlain by extremely thick feldspathic gritstone and later by a typical bimodal association of submarine lavas and tuffs,

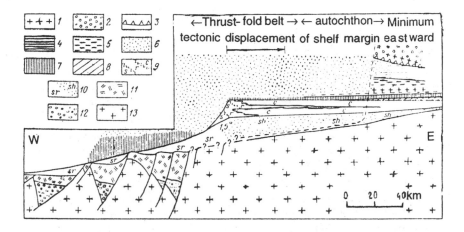

Fig. 25. Reconstruction of conditions of accumulation of the Coronation supergroup, assuming 40% compression of the sedimentary prism west of the frontal thrusts and above the basal layered deposits (after Hoffman and Bowring, 1984).

1—sills of gabbro (Morel); 2—fluvial lithofeldspathic arenites; 3—halokinetic megabreccia; 4—calcareous concretional basin shales; 5 and 6—feldspathic-lithoclastic turbidites; 7—thinly banded graphite-sulphite schists; 8—quartzite-siltstones; glauconite-dolomites.
Formations of passive margins (Apworth group): 9—dolomites; 10—quartzite-arenites; semi-lutites.
Initial rifts (Akaitcho group): 11—submarine basalts, tuffs, rhyolites, sills of gabbro ± submarine clastic sediments in the cover of reef dolomites; 12—feldspathic turbidites, semi-lutites, conglomerates ± submarine volcanics, sills of gabbro and rhyolite porphyries; 13—crystalline basement.

i.e., basalts of transitional type from continental to oceanic and subalkaline rhyolites overlain at places by reef carbonates and elsewhere by lutites.

Accumulation of the Apworth group (thickness 1–3 km), distinctly differentiated into shelf and slope (with rise ?) formations, corresponds to the stage of passive margin. In the shelf zone the sediments of the group lie directly on the Archaean basement; their lower part is made up of mature quartzite-sandstones while the upper part is made up of a carbonate (now dolomitic) platform with a distinct reef margin and stromatolitic banks in its rear. In the zone of rise and slope sediments of the Apworth group are underlain by riftogenic rocks of the Akaitcho group and are represented by 'semi-lutites' with bands of quartz-contourites and turbidites. Intrusion of gabbro sills into the peak of the ancient continental slope falling supposedly at the boundary of unreworked continental crust that has experienced extension represents an interesting feature.

Formations of the Recluse group, also differentiated into two formational zones, correspond to the foredeep stage. Dark-coloured banded shales (graphite- and pyrite-bearing), evidently of deepwater origin, are deposited at the base of the group, suggesting simultaneous subsidence of the shelf and slope. Later, a bed of turbidites of plutonometamorphic origin deposited by linear currents was formed in the zone of the inner shelf and slope. In the zone of the inner shelf, which has preserved features of autochthonous deposition, turbidites are replaced upwards by shales in a rhythmic alternation with limestones and later by continental molasse, which completes the sequence. Continental molasse is absent in the more western zone, which at this time experienced intense fold-thrust deformation with a decollement from the basement and general eastward displacement. The first phase of deformation and metamorphism of greenschist facies has been dated 1.9 b.y. on rhyolites in the upper part of the Akaitcho group and 1.88 b.y. on post-tectonic granite intrusions. It has been suggested that the first phase was caused by the collision of the Hottah block of unknown origin with the continental margin. Commencement of formation of the foredeep is placed in this phase. The second phase of deformation was manifest in the formation of cross-folds in all the zones, including the eocraton but excluding the Great Bear Lake volcanoplutonic belt. Even more tentatively, the second phase is associated with collision of the north-western zone of Churchill province (Rae block, after Hoffman, 1989) with the Slave block. This phase has been dated 1885 m.y. on intrusives and 1860 m.y. on volcanics of Great Bear Lake. The third phase of compression was manifest in formation of the conjugated north-eastern dextral and north-western sinistral strike-slip faults affecting also the volcanoplutonic belt and regarded as the result of collision on the western margin of Hottah block whose boundary, based on gravimetric data, has been put at 180 km west of the belt front. Deformation of the youngest Great Bear Lake granites occurred later, i.e., 1.84 b.y., and probably earlier at 1.81 b.y., which is the age of the post-tectonic granites in the Churchill province. Consequently, evolution of the western margin of the Slave eocraton was extremely brief: about 15 m.y. rift formation, 10 m.y. formation of continental margin sediments, 20 m.y. main epoch of deformation and 40 m.y. post-collision magmatism.

Apophyses-aulacogens set off from the Wopmay protogeosyncline north and south penetrate deep into the body of the Slave eocraton. In the north they are represented by the K i l o h i g o k (Bathurst) a u l a c o g e n, commencing in the same-named Arctic Ocean bay and extending south-south-easterly; it is intersected by the Bathurst fault (thrust) roughly along the axis. This aulacogen joined with the protogeosyncline through the latitudinal Taktu aulacogen within Coronation Gulf. In the south the A t h a p u s c o w a u l a c o g e n, originating in Great Slave Lake and extending along its Eastern Bay, represents its analogue. A major fault (thrust), Macdonald-Wilson, follows along it. The Bathurst and Macdonald-Wilson faults extend eastwards into Churchill province where they probably attenuate. Both aulacogens are filled with thick (5–7 km) beds of unmetamorphosed shallow marine carbonate-terrigenous formations, called the Goulburn and Great Slave supergroups respectively. They correlate well with shelf sediments of the outer zone of the Wopmay protogeosyncline. The internal structure of both formations is quite complex; they are formed mainly of quartzites, sandstones, shales, quite frequently red beds, limestones and dolomites, including stromatolites. Some cyclicity is noticed in the alternation of these rocks. Felsic and mafic (bimodal) volcanics are known at different levels in the Great Slave supergroup. Felsites of the Wilson group constituting the lower part of this supergroup are 1928 m.y. old (U-Pb age on zircon), granites underlying it 2175 and 2094 m.y. and intruding granites 1895 m.y. (Bewring et al., 1984). With an overall synclinal structure, formations on the north-western flank of the Athapuscow aulacogen are thrust onto the eocraton. The age of post-thrust intrusions was determined by the same method as 1865 m.y. These ages point to near synchronous development of aulacogens, at least of the Athapuscow and Wopmay.

4.1.1.2 *Transhudsonian protogeosynclinal fold belt.* This mobile belt is the largest and thus the least homogeneous structural element of the Canadian-Greenland Shield, which it intersects in a general south-west-north-easterly direction. Three major segments can be delineated in the structure of the belt. These are: 1) Churchill, west of Hudson Bay between the Slave and Superior eocratons; 2) Circum-Ungava, between Hudson Bay and the Labrador Sea (Baffin Land structure can be related to it); and 3) Central Greenland, east of Baffin Bay and Davis Strait in the Labrador Sea.

The C h u r c h i l l s e g m e n t ('province') is about 800 km in width. As already mentioned, two more eocratons (Early Proterozoic microcontinents), Rae and Hearne, are now distinguished in this part of the Transhudsonian belt (in the broadest sense here) adjoining the Slave eocraton. Felsic gneisses aged 2.9–2.8 b.y. predominate in the first of these eocratons but much older (3.1–2.9 b.y.) and much younger (2.6 b.y.) orthogneisses are also present while outliers of supracrustal rocks, relics of greenstone belts, are interspersed among the predominant gneisses. This eocraton is set off from the Slave eocraton by a fairly narrow (up to 25 km) Taltson-Thelon zone which is considered a volcanoplutonic belt aged 2.0–1.9 b.y. An even narrower mylonite-granulite Snowbird belt (1.9–1.85 and up to 1.78 b.y.) separates the Rae and Hearne eocratons. The main part of the basement of the latter is formed of Upper Archaean metavolcanics and their erosion products metamorphosed only to greenschist facies are intruded by gabbro, diorites and granites aged 2.68–2.70 b.y. At its boundary with the internal zone of Transhudsonian orogen, rocks of granulite facies are developed; zircons aged 3.2–3.1 b.y. have been encountered among gneisses. Deformed outliers of Lower Proterozoic platform cover and foredeep, particularly of the

Hurwitz group, have been preserved on the surface of the eocraton. Its analogue in the Rae eocraton is the Amery group. Their sequences are shown
in Fig. 26 and their age is 1.91–1.81 b.y. The accumulation of these sediments was preceded by deposition of molasse deposits of the Nokacho,
Martin and Baker groups and intrusion of calc-alkaline and peraluminium
plutons aged 2.02–1.92 b.y. Calc-alkaline plutons aged 1.86–1.82 b.y. are
also known while mafic volcanics and plutons formed here from 1.85 b.y.
onwards. Rapakivi granites and rhyolites are much younger, i.e., 1.76 b.y.
Finally, the relatively large superposed basins of Thelon and Athabasca
filled with practically undeformed continental clastic formations were formed
around 1.7 b.y. ago. Thus the total cratonisation of the north-western part
of Churchill province concluded only by this period even though it had commenced with the collision of the Slave and Rae eocratons in the interval
1.97–1.92 b.y. with the formation of the Taltson-Thelon granulitic and plutonic belt. In this collision, which P.F. Hoffman compares with the Himalayan,
the Slave eocraton played, in his view, the role of a foreland and the Rae
eocraton the role of a hinterland. He traces the Taltson-Thelon belt over a
distance of 3200 km from Alberta to Ellesmere Island in the Canadian Arctic
Archipelago.

In general, the north-western part of the Transhudsonian orogen,
although including major cratonised blocks even at the end of the Archaean,
experienced final cratonisation only at the end of the Early Proterozoic,
which justifies its inclusion as a constituent of the 'Churchill province'
and its orogen. The structure of the south-eastern part of the Churchill
province, the Reindeer belt, is altogether different. It adjoins with the north-
western part of the Virgin River fault and with the Superior eocraton along
the 'Thompson front' accentuated by salients of serpentinised ultramafics.
A zone (Cree Lake) of shelf development, predominantly clastic Lower
Proterozoic formations, adjoins the Virgin River fault from the south-east.
These formations are relatively weakly dislocated and metamorphosed and
lie unconformably on the Archaean basement projecting into granite-gneiss
domes. These formations, known as the Wollaston group, have the Hurwitz
group on the west coast of Hudson Bay as their north-eastern continuation.
They represent typical formations of a passive continental margin (Fig. 26)
which was evidently transformed later into a foredeep. Farther south-east,
beyond the Needle Fork zone of faults, there are no reliable signs of the
presence of Archaeans even at depth, barring the xenoliths of uncertain age
in the Wathaman granite batholith. This batholith, one of the largest in the
world, extends through the entire province with a length of up to 900 km and
width 70–130 km; its U-Pb age is 1865 ± 10 m.y. The La Ronge greenstone
belt or island arc with basic to intermediate and subordinate felsic volcanics
and quartzites, arkoses and conglomerates extends south-east, parallel to
the batholith but is separated from it by a tonalite-migmatite complex. Ages

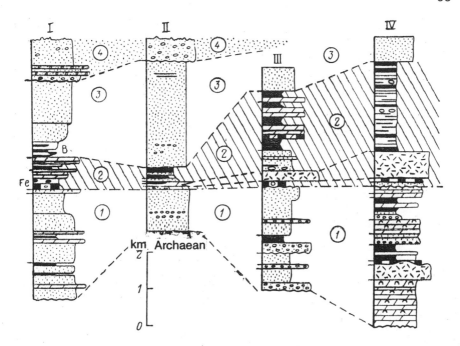

Fig. 26. Sketch showing structure of the Lower Proterozoic formations of some fold belts of the central and western parts of the Canadian Shield (after Young, 1980).

1 to 4—successive stages of sedimentation: lines joining the columns do not signify stratigraphic correlation; I—Kilohigok basin; II—Amery belt; III—Hurwitz series; IV—Belcher Island; Fe—ferruginous rocks; B—breccia.

of 1835 ± 75 and 1790 ± 35 m.y. were recorded for the volcanics and 1840 and 1825 m.y. for the granitoids intruding them. Interestingly, before radiogeochronological investigations, this zone was regarded as part of a typical Archaean granite-greenstone terrane. Kissenew gneisses are formed even more south-east of the La Ronge belt. The grade of metamorphism of these gneisses increases towards the Thompson front while granite plutons disappear (*Proterozoic Basins of Canada*, 1981; Stewart, 1976).

The C i r c u m-U n g a v a s e g m e n t, merging westwards with the Churchill segment, forms on the whole an arc bulging sharply northwards. This arc, girdling the Ungava salient of the Superior eocraton (*Proterozoic Basins of Canada*, 1981), consists of three subsegments: western, central and eastern. The western subsegment extends from James Bay south of Hudson Bay through the Belcher archipelago up to the joint with the central, latitudinal subsegment covering the extreme north of Ungava Peninsula and called the Cape Smith fold belt; the Labrador fold zone extending south-east from it occupies the western part of the same-named peninsula. In the south it is sheared by the Grenville front but initially continued up to the joint with the Penokean protogeosyncline separating the Superior and Nain-Nutak eocratons. Throughout this extension rocks of the Circum-Ungava proto-geosyncline are thrust on the Superior eocraton. The amplitude of thrust is maximum in the inlet corner of James Bay where, according to geophysical data, a series of arcuate beds gently thrust southwards on the eocraton have been delineated. The sequence of Belcher zone (see Fig. 25) with a total thickness of 7–9 km commences with the Kaseganik shelf carbonate formation containing stromatolites and evaporites. Its accumulation is succeeded by outpourings of Escimo plateau basalts aged about 2 b.y., which points to a rifting stage. This rifting not only affected Belcher zone, but also led to formation of the R i c h m o n d-B e l c h e r   a u l a c o g e n on the mainland perpendicular to the Belcher zone jutting into the Superior eocraton since analogous basalts outcrop in Richmond Bay. The basalts are overlain by the Fleurt formation comprising dolomites and ferruginous quartzites. Later, volcanism renewed with submarine eruptions predominantly of basalts but also felsic lavas; the age of the volcanics is 1760 ± 38 m.y. (Rb-Sr isochron method). This formation is interpreted as reflecting the build-up of an island arc although it is bimodal in character. In the concluding stage of development of the zone, between the arc and the continent a foredeep filled with turbidites and the red Omarolluk formation arose. All these formations were crumpled into isoclinal folds; contraction of the crust has been assessed to be 25 to 40% (in the Wopmay orogen, a minimum of 40–45%).

The C a p e   S m i t h (Hynes and Francis, 1982) or northern Quebec (Hoffman, 1989) f o l d   z o n e in the extreme north of Ungava Peninsula forms the apex of the Circum-Ungava arc. It is 50–60 km wide with an imbricated-nappe structure with a general southern vergence and thrusting onto the margin of the Superior eocraton. Gneisses and metamorphic schists of Archaean age, reworked in the Early Proterozoic, outcrop in the northern rear part of the zone along the shore of Hudson Bay. They possibly constitute the north-eastern continuation of the Rae eocraton. Peripheral nappes comprise a coarse-clastic formation predominantly of quartz composition and also flows of tholeiite-basalts of the continental type; komatiite-basalt sills are also present. In the nappes closer to the axis of the zone, the relative content of fine clastic varieties (siltstones) and mafic, less often felsic extrusions increases. It has been suggested that all these deposits were formed under the condition of a continental slope. The sequence in the central zone is almost exclusively made up of komatiite and mainly low-titanium tholeiite-basalts with sills of the same composition to a total thickness of about 4 km. Tholeiite-basalts, especially the relatively low-magnesian variety, are similar to basalts of midoceanic ridges, which permits comparing this sequence with an oceanic crust. Hynes and Francis assumed that the zone began developing with continental rifting, continued with formation of a restricted expanse with oceanic crust (Red Sea magnitude ?) and concluded with northward subduction. No magmatites have been observed in association with this subduction, however, although metamorphism, generally of greenschist facies, rises steadily northwards. The minimum scale of contraction during deformation was evaluated at 100 km. On analogy with the Belcher and Labrador zones, it has been suggested that the zone developed in the interval 1.87–1.80 b.y. The duration of spreading was established at 1871 ± 75 m.y. (U-Pb isochron on pyroxene). Hoffman (1989) has given a somewhat different interpretation of the Cape Smith zone. According to him, this zone represented a klippe, remnant of a nappe, displaced over a long distance from the north.

The third link of the Circum-Ungava system, the L a b r a d o r   f o l d   s y s t e m (New Quebec; Hoffman, 1989), extends from Ungava Bay to the Grenville front and continues south of it in a reworked form up to the joint with the Penokean system. It is 800 km long and up to 100–200 km wide. The cross-section of the system can be divided into three zones: western, central and eastern.

The western (outer) zone is made up of shelf sediments of the Knob Lake group which either unconformably lies on the Archaean basement of the Superior eocraton or is thrust on it. Sheets of trachybasalts and trachyandesites deposited at the base of this group are regarded as proof of rift formation, laying the foundation of the Labrador protogeosyncline. Volcanics are replaced by complex alternation of shallow-water and relatively deepwater clastic and carbonate sediments, among which mention should be made of ferruginous quartzites (jaspilites) forming beds of considerable commercial importance. The accumulation of this complex, exceeding 6.5 km in thickness, was interrupted by uplift and subaerial alkaline and subalkaline-basaltic volcanism; submarine eruptions of tholeiite lavas were noticed at a much lower stratigraphic level. Accumulation of the Knob Lake group concluded with the chert-clastic rock formation of the Tamarak River filling a small foredeep-type trough in front of the frontal overthrust of the system, directly north of the Grenville front; sediments similar to those of the Lower Riphean have been preserved in the axial part of this trough.

The central (axial) zone is characterised by replacement of the Knob Lake shelf formation by the deepwater, predominantly metalutite La Porte group containing mafic volcanics of tholeiite composition. This group evidently represents formations of continental slope and rise and possibly of the axial part of the basin. Eastward the metamorphism of these formations attains amphibolite facies. A trough filled with a thick (up to 5 km) pile of basalts of the Doublet group, including 3 km of pillow lava, is laid on the formations of the La Porte and partly Knob Lake groups. These earliest tholeiite-basalt outpourings were accompanied by the intrusion of sills of gabbro and ultramafics. Formation of the Doublet basalts is justifiably regarded as proof of the renewal of rifting, possibly even commencement of spreading of the type seen in the Gulf of California, for which a similar combination of effusive and hypabyssal mafic magmatism is a characteristic feature.

The eastern zone of the Labrador system is made up of metamorphics of amphibolite facies, mainly gneisses, into which a post-tectonic granite batholith has intruded. Most of the gneisses are tentatively placed in the Archaean while distinctly supracrustal rocks, i.e., marbles, quartzites and amphibolites, are regarded as Lower Proterozoic in age. They are considered analogues of the Lake Harbor formation of Baffin Land, which may thus be regarded as the northern continuation of the inner zone of the Labrador system. The corresponding zone of the Cape Smith system undoubtedly represents the north-western continuation of the same inner zone of the Labrador system.

During the epoch 1800–1750 m.y., termed *Hudsonian* in Canada, all formations of the Labrador protogeosyncline underwent intense deformation and metamorphism. The outer and central zones were crumpled into isoclinal folds and dissected by overthrusts with a western vergence while the eastern margin of the inner zone is overthrust on the Nain-Nutak eocraton.

The extreme eastern link of the Transhudsonian system is represented by the Nagssugtoqides and Rinkides of central Greenland opposite Baffin Land along the other side of Davis Strait and Baffin Bay (Esher, 1978; Kalsbeek, 1982; Kalsbeek et al., 1984). The N a g s s u g t o q i d e s form a near-latitudinal belt about 300 km wide, evidently intersecting Greenland under ice and outcropping on its eastern coast opposite Iceland. The concluding deformation of the Nagssugtoqides pertains to the Penokean tectonomagmatic epoch (1.9–1.8 b.y.). Southward they are gently thrust on the Southern Greenland eocraton (Nutak); this boundary is emphasised by deformation of dolerite dykes intersecting the basement of the eocraton and retrogressive metamorphism to amphibolite facies of Archaean rocks present in the granulite stage of metamorphism.

The R i n k i d e s represent a much younger, but also Early Proterozoic, Hudsonian proper, generation of fold structures; they are set off from the Nagssugtoqides by a thrust zone outcropping on the western coast of Greenland at Jacobshavn. According to Hoffman (1989), a north-western offshoot of an eocraton (microcontinent), Burwell (granites aged 2.8 b.y.), is wedged between the Rinkides and divides branches of the north-eastern continuation of the Transhudsonian system. One of them, Torngat, branches out into the south and separates the southern continuation of the Rae eocraton at the rear of the Labrador system from the Nain-Nutak eocraton proper. If the existence of Archaean blocks is recognised south as well as north of the Nagssugtoqidian system and the isotope data for gneisses in their central zone is also taken into consideration, this system may be regarded as a product of collision of these Epi-Archaean microplates (Kalsbeek et al., 1984).

The internal structure of the Nagssugtoqides is determined by alternation of linear zones of intense folding with a persistent east-north-easterly strike and synclinal structure with lenticular distribution of domes and basins devoid of any systematic orientation. The general structure of the zone thus recalls the structure of augen gneisses. Sections of the latter type are made up of Archaean (2.5 b.y.) orthogneisses of granodiorite-quartz-diorite composition with relicts of basic dykes. The linear belts are presumably made up of Archaean garnet-biotite gneisses with sillimanite and Lower Proterozoic marbles, quartzites, calc-silicate rocks and graphite schists, also present in the amphibolite stage of metamorphism; southward anthophyllite schists, amphibolites and marbles, probably much younger in the sequence of the zone are seen. The structure of the Nagssugtoqides in general is evidently similar to that noticed in Baffin Land and in the inner zone of the Labrador fold system.

The R i n k i d e s differ in structure. They form the western coast of Greenland north of Jacobshavn and are characterised by a much younger age of the concluding folding and metamorphism at 1.75–1.70 b.y. (K-Ar method). The structure of the Rinkides reveals two distinct complexes: gneissic and supracrustal-metasedimentary. The gneiss complex outcrops in the core of domes and the metasedimentary complex in the synforms encircles them. These synforms are laid on complex, often recumbent folds transiting into tectonic nappes. Besides granodiorite gneisses, the gneiss complex contains amphibolites with lenses of ultramafics, augen gneisses and garnet-sillimanite schists. A bed of marble up to 1.3 km in thickness is particularly interesting. The metasedimentary complex up to 8 km in thickness consists of two suites: a lower one of quartzites and garnet-staurolite schists and an upper one of much thicker greywacke schists. The rocks of this complex are metamorphosed to greenschist facies. Small plutons of syntectonic granites and a large rapakivi pluton of much later age (Riphean) are present.

The Svecofennides of the Baltic Shield may represent the continuation of the Transhudsonian system on the other side of the Atlantic (see Fig. 23).

4.1.1.3 *Penokean protogeosynclinal fold system.* This system extends in a near-latitudinal direction from the eastern slopes of the Rocky Mountains to southern Greenland, encircling the Wyoming, Superior and Nain-Nutak eocratons from the south and partly underlying the Phanerozoic cover of the Great Plains and the Grenville front. The central segment of the system situated in the region of Lake Superior and known as the Southern 'province' of the Canadian Shield has been studied best. It is here that the Proterozoic was first distinguished from the Archaean (Emmons, 1888). A series, now known as the H u r o n s u p e r g r o u p, developed on the northern coast of the same-named lake was selected as the strato-type of this Proterozoic erathem. The main sequence of this supergroup is exposed on the southern slope of the Superior eocraton where it lies with sharp unconformity and a general southward gentle dip on Archaean rocks of the basement of this eocraton metamorphosed and granitised in the Keno-ran epoch (2.7–2.6 b.y. ago). Formations of the Huron supergroup are very weakly metamorphosed (lower greenschist facies) in this region and are represented by continental-littoral and shallow-marine sediments repeated four times in a cyclic sequence. These sediments are: conglomerates, quartzites and arkoses in the lower part, greywackes, siltstones and shales in the middle, and coarse clastic rocks again in the upper. The cycles are separated by hiatuses and successively transgress northwards. The commencement of each cycle is signified by the resumption of movements along faults. Volcanics ranging from mafic to felsic are locally developed at the base of the three lower cycles, especially the basal cycle (Elliot Lake group). The basal conglomerates of the two lower cycles, Elliot Lake and Hough Lake, contain commercial uranium mineralisation while the upper two cycles, Querc Lake and Cobalt, begin with tillites. Of particular interest are the tillites in the lower part of the Gowganda upper cycle formations as they possess all the characteristics of glacial structures and pass in a northerly direction on a striated bed of Archaean rocks. Limestones with stromatolites are present in the central part of the third cycle (Querc Lake). The total thickness of the Huron supergroup in this section is about 5 km. The lower age limit of the supergroup is determined by the intrusion of dykes and sills of Nipissing diabases dated 2.1 b.y. by the Rb-Sr isochron method. Gentle folding already preceded the intrusion of these diabases. The much older Crayton granites intersecting only the Elliot Lake group are aged 2333 m.y. while rhyolites at the base of the latter are 2460 ± 20 m.y. old.

Southward the homoclinal attitude of the Huron supergroup changes into a gentle folded one. Farther south its spread is intersected by the Murray fault, following which important changes are noticed in the lithofacies composition, thickness, intensity of deformation and metamorphism of the complex (Zolnai et al., 1984), viz., alluvial and deltaic deposits south of the fault zone are replaced by deepwater turbidites, thickness increases to 10–15 km, metamorphism attains amphibolite facies and deposits appear detached from the basement and crumpled into folds with a northern vergence and cut by thrusts (Fig. 27).

As pointed out by R. Dietz and J. Holden as early as 1966, Huronian deposits accumulated under conditions of a passive continental margin. Developing this idea, Zolnai et al. (1984) regarded the basal bimodal volcanics as proof of the commencement of rift formation accompanied by the formation of listric faults merging at a depth of about 15 km into a common

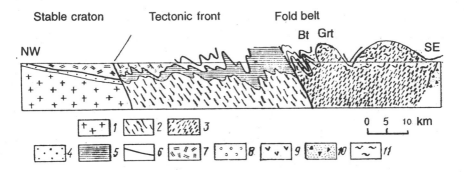

Fig. 27. Generalised geological profile through north-eastern Minnesota (after Sims et al., 1989).

1 and 2—Archaean basement, granites; 3—the same, gneisses; 4 to 10—Lower Proterozoic (6—iron ore formations, 9—basalts); 11—crystalline schistosity; Bt—biotite isograde; Grt—garnet isograde.

detachment surface. The amplitude of primary subsidence along the Murray fault has been estimated at 10–15 km based on variation in thickness and metamorphism. Faults were later transformed into overthrusts. Compression has been determined at 30%. This compression deformation is regarded as a consequence of the collision of the Superior eocraton against another continental block, situated south of present-day Manitoulin Island. The main epoch of deformation (Penokean) manifested 1.9–1.8 b.y. ago (more precisely in the interval 1875–1825 m.y., according to Van Schmus, 1980); its completion corresponds to the intrusion of the well-known nickel-bearing basic Sudbury lopolith in the extreme east of the Southern province (1850 m.y.). Thus the Penokean epoch is somewhat older than the Hudsonian (1.8–1.75 b.y.), the latter being the main one for the Wopmay and Labrador orogens and the Transhudsonian system.

A wholly similar pattern of structure and development of the Southern province has been established by American geologists in the region of Lake Superior. Here the Lower Proterozoic formations, called the Animikie group or Markett Range supergroup, emerge on both shores of the western half of the lake on the flanks of the Riphean Keweenaw aulacogen. The fault of Lake Superior traced along the southern bank of the lake represents a clear continuation of the Murray fault; similar changes occur in the sequence, structure and metamorphism of the Lower Proterozoic and volcanics begin to play a distinct role in its composition. The Animikie group, developed north of the fault on the northern shore of Lake Superior, corresponds only to the middle part of the Markett Range supergroup south of the fault. The age range of the accumulation of this supergroup is 2.05–1.9 b.y.; this signifies that, in general, it is younger than the Huron supergroup. The Markett Range supergroup, like the Huron, has a fourfold cyclic structure but the composition of the cyclothems differs from that of the Hudson supergroup. Tillites in the Markett Range lie at the base of the lower cycle (Chokalay group); ferruginous quartzites of great commercial importance (or rather, which had) played a significant role in the structure of the second (Menomikie group), third (Beraga; Menomikie + Beraga = Animikie) and fourth (Paint River) cycles. The upper part of the first cycle consists of dolomite formation up to 700 m in thickness; volcanics, mainly mafic pillow lavas, subordinate to felsic, occupy a predominant place in the structure of the third cycle. Finally, attention is drawn to the presence of cherts with well-preserved and widely known algal remnants in the formation of Gunflint horizon.

Similarity in conditions of deposition of the Huron and Animikie-Markett Range formations and presence of tillites in their composition together with distinct age discrepancies call for an explanation, even if it is assumed that the upper Huron (Cobalt group) corresponds to the lower parts of Markett Range. One possible explanation is the existence of a transverse fault (transform type ?—V.E. Khain) between Lake Superior and Lake Huron regions.

The structure of the zone south of the boundary fault of the Superior eocraton in the region of the southern coast of Lake Superior, in Wisconsin and Michigan, has been deciphered more completely than in the region of Lake Huron. Here the basement made up of gneisses and migmatites emerges on the surface from under the Markett Range formations. These gneisses and migmatites are of the same type as in the west, in the Minnesota River valley, i.e., a complex of Early Archaean age. Gneisses concealed under a Palaeozoic cover on Manitoulin Island in Lake Huron may represent an analogue of the aforesaid basement. In this case the northern and southern zones of Lower Proterozoic development possess different basements, i.e., Late Archaean and Early Archaean respectively. This confirms the assumption of a collision between two continental blocks in the Penokean epoch. These blocks originally were probably separated by a band of thinned and reworked (if not distended) continental crust. Furthermore, the northern block could have been thrust under the southern block since plutons of tonalites and granites aged 1.85–1.60 b.y. forming a chainlet parallel to the Lake Superior fault are known only within the southern block. Moreover, the Wisconsin volcanic belt extends in Wisconsin and adjoining parts of Michigan and Minnesota south of the basin filled with the Animikie supergroup and partly Markett Range formations and the Niagara fault bordering it. Volcanics are represented in the much older part (1860–1889 m.y.) predominantly by tholeiite-basalts and andesite-basalts and in the much younger part (1835–1845 m.y.) by calc-alkaline andesites, dacites and rhyolites. The following groups of granitoids are associated with them: synorogenic tonalite-trondhjemite-granodiorite aged 1840–1870 m.y., granites aged 1835 m.y. and alkaline granites aged 1760 m.y. Sims and colleagues (1989) suggested that the Niagara zone of faults represents a suture corresponding to the zone of Early Proterozoic collision of a volcanic arc with the continent.[2] In the opinion of Sims and colleagues, during formation of the ancient group of volcanics the subduction zone was inclined southwards but later, after collision at around 1860 m.y., shifted southwards and acquired a northern dip.

An analogue of the Lake Superior fault in Wyoming is the s h e a r z o n e   o f   C h e y e n n e (or Wyoming) separating the Wyoming eocraton from the zone of development of the Lower Proterozoic in the southeastern part of this state and also in the Rocky Mountains of Colorado and New Mexico (Condie, 1982; Condie and Shadel, 1984). The zone extends north-easterly, presumably continuing south of the Precambrian salient in the Black Hills of South Dakota. Like the Southern province, north of the Cheyenne fault a 13-km series (Snowy Pass supergroup) of relatively weakly metamorphosed continental-littoral, including glacial-marine, clastic deposits in the lower part and shallow-marine carbonate-clayey deposits in the upper part rests unconformably on the southern slope of the Wyoming eocraton on an Archaean basement (Karlstrom et al., 1983). These deposits are separated from the basement by an unconformity to which intrusive dykes and sills of tholeiite and subalkaline basaltoids of 2 b.y. age are confined.

Lower Proterozoic formations developed in the form of cover sags among granites south of the shear zone differ sharply from the above deposits as well as from the Lower Proterozoic formations of the Southern province of the Canadian Shield occupying a similar tectonic position. These are almost exclusively volcanic and volcanoclastic rocks metamorphosed from greenschist to amphibolite facies and relatively weakly and unevenly deformed.

The age of the volcanics directly south of the shear zone was determined as 1.79–1.78 b.y., in southern Colorado, as 1.76–1.74 b.y. and in New Mexico as 1680 m.y. (all determinations by U-Pb method on zircon) (Condie and Shadel, 1984; Nelson and de Paolo, 1984). This denotes that Early Proterozoic volcanics of the Rocky Mountains are positively younger than the Lake Superior formations deformed by Penokean folding and belong to the end of the Early Proterozoic. Mafic (tholeiites) and felsic (dacites and subordinate rhyolites) varieties are present nearly equally among the volcanics of the Rocky Mountains. Syntectonic granodiorites are comagmatic with them. In spite of the distinct bimodal nature of this association, Condie and Shadel (1984) regarded it as an island arc on the basis of the predominance of products of subaerial eruptions along with volcanoclastic turbidites. The origin of bimodal volcanics according to them, is explained either by liquation

---

[2] Serpentinites, tabular dykes, tholeiite-basalts and plagiorhyolites, i.e., elements of ophiolite complexes, were encountered along this zone.

differentiation of andesite magma or crystallisation differentiation of basalt magma with the persistence of intermediate and mafic cumulates at depth. Preferring the second mechanism, they have associated its more extensive manifestation in the Proterozoic with a heat flow considerably greater than at present.

Let us turn now to the eastern continuation of the Southern province of the Canadian Shield falling already beyond the Grenville front. In the southern exposed part of the Grenville belt analogues of the Huronian are 'recognised' at some distance from the front. These analogues also begin with coarse clastic formations but pass in the south-east into deeper water, finely layered greywackes and shales. These formations fill the basin extending parallel to the Grenville front. Farther south, coarse clastic formations are seen once again. These are thick arkoses, quartzites and ferruginous quartzites, intruded by granites aged 1.8 b.y. It is not clear whether these sediments represent the facies substitution of deepwater formations developed more to the north or are superposed on the sequence of deepwater formations.

More north-eastward, along the Grenville front, we encounter the southern, reworked continuation of the Labrador fold system and the M a k k o v i k 's u b p r o v i n c e' wedges even more north-east, on the eastern coast of the Labrador Peninsula, between the Nain eocraton bordering the Labrador system from the east and the Grenville front (see Fig. 23). This subprovince is regarded as the western tip of the southern Greenland Ketilides. The basement of this subprovince is formed by Archaean gneisses, migmatites and amphibolites (age over 2.58 b.y.). The lower Ilik group rests on them with a sharp unconformity. This group is made up of mafic volcanics and fine clastic sediments (siltstones, greywackes and shales). The age of the group by the U-Pb method on uranium smudge was determined as 1.96–1.73 b.y., most probably around 1.9 b.y. The formations of this group are pierced by the subalkaline granitoid Translabrador batholith with an Rb-Sr age 1788 ± 29 (granodiorite) to 1687 ± 30 (monzonite) m.y. The batholith extends parallel to the Grenville front which intersects it along the axis and divides into unreworked northern and reworked southern parts. Formations of the upper Ilik group, i.e., felsic volcanics and clastic sediments aged 1.75–1.65 b.y. (Rb-Sr isochron and U-Pb method on uranium minerals) are comagmatic with the batholith. At around 1.6 b.y. the rocks of the subprovince were intersected by basic dykes (the first generation of such dykes preceded the lower Ilik group) followed later by intrusion of quartz-syenites and granites (1625 m.y.). The deformation history of the subprovince includes reworking of Archaean gneisses and the dykes intersecting them before deposition of the Ilik group (about 2.1 b.y.), formation of 'tectonic interstratification' of basement gneisses and formations of the lower group before accumulation of the upper Ilik group (1.75–1.65 b.y.) and folding with vertical axial surfaces and metamorphism before intrusion of the Translabrador batholith (1625 m.y.).

As already mentioned, the K e t i l i d e s in the southern extremity of Greenland are regarded as a direct continuation of the Makkovik subprovince. Their relations with the Nutak eocraton lying northwards duplicate those described above for the Southern province of the shield: the Archaean gneiss basement of the southern margin of the eocraton pierced by basic dykes is unconformably overlain by Lower Proterozoic metasedimentary and metavolcanic formations. Southward the intensity of deformation and metamorphism rises rapidly and the Archaean basement disappears from the surface while isotope data reveals its absence even at depth. Here are also seen granite plutons aged 1.8 b.y. In the southernmost part of the island metamorphism rises to granulite facies; this metamorphism is succeeded by intense migmatisation combined with formation of a gentle isoclinal recumbent fold. Post-tectonic mushroom-shaped plutons of rapakivi granites aged 1774–1786 m.y. are also manifest here.

The problem of the conditions of development of the Circum-Ungava protogeosyncline has been under active discussion in recent Canadian literature. For its south-western segment, corresponding to the south-eastern part of the Churchill province, two plate-tectonic models have been proposed (*Proterozoic Basins of Canada*, 1981; Stauffer, 1984), with one and two island arcs and subduction towards the north-west. One model is shown in Fig. 28. The authors of the study on Belcher zone (in: *Proterozoic Basins of Canada*, 1981) also support, as mentioned before, the plate-tectonic concept while investigators of the Cape Smith zone (Hynes and Francis, 1982) have assumed its distension on the Red Sea scale with a gentle subduction northwards. A model of rifting without formation of a new oceanic crust and without its subduction has been proposed for the Labrador zone (Fig. 29).

In so far as the Penokean geosyncline is concerned, researchers have studied its western segment, presently falling within the Rocky Mountains, in terms of plate tectonics (Condie, 1982; Condie and Shadel, 1984). There are two points of view with regard to the central segment: some consider it an aulacogen relying on the existence of a continental massif even south of the fold system while others (Zolnai et al., 1984) regard it as a passive margin of an oceanic basin but evidently with a microcontinent adjoining it.

Analysing this problem as objectively as possible, the following statement can be made. The structure of the sequences of the outer zones of both protogeosynclines, as also their Wopmay analogue, is quite typical of passive rift margins of the Atlantic type and suggests that the basins accompanying them were relatively deep and wide. Hence their central parts should be underlain at least by a crust of transitional, more probably oceanic type. The absence of ophiolites (except in the Cape Smith region!) does not permit categorical affirmation that destruction here actually proceeded to the

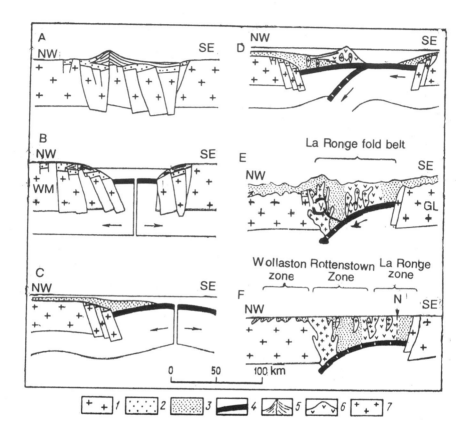

Fig. 28. Two-dimensional model showing possible development of Wollaston, Rottenstown and La Ronge zones in northern Saskatchewan (after Stauffer, 1984).

A—Late Archaean-Early Aphebian continental rifting accompanied by limited magmatic activity and deposition of supracrustal coarse clastic sediments; B—deposits of shelf and deepwater sediments along the rear edge of the Wollaston-Mudjatik continental margin; C—rotation of direction of plate movements in the Late Aphebian leading to oceanic subduction, formation of migrating island-arc La Ronge complex zone made up of metavolcanics and intrusives of early plutonic belt; D—Hudsonian orogeny—intrusion of Wathaman (W) batholith about 1865 m.y. ago along the ancient continental margin accompanied by felsic volcanics; E—closing of ocean with deformation of shelf sediments (Wollaston group) and eugeosynclinal supracrustal formations of Rottenstown-La Ronge; F—present-day profile through Wollaston, Rottenstown and La Ronge zones in northern Saskatchewan showing post-Hudsonian development of Needle Fall zone of shearing.

1—blocks of continental crust (WM—Wollaston-Mudjatik; GL—Glenny Lake); 2—coarse clastic and volcanoclastic sediments; 3—shelf deposits, continental slope and rise; 4—oceanic crust; 5—early volcanics; 6—volcanic arc formations; 7—granite plutons.

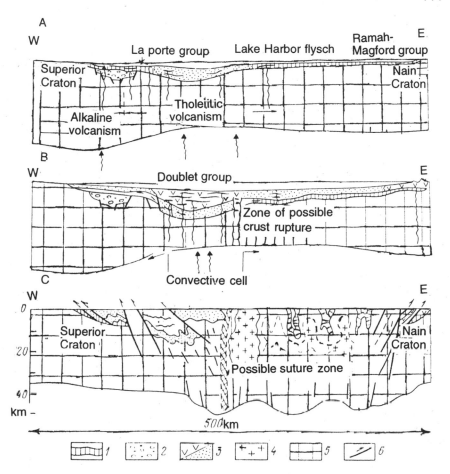

Fig. 29. Model of tectonic development of the Labrador fold system (after Wardle and Bailey, 1981).

1—coarse clastic rift facies; 2—shallow-water facies; 3—mafic flysch volcanics; 4—granites; 5—Archaean basement; 6—tectonic disturbances.
A—continental rifting and alkaline volcanism in the western trough accompanied by extensive submergence and shallow-marine transgression (La Porte group developed as facies of a much deeper basin); B—subsidence leading to flysch sedimentation and submarine volcanism (Doublet group may represent the western rim of proto-oceanic rift); C—deformation, metamorphism and granite intrusion during Hudsonian orogeny (possible suture zone may correspond to closing of Doublet rift zone).

stage of spreading and formation of a new oceanic crust on any significant scale. In fact, it may be said that conditions still did not exist in the Early Proterozoic (as in the Archaean) for obduction of oceanic crust due to the low density of the asthenosphere. At the same time, the chemical composition of tholeiite-basalts, especially in the Belcher and Cape Smith segments, points to genetic affinity with oceanic tholeiites and confirms the possibility of formation of a new oceanic crust. The overall asymmetry of fold zones and presence of batholithic bodies of granitoids in their rear (Wathaman batholith in the Reindeer segment and the Central Labrador batholith) and zones of metamorphism of amphibolite facies and relics of island arc (La Ronge zone) in the south-eastern part of the Churchill province, provide distinct proof in favour of subduction of oceanic crust. Another proof in favour of plate-tectonic models, even though on limited scale, is the significant transverse contraction of fold zones with their nappe-like overthrusting on the foreland in exactly the same manner as noticed in the Late Precambrian and Phanerozoic fold systems with transformation of listric faults of the rift stage into overthrusts in the orogenic stage. Thus the overall balance goes in favour of a plate-tectonics interpretation although with due consideration of specific features of Early Proterozoic plate tectonics, particularly the limited scale of spreading.

However, recent publication of palaeomagnetic data for the Wathaman batholith in the Churchill segment of the Transhudsonian belt (Symons, 1991) indicates that the above conclusion may prove overcautious. This data, recorded for an epoch 1865 m.y. ago, shows that an ocean (Manikevan) of up to 5000 km in width, i.e., wholly comparable with the present-day Atlantic, stretched in this epoch between the Slave (+ Rae and Hearne) and Superior eocratons. In this ocean not one (La Ronge) but two (Flin Flon is the second) volcanic arcs existed at a distance of 1700 km from each other. The Wathaman batholith marks the suture formed at the site of collision of the Slave and Superior continents.

4.1.1.4 *Hebrides massif.* Archaean history of the massif concluded 2.5 b.y. ago with the intrusion of small plutons of granitoids and general uplift. Later, tectonometamorphic reworking of the Scourian (Gelian) complex occurred in the *Inverian epoch.* This reworking is perceptible only at places where the complex had previously experienced metamorphism to granulite facies. Here it was manifest in retrogressive metamorphism to amphibolite facies, formation of zones of vertical schistosity, development of agmatites and pegmatites and, at the end of the epoch, intrusion of mafic and picrite dykes 2.4–2.2 by.y. ago and finally in the development, again in some random zones, of open folds with fractures (in: *Early History of the Earth,* 1976).

Later, mafic lavas and sediments accumulated in the southern part of the massif. These were predominantly sandy-clay with carbonates and iron ores of the Loch Mary group, which experienced folding and metamorphism to amphibolite facies in the epoch 1.9–1.8 b.y. ago called here *Laxfordian.* The Archaean Lewisian complex also underwent considerable reworking in this epoch.

4.1.2 East European Craton

4.1.2.1 *Baltic Shield.* The zone of Epi-Archaean cratonisation encompasses the north-eastern part of the shield: Kola Peninsula with continuation north-westward, Finnish Lapland and Karelia with continuation southward and Finnish and Swedish Lapland (Fig. 30). However, in the Early Proterozoic the unity of this Kola-Karelia eocraton was significantly disturbed, firstly by formation of the Murmansk and Lapland-Belomorian belts of tectonothermal reworking and secondly, by formation of rift zones (protoaulacogens) filled with clastic sediments and products of basaltoid volcanism. It is not clear what occurred south-west of the Kola-Karelia eocraton in the rest of the Baltic Shield during a significant interval in the Early Proterozoic since rocks older than 2.1 b.y. have not been established here to date. In central Sweden and central and southern Finland, i.e., in the province of Svecofennides, active protogeosynclinal development commenced only after this period.

The structure and history of the M u r m a n s k   T T R   b e l t has not been adequately deciphered to date. This is largely due to the predominant spread of Archaean soda as well as potash granitoids which have absorbed supracrustal formations, probably identical to those of the Kola series of the adjoining central Kola block. At the end of the Archaean to the Early Proterozoic, the central Kola block including Kolmozero-Voronya greenstone belt experienced underthrusting below the Murmansk belt. Remobilisation of the basement of the belt and formation of Early Proterozoic granitoids aged 2.3–2.2 b.y. are probably associated with this underthrusting.

The L a p l a n d-B e l o m o r i a n   T T R   b e l t has an even more complex history and structure. The Lapland granulite belt is delineated in it, which appears in the north-west, in Finnish Lapland, from under the Caledonides nappes and, with a gap, reaches the north-eastern coast of Kandalaksha Bay of the White Sea (Belomorie). In the Russian territory the granulite belt is substituted from the south by the Belomorides, which are traced in the east along the south-eastern coast of the White Sea after concealment beyond the Onega River under the Phanerozoic cover of the Russian Platform. The total extent of this TTR belt is 1200 km and width from 30–40 to 100 km. The belt boundaries are marked by overthrusts; the belt is distinctly thrust on the southern Lapland-Karelian megablock in the south and partly overthrust and partly underthrust below the Kola megablock in the north (Fig. 31).

The major part of the Belomorides complex falls in the Archaean as shown by Rb-Sr isochron datings of 2.95–2.7 b.y. Indirect data, i.e., the distinct difference in primary composition between Upper Archaean formations of the Kola Peninsula and Karelia, prompt consideration of the possibility of an even older age of the Belomorian complex metamorphosed to amphibolite facies with relics of granulite facies. The intrusion of dykes of basic composition into the Belomorides and later granitoids from granodiorites to

114

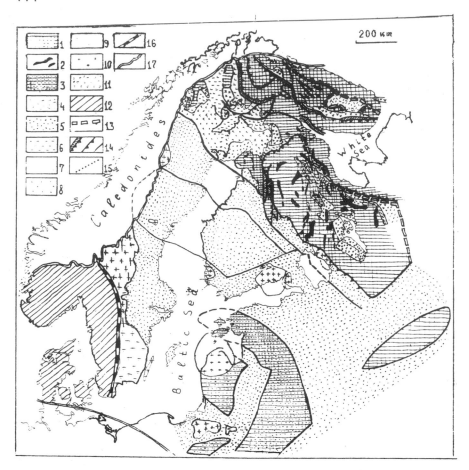

Fig. 30. Tectonics scheme of Baltic Shield and adjacent area (specially compiled for this book by A.M. Nikishin).

1—Archaean terranes; 2—Archaean greenstone belts; 3—Archaean (?) terranes reworkod in Early Proterozoic; 4—Late Archaean deformed cover (Keivy basin, 2.8-2.7 Ga); 5—Early Proterozoic terranes; 6—2.5-2.3 Ga deformed cover (Lapponian, Sumiy, Sarioliy); 7—2.3-1.8 Ga deformed cover (Jatuliy, Suisariy); 8—Archaean crust remobilised and remelted in Early Proterozoic; 9—Trans-Scandinavian magmatic belt (1.7-1.6 Ga); 10—rapakivi granites (1.7-1.6 Ga); 11—1.8-1.6 (?) cover (Vepsian); 12—Dalslandian terrane (~ 1.1-0.9 Ga); 13—Archaean suture (2.7-2.5 Ga); 14—Early Proterozoic suture zones and sutures (2.0-1.75 Ga); 15—Early Proterozoic strike-slip fault; 16—Late Precambrian suture (~ 1.1-0.9 Ga); 17—Recent East European Craton boundary.

essentially potash granites and finally pegmatite veins whose age (about 1.8 b.y.) corresponds to the end of deformation and metamorphism of the Belomorides and the final formation of their extremely complex, multifacies structure, correspond to the Early Proterozoic.

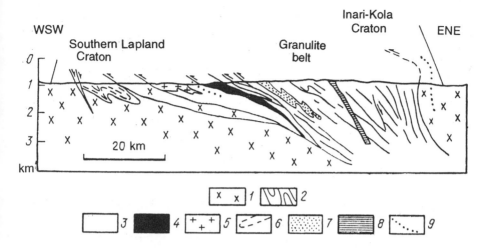

Fig. 31. Synthesised profile through Lapland granulite belt at the latitude of Inari Lake (after Barbey et al., 1984).

1—undifferentiated Archaean basement; 2—Archaean greenstone belts; 3—Tana zone; 4—Vaskoyoka anorthosite complex; 5—granites; 6 and 7—formations of granulite belt (6—khondalite suite, 7—charnockite complex); 8—Jatulian formations; 9—isograds of granulite facies.

In spite of structural unity and almost identical age of concluding deformation, the Lapland granulite belt differs significantly from the Belomorian belt in composition, age of rocks and the character of their metamorphism. Most researchers place not only the granulite metamorphism, but also the accretion of volcanics and sediments of the Lapland complex in the Early Proterozoic. According to the data of French researchers (Barbey et al., 1984), the oldest formation here is the volcanic series of tholeiitic (basalts and andesites) and later calc-alkaline (rhyodacites and rhyolites) affinity. This series is replaced by a flyschoid terrigenous formation with subordinate tholeiitic volcanism. The accretion of this entire complex occurred in the interval 2.2–2.0 b.y.; this was followed by three phases of deformation and metamorphism. Charnockites and anorthosites intruded during the main phase (2.0–1.9 b.y.). The aforesaid French researchers proposed a model of the evolution of the Belomorian belt (implying the Lapland belt) including: 1) continental rifting in the interval 2.4–2.2 b.y. simultaneous with a similar rifting in the Kola and Karelian megablocks; 2) spreading in the interval 2.2–2.0 b.y. leading to formation of the 'Belomorian geosyncline'; 3) subduction of oceanic crust of this 'geosyncline' under the Kola eocraton with granitoid plutonism in its southern extremity (2 b.y.) simultaneous with opening of the Svecofennian geosyncline; and 4) formation of the Belomorian fold belt 1.9 b.y. ago as a result of collision of the Karelian and

116

Kola eocratons, giving rise to nappe tectonics, granulite metamorphism and tholeiitic and calc-alkaline plutonism on the hanging wall (Fig. 32).

The following comments can be made about this model. Firstly, it may be relevant only to the Lapland granulite belt proper since the Belomorides, as already mentioned, are mainly made up of Archaean material. It is quite likely that the Lapland rift may have originated in the north-western continuation of the Belomorian belt as its constituent. Or, it could be assumed that the

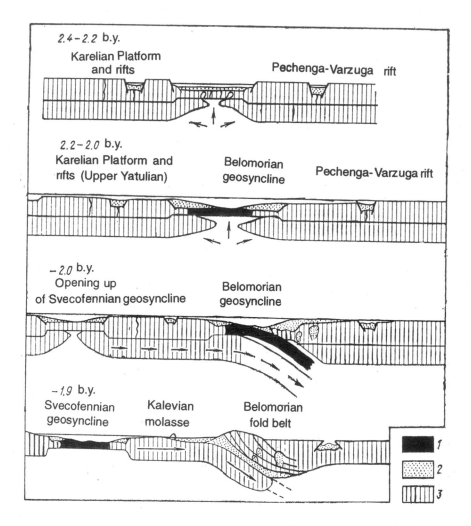

Fig. 32. Hypothetical evolution of the Belomorian orogenic belt (after Barbey et al., 1984).

1—oceanic crust; 2—volcanic and sedimentary formations; 3—continental crust.

Lapland complex is also primarily Archaean, but was rejuvenated in the Early Proterozoic, and place formation of the 'Belomorian geosyncline' in the Archaean, drawing a parallel between the Khetolamba amphibolite suite and tholeiite suite of the basement of the Lapland belt. Secondly, there is no direct proof that destruction in the Lapland belt attained the stage of formation of a genuine oceanic crust since there is no complete ophiolite association here. Thirdly, a unique proof of subduction is provided by Early Proterozoic calc-alkaline plutonism in the Kola megablock, which otherwise cannot be explained. At the same time, it is highly probable that the Lapland-Belomorian belt was formed in the process of break-up and extension of the Kola-Karelian eocraton and evidently its formation occurred as a result of the reverse process of collision of the Karelian and Kola fragments of this originally single eocraton and underthrusting of the former below the latter with thickening of the crust, nappe formation, metamorphism etc.

Yet it must also be assumed that collision between the Karelian and Kola megablocks was accompanied by some relative rotation of these blocks since their structures and especially the greenstone belts and protoaulaco-gens within them have a different orientation—sublatitudinal in the Kola megablock and submeridional in the Karelian megablock.

Based on new geochronological data, Mints (1993) advanced a new model of the Early Proterozoic development of the north-eastern part of the Baltic Shield. The basis of the new model, which differs radically from the ones suggested earlier, is the formation of a compact massif of continental crust in the region towards the end of the Archaean. Right at the beginning of the Proterozoic (2.49–2.40 b.y. ago) this crust was subjected to rifting followed by formation of the Pechenga-Varzuga and northern Karelian 'micro-oceans' with isolation of the Belomorian microcontinent between them. The lithosphere of the first basin later underwent subduction in a southerly direction with outbursts of island-arc andesite-basalts in the interval 2.36–2.22 b.y. ago. Later, 2.15 b.y. ago, subduction of the lithosphere of the southern part of the northern Karelian basin commenced south of the island arc or volcanoplutonic belt on the northern margin of the Belomorian microcontinent. This resulted in formation of an island arc or a volcanoplutonic belt already on the opposite fringe of the microcontinent. The first stage of granulite metamorphism of the Lapland belt pertains to this very period. Much later, 2.15–1.90 b.y. ago, the nappe-thrust structure of the Lapland belt was formed initially in the northern part of the microcontinent as a result of two-sided subduction under the Belomorian microcontinent. This process ceased by 1.85 b.y. in a general piling up of the rocks of the microcontinent as a result of its compression during collision with the Karelian and Kola continents bordering it. The second epoch of granulite metamorphism coincides with this event. Post-collision magmatism and doming continued up to 1.70–1.67 b.y.

In this model, as in all the preceding ones, the apparent features are rifting initially, culminating ultimately in collision of the Karelian and Kola continental blocks; a more disputable element is the scale of extension of the Pechenga-Varzuga and northern Karelian (Vetrenny Poyas) rifts.

In the central part of the Kola megablock the Pechenga-Varzuga rift system (protoaulacogen) extends roughly parallel to the Lapland-Belomorian belt. This system consists of two links—Pechenga basin and the far more extensive Imandra-Varzuga graben-trough. The total extent of the system is about 600 km at a width of up to 30–40 km. Both basins are characterised by a sharply asymmetric structure with gentle northern flanks and intensely reduced southern ones truncated by thrusts. Further, the Kola superdeep borehole intersected an overthrust of northern vergence even on the northern flank of the Pechenga basin. Thus the vergence here is inverse in relation to the Lapland-Belomorian belt but could be explained by the pressure exerted by the latter. The thickness of the volcanosedimentary filling of the basin reaches 15–20 km. The Imandra-Varzuga trough partly inherits the Late Archaean greenstone belt having been displaced slightly northwards relative to it. The sequences of the basins are cyclic in structure, such cycles being four in Imandra-Varzuga and up to six in Pechenga (Negrutsa, 1984). Sedimentary-pyroclastic rocks are deposited in the lower parts of the cycles and extrusive rocks in the upper. Moreover, the sequence is intercalated with bedded intrusives of mafic and ultramafic composition and in the Pechenga basin and the Monchetundra, at the joint between the two basins containing commercial deposits of copper and nickel sulphides. Sedimentary rocks of the lower parts of cycles are coarse and largely clastic, with textural features of shallow marine origin; beds of dolomites with stromatolites and oncolites are present. Extrusive rocks are represented by picrite-basalts, basalts and andesite-basalts with subordinate lavas of felsic and intermediate composition. Further, the increasing role of picrite-basalts on the one hand and felsic and intermediate lavas on the other, is noticed up the sequence. The metamorphism of basin filling is mainly greenschist, rising to amphibolite in the fault zones on the southern boundary. The age of metamorphism was estimated as 1.8–1.7 b.y. and of mafic and ultramafic rocks as 2.0–1.8 b.y. The volcanosedimentary formations of Pechenga and Imandra-Varzuga most probably accumulated 2.3–1.9 b.y. ago. The intrusive complex of leucocratic granites and alaskites, granodiorites and granites, alkaline granites and syenites is much younger than completion of formation of the rift system. Age values of 1840 ± 50 m.y. and 1755 ± 25 m.y. were recorded on granites of the Litsa-Araguba region by Rb-Sr and U-Th-Pb isochron methods and 1840 ± 30 m.y. for leucocratic granites and alaskites by the Pb isochron method. Even older granites aged 2095 m.y. (Umba complex) are found in the east.

A typical asymmetric synclinal structure of Keivy tundras lies north of the Imandra-Varzuga trough. It also inherited a Late Archaean trough filled with protoplatformal formation of the Lebyazhia series. Formations of the Lower Proterozoic, mature clastic formations and bimodal volcanics bear an even more platformal character. However, they experienced regional metamorphism to amphibolite facies.

The Lower Proterozoic of the Karelian megablock is distinctly divided into three parts, their age boundaries approximately 2.3 and 1.9 b.y. The lower subdivision is known in Karelia as the *Sumian* and *Sariolian*, the middle as the *Jatulian* and *Suisarian* and the upper as the *Vepsian* (*Kalevian* in Finland). The sequence is made up of alternating continental and shallow-marine clastic rocks and volcanics (Fig. 33). It is cyclic, each of the three main subdivisions representing an independent cycle and it itself consisting of several cycles of the highest order, two in the lower, three in the middle and one in the upper. This cyclicity generally corresponds to that of the Pechenga-Varzuga complex of the Kola Peninsula. Again, like the latter complex, cycles are separated by discontinuites; their lower transgressive half is predominantly sedimentary and the upper regressive half volcanic. The composition of volcanics generally reveals an antidrome direction. Thus in the Sumian volcanics are essentially felsic (quartz porphyries), in the Sariolian replaced by andesite-basalts, and in the Jatulian by tholeiite plateau basalts and gabbro-dolerites in the accompanying sills and dykes. In the Suisarian and Vepsian, along with tholeiite-basalts, picrite-basalts play an important role. Sills and dykes of peridotites represent intrusive analogues of picrite-basalts. Volcanic rocks constitute up to 50% of the Karelian sequence, its total thickness running up to 4 km at places. The thickness of volcanics alone in northern Sweden is estimated to be 5 km and of the upper series of sediments, evidently corresponding to the Vepsian, 6 km (Park and Bowes, 1983).

Apart from sills, layered intrusions of peridotite-norite formation dated 2.45–2.43 b.y. by the U-Pb method on zircons in Finland are known within the megablock under consideration. Formations of palingenetic-metasomatic granites and migmatite-granites emerging in the form of domes in southwestern Karelia and eastern Finland are younger, i.e., of pre-Jatulian, possibly pre-Sariolian age. Their Rb-Sr age in the Karelian region has been determined as 2.2–2.1 b.y. and in Finland as 2.05 b.y., possibly somewhat younger. An even younger intrusive complex, Haparanda, is known in northern Sweden. This complex is made up of rocks ranging from gabbro to granites, with a predominance of diorites and granodiorites. Their U-Pb age on zircons is 1.88–1.86 b.y. and Rb-Sr age up to 1.7 b.y. A slightly younger volcanoplutonic complex developed in the same region, but more westward, consists of a porphyry formation in the Kiruna region with which pyrite deposits and subvolcanic bodies of monzonites with perthitic feldspar

120

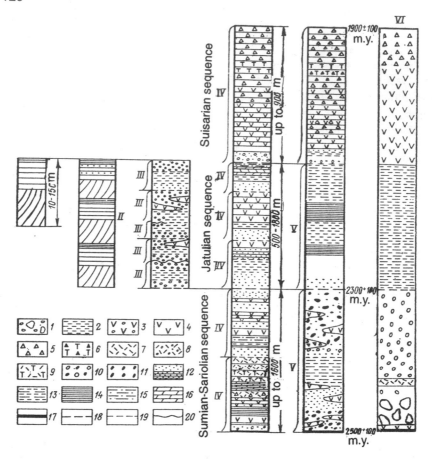

Fig. 33. Combination of volcanosedimentary cycles of a lower order (I and II) in cycles of a higher order (III to VI) in the Lower Proterozoic of Karelia (after Tikhonova, 1982).

1—continental (coarse clastic) formations; 2—basinal carbonate-terrigenous formations; 3 to 9—volcanic formations (3—extrusions of mafic composition, 4—the same, ultramafic composition, 5—lava breccla, 6—agglutinated tuffs, 7—extrusive rocks of felsic composition, 8—the same, amygdaloidal; 9—tuffs of felsic composition); 10—polymictic conglomerates; 11—the same, quartzitic; 12—sandstones, quartz-sandstones; 13—siltstones and siltolutites; 14—schists; 15—schungites; 16—marls; 17—structural unconformity; 18—marginal unconformity; 19—erosional disconformity; 20—hiatuses.

are associated. The prophyry formation consists of volcanics ranging from subalkaline basalts and andesites to calc-alkaline dacites and rhyolites. The age of granite from the formation of 'perthitic monzonites' was determined by the U-Pb method on zircon as 1863–1826 m.y. and by the Rb-Sr method on whole rock as 1725 m.y. (Skiold and Clift, 1984). Some similarity of 'perthitic monzonites' (zonal feldspar) with rapakivi granites has been noticed.

Metamorphism of the volcanosedimentary protoplatform cover of the Karelian megablock is predominantly greenschist but attains amphibolite facies close to intrusions and faults.

Localisation of the thickest sediments and volcanics within the megablock is undoubtedly associated with rifting. Three main belts of volcanotectonic basin-grabens are noticed: 1) eastern Karelian along its boundary with the Belomorian belt (one of the largest basins, Vetrenny Poyas, falls in this zone); 2) Central Karelian, extending for 500 km from Lake Onega in a north-westerly direction; and 3) Finnish, falling in the east in the Russian Karelian territory. These rift zones possess many common features: their form in plan is uneven; considerable dislocation at places, development of reverse faults, even overthrusts, schistosity and strike-slip faults at places enabled Tikhonova (1982) to consider, wholly justifiably, these zones connected in the initial stages of their development to strike-slip faults (with a predominance of sinistral displacement). The basins themselves in configuration, and evidently in their nature, fall in the category of extension basins confined to major strike-slip faults; this type of trough is now widely known as a pull-apart basin. In the concluding stages of development such basins underwent compression, especially basins of the Finnish belt situated close to the overthrust front of the Svecofennides.

During the Early Proterozoic the Karelian megablock experienced progressive cratonisation and Vepsian formations represent a nearly undeformed and unmetamorphosed platform cover.

The Svecofennlan[3] protogeosynclinal fold system tectonically represented the most active constituent of the future Baltic Shield in the Early Proterozoic. The formations constituting it are distinctly thrust onto the south-western margin of the Karelian protoplatform. The overthrust front has been traced from Lake Ladoga to the apex of the Gulf of Bothnia; it later extended through northern Sweden (where it is poorly manifest), concealed under the Caledonides and should have emerged in the Sea of Norway south of the Lofoten Archipelago. A zone of dextral strike-slip displacement evidently runs parallel to it in the rear.

In the south-west the Svecofennides are bound by the much younger (1.75–1.65 b.y.) Trans-Scandinavian volcanoplutonic belt, possibly partly superposed on an Archaean (pre-Gothian) gneiss basement. The province of Epi-Archaean cratonisation restricts the Svecofennides from the south-east too, judging from the presence of Archaean zircons in the Vyborg rapakivi massif (according to Gaal, 1982). But in the south-west the Svecofennides extend under the cover of the Russian Platform. In the north-west the

---

[3] In Swedish-Finnish literature the terms Svecofennokarelides or simply Svecokarelides and their adjectival forms are frequently used. The latter term presents a contradiction, however, and should justifiably be rejected (Gaal and Gorbatschev, 1987).

Svecofennides are traced in the tectonic windows of the Scandinavian Caledonides.

Accumulation of the volcanosedimentary formations constituting the Svecofennides commenced not earlier than 2.0–1.9 b.y. ago and the nature of much of the basement on which it occurred is not well understood. However, outcrops of ultramafics, gabbroids, trondhjemites, basaltic pillow lavas and cherts along the Bothnian-Ladoga zone of frontal overthrusts in the nickel-bearing Katalahti belt may be regarded as elements of ophiolitic association (Gaal, 1982). A similar rock assemblage was detected in yet another such belt, the Kilmekoski, in south-western Finland. Proto-ophiolites of the Katalahti have been dated 2.1–1.9 b.y. Kontinen (1987) described quite typical ophiolites in the Jormua region; these have been dated 1.97–1.96 b.y. and represent an excellent example (if not the most outstanding) of Early Proterozoic ophiolites. This data, together with isotope studies of granitoids in Sweden (Wilson et al., 1985), quite positively favours formation of a significant part of the Svecofennides on an oceanic crust. However, the origin of the immense (400 km in cross-section) Central Finland batholith present in the centre of the Svecofennian province is still an open question.

The Svecofennian protogeosynclinal complex comprises two significantly different types of formations. One of them, terrigenous, consists of black shales and greywackes and is correlated with the Jatulian (Middle and Upper) of the Karelian megablock in which black shales are also present in subordinate quantities. In addition to black shales, metagreywackes and metabasalts are also developed; with growing metamorphism, lutites transformed into micaschists and greywackes into gneisses. The second type of Svecofennian complex is significantly volcanic. Volcanics range from mafic to felsic. That felsic metavolcanics, called leptites in Swedish literature (they form predominantly after tuffs and ignimbrites), predominate in the sequence is of particular interest. Turbidites and conglomerates are associated with volcanics at places. This type of deposit was evidently formed in an island-arc environment. One such arc, Scellefto, is situated in northern Sweden; another, Bergslagen, girdles the Stockholm region from the west and its northern branch extends through the Åland islands in southern Finland. The age of the leptites in central Sweden determined by the U-Pb method on zircons is 1900 ± 19 m.y. and of mafic volcanics of the Kiruna region in the northern part of the country 1932 ± 45 m.y. (Sm-Nd isochron method); quartz-porphyries of the same region are $1909 \pm 16\,5$ m.y. old (Rb-Sr method on zircons). The main zone of development of the schist-greywacke formation falls between volcanic arcs; this zone is called the Nurrland or Bothnian geosyncline in Swedish literature (see Fig. 29). It is evidently a marginal or interior sea in which deposits reach 10 km in thickness. Another section in which the same formation prevails lies within the Bergslagen arc.

K a l e v i a n represents a much younger member of the Svecofennides section; it is a terrigenous flysch formation developed in a marginal trough along the Svecofennides overthrust front. In the territory of Russian Karelia the L a d o g a s e r i e s corresponds to the Kalevian; like the Kalevian, it lies unconformably on underlying formations and contains olistostromes (Negrutsa, 1984). Gaal and Gorbatschev (1987) have given a totally different interpretation of the Kalevian. According to them, the unconformity between the Jatulian and the Kalevian corresponds to the commencement of formation of a passive margin of the Svecofennian basin due to rifting of the Epi-Archaean Karelian craton. Thus the Kalevian, which has been assigned an age of 2.0–1.9 b.y. (Ward, 1987), should represent not the concluding, but an initial member of the Svecofennian complex. The question continues to be debated, however; the presence of slices of ophiolites runs more in favour of the first interpretation. In northern Finland the Lower Proterozoic sequence ends in the variegated Kumpu (= Vepsian) molasse formation.

The structure of the north-eastern front of the Svecofennides is highly complex (Martynova, 1980; Park and Bowes, 1983). Geosynclinal Svecofennian formations containing ophiolites are thrust here on their Jatulian platform analogues, which are underlain by Archaean gneisses and granites. The latter form classic mantled granite-gneiss domes; along their margins are rocks of an autochthonous cover and a geosynclinal allochthon with Outokumpu ophiolites at their base. The structure is additionally complicated by slicing of the western flanks of the domes causing tectonic interlayering of the basement and cover.

Deep within the Svecofennian province the degree of deformation—isoclinal folding, overthrusts etc.—evidently diminishes slightly but nevertheless remains quite high. Folds run around the Central Finland batholith and form an arc in central Sweden. The main epoch of deformation began 1.98 b.y. ago and its first phase, up to 1.85 b.y., was accompanied by the intrusion of syntectonic soda granitoids of mantle origin. Svecofennian tectonics and plutonism culminated 1.83–1.77 b.y. ago. Late orogenic granitoids, unlike early orogenic granitoids, are of crustal, anatectic origin. They are accompanied by migmatites and abundant pegmatites. These are succeeded by the intrusion of minor post-orogenic granites aged 1.7 b.y. and younger. In the concluding phase of deformation displacement occurred along strike-slip faults, in particular along the major Ladoga-Bothnian zone. In such fault zones metamorphism increases at places up to granulite facies while amphibolite facies of metamorphism predominates in the rest of the area of the Svecofennides.

After 1.78 b.y., the T r a n s-S c a n d i n a v i a n v o l c a n op l u t o n i c b e l t with large diapiric plutons of granitoids of mantle origin (according to Sr isotope data) lay unconformably on the western margin of the Svecofennides. This belt developed until 1.6 b.y. (see Sec. 4.3).

Geodynamic interpretation of the evolution of the Svecofennides varies somewhat. Most researchers regard the north-western segment of the Kola-Karelian eocraton as an active continental margin with the Scellefto volcanic arc and a somewhat younger (1.89–1.84 b.y.) marginal volcanoplutonic belt, at the rear of which calc-alkaline magmatism was replaced by subalkaline and alkaline. Thus subduction north-east under the Kola-Karelian continent is evident here. Less distinct is the direction of subduction in the Finnish-Karelian segment on its boundary. Some researchers (G. Gaal, and others) assume the same direction of subduction here but we see more obduction than subduction in this region and there is no marginal magmatism. Bowes (1980) assumed subduction to the south-west here, which could explain in part the formation of the Central Finland batholith. Regarding the Bergslagen volcanic arc, it is quite evident that subduction should have been towards the east in Sweden and towards the south in Finland but the latter is poorly correlated with the formation of the same Central Finland batholith. In fact, this batholith may have formed on the site of a microcontinent due to remobilisation of the ancient sial. Either way, the origin of the Central Finland batholith is, in our opinion, at the root of the entire problem of the geodynamics of the Svecofennides.

Canadian geologist A.F. Park (1991) recently proposed a significant new model of the evolution of the western and central parts of the Baltic Shield in the Early Proterozoic (2.50–1.75 b.y.) based on the terrane concept. Within the Svecofennides and their boundaries, Park has differentiated eight terranes: Kuhmo and Insalmi with Archaean continental crust, allochthonous 'hybrid' Lapland and Savo; young island-arc type Scellefto-Savonlinna in the north and southern Finland-central Sweden, back-arc or island-arc Outokumpu tectonic nappe and Jormua ophiolite nappe. Moreover, the Svionian magmatic arc, Kalevian flysch and Bothnian basins are differentiated. The collage of all these terranes occurred over several phases of accretion, deformation and magmatism 1.95, 1.90, 1.88–1.87 b.y. ago and concluded 1.85 b.y. ago with the continuation of magmatic and thermal episodes in the interval 1.85–1.75 b.y.

4.1.2.2. *Ukrainian Shield and basement of the Russian platform.* By the end of the Archaean the entire Ukrainian Shield and probably much of the basement of the Russian Platform had undergone cratonisation but at the commencement of the Early Proterozoic, as everywhere else, this gave way to destruction. The consequence of the latter was the formation of mobile protogeosynclinal systems, of which the most prominent is the Kursk-Krivoy Rog exposed within the Ukrainian Shield (Fig. 34) and explored by drilling in the Voronezh massif.

The Kursk-Krivoy Rog protogeosyncline has been traced in the north and south along magnetic anomalies over a stretch of about

125

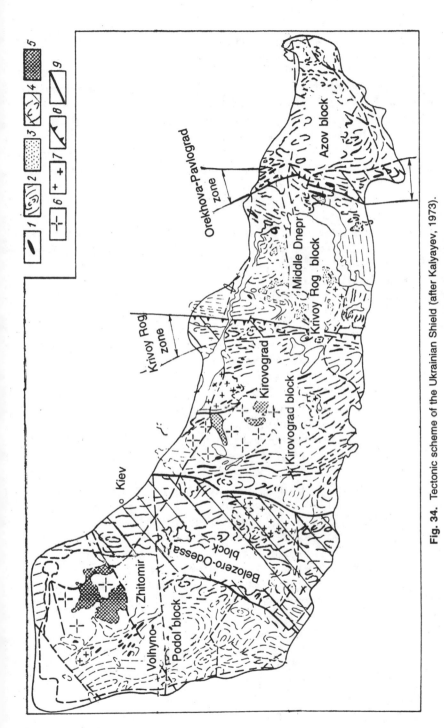

Fig. 34. Tectonic scheme of the Ukrainian Shield (after Kalyayev, 1973).

1 to 3—Early Archaean and Archaean (1—Aulian series, 2—gneisses and granite-gneisses, 3—greenstone belts); 4—protogeosynclinal systems of Early Proterozoic; 5—mafic intrusives; 6—rapakivi granites; 7—microcline granites; 8—overthrusts; 9—other faults.

1000 km in a meridional direction, its width not exceeding 75 km. The Krivoy Rog synclinorium proper of the Ukrainian Shield is a narrow asymmetric trough complicated by folds and overthrusts of higher order. The more quiescent eastern limb lies unconformably on the Saksagan gneiss-migmatite dome of the Middle Dnepr eocraton. The western limb is underlain by an overthrust which separated it from the Ingulets gneiss swell (a marginal element of Kirovograd megablock of the shield) with Archaean basement and Lower Proterozoic paragneiss cover of the Teterev series metamorphosed together (repeatedly in respect to the Archaean) to amphibolite facies and intruded by plagiogranites in the epoch 2.2–2.0 b.y. Thus the vergence of the Krivoy Rog synclinorium is directed from the younger Kirovograd block towards the more ancient Middle Dnepr block.

The Krivoy Rog series filling the same-named synclinorium has a three-tiered structure. It commences with conglomerates and contains in the lower part gritstones, quartzite-sandstones, quartz-sericite, chlorite and graphite schists; in the central part schists alternate rhythmically with bands of ferruginous quartzites forming the main Krivoy Rog iron ore basin deposits. The upper part of the series is retrogressive in character—quartzite-sandstones and schists with bands of dolomite-marbles are replaced by coarse clastic rocks. The total thickness of the series reaches 8 km; it is metamorphosed under conditions of high grades of greenschist and low grades of amphibolite facies.

The Ingulo-Ingulets suite, an analogue of the upper suite of the Krivoy Rog series, is developed on the slope of the Kirovograd megablock, west of the main overthrust of the Krivoy Rog synclinorium and the Ingulets swell. This formation lies unconformably on underlying rocks, is relatively weakly deformed but metamorphosed to amphibolite facies. It is intruded by plagioclase-microcline, aplite-like and pegmatoid granites aged 2.0–1.8 b.y. (Verkhoglyad, 1985), constituting the second phase of the Kirovograd-Zhitomir complex (first phase—plagiogranites intruding the Teterev series, probably an analogue of the lower suite of the Krivoy Rog series).

In the Voronezh massif, in the region of the Kursk Magnetic Anomaly (Krestin, 1980), the Krivoy Rog series, although not so thick, is close in composition to the Kursk series recovered by the schist-carbonate Oskol' series. A volcanic suite, felsic below but mafic and intermediate above, is deposited on the Oskol' series. The Stoilensk-Usman gabbro-plagiogranite and the much younger Pavlovsk complexes of normal granites (2.0–1.95 b.y.) are probably partly comagmatic with the volcanic suite. The Vorontsovo series (possibly an analogue of the Oskol' series), much younger than Kursk, is developed in the eastern part of the massif, separated by a major fault from the western part. This is a flyschoid terrigenous series; in age and structural position it resembles the Ladoga series of the Baltic Shield and it is not by mere chance that numerous layered mafic-ultramafic intrusions with copper-nickel-sulphide mineralisation (Trosnyansk-Mamonov complex) are also well known in this area. It is therefore possible that the Bothnian-Ladoga frontal zone of the Svecofennides has its continuation here. Its further continuation to the south-south-east might be noticed in the eastern part of the buried Rostov salient of the Ukrainian Shield. From the east this zone is bound by, probably thrust on, an extensive block of Epi-Archaean consolidation encompassing the entire Volga-Urals region and evidently joining with the Kola-Karelian eocraton in the north.

In the region of the Kursk Magnetic Anomaly another mobile zone of the same type and age branches southwards from the northern part of the Kursk-Krivoy Rog protogeosyncline. On the Ukrainian Shield this zone (system), called O r e k h o v o-P a v l o g r a d, wedges between the Middle Dnepr and Peri-Azov megablocks. It experienced intense compression and tectonic reworking and hence only some odd synclinal remnants have remained of the analogues of the Krivoy Rog series, including its central iron ore suite. Formations corresponding to the upper suite of the Krivoy Rog series and consisting of diverse paragneisses with beds of calcareous quartzites and dolomite-marbles are most widespread; a suite of mafic and ultramafic volcanics lies above, as in the Voronezh massif. Like the Krivoy Rog synclinorium, the Orekhovo-Pavlograd synclinorium is complicated to an even greater extent by overthrusts whose surfaces dip eastwards under the Peri-Azov megablock. In structure, the Peri-Azov megablock is similar to the Kirovograd in that the former is mainly made up of Archaean formations but experienced intense reworking and repeated granitisation in the first half of the Early Proterozoic (2.3–2.1 b.y. ago). Glevassky (1989) noticed a definite polarity in the change of composition of granitoids in the latitudinal section of the block with increasing distance from Orekhovo-Pavlograd overthrusts, which are interpreted as a former zone of subduction. A much younger complex of alkaline granitoids aged 1.7 b.y. is known in the eastern part of the megablock.

Yet another Early Proterozoic protogeosynclinal fold system—O d e s s a-K a n e v (Belaya Tserkov)—extends between the Kirovograd megablock of the Ukrainian Shield in the east and the Volhyno-Podol megablock in the west. The latter, especially in its northern, Volhyn' section, is similar in structure to the Kirovograd and the Peri-Azov; like them, it underwent tectonothermal reworking 2.3–2.1 (or 2.2–2.0) b.y. ago with the intrusion of major granite plutons. The Odessa-Kanev system is itself poorly exposed and traced on the surface along synclinal outliers of ferruginous quartzites and, outside the shield, along linear magnetic anomalies. This characteristic and also increase in crustal thickness, which is characteristic of all protogeosynclinal systems of the Ukrainian Shield, are observed in another such system westwards, beyond the limits of the shield. This system runs southwards under the Carpathians and could be recognised geophysically under the Carpathian arc (Sollogub and Chekunov, 1985).

The geodynamic interpretation of the evolution of the Ukrainian Shield from the viewpoint of plate tectonics was recently adopted by G.I. Kalyayev and E.B. Glevassky. However, we do not wholly subscribe to this view for the formation in the Early Proterozoic at the site of protogeosynclinal systems of considerable expanses with new oceanic crust. The absence of ophiolites and extremely poor development of even mafic magmatites do not suggest that destruction here reached the stage of spreading on the oceanic scale. There is no doubt, however, that continental crust at the base of proto-geosynclines should have undergone significant extension and reworking and was probably replaced by a transitional type crust. The formation of small oceanic basins of the Red Sea type is also quite possible. The crust of such basins, transitional or simatic, could have subducted under the Peri-Azov, Kirovograd and Volhyn' megablocks and thus explain thickening of the crust in the zones of former protogeosynclines and granitoid magmatism on the hanging walls of the ancient zones of subduction. As for the width of these basins, palaeomagnetic data is needed for objective determination.

Besides the Ukrainian Shield and the Voronezh massif, Early Protero-zoic protogeosynclines have recently been established within the limits of Belarus', NE Poland and the Baltic states according to geophysical and drilling data and radiometric determinations of rock ages. However, only the strip along the southern coast of the Gulf of Finland can be confidently placed in this category of structures, since it undoubtedly represents the marginal zone of the Svecofennides. It is bound from the south by the Latvia-Estonian Archaean massif. In so far as the more southern meridional zones of development of amphibolite-gneiss (metaextrusive greywacke) and gneiss-schist (metaflyschoid and volcanosedimentary, including iron ore) complexes extending from Latvia through Lithuania towards Belarus' and Poland (Mazovets-Inchukalin and Lithuanian-Belarus' zones) are concerned, isotope data obtained for these complexes suggests that they belong to Early Proterozoic protogeosynclines.

In general, the eastern part of the East European Craton east of the line running through the apex of the Gulf of Bothnia and Lake Ladoga to Ros-tov district represents mainly the area of Epi-Archaean cratonisation which maintained relative stability in the Early Proterozoic. The western part of the craton, however, underwent considerable destruction at the beginning of the Early Proterozoic with formation of a system of protogeosynclines which degenerates southwards.

4.1.3 Mediterranean Geosynclinal Belt

Numerous outcrops of the Precambrian basement are known within this belt but only a few have proven or assumed Early Precambrian age with the presence of Lower Proterozoic established or suggested only in certain cases. This pertains predominantly to the more eastern part of the belt lying east of the Ural-Oman lineament. In the west the Lower Proterozoic has been most firmly established in the Palasu-Mare borehole drilled in southern Dobrogea near Konstantsa. This age is based here on radiometry (1.7 b.y.) and from the similarity of rocks (ferruginous quartzites) with those of the Krivoy Rog series. Early Proterozoic ages were recently determined for eclogites of southern Brittany (2208 ± 30 m.y.) by the U-Pb method on zircon. Early Proterozoic age can very tentatively be assigned to the lower part of the Rhodope crystalline complex consisting of gneisses, crystalline schists, amphibolites and marbles with ferruginous quartzites.

In the eastern segment of the Mediterranean belt, in Badahshan and the Pamirs, Hindu Kush, Karakoram, Hazara and Himalayas, outcrops of the crystalline basement now integrated in its complex nappe structure are quite extensively developed. Karapetov (1979), who correlated the entire data on the structure of this basement, distinguished two complexes here, of which the lower is predominantly gneissic and granite-gneissic, metamorphosed to amphibolite, partly granulite facies and migmatised, and has been placed in the Archaean; the upper is crystalline schist with gneisses, quartzites, marbles and amphibolites, metamorphosed to epidote-amphibolite facies and intruded by small plutons of granite-gneisses, plagiogranites, gabbroids and veins of pegmatites and has been placed in the Early Proterozoic. The first of these complexes was radiometrically dated only in the south-western Pamirs and the second in the Himalayas (1.8 b.y.). At the same time, in the central and western segments of the belt (the Balkans, Carpathians and Alps) rows of analogous composition and degree of metamorphism have been placed presumably in the Lower and Middle Riphean, having experienced Grenville thermotectonics. But it is possible that a Riphean age could be considered for these rocks as a result of later reworking. If placing this complex in the Early Proterozoic is correct, its tectonic nature can be determined based on the development of quartzites, marbles and amphibolites as cover formations (deposits of epicontinental basins or shelf).

4.1.4 Siberian Craton and Its Framing

As noted in the preceding chapter, the build-up of the Siberian Craton was almost completed by the commencement of the Proterozoic. It was already in the form of either an eocraton or protoplatform in the Early Proterozoic. This is well seen on the example of the Aldan Shield; in its western part, the typical protoplatform Kodar-Udokan basin filled with the very thick clastic Udokan series lies with a sharp unconformity on the Archaean basement, which includes Late Archaean greenstone belts extending meridionally.

Assuming cratonisation of the basement of the Siberian Craton already at the end of the Archaean, it would be incorrect to reject the assumption that it maintained considerable mobility in the Early Proterozoic, however. For example, the Anabar massif, which has been well studied (Markov,

1988) shows that considerable mobility occurred in the Early Proterozoic along faults with the rise of fluids, diaphthoresis of Archaean granulites and formation of anatectic granites.

4.1.4.1 *Kodar-Udokan and other basins of the Aldan Shield.* The Kodar-Udokan basin (Biryul'kin et al., 1983; Fedorov, 1985) extends for 250 km in a latitudinal direction and is up to 100 km wide. The Udokan series filling it represents a clastic formation with carbonate layers. This series of fluvial-lacustrine and shallow-marine origin has a distinct cyclic structure with a generally retrogressive tendency in development. Quartzites, metasandstones and metasiltstones with bands of lenses or gritstones and conglomerates were deposited at the base of the cycles. Along with these rocks, dolomite and limestone marbles are seen in the middle of the cycles. The upper parts of the cycles are made up of sandstone, siltstone and shale with traces of erosion, ripple marks and shrinkage cracks. Sandstones of the third cycle are copper-bearing (Udokan copper ore) from below. The total thickness of the series is 8–12 km. Metamorphism is zonal, generally within amphibolite facies; it rises steadily towards the periphery of the basin where synmetamorphic granite-gneisses are seen. They form a belt of domes, swells and other forms encircling the basins and arose due to remobilisation of basement rocks. Moreover, the basin houses an immense post-metamorphic Kodar-Kemen granite lopolith; granites are proximate to rapakivi granites in chemical composition and porphyric structure. The age of these granites was determined as 2.0–1.8 b.y. and of synmetamorphic granitoids as 2.1–1.9 b.y. These ages fix the lower limit of the age of the Udokan series. Its uneven fold dislocation is associated with the intrusion of granites as well as with the build-up of granite-gneiss domes (in the marginal parts of the basin) and, finally, with the reflection of faults along the margins of buried Archaean greenstone belts. The Udokan series is overlain by the Kebeta suite, which concludes the sequence of the basin. This suite is made up of red continental sandstones and siltstones with lenses of felsic volcanics. It lies unconformably on the Udokan series and is almost unaffected by metamorphism. There is good reason for assuming that this suite is younger than the Lower Proterozoic.

Apart from the Kodar-Udokan basin, the Chara block of the Aldan Shield houses some similar basins but of considerably smaller dimensions. The existence of such basins has been suggested even under the cover of the Central Siberian Platform on the basis of geophysical data. East of the Aldan Shield lies the large Ulkan trough, also filled with formations of Lower Proterozoic age but almost wholly subaerial-volcanic (volcanics of felsic composition, thickness up to 4 km). Granites intruding them are aged 1900 ± 100 m.y., i.e., similar in age to the Kodar-Kemen pluton. In the Ulkan and, to a lesser extent in the Bilakchan troughs, subsidence continued even later with the accumulation of a red-coloured clastic formation, an analogue of the Kebeta series. The age of the Ulkan laccolith of alkaline granitoids is 1660 m.y.

4.1.4.2 *Baikal-Vitim protogeosynclinal system and Northern Baikal volcanoplutonic belt* (Fig. 35). The zone west of the Aldan Shield, and the Zhuia fault restricting it, and south of the present-day southern margin of the platform in the Baikal region maintained high mobility in the Early Proterozoic. This region, which corresponds today to the Baikal-Patom upland, forms a re-entrant into the body of the Siberian Craton. It is extremely complex in structure and its interpretation, especially of the southern part, is rendered additionally difficult due to extensive development of granite intrusions forming in general an immense polyphase Barguzin (Angara-Vitim) batholith.

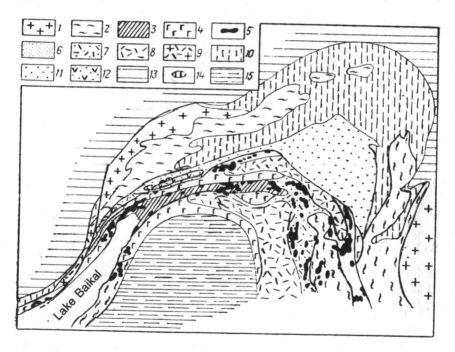

Fig. 35. Sketch showing the Baikal-Muya (Vitim—V.E. Kh.) ophiolite belt (after Dobretsov, 1982).

1—Archaean rocks; 2—sialic crystalline basement in structures of the Baikalides; 3 and 4—Baikal-Muya belt (3—inner zone, 4—outer zone); 5—'boudinage pluton' of gabbro and ultramafics; 6—metamorphosed olistostromes; 7 and 8—regions of development of Kilyana series (7—with olistostrome horizons, 8—with felsic metavolcanics); 9—Peri-Baikal (Akitkan—V. Kh.) volcanic belt; 10—terrigenous strata of marginal troughs; 11 and 12—Olokit-Bodaibo suite of inner trough (11—metaterrigenous, 12—volcanic); 13—Siberian Platform cover; 14—olivinite-peridotite-troctolite (Dovyren) complex; 15—inner Baikalides zone.

Dislocated Riphean formations are extensively developed in the northern part. An interrupted arc of uplifts-anticlinoria consisting of crystalline schists and granites of Lower Proterozoic age extends amidst this field of Riphean strata conformably to its boundaries.

The Northern Baikal (Akitkan) volcanoplutonic belt, delineated and described by Bukharov and colleagues (1985), extends along the western flank of the re-entrant at the very boundary of the craton. It forms the Baikal Range on the western bank of the lake. The belt is made up of the volcanosedimentary Akitkan series and Irel' intrusive complex associated with this series. Variegated clastic rocks ranging from conglomerates to siltstones with subordinate sheets of felsic volcanics predominate in the lower parts of the Akitkan series. Its upper part is made up of volcanics of the calc-alkaline and alkaline series with a predominance of felsic varieties, including ignimbrites. Volcanic centres and subvolcanic formations have been identified. Diorites, granodiorites, granites, granosyenites and quartz-monzonites are found in the comagmatic Irel' intrusive complex. An age of 1700 ± 100 m.y. has been established for granites of the late phase of the complex and 1700 ± 35 m.y. (lower section) and 1630 ± 40 m.y. (upper section) for lavas of the upper part of the Akitkan series.

The geodynamic environment of formation of the Northern Baikal volcanoplutonic belt has not yet been thoroughly understood, largely due to non-availability of strontium and neodymium isotope data. The disposition of this belt strictly at the boundary of the craton (even considering that its rocks are discovered by drilling more towards the north under the platform cover) suggests that it may be a belt of the Andean type; this was the view of Zonenshain and colleagues (1990). But then, where are the relics of the corresponding adjoining oceanic basin? These could be the metaophiolites of the Baikal-Vitim belt recently discovered by Konnikov (1991).

Another interpretation of the Northern Baikal belt is possible: it might be a post-collision formation, keeping in view the collision of the Early Proterozoic Baikal (Chuy-Tonod) arc with the Epi-Archaean Siberian Craton.

4.1.5 Ural-Okhotsk Geosynclinal Fold Belt

Numerous outcrops of Early Precambrian crystalline basement are seen within the Ural-Okhotsk belt throughout the Ural Mountains from Marunkeu and Kharbey uplifts in the Polar Urals to Mugodzhary. These are concentrated in the central and southern Urals in the east Uralian and Preuralian anticlinoria where cores of granite-gneiss domes surrounded by Riphean and Palaeozoic strata are found. Some researchers (A.I. Rusin, among others) oppose the domal nature of these structures, pointing out that Precambrian rocks are quite often in contact with Palaeozoic formations along fractures, with the latter revealing no perceptible metamorphism. It is probable that such contacts represent secondary phenomena. The Early

Precambrian age of these formations has been confirmed by a few isotope datings, which points to the presence of the Archaean (see Sec. 3.1.5) and the Lower Proterozoic. The following are the most reliable Early Proterozoic datings: 1) 2 b.y. for gneisses of the Kharbey uplift (U-Pb method on zircon) and 2) 1850 m.y. (U-Pb and U-Th-Pb methods) for the Selyankin complex of the Ilmenogorsk uplift in the southern Urals (formerly, for these same Selyankin gneisses, Rb-Sr and Pb isotope methods gave values of 2100 ± 200 and 1980 ± 100 m.y.). For Mugodzhary (in: *Precambrian of Central Asia,* 1984), values of 1730 ± 120 and 1580 ± 65 m.y. have been cited (Pb isotope method on zircons). In composition, Lower Proterozoic formations of the Urals are represented by gneisses, crystalline schists and amphibolites with bands of quartzites, metamorphosed to amphibolite or epidote-amphibolite facies, quite often migmatised and intruded by granites.

In c e n t r a l K a z a k h s t a n the most significant outcrops of the Lower Proterozoic are known in the Ulytau Range in the Kazakhstan-Northern Tien Shan median massif (microcontinent). The most characteristic element of the Lower Proterozoic sequence here is the iron ore K a r s a k p a y s e r i e s comprising four-tiered cyclically recurrent formations. The lower part of these formations comprises basaltic and andesite-basaltic metavolcanics and the upper part phyllites, ferruginous and barren quartzites and marbles. The thickness of the series has been estimated as 4 km and metamorphism as greenschist facies. A similar series is known farther north in the Kokchetav massif and south in Betpak-Dala. The Karsakpay series as such has not been radiometrically dated but it is unconformably overlain by the molasse-porphyry formation of the Maityube series (see Sec. 4.3); granite-gneisses associated with the latter are aged 1.8 b.y. A view has been expressed (Makarychev et al., 1983) about the greater antiquity of the Maityube series compared to the Karsakpay but hardly accords with the typical (jaspilites!) composition of the latter. Most researchers regard the Karsakpay series as underlain by the Aralbay series of volcanic composition. Metadacites and metarhyolites and their tuffs and tuffites with sodium specialisation predominate among the Aralbay volcanics. Along with them are also contained sericitic (with albite, chlorite and quartz) schists, phyllites, marbles and thin bands of ferruginous quartzites. The Ulytau system, traced along the Karsakpay iron ore series for at least 500 km (southward under the Dzhezkazgan basin), represents a fairly typical protogeosyncline bound most prominently in the west by the Archaean protoplatform under the Turgay trough.

North-east of Ulytau the Lower Proterozoic is exposed at places in the Yerementau-Niyaz anticlinorium. According to German (1986), it is made up here of metaterrigenous rocks with high-grade greenschist facies and bands of quartzites, marbles and metatuffaceous schists; the exposed thickness exceeds 3 km. These formations are supposedly underlain by Archaean gneisses and amphibolites forming dome structures. It is possible that these formations represent an Early Proterozoic protoplatform cover.

Lower Proterozoic formations of similar composition are exposed in the Makbal uplift (dome) of the Kirghiz Range of Northern Tien Shan and in Kasan uplift (also a dome) of central Tien Shan (according to Ges', 1988). In the former the Keldin suite of the Kirghiz series, dated 2010 ± 100 and 1920 ± 50 m.y. on zircon, is made up of garnet-muscovite-quartz schists with bands of amphibolites and marbles. The underlying Makbal suite consists of quartzites with graphite and mica and is dated 2075 ± 10 m.y. In the Kasan uplift the Lower Proterozoic is represented by the Tereksay suite (garnet-biotite and binary mica schists with bands of amphibolites and marbles). In both regions metamorphism is of epidote-amphibolite facies and metamorphic schists are present in tectonic contact with serpentinites, metagabbroids and metabasaltoids which Makarychev and Ges' (Makarychev et al., 1983) regard as relics of an ophiolite association.

For gneisses and crystalline schists of the Chu-Ili system, Kazakhstan geologists obtained the following Early Proterozoic ages (in: *Precambrian in Phanerozoic Fold Belts,* 1982; *Precambrian of Central Asia,* 1984): Sarybulak suite of Kendyktas (1877 ± 8 m.y.), Anrakhay (1722 ± 14 m.y.), Karakamys (1750 and 1800 ± 100 m.y.) and Ogiztau (1755 ± 100 m.y.) suites (Pb isochron method on zircon).

Outcrops of the Early Precambrian disappear farther east in the northern part of the Ural-Okhotsk belt, evidently due to the development of much younger oceanic crust. They reappear along the Siberian Craton, in the Yenisei Range and in Eastern Sayan. In the northern part of the Y e n i s e i R a n g e the Teya series of high-alumina gneisses, amphibolites and marbles in the lower part and marbles and metamorphic schists (with epidote, chlorite, biotite and garnet) in the upper part lying on the Tarak complex of granitoids aged 1850 ± 100 m.y. belongs to the upper part of the Lower Proterozoic. Rocks of this series are exposed in the cores of granite-gneiss domes and swells and have been dated 1.6–1.5 b.y. in their upper part (K-Ar method). Southward, in the central zone of the Eastern Sayan, an analogue of these series is the Derbin series formed of gneisses, amphibolites, quartzites and marbles (Alykdzher suite) in the lower part and of marbles with graphite (Derbin suite proper)[4] in the upper part. Rocks of the Alykdzher suite are intruded by granites aged 1.6–1.5 b.y. Evidently the Derbin suite represents a protoplatform cover.

Formations such as the Derbin protoplatform complex are extensively developed south-west and south-east of Eastern Sayan, in the Tuva-Mongolia median massif, Malkhan 'promontory' of the Aldan Shield, Kerulen-Argun massif, and in the small blocks of the Early Precambrian continental

---

[4] Volobuyev and colleagues (1980) place these formations already in the Riphean.

crust within the eastern half of the Ural-Okhotsk belt. Among them in particular are the Moren series of the Sangilen upland, made up of gneisses (biotitic and garnetiferous) with bands of marbles, amphibolites and jaspilites intruded by granitoids aged 1.9–1.8 b.y. and possibly the Balyktygkhem series of marbles with bands of gneisses; the Malkhan series of the same-named zone with a predominance of gneisses and crystalline schists and bands of amphibolites, quartzites and marbles; the Bumbuger complex of the Baidaryk (Baidaragyn) block; and the Yesenbulak complex of the Gobi-Altay block. These complexes consist of gneisses, amphibolites, marbles and quartzites; the Bumbuger complex is intruded by granitoids aged 1.9–1.7 b.y. (Pb-Pb thermoisochron and K-Ar methods). Somewhat north of the latter two blocks in western Mongolia, in the Dzabkhan and Khunguy-Gol interfluve, Makarychev (1992) identified a complex of ultramafic gabbroids and amphibolites regarded as an ophiolite association underlying the formations of the Bumbuger-Yesenbulak type. However, other scientists dispute the affinity of these rocks to ophiolites as well as their Early Proterozoic age.

The Lower Proterozoic may still be present in the southernmost part of Mongolia, in the Ulanul zone (microcontinent), constituting a part of the Southern Mongolia Hercynian fold system but has not yet been adequately studied.

4.1.6 Sino-Korea Craton

The beginning of the Proterozoic here, as everywhere else, was a period of destruction of recently formed monolithic continental crust, isolation of its remaining unreworked blocks, i.e., protoplatforms (eocratons) and the formation of protogeosynclines between them (Fig. 36). The more prominent platforms number five, including the Tarim eocraton whose core was also formed by consolidated Archaean emerging into its southern periphery and in the north into Kuruktag, and the Sichuan eocraton, later entering the structure of the South China Craton.

In China the Lower Proterozoic is distinctly subdivided into two parts: from 2.5 (2.6) to 2.3 (2.2) b.y., represented by formations of the Wutai group whose accumulation culminated in the same-named epoch of diastrophism and from 2.3 (2.2) to 1.9 (1.8) b.y., represented by formations of the Hutou group and its analogues deformed in the Lulian (or Jongtiao) epoch of diastrophism (Cheng Baijin and Sun Dazhong, 1982; Ma Xingyuan and Wu Zhengwen, 1981; Wang et al., 1985).

An axial trough with intense manifestation of volcanism and peripheral zones practically devoid of such manifestation can be distinguished within much of the Wutai protogeosynclines; they exhibit similarity with eu- and miogeosynclines respectively, and are sometimes so called. In addition, protogeosynclinal systems reveal geoanticlines made up of Archaean

136

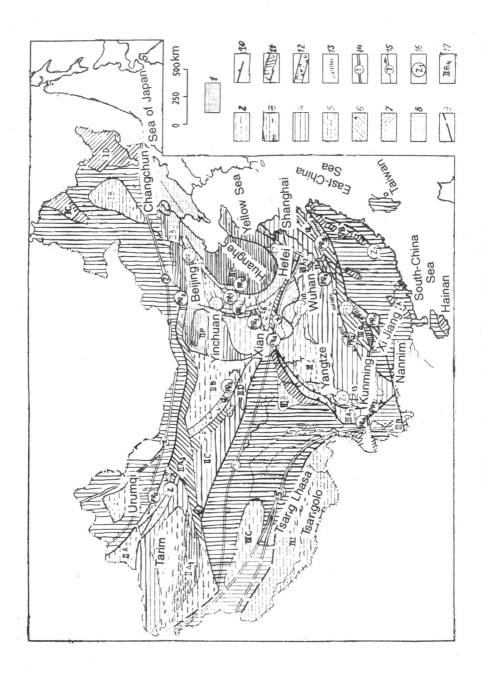

rocks which subdivide them into individual troughs. The well-studied protogeosyncline Jinshan extends in a north-easterly direction from southeastern Shanxi to Futian, separating the Ordos and Hehuai protoplatforms. In its axial Wutai-Luliangshan trough a thick suite of volcanics was deposited, ranging from mafic to felsic, partly spilite-keratophyre, with bands of greywackes, turbidites and ferruginous quartzites metamorphosed to greenschist, partly amphibolite facies, migmatised and intruded by granites (some granites intrude only the lower part of the sequence). The thickness of formations reaches 6–7 km. Some similarity is noticed with greenstone Archaean belts together with differences, i.e., higher content of sedimentary rocks and absence of true komatiites.

The following age data is presently available for formations of the Wutai group: in eastern Hebei 2403 and 2193 m.y. by the Rb-Sr isochron method on whole rock; in the hills of Wutaishan (Shanxi) 2350 ± 137 and 2392 m.y. by the same method and 2508 ± 9 and 2521 ± 17 m.y. by the U-Pb method on zircon from granites and keratophyres.

The M a c h h o l l e n p r o t o g e o s y n c l i n e in North Korea is another typical protogeosyncline on the eastern extremity of the Sino-Korea Craton. It has a submeridional trend and is filled with thick (up to 10–12 km) volcanosedimentary formations. Mafic, partly felsic volcanics predominate in

---

Fig. 36. Schematic map of the structure of the Precambrian basement of China (after Wang and Qiao, 1984)..

Continental crust (basement of cratons and median massifs): 1—pre-Fuping (2600 m.y.); 2—pre-Luliang (1850 m.y.); 3—pre-Chong'yan (1700 m.y.); 4—pre-Jiannan (1000–850 m.y.); 5—undifferentiated Precambrian.

Crust of transitional type (island arcs, marginal seas etc.): 6—mainly 1000–850 m.y. (Jiannan); 7—mainly 850–550 m.y. (Xinkai); 8—open seas, mainly oceanic crust.

Other formations: 9—passive continental margins; 10—accretionary zones of crust absorption; 11 and 12—aulacogens (11—intracontinental, 12—continental marginal); 13—ophiolite complexes; 14—successive convergent zones of subduction (1—Aiby-Zhduiyan, 2—Suslun-Xilamulun, 3—Chyugou-Machin, 4—Shan'yang-Tongcheng, 5—Banggong-Nujiang); 15—successive transregional faults (6—Altin, 7—Tangong-Lujiang); 16—age of diastrophism (Z—Sinian); 17—geotectonic elements and structures of the basement. IA—Junggar massif; IB—Songlyao massif; IC—Iilchuli; ID—Xinkai uplift. II—North China continental province; IIA—Tarim Craton (IIA$_1$—south Tarim core, IIA$_2$—Iinin massif, IIA$_3$—Beishan uplift); IIB—North China Craton (IIB$_1$—Ordos core, IIB$_2$—Yan-Liao core, IIB$_3$—Hehuai core, IIB$_4$—Alxa massif and IIB$_5$—Dzhailyao uplift); IIC—Qaidam massif; IID—Lanzhou-Xinin massif; IIE—central Qilian uplift. III—South China continental province; IIIA—Yangtze Craton (IIIA$_1$—central Sichuan massif, IIIA$_2$—Dabei massif, IIIA$_3$—Kam-Yunnan uplift, IIIA$_4$—Jiannan uplift); IIIB—Songpan massif; IIIC—Qtantong massif; IIID—Linqang massif. Southern (Gondwana) continental region; V—East China marginal-continental province; VA—Yunkai uplift; VB—Dzhinu uplift. Aulacogens: 1—Hutou; 2—Gantoahe; 3—Jongtiao; 4—Bayan-Obo; 5—Yanshan-Taishan; 6—Luliang Xioner; 7—central Tien Shan; 8—Quruktagh; 9—northern Qilian; 10—Shennungjia; 11—Luokedong; 12—Shangshu; 13—Suchong.

138

its lower part and jaspilites are developed although carbonates predominate in the upper part. Folding is linear and metamorphism is to amphibolite facies. Intrusions of granites are known. Westward the Machhollen fold system adjoins the Archaean Nannim block of the Sino-Korea Shield along a wide (up to 4 km!) zone of faults accompanied by mylonites.

Wutai tectonics promoted a significant growth of the finally cratonised area of the Sino-Korea as well as South China cratons. Fig. 36 shows the blocks of consolidated continental crust in Manchuria, Junggar, Qaidam and Tibet. The Sino-Korea Craton was formed as a single continental megablock from North Korea to Tarim inclusive. Two latitudinal protoaulacogens arose along the northern and southern peripheries of this craton. These are the Yanshan-Taishan in the north and the Sinyan-Fozilin in the south. Another aulacogen with a similar trend is seen along the northern periphery of the Tarim massif. These aulacogens continued to grow in the Late Proterozoic (Riphean), having been filled with conglomerates, quartzites, calc-clay fly-schoid formations and dolomites with a cyclic sequence. Initial rifting was accompanied by lava extrusions of mafic and intermediate composition.

Along with the formation of aulacogens, some protogeosynclines continued to grow, especially the Ji-Jinshan. Coarse clastic, sandy-clay rocks, stromatolitic dolomites of the Hutou group with subordinate sheets of andesites and basalts accumulated in these protogeosynclines to a thickness of over 10 km. South-easterly the geosynclinal complex gradually passed into a platform cover. Geosynclinal formations experienced metamorphism to greenschist facies and locally were migmatised during the first phase of Luliang diastrophism 1.8 b.y. ago. Deposition of coarse clastic molasse sediments commenced in the intermontane troughs which were formed after this phase and deformed and metamorphosed during the second phase of Luliang diastrophism (1.7 b.y. ago). The folds are intensely compressed, up to isoclinal, and complicated by break-up.

Final consolidation of the Sino-Korea Craton, probably including the Tarim and Qaidam massifs, occurred as a result of Luliang tectonics. This was succeeded by the intrusion of huge swarms of dolerite dykes (in the Wutai-Taishan region one such swarm was 150 km long and 100 km wide) and build-up of plutons of anorthosites and rapakivi granites which extended to the entire Early Riphean (see Sec. 5.1.4).

4.1.7 South China (Yangtze) Craton and Adjoining Massifs[5]

Lower Proterozoic formations are developed more extensively in this region than Archaean and constitute the bulk of the basement of the South China (Yangtze) Craton, Indosinian and Sinoburman massifs and some

---

[5] After Fan Chyong Thi (1981), Le Zui Bath (1985) and *Metamorphic Complexes of Asia* (1977).

small uplifts in the northern part of Vietnam and in the Viet-Lao Phanerozoic fold system.

4.1.7.1 *Sikan-Yunnan Range.* In the sequence of the Sikan-Yunnan Range much of the K a n n g i n g   c o m p l e x falls in the Lower Proterozoic and is divided into three parts. The lower part is made up of gneisses and plagiogneisses hosting plagiogranites and migmatites dated 2.1 b.y.; the middle section is made up of gneisses with lenses of marbles and amphibole-magnetite rocks while the upper part is made up of quartz-mica schists with graphite and intercalations of ferruginous quartzites. The complex metamorphosed to amphibolite facies, underwent granitisation and is unconformably covered by Riphean formation.

An analogue of the Kannging complex is the Ailaoshan complex exposed in a narrow suture horst of the Danjianshan and Ailaoshan mountains into China and Fansipan farther east-south-east in Vietnam in the boundary zone between the South China Craton and the Viet-Lao fold system. The Suanday series, representing a complex equivalent to Ailaoshan in Vietnam, consists of two formations. The lower formation (thickness 5–5.5 km) is represented by gneisses and amphibole-biotite plagiogneisses with intercalations of amphibolites, quartzites, amphibole-magnetite rocks and magnetite-quartzites. It was intruded initially by plagiogranites and trondhjemites and later by microcline migmatites and pegmatite-granitoids. The upper part (thickness up to 2.5 km) has been granitised only at the bottom; it is made up of garnet-mica and garnet-graphite schists and quartzites with bands and lenses of amphibolites, haematite-magnetite-quartzites and marbles. The thickness of the Ailaoshan complex in China is 9 km. The age of the Suanday series has been estimated as 2.3–2.07 b.y. by radiometric dating of plagiogneisses of the lower formation and at 1386 m.y. for pegmatites intersecting it.

In the Jiannan uplift of the Yangtze Craton (core of same-named anteclise), the Lower Proterozoic is represented by biotite and hornblende gneisses and higher by biotite schists with bands of quartzites and marbles and bodies of hypersthene-granites and amphibolites. Similar formations (gneisses, crystalline schists and amphibolites) are also known within the Caledonian fold system of south-eastern China (Cathaysia) in the Uishan Range where they are intruded by granites aged 1.9–1.7 b.y. Rocks of proximate age (1.8 b.y.) were reached by drilling in the Sichuan syneclise in the western part of the South China Craton and are exposed in its northern framing.

4.1.7.2 *Indosinian massif.* Much of the Kontum salient of this massif is made up of the Lower Proterozoic Ngoklinh series consisting of two formations. The lower, up to 4–5 km thick, is formed of gneisses (in the lower part), amphibole, biotite, biotite-sillimanite schists, amphibolites and quartzites with marbles on top. The upper formation is up to 3 km thick and consists of sillimanite-mica-graphite schists and quartzites with graphite. Both formations are metamorphosed to amphibolite facies; the lower formation is intensely granitised while the upper is weakly affected by ultra-metamorphic processes. In primary composition the lower formation represents an alternation of mafic volcanics and carbonate-terrigenous sediments while the upper is wholly made up of terrigenous sediments. The age of the gneisses in the lower formation was placed at 2.3 b.y. by the lead isochron method.

North of the massif, Lower Proterozoic formations of similar composition outcrop in the dome-shaped Phu Hoat uplift in the Viet-Lao fold system. These are amphibole gneisses, amphibole and biotite schists, garnetiferous amphibolites and sillimanite-mica-graphite schists. The age of the gneisses was determined as 1.6–1.51 b.y. by the K-Ar method. The complex is unevenly granitised and migmatised. The Lower Proterozoic complex exposed in the Red River fault and named the Shong Hong complex, exhibits some peculiar features. It consists of high-alumina gneisses with garnet, sillimanite, biotite, ortho- and para-amphibolites and contains intercalations of marbles in the lower part for which an age of 2.3 b.y. was determined by the lead isochron method. The visible thickness of the complex is of the order of 6 km. All the rocks in it are highly migmatised with the formation of pegmatites and porphyric granitoids. The structure is characterised by the development of narrow, intensely compressed folds with a persistent north-western orientation. The zone of development of the Shong Hong complex coincides with the contemporary Red River fault zone. The Hanoi rift is developed in the eastward continuation of this fault zone. The linearity, intense tectonics, metamorphism to high grade amphibolite facies with transition to granulite facies in the lower part make the Shong Hong zone similar to the granulite belts of the Precambrian and permits regarding it perhaps as a zone of collision of the South China and Indosinian Early Precambrian megablocks, like the Limpopo Archaean zone in South Africa.

4.1.8 South American Craton

Like other cratons, this craton too possessed a mature continental crust by the end of the Archaean but destruction in the Early Proterozoic was extremely deep here and only small massifs survived. Contouring of these small massifs has proved difficult to date. They have been separated by extensive protogeosynclines whose development ceased 2.0–1.9 b.y. ago as a result of intense diastrophism, termed the T r a n s a m a z o n i a n o r o g e n y. This diastrophism played an important role in consolidation of the South American Craton.

4.1.8.1 *Guyana Shield.* Three residual cores of Epi-Archaean cratonisation are distinguished within the Guyana Shield. These are Imataca,

Pacaraima and Xingu; the last extends southwards along the other side of the Amazonian syneclise into the Central Brazilian (Guapore) Shield but like the Pacaraima, is only tentatively delineated. A cover of felsic, subaerial volcanics aged 1.9–1.75 b.y. is formed on the Pacaraima massif and underlain at places by continental clastic formations. The space between the massifs represents a granite-greenstone terrane, delineated in the east (Peri-Atlantic region) as the M a r o n i-I t a c a y u n a s b e l t with a south-easterly trend. Its north-western branch separates the Imataca and Pacaraima massifs while the western one separates the Pacaraima and Xingu. In the southeast the belt intersects the Amazonian syneclise at depth and extends into the north-eastern part of the Central Brazilian Shield. The total extent of the Maroni-Itacayunas belt is 2500 km and its width reaches 500 km. Further, it must be remembered that before the opening of the Atlantic, it extended into Western Africa (Choubert, 1969). We revert to this subject in Section 4.1.9.3

Greenstone belts of the Guyana Shield comprise the B a r a m a- M a z a r u n i c o m p l e x (Gibbs and Barron, 1983) with a thickness of up to 8–10 km. It consists of metabasalts, komatiites and small amounts of meta-andesites, dacites, rhyolites and also metagreywackes (often turbidites), phyllites and conglomerates formed by erosion of volcanic rocks. Manganiferous and ferruginous quartzites, cherts and carbonates are also present. Volcanics pertain to the tholeiitic, calc-alkaline and magnesian series. Subvolcanic rocks of felsic composition are often encountered. Homodromous succession of volcanics and topping of the sequence with clastic sediments are common. Metamorphism varies from greenschist to amphibolite facies, increasing towards the granite-gneiss periphery of the belt.

The M a r o v i j n e g r o u p in northern Surinam extending in the east into French Guiana represents an analogue of the Barama-Mazaruni complex (Bosman et al., 1983). It has the three-tiered structure typical of greenstone belts. The lower formation, Paramaka, is mainly made up of mafic metavolcanics with intrusions of metagabbro, replaced by volcanics of intermediate composition with bands of metacherts and phyllites increasing upwards in the sequence. The middle formation, Armina, is formed of cyclic alternation of metagreywackes with gradational texture and phyllites (evidently flysch) while the upper formation, Rosabel, consists of metasandstones and metaconglomerates. The total thickness of the group is roughly estimated at 10 km. The age of basalts and quartz-andesites of the lower formation was determined as 1950 ± 50 and 1955 ± 60 m.y. respectively by the Rb-Sr isochron method at $^{87}Sr/^{86}Sr = 0.702$ in the first case. The type of metamorphism is the same as in Venezuela and Guiana. In eastern Venezuela corresponding formations have been distinguished in the Pastora supergroup, while in northern Brazil they are known as the Amapa, Vila Nova and Andorinhas groups.

The basement of the Guiana greenstone belts is a highly disputed subject. The oldest age values of gneisses are 2.25 and 2.05 b.y., the former being proximate to the age of volcanics of the belts themselves and the latter corresponding to Transamazonian diastrophism. This together with the absence of erosion products of rocks of continental crust in the composition of sediments prompts some researchers to support the possibility of an ensimatic origin of the belts (Gibbs and Barron, 1983). The age of rocks metamorphosed to granulite facies and encountered in the central part of the shield is not clear, however. These rocks also include orthoquartzites of continental origin. Other researchers (Bosman et al., 1983) assume the existence here of an Archaean folded basement. Transamazonian diastrophism was manifest in intense crumpling of the greenstone belts conformably with their gneissic frame and was accompanied by intrusion of numerous plutons, mainly of the calc-alkaline series, ranging from gabbro-diorites and quartz-monzonites to tonalites and adamellites. At least part of these rocks arose as a result of anatexis of the rocks of greenstone belts.

Two generations of granitoids are distinguished in Surinam (Bosman et al., 1983). The much older ones (tonalites and trondhjemites) form diapiric intrusions into the lower formations of the greenstone belts while the much younger ones form diapiric intrusions of binary micaceous leucocratic granites intruding even into the middle formation of the Marovijne complex. Moreover, granites are extensively developed beyond the greenstone belts; with them is associated the subaerial volcanic calc-alkaline Dalban formation of predominantly rhyolite and dacite composition with ignimbrites. The age of this volcanoplutonic association has been estimated at 1874 ± 40 m.y. by the Rb-Sr method. In southern Guiana and in Surinam a complex of small intrusive bodies of gabbro and ultramafics is of similar age (1818 ± 165 and 1845 ± 285 m.y.).

4.1.8.2 *Central and Eastern Brazilian Shields.* Relics of an Epi-Archaean cratonisation area are preserved here in Bahia (Brumado) and Goias (Crixas) states; in the contemporary structure they reveal affinity to the Sao Francisco eocraton and Goias median massif in the Brazilides system. South of the Sao Francisco eocraton the Lower Proterozoic iron ore formation of Minas also evidently represents a protoplatform cover. The first of these massifs in the north is partly overlain by Chapada-Diamantina protoplatform cover. The largest Early Proterozoic protogeosyncline was the meridional system extending for 2000 km from Ceara in the north to Sao Paolo in the south. This system, of ensialic origin, consists of greenstone belts similar to those of the Guianian system. Parallel to this large protogeosyncline, some very narrow fold zones of Transamazonian age extend westwards. Transamazonian metamorphism attains granulite facies while granitisation, largely anatectic, and migmatisation as well as deformation are very intense.

Tentatively, an Early Proterozoic age is ascribed to the even thicker volcanosedimentary rocks constituting the meridional zone within the Goias median massif. It is made up of basalts, terrigenous and siliceous sediments metamorphosed to amphibolite facies; layered intrusives of gabbro-anorthosites and mafic-ultramafics with sulphides of copper, nickel and cobalt mineralisation are presumably associated with this zone. In the same region less thick and weakly metamorphosed clastic-carbonate (with cherts and jaspilites) formations of the cover type rest unconformably on the Archaean and clastic pyrite, gold and uranium occur in the basal conglomerates. These formations are intruded by tin-bearing granites whose age has not been accurately determined but possibly fall in the Early or Middle Proterozoic.

On the whole, the data for the South American Craton shows that it experienced extremely significant reworking in the Early Proterozoic, which was manifest in different forms under conditions of extension (formation of proto-geosynclines and greenstone troughs) or compression (development of belts of tectonothermal reworking, i.e., granulite belts). These processes evidently concluded even before the end of the Early Proterozoic, in the Transamazonian epoch of diastrophism, not later than 1.9–1.8 b.y., and were succeeded by processes of cratonisation accompanied by magmatism specific for them (felsic volcanics, anorthosites etc.) (see Sec. 4.2).

4.1.9 African Craton

A significant part of this craton escaped destruction at the beginning of the Proterozoic and survived in the form of protoplatforms, among which the Kalahari was the largest and most stable.

4.1.9.1 *Kalahari protoplatform* (Tankard et al., 1984). This protoplatform was formed at the end of the Archaean as a result of completion of development of the Limpopo mobile belt which fused the Kaapvaal and Zimbabwe eocratons into a single entity. Here we find platform-type formations that are unique in stratigraphic completeness and thickness. These formations encompass the entire Lower Proterozoic and the underlying proximate type 'Witwatersrand triad' now placed in the Upper Archaean and described in Section 3.1.9.1. They constitute the Transvaal syneclise in the northern part of the Kaapvaal eocraton. Two complexes are distinguished in the Lower Proterozoic sequence of this syneclise (Fig. 37).

The lower complex of the Lower Proterozoic of South Africa, i.e., the T r a n s v a a l s u p e r g r o u p, consists of epicontinental marine formations up to 12 km thick formed in the interval 2.35–2.05 b.y. Continental clastic sediments with fairly thick sheets of basalts and felsites are lying at

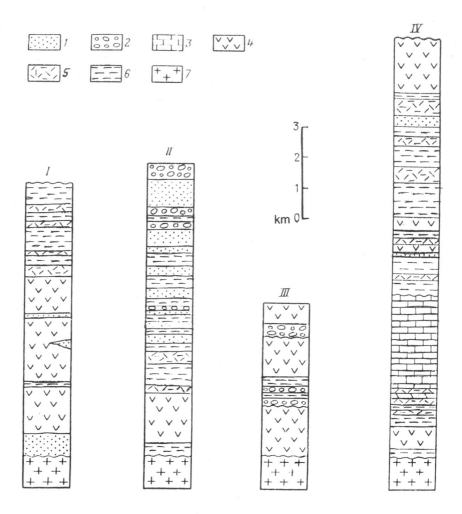

Fig. 37. Stratigraphic sequences of (I) Pongola, (II) Witwatersrand, (III) Ventersdorp and (IV) Transvaal basins (after Tankard et al., 1984).

1—sandstones; 2—conglomerates; 3—carbonates; 4—mafic extrusives; 5—felsic extrusives; 6—shales; 7—Archaean basement.

the base of this formation. These are gradually replaced by much thinner marine sediments and later by a massive dolomite formation, abounding in stromatolites and, in the upper part replaced by a ferruginous formation. The upper part of the complex, the Pretoria group, resting unconformably on the lower formation, consists of alternation of quartzites and clay-silt sediments with bands of dolomites and mafic volcanics of a type transitional from continental to oceanic tholeiites. Clastic material entered in the basin in large quantities from its northern framing, coinciding with the Archaean Limpopo belt which has evidently preserved high mobility.

Further accumulation of sediments and volcanics in the Transvaal syneclise was interrupted by formation of the largest (66,000 km^2, 350 km across) Bushveld layered ultramafic-mafic-granite multiphase and complex lopolith extending in a latitudinal direction. This lopolith intruded along the surface of unconformity, separating the quartzites of the top of the Pretoria group from the Rooiberg felsites. It factually represents the sum total of intrusive bodies of arcuate form formed at different periods. Each of these bodies had its own incurrent canal, as demonstrated by the existence of positive gravitational anomalies.

Mafic rocks constituting a stratified Rustenburg suite consist of five zones. The lowermost marginal zone is of norite and is replaced above by the lower zone of pyroxenites and harzburgites. Above them, in the pyroxenites, are seen deposits of chromites forming together the so-called Critic zone. Higher, pyroxenites begin to alternate cyclically with norites forming a cumulative pair. Further, norites are replaced by anorthosites. In these uppermost cycles of the Critic zone, especially in the penultimate Merensky cycle, pyroxenites contain a concentration of platinum metals. The main zone, consisting of norites and gabbro with subordinate amounts of anorthosites, lies above the Critic zone. The main zone is replaced by the Upper zone of gabbro containing up to 18 bands of titanomagnetites enriched with vanadium. Granites with which tin ores and fluorite deposits are associated represent the youngest intrusives.

The Bushveld lopolith is known for its rich mineral resources including gold, platinum group metals, sulphides of nickel and copper, chromites and ores of vanadium, tin and fluorite (the last two in granites).

A swarm of granite-gneiss domes occurs along the periphery of the Bushveld pluton, of which the largest is the Vredefort. The age of mafic rocks in the lower part of the pluton was determined as 2095 ± 24 m.y. and of granites of the upper part as 1920 ± 40 m.y., both by the Rb-Sr method.

The youngest Early Proterozoic complex of South Africa is the red clastic complex, its spread extending far beyond the boundaries of the Transvaal syneclise and known under different names: Matsap in Cape Province of the Republic of South Africa, Waterberg and Southpansberg in Transvaal, Palavie in Botswana, and Umkondo in Zimbabwe. The average age of these formations is 1.8 b.y.

The W a t e r b e r g g r o u p (thickness up to 5 km) lies with an erosional disconformity on the Transvaal supergroup of formations or granites of the Bushveld pluton. It is made up of red conglomerates and sandstones (latter predominate); trachyte and quartz-porphyry lavas are present at the base. The conditions of accumulation of the Waterberg group of sediments are interpreted as an environment of littoral plain and shallow sea (or lacustrine basin).

Accumulation of the S o u t h p a n s b e r g   g r o u p proceeded in a significantly different environment. This group is confined to a complex broad graben (aulacogen) superposed on the Limpopo belt, with half of the group made up of volcanics, predominantly basalts; extrusions were along fissures and subaerial. The sediments are regarded as alluvial. The aulacogen regenerated at the end of the Palaeozoic, in the Karroo epoch.

The U m k o n d o   g r o u p (thickness about 2 km) falling in the eastern part of Zimbabwe close to the boundary of the Mozambique mobile belt, consists predominantly of clastic rocks with thin bands of carbonates in sands. Erosion proceeded from the Zimbabwe massif into the marine basin opening eastward.

The M a t s a p   g r o u p (thickness 4.8 km) formed on the opposite, south-western rim of the Kalahari Platform. Apart from the predominant clastic and clay rocks, dolomite-limestones and volcanics are present in it. Towards the south-west it passes into the Namaqualand protogeosyncline.

4.1.9.2 *Congo protoplatform.* The rest of the African protoplatforms are delineated based on negative features: absence of Lower Proterozoic formations and radiometric data pointing to Early Proterozoic reworking. An exception is the C o n g o   p r o t o p l a t f o r m within which the F r a n c e v i l l i e n   g r o u p of formations was formed in the territory of the Congo and Gabon. This group is remarkable firstly for the total absence of metamorphism (these are the oldest unmetamorphosed sedimentary formations) and secondly for its rich uranium and manganese mineralisation. This is a red clastic formation with black shales, dolomites and ignimbrite tuffs. It lies unconformably on an Archaean (2.7 b.y.) crystalline basement; its lower part precedes the intrusion of syenites aged 2140 ± 70 m.y. Diagenesis of sediments of the group occurred between 2.05 and 1.87 b.y. (Bonhomme et al., 1982). Westward, analogues of Francevillien formation, the Ogooue group, reveal already a significant degree of dislocation and metamorphism. They represent the northern continuity of the Kimezien metamorphic complex extending parallel to the Atlantic coast from Gabon to northern Angola. The 'intermediate series', formed northwards in Cameroon, Republic of Central Africa and north-western Zaire, could be an analogue of Francevillien but, unlike it, experienced metamorphism to greenschist facies. These intermediate series are made up of conglomerates, quartzites, metalutites and volcanoclastic rocks.

4.1.9.3 *Mobile (protogeosynclinal) belts.* Among the Early Proterozoic mobile belts of Africa, foremost for consideration is the B i r r i m i a n (Eburnean) p r o t o g e o s y n c l i n e (Fig. 38). It extends as a broad

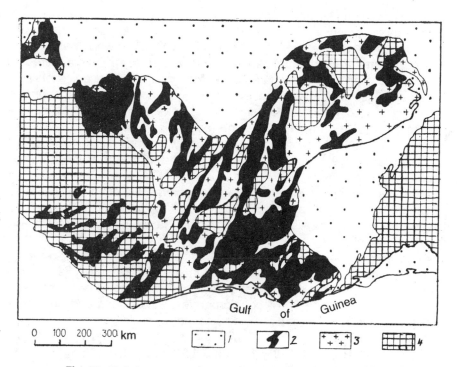

Fig. 38. Birrimian protogeosynclinal fold system of Western Africa.

1—Phanerozoic formations; 2—volcanosedimentary Birrimian complexes; 3—Early Proterozoic granitoids; 4—Archaean substratum.

band in a meridional direction through the Ivory Coast, Ghana and Burkina Faso and is concealed more to the north under the Late Proterozoic-Phanerozoic cover of the Taoudenni syneclise but its probable northern continuation can be seen in the south-western part of the Ahaggar massif (In-Quzzal), eastern part of the Reguibat massif and western half of Anti-Atlas. According to mobilistic reconstructions, the Birrimian protogeosyncline represents a direct continuation of the Maroni-Itacayunas belt (Guyana Shield, South America). In fact, it is very similar to the latter in structure, representing an analogous granite terrane, and an age of concluding diastrophism (2.18–2.15 b.y.) termed Eburnean (= Transamazonian) by French scientists. Westward the Birrimian protogeosyncline is bound by the Epi-Archaean Leon-Liberian massif with continuation into the western part of the Reguibat massif, in the east by the Late Proterozoic (Pan-African, see below) Libya-Nigerian belt formed on an Early Proterozoic basement. The structural plan of the Birrimian belt is characterised by a complex pattern with alternation of comparatively narrow sinuous bands of supracrustal rocks and more extensive fields of granites of tonalite-trondhjemite composition. Greywacke flysch and fine-grained tuffaceous sediments predominate in the metasedimentary complex while manganese-bearing cherts and intraformational conglomerates are encountered in subordinate amounts. Carbonate rocks are negligible. Volcanics play an extremely significant role in the Birrimian sequence. They correspond to the tholeiitic (45%) and calc-alkaline (55%) series. Quartz-tholeiite-basalts predominate (91%) in the former. High-magnesian basalts and rhyodacites are insignificant. In the calc-alkaline series 40% are andesites, 48% dacites and rhyolites and 9% calc-alkaline basalts. The thickness of the Birrimian deposits reaches 7.5 km.

The deposits are deformed into highly compressed, often isoclinal folds, frequently complicated by imbricated thrusts. Metamorphism is weak, of lower greenschist facies, rising to amphibolite facies only at contacts with intrusive granitoids.

The post-Eburnean age in this fold system applies to the T a r k w a m o l a s s e f o r m a t i o n developed in southern Ghana. It lies unconformably on Birrimian formations and was deformed together with the latter. This formation is made up of conglomerates, sandstones and quartzites, while its upper part consists of much finer terrigenous sediments, right up to schists (phyllites). The formation is pierced by sills and dykes of granophyres, dolerites, gabbro and norites; metamorphism reaches greenschist facies only at places. Tarkwa is known for its gold mineralisation; for this reason Ghana was known as the Gold Coast in the colonial past.

The Birrimian belt survived at least three phases of deformation and two main phases of granite formation (Cahen et al., 1984)—syntectonic at the level of 2125 ± 40 to 2137 ± 50 m.y. (Baoule-Ivory Coast and Cape Coast-Ghana batholiths) and post-tectonic at 2037 ± 45 m.y. (minor intrusions of Boundoukou-Ivory Coast and Dixcove-Ghana). The Tarkwa molasse was intruded by Kinkene granite 2047 ± 97 m.y. ago (Vachette, 1964 in Goodwin, 1991).

The geodynamic conditions of formation and development of the Birrimian mobile belt, like the analogous Maroni-Itacayunas belt in South America, continue to be debated. There is no doubt that these belts were formed on an Archaean sialic basement and that this was accompanied by destruction of the basement and its extension but the magnitude and consequence of rifting are not clear. According to one of the models suggested (Leube et al., 1990), the process was restricted to the formation of parallel rift systems with general thinning of the crust followed by the birth of volcanic arcs, general compression and intrusion of granite plutons and, finally, formation of secondary rifts filled with Tarkwa molasse. Another model (Ledru et al., 1989) suggests development of the belt conforming to a plate-tectonics scenario with collision of several microcontinents. The first model is supported by the multiplicity of greenstone belts (if this is regarded as primary) and more importantly by the absence of ophiolites; the second model is supported by the similarity of basalts with midoceanic ones, development of volcanic arcs, compressive deformation, metamorphism and granitisation. Hence the plate-tectonics model enjoys preference albeit further substantiation is required.

These conclusions should also evidently be applied to the South American continuation of the Birrimian belt.

As pointed out above, the Birrimian belt continues northwards under the cover of the Taoudenni syneclise into the eastern part of the Reguibat massif. However, the analogues here of Birrimian have a significantly different lithological and tectonic expression. Volcanoclastic formations of the Yetti group lie here unconformably on an Archaean basement. These formations are unevenly dislocated and metamorphosed into greenschist facies including rhyolites, ignimbrites, sills, dykes and plutons of granitoids aged 2.03 b.y. Apart from normal granitoids, intrusives of alkaline granites and nepheline syenites are also present. Intermediate and felsic Eglab volcanics with which granites aged 2.05–1.8 b.y. are associated lie unconformably on top. These formations in turn are overlain by unmetamorphosed and undeformed Guelb-El-Hadid molasse.

East of the Reguibat massif the Lower Proterozoic volcanosedimentary formation of the shelf type, known as Suggarien, metamorphosed to granulite facies and participated in the build-up of the Ahaggar massif. It lies unconformably on Archaean formations and houses granitoid intrusives. The age of metamorphism is 1975 ± 20 and of granites 1855 ± 80 m.y.

The northern continuation of the Reguibat Lower Proterozoic is seen in the western part of Anti-Atlas and the southern continuation of the same Ahaggar formations in the Benin-Nigerian massif. Rocks of this age are represented in the latter by a gneiss-migmatite complex of amphibolite facies, metasedimentary complex of greenschist facies and granitoids aged 2.1 b.y.

Early Proterozoic formations also outcrop east of Ahaggar, firstly in the small Oweinat massif at the confluence of the boundaries of Libya, Egypt and Sudan in the form of gneisses, amphibolites and granites ($\sim$ 1.8 b.y.), and secondly in the eastern part of the Arabia-Nubian Shield, as granodiorites aged 1.8–1.6 b.y.

In Equatorial Africa the R u z i z i-U b e n d i-U s a g a r a is the largest Early Proterozoic protogeosynclinal system; it extends fairly much south-easterly from the eastern part of Zaire (Kiwu province) through Burundi and Tanzania to the northern part of Malawi, then turning north-east encircles the southern tip of the Tanganyika eocraton. The name Ruzizi is applied to its north-western segment up to Lake Tanganyika; between Tanganyika and Malawi lakes it is known as Ubendi 'and in the north-eastern segment as Usagara. The total extent of the system is about 1500 km and its width

200 km. This protogeosyncline separates the Central African (Congo) pro-toplatform from the Tanganyika protoplatform (eocraton). In the north-west it evidently attenuates somewhere into the base of the Congo syneclise and in the south-east is sheared by the tectonic front of the Mozambique belt, prob-ably continuing within it in a reworked form. The ensialic and rift origin of this zone is evident since its formations at several points were deposited on Archaean granites or gneisses that have been dated radiometrically. However, mafic and ultramafic rocks, possibly forming an ophiolite (proto-ophiolite ?) association, are known in the Ubendi segment. Primary sedi-mentary rocks predominate in the composition of the geosynclinal complex, namely quartzites, lutites and carbonates transformed into gneisses, amphi-bolites and marbles. Volcanics, predominantly tholeiite-basalts, less often (in the upper part) felsic lavas (dacites and rhyolites) are also encountered. The complex is deformed into narrow straight folds and intensely metamor-phosed (amphibolite facies transiting into granulite facies at the joint with the Mozambique belt), migmatised and intruded by granites aged 2.0–1.7 b.y. (Rb-Sr isochron method). Evidently there should be manifest two different phases of tectonics, of which the earlier corresponds to the Eburnean (Nyika granites in Malawi aged 2.05 b.y. by the U-Pb method on zircons) and the latter to post-Tarkwa 'Mayombe'. Here and there a series of terrigenous rocks and felsic volcanics lies unconformably on the main geosynclinal complex; this series evidently corresponds to the interval between these two phases, i.e., represents a homologue of the Tarkwa series although it does not exhibit such a distinctly molassic character.

At the latitude of the northern coast of Lake Victoria, a similar Early Proterozoic f o l d s y s t e m, the K i b a l i-T o r o-B u g a n d a (or R u w e n z o r i) extends east-south-easterly from Zaire through Uganda into Tanzania. It is made up of metaterrigenous sediments—schists (micaceous), quartzites (at places ferruginous) and also mafic and ultramafic volcanics and hypabyssal intrusives regarded by some researchers as ophiolites. They lie with sharp unconformity on an Archaean basement in the north. The intensity of folding, migmatisation, granitisation and metamorphism of the system is quite high in the west but progressively diminishes eastwards, i.e., towards Lake Victoria, where the folds become straight, open and inclined, and metamorphism is to greenschist facies. The age of the Kibali group is 1850 m.y. for pegmatites and of the Buganda group 1807 ± 60 m.y. for Mubende granite (K-Ar method) (Cahen and Snelling, 1984).

In the Peri-Atlantic zone of Equatorial and South Africa (southern Zaire, Angola, Namibia and Republic of South Africa, Cape province) another protogeosynclinal system is seen. It encircles the Congo and Kalahari protoplatforms from the west, partly dividing them. In south-western Angola (Carvalho, 1983) the geosynclinal complex of this P e r i-A t l a n t i c s y s t e m is represented by the volcanosedimentary U a m b o   g r o u p lying unconformably on Archaean gneisses and migmatites aged about 2.7 b.y. The composition of this group consists of schists, greywackes, mafic volcanics including pillow lavas, phtanites and ferruginous quartzites. The Uambo group is unconformably overlain by another volcanosedimentary S h i v a n d a   g r o u p containing, among other rocks, gold-bearing conglomerates and itabirites. The sequence of the complex closes with the clastic, molassic type O n d o l o n g o   g r o u p, also lying unconformably on underlying formations. The complex forms synforms with a submeridional trend, is metamorphosed to greenschist facies and intruded by granites aged 2.2 and 1.8 b.y. (Rb-Sr method). Moreover, the Uambo group hosts gabbro intrusions comparable to the major gabbro-anorthosite pluton of south-western Angola. It is quite possible that the Uambo group is still of Late Archaean age and represents the filling of a greenstone belt, albeit in Angola much older formations of the same type, more intensely metamorphosed, are known (see Sec. 3.1.6).

Lower Proterozoic formations of similar composition are known even north of the south-western part of Angola, on one side in the north-western part of this country, in lower Zaire, Congo and Gabon and, on the other, in north-eastern Angola and southern Zaire in Shaba and Kasai provinces. The age of these formations is determined by their unconformable deposition on the Archaean, more highly metamorphosed basement, metamorphism being 2.2–1.9 b.y. old and intrusion of granites and pegmatites 2.0–1.9 b.y. The northern band of these formations projects into the western Peri-Atlantic rear of the Late Proterozoic fold system of the Western Congolides (see Sec. 7.1.15.5), constituting the Mayombe Hills and intersecting the lower course of the Congo River. These formations are known under the common name Kimezien and are represented by intensely deformed terrigenous and carbonate formations metamorphosed to greenschist (in the east) and amphibolite (in the west) facies. The age of the complex has been determined for migmatites, gneisses and granites as 2126 ± 39 to 1940 ± 107 m.y. In southern Zaire and north-eastern Angola these micaceous quartzites, mica schists and felsic lavas overlain by mafic lavas and then by chlorite and sericite schists are termed the Zadinian group. In southern Zaire the lower part of the Lower Proterozoic is known as the Luiza group (quartzites, mica schists and ferruginous quartzites) and its age of metamorphism is 2.2–2.0 b.y.; the upper part, also metasedimentary, with manganese ores, is known as the Lukoshi group (Lavreau, 1982).

In the south, in Namibia, and in the western part of the Republic of South Africa (Tankard et al., 1984), tectonic conditions are sharply complicated and the degree of metamorphism increases; it is therefore difficult to delineate Lower Proterozoic formations here. They are most decisively established in the Orange River basin where the lower limit of their age was determined from the period of formation of the major and complex Vioolsdrift batholith, some phases of which provide an entire gamut of transition from ultramafics and mafics aged 2.0–1.9 b.y. to tonalites and granodiorites aged 1.93–1.87 b.y. and finally, granites aged 1.73 b.y. The batholith is distinguished by a low strontium isotope ratio of 0.7031 and copper porphyry and molybdenum mineralisation. The Orange River group intruded by the batholith consists (bottom upwards) of intermediate and felsic volcanics; quartzites, ferruginous quartzites, chlorite schists and metalutites; and felsic and mafic lavas of the calc-alkaline type (possibly this is a tectonic repetition of the lower part of the sequence).

Northward in Namibia two types of supposedly Lower Proterozoic formations are known. One (Gariep) is mainly represented by dolomite marbles with subordinate quartzites and lutites. It is regarded as shelf facies of the adjoining platform Matsap group (Kalahari Craton). The other type, metalutite-gneisses with bands of quartzites and graphitic schists, is regarded as possibly a deepwater formation of the same age. All these formations are complexly deformed; intrusion of granodiorites aged 2040 ± 40 m.y. occurred after the first phase of deformation. The complex of granite-gneisses and augen-gneisses aged 1780 ± 35 m.y. is correlated with the second phase of deformation. Granite-gneisses of granodiorite or quartz-monzonite composition are dated 1755 m.y. (Pb-Pb method on zircon). Metamorphism attains granulite facies here and there with manifestation of charnockites and enderbites.

Volcanics (from andesites to rhyolites of the calc-alkaline series) in the erosion windows of Francefontein and Grootsfontein and white Khoabendus quartzites intruded by porphyric granites are very similar to the Orange River group. Similar rocks are likewise traced up to Kunene River and farther in southern Angola.

Early Proterozoic geosynclinal formations extend from Namibia into Botswana and the north-western part of Zimbabwe, thrusting directly onto the Zimbabwe and Kaapvaal eocratons and shearing the Limpopo belt. The zone of development of these formations is divided into two belts: M a g o n d i[6] in the north-western framing of the Zimbabwe eocraton with a north-easterly trend and K h e i s on the western periphery of the Kaapvaal eocraton with a meridional trend. In the first belt two structural-formational zones are distinguished: carbonate-shelf and geosynclinal schist-greywacke; the age of deformation ranges from 2.3–2.2 to 2.05–1.76 b.y. Formations of the Kheis belt are metamorphosed and weakly deformed; they are intruded by a granitic batholith aged 1.8 b.y. (Stowe et al., 1984).

---

[6] Includes the Deweras, Lomagundi and Piriwiri groups, the latter two probably being equivalent.

In the opposite, southerly and south-easterly directions, possible Lower Proterozoic formations represented by gneisses emerge in Bushmenland and Namaqualand and later, after an interval, with less probability in Natal. In South Africa as a whole the emergent picture is framing of the Kalahari protoplatform from the north-west, south-west and south-east by an extremely complex polymetamorphic fold-overthrust belt. Its outer zone is represented by the Magondi and Kheis systems, constituting respectively the distal continuation of Umkondo and Matsap cover formations. Partly together with the underlying rocks of the Archaean basement, these formations are involved in imbricate-overthrust and nappe structures with a distinct vergence towards the platform. Here the main age of deformation, metamorphism and granitisation falls towards the end of the Early Proterozoic. The central zone is characterised by participation in its structure not only and even, possibly, not so much Early Proterozoic as Middle Proterozoic formations and by the high level of metamorphism to the granulite stage; the latter feature helps interpret the central zone as a fairly typical granulite-gneiss belt. This zone extends through Namaqualand and farther into Natal. It is devoid of a much older (older than 2.0 b.y.) basement and ophiolites have been observed in it (in Natal). Finally, the very peripheral zone is distinguished by development and folding on the much older sialic basement of Upper Proterozoic formations at the end of the Proterozoic. This zone extends along or parallel to the boundary of the Orange River and Cape provinces of the Republic of South Africa. We shall revert to a description of the latter two zones of the Circum-Kalahari belt later.

A branch running from the Peri-Atlantic Early Proterozoic protogeosynclinal system into the central part of Namibia towards the north-east separates the Kalahari and Congo protoplatforms and probably runs to the confluence with the Ruzizi-Usagara protogeosyncline through the basement of the Bangweolo massif (Stowe et al., 1984). It is partly inherited by the Mid-Proterozoic intracratonic Kibarides geosyncline. This branch is traced in Namibia along outcrops of Ababis gneisses aged 2.0-1.8 b.y. and in the north-east along outcrops of felsic volcanics, quartzites and muscovite and chlorite schists intruded by granites aged 1930 ± 30 m.y. into the Bangweolo massif at the joint of the boundaries of Zambia, Tanzania and Zaire. The age of the volcanics is 1.82 b.y. and their analogues are known in Shaba in southern Zaire.

Among Early or Middle Proterozoic formations of the geosynclinal type are also the quartzite-schist and cipolin (carbonate) series of the central part of Madagascar intruded by gabbro-norites, granites and granodiorites of the Ilaka complex (1538-1400 m.y.) and, even more tentatively, metavolcanics of the Daraina and Milanoa series and carbonaceous shales of Andrarupa in northern Madagascar intersected by pegmatites aged 1600-450 m.y. These formations are metamorphosed to greenschist or epidote-amphibolite facies. Folding is intense only close to fracture zones.

Thus the evolution of the future African Craton in the Early Proterozoic is wholly similar to the history of other cratons: break-up of the Epi-Archaean continental crust; isolation of protoplatforms; formation of mainly ensialic protogeosynclines accompanied by extension and volcanism, at places with some signs of formation of new oceanic crust; transition to compressive conditions commencing 2.2–2.0 b.y.; two or more epochs of deformation, metamorphism and granite formation; orogeny with local molasse-type formations; and ultimately restoration of a compact and more consolidated crust by commencement of the Middle Proterozoic.

4.1.10 Indian Craton

Much of this craton was finally stabilised by the Early Proterozoic (Fig. 39). This applies primarily to the whole of its southern peninsular part (South Indian eocraton, protoplatform), which in the Early Proterozoic evidently experienced a general uplift, more intensely in the east (Eastern Ghats) and southern extremity in the region of Late Archaean granulite metamorphism. In the northern part of the craton conditions were less

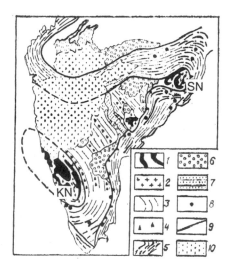

Fig. 39. General geological map of the Indian peninsula (after Radhakrishna and Naqvi, 1986).

KN—Karnataka nucleus; SN—Singhbhum nucleus; 1 to 4—Archaean (1—schist belts within the nucleus, 2—tonalite-gneisses, 3—granodiorites-granulites, 4—potash-granites); 5—granulites and gneisses of Early Proterozoic; 6—Middle Proterozoic sedimentary basins; 7—Gondwana sediments of Godavari rift valley; 8—anorthosites along the Archaean-Proterozoic contact; 9—overthrusts of Eastern Ghats, Sukinda-Singhbhum; 10—Deccan traps.

quiescent; here the Epi-Archaean continental crust was broken into individual blocks and separated, as in most of the cratons studied thus far, by mobile systems, i.e., protogeosynclines. The A r a v a l l i   s y s t e m represents the largest and best known of the protogeosynclinal systems of India and coincides with the same-named mountain range extending for 750 km almost from the Narmada valley in the south to the alluvial plains of the Ganges in the north, initially north-westerly then north-easterly, with a width of about 200 km. The eastern boundary of the Aravalli protogeosyncline is represented by an Archaean massif (protoplatform) made up of Bundelkand granites and 'banded gneisses' intruded by Archaean Berach granites. Some of these 'banded gneisses' have been proven to be products of deep metamorphism of rocks of the Aravalli protogeosyncline.[7] At its boundary with the Archaean block, a melange zone has been established in which talcose mafics and ultramafics are present. These probably represent elements of an ophiolite association formed at the base of the protogeosyncline. This assumption is supported by the fact that mafic wholly crystalline rocks transformed into amphibolites and hornblende slates, ultramafics in talc and serpentinites as well as dolerite dykes are known in the composition of the lower complex of the filling of the protogeosyncline (Aravalli supergroup). Some researchers already regard these formations as ophiolites.

Two structural-facies zones have been distinguished in the structure of the Aravalli protogeosyncline: the eastern shelf and western relatively deep-water zone. Sediments of the shelf zone were deposited in grabens, separated by horsts of the Archaean basement; their unconformable deposition on the basement is observable at places. Sheets of tholeiite-basalts are present at the base of the shelf sequence; it has been stated that these basalts extruded along faults separating horsts and grabens. All this evidently characterises the rift stage of development of the passive margin of the Aravalli basin. The main part of the shelf sequence is made up of quartzites, carbonates with stroma-tolites and lutites. It is subdivided into two groups, lower and upper, separated by an interval and unconformity. Basal conglomerates of the upper group are similar to diamictites at places. A division into three groups with two levels of unconformities is adopted in some works.

The deeper water deposits of the western zone are represented by mica schists and phyllites with rare bands of quartzites (turbidites ?) forming a continuous succession. In the eastern zone replacement of shallow-water deposits by deeper water clay types, as developed in the western zone, is noticed in the upper part of the sequence.

Three phases of deformation are distinguished in the Aravalli system: the first phase, giving rise to isoclinal folding and schistosity, was the most

---

[7] The name Mewar gneisses was therefore proposed for true Archaean gneisses.

important. Simultaneous with it occurred the intrusion of the Darval granites aged 1900 ± 80 m.y. and metamorphism, very weak in the east and somewhat more intense westward. The formation of zones of plastic strike-slip faults and brittle faults is associated with the subsequent phases of deformation.

Accumulation of the Aravalli supergroup of deposits concluded with deposition of carbonates and volcanics (continental tholeiitic type) of the Rayalo group, known as a source of white marble widely used in building practice, especially in the famous monument Taj Mahal. This group had earlier been placed in the Delhi supergroup (some researchers hold this view even today). The Aravalli protogeosyncline was undoubtedly formed on an Archaean sialic crust accompanied by its extension, thinning and volcanism. The degree of this extension and of crust transformation is variously interpreted. Some researchers assume that the process was restricted to continental rifting while others, on the basis of the presence of ophiolites (although not acknowledged by everyone), assume the transition of rifting into spreading, albeit on a limited scale. This latter view appears more probable.

Northward the Aravalli system is sheared unconformably in the plan by the eastern boundary of the much younger D e l h i   f o l d   s y s t e m with a more north-easterly strike. This feature hinders recognition of the natural western boundary of the Aravalli system and thus makes geodynamic interpretation difficult.

The Delhi fold system consists of the same-named supergroup of formations, traditionally divided into three parts: basal conglomerates, quartzites, arkoses, gritstones (Alwar group), then limestones (Kuchalgarh group) and finally clay rocks transformed into phyllites, biotitic schists and even gneisses (Ajabgarh group); the total thickness is about 6 km. It has now been demonstrated that, like the Aravalli protogeosyncline, the Delhi protogeosyncline consisted of two trough zones separated by an uplift of Archaean basement. The north-eastern trough is devoid of volcanics; its filling consists mainly of carbonates, lutites and greywackes. Contrarily, volcanics, predominantly low-potash tholeiites and partly rocks of intermediate and felsic composition, predominate in the western trough. Here ophiolites (but intensely tectonically dismembered) and also glaucophane-lawsonite schists are presently known. This feature undoubtedly favours the plate-tectonics model of development of not only the Aravalli, but also the Delhi system, although like the former, it too should have originated under conditions of continental rifting.

The rocks of the Delhi system underwent intense and polyphase deformation, experienced break-up and metamorphism to amphibolite facies at relatively low pressure and high temperature, with the formation of andalusite, staurolite and garnet. Folding exhibits a general eastern vergence. Bairath granites intruding the Delhi group are dated 1.66 b.y., which points to the affinity of this group to the uppermost part of the Lower Proterozoic.

In the Riphean the Aravalli fold system experienced repeated tectono-magmatic reactivation, evidently under the influence of processes in the volcanoplutonic belt adjoining it from the west and forming the north-western part of the Indian Shield. The primary boundary of the Aravalli-Delhi protogeosyncline remains concealed under this belt.

A second protogeosynclinal system is seen in the northern frame of the Singhbhum Archaean arch (a granite-gneiss dome) between it and the Chitradurga massif of gneisses and granites, also probably of Archaean age. This N o r t h e r n   S i n g h b h u m   p r o t o g e o s y n c l i n e has a general latitudinal strike and forms a constituent of the belt formerly delineated as the Satpura, coinciding with the Satpura range. North of the Singhbhum dome, formations of this protogeosyncline constitute an arc weakly bulging northwards and enveloping the dome and overthrust on it along the well-known zone of the 'Copper Belt overthrusts'. In the zone of maximum compression against the northern apex of the dome, the width of the protogeosynclinal system is only 48 km but enlarges perceptibly west and east. The exposed extent of the system is about 200 km. It is made up of an extremely thick (9.1–11.6 km) formation of terrigenous flysch (mainly clayey) metamorphosed to greenschist and amphibolite facies and consisting presently of phyllites and mica schists with bands of quartzite. The age of this bed has been estimated as 2.02 to 1.7 b.y. (Sarkar, 1982).

South of the overthrust zone of the Copper Belt, flysch is replaced by sediments of a protoplatform cover consisting of argillites, ferruginous rocks and subordinate sandstones and conglomerates. A narrow band of outcrops of metabasalts with tectonic inclusions of mafic and ultramafic wholly crystalline rocks and bound on both sides by thrusts with a northerly dip runs roughly in the middle of the flysch zone. Here and there metabasalts reveal the pillow variety and are proximate to midoceanic tholeiites in petrochemical composition. With them are associated carbonaceous and red phyllites and layered cherts. Sarkar (1982) considered all these together under the name D a l m a   o p h i o l i t e s. Ophiolite-clastic conglomerates are deposited on them while numerous olistolites of mafics and ultramafics, evidently from this same ophiolite complex, are encountered in the flysch, mainly in its northern band. The origin of this complex clearly preceded the accumulation of flysch formation. The latter subsequently underwent several phases of deformation, of which the first was manifest as isoclinal folding and cleavage of the axial surface; commencement of metamorphism coincided with it. This process was a consequence of the collision of the southern and northern platforms with obduction of ophiolites and flysch on the southern platform. Accumulation of molasse and extrusions of continental basalts occurred at this same time, dated by Sarkar at 1.7–1.55 b.y., i.e., commencement of the

Riphean, south of the overthrust of the Copper Belt in the north-western frame of the Singhbhum dome. Subsequent events in this region continued already into the Riphean, right up to 850 m.y. (two more phases of deformation and metamorphism).

The S a t p u r a p r o t o g e o s y n c l i n e extending in the same WSW-ENE direction and separating the North Indian protoplatform from the South Indian one, represents the western continuation of the Northern Singhbhum protogeosyncline and a probable connecting link between it and the Aravalli-Delhi system.

Three groups (supergroups) of deposits constitute the geosynclinal complex of the Satpura protogeosyncline, namely the Dongargarh, Sakoli and Sausar. The D o n g a r g a r h g r o u p in turn consists of three groups: 1) Amgaon—schists, quartzites, gneisses and amphibolites; 2) Nandgaon—bimodal volcanics; and 3) Khairagarh—schists, sandstones and basic volcanics. The first group was metamorphosed and granitised about 2.3 b.y. ago; the second group consists of volcanics aged 2180 ± 25 m.y. and intruded by granites aged 2270 ± 90 m.y. (Rb-Sr method).

The S a k o l i g r o u p consists of a large synclinorium and is formed mainly of metalutites, amphibolites and jaspilites with subordinate amounts of ultramafics and mafic volcanics. Jaspilites form very large deposits of great economic importance.

The S a u s a r g r o u p is formed of metamorphosed sandy-clay and carbonate rocks containing large deposits of manganese ores. Metamorphism ranges from greenschist to amphibolite, at places even granulite. Rocks are intensely dislocated, right up to the formation of overthrusts and nappes with a southern vergence and are intruded by potash granitoids. The age of these rocks is not wholly clear. Evidently rejuvenated age determinations on mica gave values of 1000 to 850 m.y.

4.1.11 Australian Craton

Development of the Australian Craton in the Early Proterozoic did not differ from that of other cratons described thus far. Break-up of the Epi-Archaean continental crust, formation of protogeosynclines and isolation of protoplatforms occurred here too (Fig. 40).

4.1.11.1 *Western Australia.* The P i l b a r a (in the north) and Y i l g a r n (in the south) e o c r a t o n s described in Western Australia in the preceding chapter emerged as protoplatforms. The A s h b u r t o n-G a s c o y n e protogeosyncline lies between them and cover formations, the H a m e r s l e y (N u l l a g i n e) and N a b b e r u b a s i n s, accumulated on their slopes.

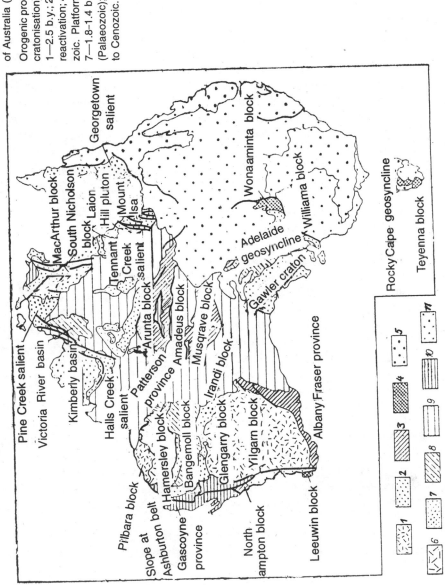

Fig. 40. Scheme of main tectonic provinces of Australia (after Plumb, 1979).

Orogenic provinces (age of orogeny and cratonisation):

1—2.5 b.y.; 2—2.0–1.4 b.y.; 3—1.9–0.9 b.y., reactivation; 4—900–650 m.y; 5—Palaeozoic. Platform covers: 6—2.8–2.4 b.y.; 7—1.8–1.4 b.y.; 8—1.2–0.9 b.y.; 9—900 m.y. (Palaeozoic); 10—Phanerozoic; 11—Permian to Cenozoic.

The first of these basins developed mainly during the Late Archaean and its filling has already been described (see Sec. e.1.11.2). We shall revert to the second basin after describing the Ashburton-Gascoyne system.

The Ashburton-Gascoyne fold system is also known as the C a p r i c o r n   o r o g e n (from its disposition on the Tropic of Capricorn). In the present-day structure the superposed Bangemoll Riphean syneclise divides the Ashburton and Gascoyne systems. Their exposed length is about 500 km. The ensialic formation and rift origin of these protogeosynclines are beyond doubt. In any case they are evident for the northern zone of the Ashburton system and southern and central zones of the Gascoyne system where the remobilised Archaean basement corresponding to the Pilbara and Yilgarn blocks emerges on the surface in mantled domes and cores of folded uplifts fringed by granite-gneisses while the sedimentary cover is represented by continental-marginal and shallow-marine sediments and sheets of bimodal volcanics. This pattern changes on transition to the axial trough of the Ashburton system and northern zone of the Gascoyne system. In the former a thick (up to 7.2 km) sequence of lutites and greywackes with subordinate quantites of sandstones, congolomerates, jaspilites, dolomites and volcanics lies atop the aforesaid deposits; in the latter system (Gascoyne) a more deeply metamorphosed and intensely granitised series of rocks represents an analogue of all the formations described for the first system. The total thickness of the volcanosedimentary complex has been estimated to be 12–14 to 20 km. A sialic basement is not known in these zones.

The structure of the Ashburton-Gascoyne system possesses a bilateral symmetry. The northern zone, Gascoyne, is characterised by intense folding, cleavage, metamorphism from amphibolite to granulite facies and a large number of anatectic granitoid bodies aged 1.7 b.y. Filling of the Ashburton trough likewise experienced intense folding and cleavage. Migmatites aged 2.0 b.y. form the basement of the geosynclinal complex here. Thus the protogeosyncline grew during the second half of the Early Proterozoic. In its northern (Ashburton) and southern (Glengarry) periphery metamorphism drops to greenschist facies, the Archaean basement appears and the structure is simpler. Gradual transition is noticed in the south into thick (up to 19 km) clastic and carbonate platform formations of the N a b b e r u b a s i n located on the northern slope of the Yilgarn eocraton affected by small basic intrusions. The fold system disappears in the east under the Bangemoll syneclise. The filling of the Nabberu basin was deformed and metamorphosed very unevenly under the influence of the movements of the remobilised Archaean basement, intrusion (at depth) of granitoids and action of the adjoining orogen.

Development of the Capricorn orogen concluded with the formation of grabens filled with continental molasse (sandstones and conglomerates) and intrusion of post-tectonic granites in the interval 1.7–1.5 b.y., i.e., already in the Early Riphean. Molasse passes southward into the upper complex of the Nabberu basin cover, consisting of 6 km of shelf sediments that accumulated between 1.8 and 1.7 b.y. ago.

The above description of the Capricorn orogen shows that its formation should have been accompanied by very significant extension and reworking of a very old continental crust. J. Muhling (1985) pointed out just such a possibility, having assumed the existence of a basin of oceanic type between the Pilbara and Yilgarn eocratons in the Early Proterozoic and subsequent formation of the Capricorn orogen as a result of the collision of these cratons. Herein lies an explanation for the great dissimilarity of the internal structure of cratons.

4.1.11.2 *Northern Australia.* Within Northern Australia, the Archaean basement underwent reworking and almost overall destruction; its primary structure was fairly preserved only in a small section in the north in the region of Pine Creek and possibly the basement of the Kimberley basin filled with platform formations commencing 1.8 b.y. ago. Elsewhere, proto-geosynclines of two generations developed (after Plumb, 1979): 2.05–1.85 and 1.9–1.7 b.y.

One of these protogeosynclines is the K i n g L e o p o l d p r o t o g e o s y n c l i n e extending south-east from the Indian Ocean coast between Kimberley basin and the Phanerozoic Canning syneclise whose formations partly overlie this protogeosyncline. The Canning syneclise separates the King Leopold system from the Pilbara massif; in the east this system gently joins another protogeosynclinal system, H a l l s C r e e k, with a north-north-easterly trend. In the sequences of both the protogeosynclines, greywacke and quartz-turbidites and metalutites (phyllites, sericite and quartz-mica schists) predominate to a thickness exceeding 7 km; carbonate-siliceous rocks are present in very subordinate amounts. In the Halls Creek system, moreover, basaltic and, to a lesser extent, felsic lavas were formed. Folded bodies of mafic and ultramafic rocks (from dolerites to peridotites) originally representing sills and dykes are also present here. In this same system the main geosynclinal complex experiencing zonal metamorphism to granulite facies in the interval 2200–1920 m.y. is unconformably overlain by thick (up to 12 km) Whitewater felsic volcanics (1912 ± 107 m.y.) with subvolcanic porphyries. Batholiths of porphyric Tau River granites and Machale granodiorites aged 1840 to 1815 m.y. (Rb-Sr method) are comagmatic with the above subvolcanic porphyries. All of them evidently form a volcanoplutonic belt at the boundary with the Kimberley block. Thus the Halls Creek protogeosyncline and probably the King Leopold system congruent with it belong to the earlier generation of Early Proterozoic geosynclines of Australia deformed in an epoch termed the

B a r r a m u n d i   o r o g e n y (1880–1840 m.y.) by Page and Williams (1988). Correspondingly, the sedimentary cover of the K i m b e r l e y s y n e c l i s e had commenced already with Lower Proterozoic formations intruded by dykes and sills of dolerites aged 1762±15 m.y. (Rb-Sr method). These formations are represented by quartz and arkose sandstones, variously coloured siltstones and shales, sheets of amygdaloid-basalts, bands of algal dolomites and, in the upper part, by tillites. Their thickness reaches 5 km. Along the periphery of the basin, formations affected by thrust-shear dislocation of the adjoining fold systems have been deformed and locally affected by metamorphism attaining amphibolite facies.

In the north, in the region of Darwin on the coast of the Arafura Sea, the Halls Creek system joins with the similar P i n e   C r e e k system in the central part of which Archaean rocks outcrop in granite-gneiss domes. This confirms the ensialic origin of the Pine Creek system. The geosynclinal sequence commences with conglomerates and includes quartz and greywacke turbidites, siliceous and carbonate (dolomites and marls) rocks. The latter are developed in the axial, deeper part of the basin together with black shales and ferruginous siltstones while algal barrier reefs rising up to 300 m in height extend along its periphery. The sequence ends in thick turbidites. Volcanics of mafic composition are developed locally in the upper part of the sequence; their age (U-Pb method on zircon) is 1877 ± 11 and 1884 ± m.y. The geosynclinal complex runs to 10 km in total thickness. It is unevenly deformed into linear, up to isoclinal and recumbent folds and regionally metamorphosed to amphibolite facies at places. The Nimbuwak granites, probably coeval to metamorphism, are dated 1886 ± 5 and 1866 ± 8 m.y. The Pine Creek geosynclinal complex is well known for its rich uranium mineralisation.

The geosynclinal complex is unconformably overlain by a plutonovolcanic molasse rock association. This complex includes subaerial andesite-dacite-rhyolite lavas, ignimbrites and tuffs of Edith River, subvolcanic bodies of porphyries, batholiths and smaller bodies of granitoids (from tonalites to adamellites), greywacke and arkose sandstones, less often conglomerates and breccia. The age of these post-tectonic volcanics and granites is 1.9–1.8 b.y. and of much younger dolerites constituting a major layered differentiated lopolith 1688 ± 13 m.y. (Rb-Sr method).

South of the Pine Creek protogeosynclinal system and south-east of the King Leopold system, an Archaean block (hypothetical Sturt block) might possibly lie concealed in the northern central part of the continent under the Riphean-Lower Palaeozoic cover. Protogeosynclinal formations outcrop again in different districts called Granites Tanami, Tennant Creek and Davenport.

The   G r a n i t e s   T a n a m i   f o l d   z o n e   consists of two complexes, the much older metamorphic, correlated with the geosynclinal complex of the Halls Creek system and placed in the interval 2.2–1.92 b.y. and a much younger volcanoplutonic complex consisting of felsic volcanics of Mount Winnecke and hypabyssal granites with respective ages of 1770 ± 15 m.y. ($^{87}Sr/^{86}Sr$ = 0.705) and 1764 ± 15 m.y. Much younger granites here have an Rb-Sr isochron age of 1740–1685 m.y. ($^{87}Sr/^{86}Sr$ = 0.709 to 0.707).

The T e n n a n t C r e e k f o l d z o n e comprises the terrigenous flyschoid Warramunga complex consisting of greywacke, quartz turbidites and shales, and up to 5–6 km in thickness. Folding is linear (from moderate to intense) and cleavage evident. Numerous bodies of granitoids and felsic volcanics are present. The age of the complex is estimated from volcanics (U-Pb dating on zircon) at about 1870 m.y., i.e., roughly the same as the Pine Creek complex. The first epoch of granite formation is close to this period at 1869 ± 20 and 1846 ± 8 m.y. (U-Pb method on zircon) while the much later epoch falls at 1650 m.y. (Rb-Sr method).

The D a v e n p o r t f o l d z o n e, extending on the south-eastern flank of the Tennant Creek system is made up of variegated sandstones, siltstones, shales with bands of conglomerates and volcanics in the upper part (from mafic to felsic) and exceeds 6 km in thickness. This volcanic molasse, Hatches Creek series, lies unconformably on the Warramunga series, crumpled into narrow linear folds and intruded by small bodies of potash granites and quartz porphyries aged 1.8–1.7 b.y.

Early Proterozoic formations outcrop also north-east of the Tennant Creek zone in the M u r p h y r i s e, represented by metamorphites, volcanics and granites aged about 1.87–1.85 b.y. The Nicholson granite complex here has been dated at 1804 ± 83 m.y. (Wyborn, 1988). These Early Proterozoic formations are overlain by the platform cover of the MacArthur basin (syneclise); glauconite in its lower part may reach the age of 1.8 b.y., as in the Kimberley basin.

South of the Murphy rise and east of the Tennant Creek zone, the M o u n t I s a f o l d s y s t e m extends in a meridional direction. This system commenced evolution in the Early Proterozoic and was completed in the Early Riphean. Lower Proterozoic formations represented by felsic volcanics metamorphosed to amphibolite facies and batholiths and bodies of granitoids outcrop in a stretch of about 300 km of the Calcadoon-Leinhardt horst rise, separating the Early Proterozoic intracratonic geosyncline into two troughs, western and eastern. This metamorphic complex is complex in structure and paragneisses with an age of metamorphism at 1885 ± 10 m.y. and with clastic zircons aged 2.55–2.3 b.y. have been detected in it. The oldest Leinhardt volcanics are 1870–1850 m.y. and the youngest 1720 ± 7 and 1678 ± 6 m.y. and tuffs 1670 ± 19 m.y. (U-Pb method on zircon). Much younger, already Riphean metamorphites and magmatites are also present (see Sec. 5.1.8).

The Mount Isa system is bound in the east by the G e o r g e t o w n
m a s s i f taking part in a meridional tectonothermally reworked belt in the
extreme north-eastern part of the Australian Craton. Gneisses with Sm-Nd
age of 2200 and even 2490 ± 70 m.y. are known in this massif. These
gneisses were formed after shallow-marine sediments and mafic volcanics
while dolerites older than 1.8 b.y. are known in the more northern Cohen
uplift. The main deformation, metamorphism and granitisation occurred here
already in epochs 1570 and 1470 m.y., however, i.e., in the Early Riphean,
and hence we shall revert to this massif in the next chapter (Sec. 5.1.8).

4.1.11.3 *South Australia.* The G a w l e r b l o c k represents the
major salient of the Precambrian basement in this part of the craton. The
basement of this salient is made up of Archaean formations metamorphosed
to granulite facies (see Sec. 3.1.8). The Lower Proterozoic cover lying
unconformably on Archaean formations consists of epicontinental sediments
(thickness 1.5–2.0 km) comprising quartzites, dolomites, jaspilites and basic
lavas. Commencement of accumulation of this cover has been estimated
at 1940–1847 m.y.; during the period 1.84 and later 1.68–1.6 b.y. ago, it
experienced several phases of deformation, metamorphism from greenschist
to amphibolite and here and there even granulite facies and granite intrusion.
Later, already in the Early Riphean (see Sec. 5.1.8), the Gawler Range
volcanoplutonic and molassic belt arose on the eastern margin.

The W i l l i a m a or B r o c k e n H i l l b l o c k, set off from the
Gawler block by the Late Riphean-Cambrian Adelaide fold system, represents the easternmost salient of the basement in the southern half of the
Australian Craton. Paragneisses and granite gneisses with model Sm-Nd
ages in the range 2.2–2.05 b.y. are present in the Williama block. On top of
them is a thick (5–7 km) suite of psammites and lutites of the turbidite type
metamorphosed to amphibolite facies and transformed mainly into sillimanite
gneisses with subordinate bimodal volcanics and jaspilites. The suite comprises bedded deposits of ferrous and non-ferrous metal sulphides. The age
of the upper levels of the sequence has been established as 1820±60 m.y.
and of main deformation, metamorphism and granitisation as 1660±10 m.y.
There are also much younger granites (1490 ± 20 m.y.) as well as manifestation of Delamerian orogeny at around 520±40 m.y. (Stevens et al., 1988),
reflecting the marginal position of the block relative to the craton as a whole
at the boundary with the Tasman Palaeozoic mobile belt. Similarity is quite
evident with the Georgetown-Cohen massif whose southern continuation is
probably the Williama block.

4.1.11.4 *Central Australia TTR province.* There are two major uplifts
of Precambrian basement right in central Australia—Arunta (northern)
and Musgrave (southern)—separated by the latitudinal Riphean-Palaeozoic

Amadeus aulacogen. Both rises belong to an extremely typical, recurrent TTR belt, manifest in multiphase deformation, metamorphism and granitisation. This complex history is most completely revealed in the Arunta massif where the oldest Sm-Nd datings of granulite gneisses are 2015–1980 m.y. (2070 ± 125 m.y. according to other sources) and the youngest event dated by radiometry is the Early Carboniferous epoch of Alice Springs diastrophism in common with the adjoining Amadeus aulacogen. In the interval an epoch of granulite metamorphism around 1800 (1790 ± 35 m.y.) and an epoch of metamorphism of amphibolite facies 1.7–1.65 b.y. ago have been established. An intrusion of granites between 1.77 and 1.5 b.y. also occurred within the confines of the Early Proterozoic.

In so far as the M u s g r a v e   b l o c k, representing the reworked northern margin of the Gawler eocraton is concerned, the oldest age values for granulite rocks in the southern part of the block are 1615±170 m.y.; this dating is interpreted as the age of metamorphism; consequently, the rocks themselves may be of Lower Proterozoic age, like the cover of the Gawler eocraton.

The south-western continuation of the Musgrave block lies, after an interval associated with spread of the Palaeogene cover, in the A l b a n y-F r a s e r belt superposed on the southern part of the Yilgarn eocraton and set off from it by the same-named fault zone. Directly north of this zone and partly south, Upper Archaean rocks are developed and, at some distance from them, Lower Proterozoic rocks aged 2.0 and 1.7 b.y. (granites and gneisses respectively; Plumb, 1979). Farther south they are replaced by metamorphites and granites of essentially Riphean age.

Directly west of the Musgrave block and east of the Pilbara block is another small Paterson salient of basement associated with the former block based on geophysical data. The Paterson basement reveals a similar history of metamorphism and granitisation but commencing only from 1.33 b.y. However, the high primary strontium isotope ratios tend to suggest that the corresponding gneisses may be of Early Proterozoic age (Plumb, 1979).

Concluding the overview of data on the Early Proterozoic of Australia, an important point should be emphasised: all concrete data on sedimentation, magmatism, metamorphism and deformation pertains only to the second half of the Early Proterozoic commencing from 2.0 b.y. Thus the destruction of Epi-Archaean continental crust commenced here only from this period.

4.1.12 Antarctic Craton

As pointed out in Section 3.1.12, at the end of the Archaean aeon the Antarctic Craton underwent total cratonisation accompanied in many regions by the intrusion of swarms of basalt dykes, as noticed in other continents. Subsequent Early Proterozoic development was, however, nowhere wholly platformal, as supported primarily by the absence of protoplatform sedimentary cover. Nearly all sections of the craton experienced tectonothermal reworking at this time but it was particularly intense within the W e g e n e r-M a w s o n T T R b e l t delineated by Kamenev (1991). This belt intersects the entire eastern Antarctic in a latitudinal direction. Reworking included recurrent metamorphism to granulite of amphibolite facies and intrusion of granitoids, evidently anatectic. The Wegener-Mawson belt, like other belts of this type, had a long history. Its first phase ended in a spur of endogenic activity 1700 ± 200 m.y. ago and subsequent ones around 1000 and 550 ± 100 m.y. ago. Diastrophism at the end of the Early Proterozoic in several regions was accompanied by formation of new swarms of basalt dykes and here and there pegmatites as well.

The presence of supracrustal deposits of Early Proterozoic age could be suggested only in a few regions. With utmost justification, it is applicable to Dronning Maud Land and the montane framing of Lambert and Amery glaciers (Prince Charles Mountains etc.). In the first of these regions the so-called Inzelian complex was distinguished. It consists of various gneisses and migmatites with lenses and bands of quartzites, marbles, amphibolites, metamafics and metaultramafics. It is intersected by mafic intrusives aged 1700 m.y.

In the second region metavolcanics represent the predominant rocks and form a full range from komatiite-basalts to dacites and rhyolites. Quartzites, metaconglomerates and crystalline schists are of subordinate importance in this complex. E.N. Kamenev suggested that these formations belonged to a greenstone belt. They were abundantly intruded by plutons of gabbroids and more felsic rocks, right up to granites and syenites.

Adele Land and Shackleton Range represent two more regions of probable development of metamorphic supracrustals of Lower Proterozoic age. These regions are essentially no different from those described above in composition, degree of metamorphism and deformation.

4.2 TECTONIC REGIMES, MAIN STRUCTURAL ELEMENTS AND THEIR DEVELOPMENT IN THE EARLY PROTEROZOIC

The Archaean aeon thus ended with the formation of continental crust over much, if not the whole area of present-day ancient platforms and partly beyond their limits (see Fig. 22). Late Archaean cratonisation commenced already at the verge of the Early and Late Archaean (3.0 b.y. ago) and was completed 2.7–2.5 b.y. ago, as supported by the age of the oldest formations of platform cover commencing with the Pongola supergroup of South Africa. Another very important evidence is the extensive spread of swarms of mafic dykes, known in almost all the eocratons from Greenland to the Antarctic (Amundsen dykes; Sheraton et al., 1980). Towards the end of the Archaean to commencement of the Proterozoic, the first of steady faults originated, as for example in South India (Drury et al., 1984). This points out that the continental crust, at least its upper part, cooled perceptibly compared to the Archaean and became brittle. The geothermal gradient of the Early Protero- zoic has been estimated at 47°C/km compared to 54°C/km for the Archaean and the pressure in the lower parts of the crust at 430 MPa versus 410 MPa (Grambling, 1981).

The already wide development of dykes suggests that commencement of the Proterozoic was a period in which extension prevailed, possibly due to some increase in volume of the Earth as a result of phase transformation in the mantle stimulated by accumulation of heat under the thermally insulating layer of continental crust. It is still not quite clear whether this crust formed a continuous shell on the Earth or a general piling up at the end of the Archaean led to its concentration and formation of a sialic agglomeration in the form of the Pangaea supercontinent in one hemisphere, promoting formation of the primordial ocean in the other hemisphere. However, per- ceptible signs of the existence of such an ocean appear only much later, at the end of the Early Proterozoic (see Sec. 4.3).

The formation of dykes was succeeded by more intense destruction which led to the birth of protogeosynclines and divided Pangaea 0 into a large number of protoplatforms and, in an abortive form, protoaulacogens within the latter, often forming triple junctions with protogeosynclines (see Sec. 4.1.1). In some regions, especially in Australia and Africa, partly India, i.e., mainly within the future Gondwana but also in the Baltic Shield (Sve- cofennides!), protogeosynclines appear only at the end of the first half of the Early Proterozoic (about 2.1–1.9 b.y. ago); further, their origin coincided with orogeny in the first generation of protogeosynclines (Eburnean and Transamazonian epochs and their equivalents). This points out that at this time conditions of compression in some mobile belts and extension in others prevailed simultaneously and compensated each other; compression proba- bly predominated. Conditions of plate compression immediately replaced the initial extension regime in TTR belts. But the total predominance of compres- sion within the present-day cratons set in at the end of the Early Proterozoic when protoplatforms were firmly soldered by protogeosynclinal fold systems and TTR belts with regenerated and accreted continental crust.

Let us now examine the main types of Early Proterozoic structural ele- ments of the crust individually and the characteristics of their development.

4.2.1 Protoplatforms (Eocratons)

Protoplatforms represent rounded-polygonal, mostly isometric blocks of continental crust, i.e., fragments of an Epi-Archaean supercontinent from a few hundred to more than a thousand kilometres in cross-section with an area of thousands or even a few million square kilometres. The largest protoplatform is perhaps the Central Siberian eocraton, with the Antarctic

eocraton probably a close second. There were no less than four protoplatforms in North America, four or five in Africa, two or three in India, four in Australia, five in South America and six (or many more) in Eastern Europe. Thus
the total number of protoplatforms exceeds, possibly far exceeds, 30 with
a total extent of 60 million km^2 (see Fig. 22). As pointed out in the preceding chapter, the first protoplatforms manifest already in the Late Archaean
when their covers also began to accumulate (South Africa and Western Australia). The description that follows thus pertains also to the Late Archaean
protoplatforms.

Archaean granite-greenstone terranes and only partly granulite-gneiss
belts mainly constitute the b a s e m e n t   o f   p r o t o p l a t f o r m s.
Over much of the area of the cratons the basement is exposed, the rest
being overlain by volcanosedimentary cover.

Structures of three types constitute the p r o t o p l a t f o r m
c o v e r: protoplatform slopes, associated with the volcanosedimentary
filling of protogeosynclines; flat basins, often of large size, i.e.,
protosyneclises; and grabens, i.e., protoaulacogens. The Huron supergroup
homocline at the border of the Lake Superior eocraton and its transition to
the Penokean protogeosyncline represents a structure of the first type. An
example of the second type is the Transvaal syneclise and of the third the
Bathurst protoaulacogen of the Slave eocraton or the Pechenga-Varzuga of
the Baltic Shield.

Three main types of sedimentary formations participate in the composition of protoplatform cover formations. The first type is the grey-coloured,
predominantly continental clastic formation including sediments of alluvial
and delta plains and lakes, partly shallow-marine basins. This formation
was deposited usually at the base of cover sequences. Pongola, Dominion Reef, Witwatersrand (South Africa), Sumian-Sariolian (Karelian) groups
and lower parts of the Udokan series and Huron supergroup belong to this
type. In South Africa and Canada this type is remarkable owing to abundant uranium and in Africa additionally gold mineralisation. Shallow-marine
sediments of shelf slopes of eocratons and epicontinental seas predominate in the central part of the cover. Dolomites, often forming thick beds
and abounding in stromatolites (for example, in the Transvaal group of
South Africa), play a significant role in the constitution of this shallow-marine
carbonate-terrigenous formation. Jaspilites as in the Hamersley (Australia),
Transvaal (South Africa) or Huron supergroup (Canada) represent another
important type of rock. Carbonaceous shales or carbonate rocks (for example Jatulian shungites of Karelia) are known here. Red or variously coloured
lagoon-continental coarse clastic formations of the molasse type complete
the sequence of the platform cover. To this type belong Matsap, Waterberg,
Southpansberg, Umkondo (South Africa) and the upper part of the filling of

the Kodar-Udokan basin containing copper-bearing sandstones of considerable commercial importance. The red colour of this formation (1.9 b.y.) points to the presence of free oxygen in the atmosphere but the early signs of these changes in the composition of the atmosphere are noticed only 2.2–2.1 b.y. ago. That the chemical composition of the hydrosphere underwent change even earlier is suggested by the extensive growth of stromatolites and the presence of evaporite minerals—anhydrite and gypsum—after which pseudomorphs begin to be encountered in the sediments. Tillites are noticed already in the uppermost Archaean (Witwatersrand); they are known in the interval 2.3–2.1 b.y. in South Africa and in Canada (Huron).

Sequences of protoplatform cover have a cyclic structure; this has also been noticed for the Udokan (four cycles), Karelia (five cycles) and Transvaal sequences; further, this cyclicity is associated with the overall retrogressive tendency manifest in the increase in role of coarse clastic rocks towards the end of the Early Proterozoic.

In addition to sedimentary rocks, volcanics sometimes forming independent suites several kilometres in thickness (for example Ventersdorp of South Africa or Fortescue in Western Australia) participate significantly in the structure of protoplatform covers. Among volcanics, continental tholeiite-basalts predominate; apart from them, picrites (Karelia) and felsic lavas (South Africa and Karelia) are encountered and form bimodal associations. Volcanism in some regions is distinctly associated with fracture tectonics, as for example in Karelia (Jatulian and Suisarian) or in South Africa where it preceded Ventersdorp volcanism.

Intrusive magmatism is most often manifest in the form of dykes, sills and bodies of gabbro-diabases, gabbro-dolerites and gabbro-anorthosites but large layered plutons are also noticed here and there. Prominent examples of the latter are lopoliths such as the Bushveld, Kodar-Kemen, Sudbury and Stillwater. Granites of the Kodar-Kemen lopolith (1.9 b.y.) belong to the rapakivi type, most widely developed at the verge of the Early and Middle Proterozoic.

A fairly characteristic phenomenon associated with remobilisation of Archaean sialic crust is the formation of mantled granite-gneiss domes. Such domes complicate the cover structure in the Kodar-Udokan and Transvaal basins as well as the Hamersley and Nabberu basins in Australia.

The thickness of volcanosedimentary cover reaches 10–12 km or more (up to 20 km), which is not very much considering the immense duration of the Early Proterozoic extending for almost a billion years; the average deposition rate is less than 10 m in one million years.

Large basins within protoplatforms, Kodar-Udokan in Siberia, Transvaal in South Africa and Hamersley (Nullagine) in Australia, may be called protosyneclises although they are of smaller size (a few hundred kilometres) compared to Phanerozoic syneclises (several hundred or even more than a thousand kilometres across). As a rule, basins of a protoplatform cover are not inherited by analogous structures of a 'true', i.e., Riphean-Phanerozoic cover.

Cover deposits are unevenly deformed. Deformation is most often associated with block movements of the basement with displacement along strike-slip faults and with intrusion of granite-gneiss domes and lopoliths. A uniformly developed regional linear folding has hardly ever been observed. Dislocation is perceptibly intensified close to faults, like metamorphism. The latter too is highly uneven. It attains amphibolite facies (even granulite facies as an exception), weakening upwards along the sequence and from the periphery to the central parts of the protoplatform while it is altogether absent in some regions (Francevillien series of Equatorial Africa, southern flank of Nabberu basin in Western Australia and upper part of the sequence of the Aldan Shield).

4.2.2 Protoaulacogens

Structures of the aulacogen type were comparatively less expanded in the Early Proterozoic. Among them are noticed two varieties: with a purely sedimentary and with a volcanosedimentary filling. Aulacogens of the Canadian Shield belong to the first type. The Pechenga-Varzuga system of the Kola Peninsula and probably the structure of the Vetrenny Poyas belt of Karelia to the second type while the Southpansberg aulacogen in South Africa combines the features of both types. Sedimentary formations are represented in the first type, in addition to terrigenous, by carbonate (dolomite) formations and in the second type mainly by clastic formations. Volcanics are represented by tholeiite and subalkaline basalts and also by picrites (Baltic Shield); sills of gabbro-diabases are extensively developed in this type. Dislocation and metamorphism increase sharply close to zones of marginal faults; further, linear folding as well as overthrusts pointing to general compression during partial inversion of corresponding aulacogens are also noticed here.

4.2.3 Protogeosynclines

These are far more extensively developed and are fairly diverse in structure and evolution. Their extent ranges from a few hundred to several thousand kilometres, e.g., the Transhudsonian system has been traced for 3000 km with a width of a few hundred kilometres. All were formed on an Epi-Archaean continental crust by rifting but the degree of destruction of this crust varied markedly. One type of protogeosyncline, in the form of m u l t i p l e   t r o u g h s, is close to Archaean granite-greenstone terranes and the corresponding structures may also be called greenstone belts. They differ from classic Archaean belts as follows: 1) komatiites are absent or rare; 2) volcanics have an almost exclusive bimodal composition; 3) sedimentary rocks play a more significant role in the sequence than in Archaean belts; 4) among clastic rocks, a perceptibly high percentage belongs to quartz and arkose varieties compared to greywackes that predominated in the Archaean; 5) the role of turbidites increases and many formations have a flysch appearance; 6) carbonates are sometimes present in perceptible amount; and 7) genuine molasse is seen in the upper part of the sequence at places. The thickness of the filling reaches several kilometres. This type is represented by the Maroni-Itacayunas province of the Guyana Shield and the Birrimian of West Africa as also the axial zone of the Transhudsonian belt.

Formations of Early Proterozoic greenstone belts, like their Archaean analogues, are complexly dislocated, metamorphosed to greenschist and amphibolite, and sometimes even granulite (Guyana Shield ?) facies and intruded by numerous granitoid plutons, partly autochthonous (anatectic) and partly allochthonous (melting of Archaean crust).

The geotectonic conditions of formation and evolution of greenstone belts do not call for special analysis here since this subject has already been dealt with in Section 3.2 in the context of their Archaean homologues. It may simply be pointed out that in the Early Proterozoic we have almost exclusively to deal with a subtype of these belts (see below), which for the Archaean were called belts of incomplete development with bimodal volcanism. It may be recalled that these belts were assumed to have formed on continental crust that experienced extension, thinning and reworking, without a break in continuity or, in any case, without significant extension, which was replaced by compression and some underthrusting of one flank under another (subduction) in the concluding stage of evolution. The latter is an essential assumption in explaining granitoid plutonism and fold-overthrust deformation at the end of the evolution of these belts.

The second type of protogeosynclines, monotroughs, is the most widely distributed and differs from the first type in smaller width, somewhat lesser extent and consists of an axial trough with abundant, usually bimodal volcanism, relatively deep, quite often flysch-type deposits, sometimes with olistostromes (northern Singhbhum etc.) and peripheral zones with accumulation of shallow-water deposits, including quite often thick bands of shelf carbonates, sometimes even barrier reefs (Halls Creek system in Northern Australia) on transition to a deeper part of the basin. In fact, the very existence of such basins is possible only on oceanic or suboceanic crust.

Two subtypes are distinctly decipherable in this type: with ophiolites ('proto-ophiolites') and without them, or ensimatic or ensialic. Four protogeosynclinal systems belong to the first subtype, namely Transhudsonian, Svecofennian, Satpura (including northern Singhbhum) and Aravalli-Delhi.

The second type includes the Krivoy Rog-Kursk system and systems parallel to it in the southern Russian Platform and protogeosynclines of Australia (Ashburton, Gascoyne, Halls Creek, Pine Creek etc.) and South Africa (Namibia-Namaqualand).

The difference between these two subtypes, as in the case of subtypes of greenstone belts, evidently lies in the degree of extension and destruction of the continental crust on which they were formed. The presence of ophiolites positively demonstrates that destruction proceeded until commencement of spreading although the scale of the latter was probably limited and proximate to that of the Red Sea. It is quite possible that in future at least a part of the protogeosynclines of the second type will have to be transferred to the first type. The example of the Transhudsonian system suggests in the light of new data that spreading can be manifest on a rather wide scale.

The concluding phase of development of protogeosynclines is characterised by the formation of felsic, calc-alkaline volcanoplutonic association (the much earlier granitoids were of the sodic type and later ones the potash type) associated with coarse clastic formations of the molassoid type. This was preceded by intense deformation and regional metamorphism reaching amphibolite facies.

The t h i r d t y p e o f E a r l y P r o t e r o z o i c g e o-s y n c l i n e s, z o n a l-p o l a r, is distinctly characterised by transverse zoning and polarity and stands closest to the Late Proterozoic-Phanerozoic geosynclines. This third type is well represented in the Canadian Shield by the Wopmay, Transhudsonian partly, Labrador and Penokean systems. The outer zone of these systems (miogeosynclinal) is made up of shelf sediments, usually containing carbonates (dolomites), distinguished only by a very large thickness and intense dislocation from the formations of eocraton slopes primarily associated with these shelf deposits. Sheets of tholeiite-basalts of the continental type are encountered. In the epoch of concluding orogeny, the cover of the outer zones experienced detachment from its continental base and was thrust in the form of slices on the margin of the eocraton. The central zone corresponds to the continental slope and rise of protogeosynclinal basins and possibly also to their axial part. The lower parts of the sequence are predominantly lutitic in composition; at this level tholeiite-basalts proximate to the midoceanic type play a significant role in the sequence with flysch superposed. In the orogenic stage flysch was replaced by molasse, which is deposited on the formations of the outer zone or even on the slope of the protoplatform (Wopmay). Formations of the outer zone are usually weakly metamorphosed; metamorphism increases perceptibly within the central zone, which evidently developed on suboceanic or oceanic crust and was later thrust on the outer zone. The innermost zone is formed by a volcanoplutonic belt consisting of large batholiths of granitoids and superposed on an intensely reworked ancient continental base. Thus fold systems are formed on the site of protogeosynclines of this type as a result of the collision of blocks of much older continental crust, i.e., eocratons, with underthrusting (subduction) of suboceanic or oceanic crust and one eocraton under another (collision).

All the types of protogeosynclines described above should represent relatively deepwater basins, judging from the extensive development of flysch or flyschoid formations and at places olistostromes and barrier reefs. The width of these basins should have been at least one-and-a-half times more than that of fold systems formed *in situ,* judging from the magnitude of compression during fold-thrust deformation. Considering the area occupied by these structures on present-day continents, they should have held a fairly large volume of seawater at a depth of 3 km. This conclusion is important in a later discussion (see Sec. 4.3).

The process of 'closing' of protogeosynclines, accumulation of their volcanosedimentary filling and metamorphism and granitisation of the filling should have been a consequence of subduction of oceanic or suboceanic ('transitional') crusts as assumed above. However, a wide recognition of Early Proterozoic subduction is contradicted by the fact of extreme rarity in the Early Proterozoic of continuously differentiated calc-alkaline volcanic series, regarded as typical products of subduction of oceanic crust, and also the absence of manifestation of low-temperature and high-pressure metamorphism, typical of the Phanerozoic Benioff zone. These arguments are only of relative importance, however; calc-alkaline series are known in several regions (Canadian Shield Wopmay orogen and La Ronge zone in Transhudsonian orogen; Penokean orogen; and Lofoten Island on the Norwegian coast). While there are no volcanics of intermediate composition in other regions, granitoids are abundant (Labrador system). Kyanite schists could have played the role of glaucophane schists in the Early Proterozoic under conditions of very high heat flow, all the more so since subduction zones should have been less deep and hence pressures not very high, while the activity of sodium-bearing fluids, assigned much importance by some Russian geologists (A.A. Marakushev), was more significant (lesser depletion of the upper mantle). According to others (Moskovchenko, 1983), eclogites could have served as Early Proterozoic representatives of high-pressure metamorphics of subduction zones. Further, glaucophane-lawsonite schists of Early Proterozoic age have been detected in just two regions of the world, i.e., the Aravalli system of India (Deb and Sarkar, 1990) and the Cathaysian system of China (Wu Geyao, 1986).

These facts (scarcity of andesites and glaucophane schists) should be studied and their explanation sought. One explanation was given by Hynes (1982) as follows. Since the temperature of the asthenosphere in the Early Proterozoic was $100^\circ C$ more than at present, the lithosphere (tectosphere according to T. Jordan) rapidly lost its buoyancy and underwent subduction at a very young age (60–75 m.y.) while volatiles were discharged closer to the trench which prevented magma formation. In our opinion, this hypothesis is somewhat artificial; it does not take into consideration the fact that felsic magma was abundantly produced in the concluding stage of the evolution of many protogeosynclines. The scarcity of volcanics of intermediate composition is better explained by the ensialic nature of many protogeosynclines causing melting of large masses of Archaean continental crust. The latter is confirmed by a predominance of high strontium isotope ratios in Early Proterozoic felsic volcanics and granites although much lower ratios are also encountered, pointing to some inflow of juvenile magma from the mantle.

On the whole, in the Early Proterozoic tectonics of minor plates prevailed (Goodwin, 1985) with manifestation of disperse spreading and subduction on a restricted scale in each of the mobile zones. At the same time, as a result of a large number of such zones, the overall global effect of these processes could be no less, or even more, than in the Neogäikum and this (especially disperse spreading) ensured heat loss by the Earth equivalent to its very high heat flow compared to the present. According to Abbot and Menke (1990), the overall length of the axes of spreading 2.4 b.y. ago was more than double (2.2 times) the present-day extent. A particularly high rate of spreading is thus not required, which as pointed out by A. Kröner (in: Powell et al., 1983) contradicts palaeomagnetism as well as geological data.

4.2.4 Tectonothermally Reworked (TTR) Granulite-Gneiss Belts

These belts are encountered in nearly all the continents. They first appeared by the Late Archaean (classic example, Limpopo belt) and peaked by the Late Proterozoic (Riphean). The Early Proterozoic occupies an intermediate position in this respect.

Among the Early Proterozoic TTR belts, three subtypes can be distinguished. One subtype, seen in the Lapland-Belomorian and Stanovoy belts, commenced and completed its development in the Early Proterozoic. Other belts continued to evolve actively in the Middle and Late Proterozoic; these are the Australian belts, Wegener-Mawson in the Antarctic region, Inner Mongolia (Liaoning) belt and Atlantic belt of Brazil. Both these subtypes are made up of Archaean material, primarily metamorphosed from greenschist (former greenstone belts) to granulite facies. In the Belomorian and Stanovoy belts granulite Archaean was subjected to retrogressive metamorphism in the Early Proterozoic to amphibolite facies and intensely granitised; supracrustal Archaean formations involved in the process of crustal subduction probably served as a source of water. In the Lapland granulite segment of the Lapland-Belomorian belt, progressive granulite metamorphism of high pressures has been dated Early Proterozoic but was probably preceded by a similar Archaean metamorphism of moderate pressure. Belts of the second subtype also reveal metamorphism of granulite facies of Early Proterozoic or even much younger age, Middle to Late Proterozoic (Australian and Antarctic regions).

Delineation of the third subtype of Early Proterozoic TTR belts is tentative. Belts of this subtype are made up of Archaean and even Early Archaean substratum while Early Proterozoic supracrustal formations are absent in them and radiometric age determinations point only to Late Proterozoic activity. Hence it is not clear what happened at the site of these belts in the Early Proterozoic but most probably a moderate uplift. An example of this subtype of TTR belts is the Eastern Ghats belt of the Indian peninsula where the main epoch of tectonothermal reworking set in 1.6–1.5 b.y. ago.

What were the geodynamic conditions in which TTR belts developed? Three elements are quite evident: high heat flow, intense tangential compression and active uplift; the latter would be relevant only in the concluding phase of their development (for example in the case of the White Sea). Compression is demonstrated by the thrusting of TTR belts on adjoining granite-greenstone Archaean terranes or, more correctly, underthrusting of the latter beneath the TTR belt (Karelian megablock beneath the Lapland-Belomorian belt and Aldan megablock beneath the Stanovoy); in other cases, thrust displacements predominate at their boundary (Musgrave block and Albany-Fraser belt). Within the much older, Late Archaean tectonothermal reworking, even rifting could have occurred in the Early Proterozoic, an example of which is the Southpansberg rift in the Limpopo belt.

4.3 END OF EARLY PROTEROZOIC—TOTAL CRATONISATION AND ORIGIN OF PANGAEA I SUPERCONTINENT

Evolution of the Early Proterozoic protogeosynclines concluded essentially 1.8–1.75 b.y. ago; from that time a continental regime was established in the region of present-day ancient platforms-cratons and their miogeosynclinal framework. This regime resulted in extensive manifestation of subaerial volcanism and typical gabbro-anorthosite-rapakivi, sometimes granite plutonism. This magmatism encompassed the period 1.8–1.6 b.y., i.e., the transitional period from Early to Middle Proterozoic, at places continuing up to 1.5 and even 1.4 b.y.; it calls for a special examination.

4.3.1 North American Craton

On the north-western margin of the Canadian Shield, in the region of Great Bear Lake, a marginal Andean type volcanoplutonic belt with formation of ignimbrites of intermediate composition and mantle granite plutons aged 1.87–1.86 b.y. arose in the concluding stage of evolution of the Wopmay protogeosyncline; a little later (1.86–1.85 b.y.) potash granites manifested (Hoffman, 1989).

Development of another protogeosyncline south of the Wyoming Archaean eocraton concluded with formation of a volcanic arc and calc-alkaline magmatism by 1.7 b.y. Following this a volcanoplutonic belt aged 1.6 b.y. formed in the south and south-east. A similar picture is observed in the southern framing of Superior (Lake Superior) eocraton where a volcanic arc aged 1.9–1.8 b.y. arose at the site of inner zones of the Penokean protogeosyncline. South of it, in the region of Lake Michigan, felsic volcanics aged 1.76 b.y. are known; in Wisconsin a pluton of rapakivi granites and small bodies of anorthosites aged 1.5 b.y. are known. In the more southern regions of the Midcontinent (USA), according to drilling data, a sheet of felsic subaerial volcanics intruded by epizonal granite plutons aged 1.5–1.35 (1.5–1.45) b.y. is widely distributed.

4.3.2 East European Craton

On the north-western margin of the Baltic Shield, calc-alkaline felsic volcanics, including abundant ignimbrites and granitoid plutons (comagmatic with ignimbrites) of batholith proportions, are replaced by Svecofennian granitoids aged 1.75 b.y. These volcanics form the Trans-Scandinavian volcanoplutonic belt with a north-north-easterly strike continuing under the Caledonides, best manifest in the central part of Sweden (Smäland-Värmland). This belt continued to develop in the south up to 1.6 b.y. and in the north up to 1.5 b.y. Strontium isotope ratios are characterised by low values in the south and high values in the north, pointing correspondingly to mantle and crustal origin of magma. According to many researchers, for example Nyström (1982) and Wilson (Wilson et al., 1985), this belt is part of the series of marginal volcanoplutonic belts of the Andean type. However, there is no clear proof of the existence of an oceanic basin west of this belt, possibly because the belt marks a suture at the site of final closing of the Svecofennian protogeosynclinal basin.

Nevertheless, based on the absence of traces of a very old sialic basement in the western part of the Sveconorgides and the island-arc nature of intermediate and felsic volcanics and associated sediments and their alternation with sequences made up of greywackes, schists and basalts, Gaal and Gorbatschev (1987) hypothesised that volcanic arcs and the basins between them developed here, evidently on oceanic crust. Given this data and these concepts, the hypothesis regarding the affinity of the Trans-Scandinavian belt to the Andean type is quite probable.

A sharp change in tectonic regime occurred at around 1550 ± 100 m.y. in Sweden, i.e., compression was replaced by extension (Wilson, 1982). Roughly in the same belt, as in the much older volcanoplutonic belt and also the latitudinal belt west of Stockholm, a swarm of basic dykes intruded. Small plutons of gabbro and rapakivi granites formed more to the north.

In the epoch 1.65–1.50 b.y. far larger massifs of gabbro-anorthosites and rapakivi granites formed in southern Norway, southern Finland and Karelia

(Vyborg and Salma), in the Peri-Baltic region (Riga pluton) and in north-eastern Poland and the Ukraine (Korosten' and Korsun'-Novomirgorod plutons), revealing high magmatic activity of the entire western periphery of the East European Craton. West of the boundary of the craton, throughout Central and Western Europe, outcrops of rocks of corresponding ages are not known and hence the tectonic conditions of this magmatism and the reason for its localisation at the western margin of this craton are not quite clear. An analogy can only be assumed with the Trans-Scandinavian belt whose southern continuation is represented by these magmatites and an association with the future formation of the Mediterranean geosynclinal belt—the Proto-Tethys Ocean.

4.3.3 Siberian Craton

Magmatic activity of this epoch is most distinct in the Peri-Baikal region (see Sec. 4.1.3) and manifest in the Northern Baikal (Akitkan) volcanoplutonic and molassoid belt; the belt was formed in the interval 1700 ± 100 to 1630 ± 40 m.y.

As pointed out in Section 4.1.4, the same double interpretation as in the case of the Trans-Scandinavian belt, i.e., either it is a post-collision belt or a belt of the Andean type, is possible with regard to this Siberian belt. Accepting the latest interpretation of both belts, an extremely important conclusion follows that the boundaries of the East European and Siberian cratons became perceptible already at the very end of the Early Proterozoic and commencement of the Middle Proterozoic. In fact, rapakivi granites and the rather younger Berdyaush pluton are known east of the East European Craton and alkaline granitoids aged 1.66 b.y. east of the Aldan Shield. The granite-rhyolite belt in the southern part of North America and the volcanoplutonic belt of Erinpura-Malani in north-western India, the latter much younger at 1.0–0.7 b.y., could be of similar significance.

4.3.4 Ural-Okhotsk Belt

In this belt the Kazakhstan-Northern Tien Shan and Balkhash microcontinents exhibit quite extensive development of the felsic volcanoplutonic association characteristic of the epoch under consideration. Such an association in Ulytau is the Maityube and Zhiida series, in Aktau-Mointy uplift the upper Atasu series and in Terskei-Alatau the Karasaz suite. A typical feature of all these formations is the predominance of porphyroids formed after lavas, tuffs and ignimbrites of rhyolite composition. Volcanics are intercalated by thick bands of clastic rocks of predominantly quartz composition. Granite gneisses aged 1.8–1.7 b.y. are closely associated with them (in: *Precambrian in Phanerozoic Fold Belts,* 1982; *Precambrian of Central Asia,* 1984; Zaitsev, 1984). But a very similar magmatism continued here even in the Early Riphean (see Sec. 5.3.5).

4.3.5 Sino-Korea Craton

Cratonisation here concluded in the interval 2.0–1.7 b.y. in two phases of Luliang tectonics, the main phase at the level of 1.8 and the concluding phase at 1.7 b.y. (Ma Xingyuan and Wu Zhengwen, 1981; Wang et al., 1985). This was followed by intrusion of plutons of rapakivi and anorthosites along the northern periphery of the craton with ages of 1638, 1624 and 1433 m.y., preceding Riphean rifting in this belt and once again marking the northern boundary of this craton and also extensive (up to 150 × 100 km) swarms of dolerite dykes in the more southern regions.

4.3.6 South American Craton

This craton, especially its north-western part, represented an arena of intense magmatic activity at the end of the Early Proterozoic. In this epoch the thick volcanoplutonic Rio Negro-Juruena belt with a north-west—south-easterly strike was formed in the western part of the Guyana Shield and in the western-central part of the Central Brazilian Shield, which in this epoch constituted a single Amazonian Craton. The period of its development has been estimated at 1.75–1.55 b.y. and the main rocks are granites and granodiorites. On the Guyana Shield, the epoch of cratonisation commenced with felsic volcanics of the Uatuma calc-alkaline (andesite to rhyolite) and alkaline (from trachytes to alkaline rhyolites) series. Granitoids are represented by granites, granophyres, adamellites and granodiorites aged 1.9–1.8 b.y. (Rb-Sr and K-Ar methods) with high strontium isotope ratios pointing to the crustal origin of the rocks. The clastic Roraima group and its analogue Urupi are much younger, with their age estimated to be about 1.8 b.y. This was followed by intrusion initially of mafic rocks and later granites, in particular Paraguaçan granites aged 1.55–1.45 b.y. Again, on analogy with the Baltic Shield and the Siberian Craton, it is not known what could have fringed the Rio Negro-Juruena belt but its gravitation towards the western boundary of the craton is clearly expressed.

We must also recall the affinity of the volcanosedimentary belt of the lower part of the Espinhaco supergroup on the eastern boundary of the Sao Francisco eocraton in Bahia and Minas Gerais states of Brazil to the period under consideration.

4.3.7 African Craton

In the north-western part of the craton the largest field of felsic volcanics and granites of the end of the Early Proterozoic is situated in the eastern part of the Reguibat massif. The lower member of the sequence of sedimentary-volcanoplutonic association of this age here is the Oued-Sous series with volcanics including ignimbrites of andesite-rhyolite composition and also layers of clastic rocks. This series is intruded by Aftout granites aged 1850 ± 75 m.y. (Rb-Sr method on biotite) and overlain by a new suite of felsic (and intermediate) Eglab volcanics transiting upwards along the sequence into the clastic Guelb-El-Hadid series intruded by dolerites aged (Rb-Sr method) about 1.6 b.y.

In Equatorial Africa felsic volcanics and granites of proximate age are distributed in the zone of confluence of the boundaries of Zaire, Tanzania and Zambia, within the Bangweolo block, mainly along its north-western periphery where a volcanoplutonic belt with a north-easterly trend is noticed. It is made up of volcanic tuffs and ignimbrites of rhyolite, rhyodacite, dacite and less often andesite composition as well as granites; the Rb-Sr age of these magmatites is 1.83 b.y. A suite of continental clastic Mpokorozo rocks with a thickness of up to 5 km lies on top.

It must be further noted that granodiorites aged 1.63 b.y., as determined by several methods (U-Pb method on zircon, Sm-Nd and Rb-Sr methods on whole rock), were recently discovered in the extreme north-eastern part of the Arabia-Nubian Shield within Saudi Arabia. Lead isotope studies showed that Early Precambrian rocks may be notably developed in this part of the shield.

4.3.8 Australian Craton

Orogenic volcanoplutonic associations concluding the development of protogeosynclines (Pine Creek, Halls Creek, Granites Tanami, Tennant Creek and Mount Isa) and formed in the period 1.9–1.65 b.y. (mainly 1.8 to 1.7 b.y.) were already mentioned in Section 4.1.8. Formation of the volcanoplutonic Gawler Range belt on the eastern margin of the Gawler eocraton in the epoch 1592–1575 m.y. is most remarkable in the period under consideration (1.6–1.5 b.y.). A much earlier epoch of Kilban granite formation affecting even the central portion of the eocraton and coinciding with regional metamorphism has been dated 1.68–1.6 b.y. This endogenous activity manifested in the form of metamorphism in the Williama block (Brocken Hill) 1.66–1.5 b.y. ago. That both the structures lie in the immediate proximity of the Tasman geosynclinal belt, which commenced development much later, in the Late Riphean, is of interest. However, even the more interior regions of the Australian Craton, Arunta and Musgrave blocks, represented an arena of metamorphic transformation to granulite facies, migmatisation and anatectic granite formation in the epoch 1.8–1.75 b.y., to amphibolite facies in the interval 1.7–1.55 b.y. in the Arunta block and granulite facies in the interval 1.6–1.5 b.y. in the Musgrave block (the latter age is disputed; it probably corresponds to the age of primordial crust formation). These two blocks, separated by the superposed Amadeus aulacogen, represent the exposed segment of an extended polymetamorphic TTR belt intersecting the Australian Craton in a latitudinal direction and actively developing right up to 1.0 b.y.

4.3.9 Antarctic Craton

The epoch of magmatism under consideration was represented here by differentiated volcanics and subvolcanic intrusions aged 2.2–1.7 b.y. (Rb-Sr and K-Ar methods) in the western part of the craton on Reacher plateau close to the Weddell Sea (see Sec. 4.1.19). This magmatism extends into the Middle and Upper Proterozoic.

4.3.10 Some Conclusions

The above discussion shows that the end of the Early Proterozoic is characterised by a definite set of common conditions in all the continents. By around 1.7 b.y., nearly all the protogeosynclinal basins had disappeared, all eocratons joined and a single large massif of continental crust, a supercontinent, should have formed. This supercontinent can logically be designated as Pangaea I, as distinct from the much later Wegener Pangaea II. According to palaeomagnetic data (Piper, 1982), such a protocontinent could have formed already at the end of the Archaean 2.7 b.y. ago (Pangaea 0) but during the Early Proterozoic it underwent destruction with disintegration into several blocks or protoplatforms. Protogeosynclines arose on crust of suboceanic ('transitional') or even oceanic type in the rift zones between the protoplatforms. However, the common trajectory of apparent wandering of the pole derived from palaeomagnetic data shows that the scale of extension could not have exceeded 1000–2000 km and that during the closure of palaeo-ocean basins, their flanks should have reverted roughly to the prespreading position. As pointed out before, geological data does not contradict but rather confirms more readily these conclusions as well as some other geophysical considerations (fast subduction of young oceanic crust as a result of low viscosity of the asthenosphere).

It should concomitantly be pointed out that the concept of a single protocontinent being dismembered commencing from the end of the Archaean is not shared by all specialists of palaeomagnetism. In particular, data (McElhinny and McWilliams, 1977) has been given which points to differential movement in the Proterozoic between Laurentia, Africa (+ South America ?) and Australia, i.e., Western and Eastern Gondwana. Moreover, the palaeomagnetic data presented by these authors for the Transhudsonian belt points to the possibility of Early Proterozoic ocean basins attaining a width comparable to that of the present-day Atlantic. Thus the question of scale of destruction of Pangaea 0 in the Early Proterozoic remains unresolved.

Emersion of most marine basins at the end of the Early Proterozoic, which held, as pointed out above, a significant volume of water, poses the question of drainage of this water beyond the contemporary continental hemisphere, i.e., about the existence (formation ?) already at that time of

Panthalassa, the possible prototype of the future Pacific Ocean, along with Pangaea. Salop (1984) assumed that such desiccation occurred repeatedly commencing right from the end of the Early Archaean, namely at the verge of the Early and Middle Archaean, Middle and Late Archaean and Archaean and Proterozoic. It must be pointed out, however, that no adequate grounds exist for drawing such conclusions as there is no proof of the simultaneous elimination of all marine basins in these periods. Moreover, available data points to the succession of such basins between Early and Late Archaean and Archaean and Proterozoic. Actual proof of general desiccation is noticeable only for the end of the Early Proterozoic, which determines the minimum age of Panthalassa. It is confirmed by manifestation of early volcanoplutonic belts which can be regarded as marginal, namely Bear belt in Wopmay orogen on the north-western Canadian Shield, eastern Australian belt from York Peninsula to the west coast of Spencer Bay and also the ophiolite belt in south-eastern China already in the Early Riphean (Zhai et al., 1985).

Thus it may be assumed that polarisation of the Earth's surface, crust and lithosphere into continental and oceanic segments was complete by the end of the Early Proterozoic. This polarisation occurred by fusion and agglomeration of continental massifs (eocratons-protoplatforms) along protogeosynclinal sutures into a single supercontinent. But, at least theoretically, Panthalassa could have arisen even earlier. The most important indirect evidence of this is the analogy with other planets of the Earth's group and Moon which also reveal characteristic asymmetry with separation into 'oceanic' and continental hemispheres. Such a comparison shows that the maximum age of Panthalassa could go up to 4.0 b.y. or more. But what was the mechanism in this case that caused the separation of crust (and lithosphere) into two dissimilar segments? One explanation, as pointed out before (see Sec. 1.1), was offered by Elsasser (1963) who pointed to the possible asymmetrical inflow of mantle iron into the core and, correspondingly, core formation caused by the gravitational influence of the Moon which was closer to the Earth at that time.

Such a mechanism has been strongly undermined by modern concepts regarding formation of the core even as a result of heterogeneous accretion. Another mechanism could be based on the assumption of predominantly meteorite bombardment of one of the hemispheres or the fall of a particularly large asteroid on one side of the Earth, forming on that side a huge pocket of fusion ('magmatic ocean') and causing commencement of a vigorous differentiation of the upper mantle, leading ultimately to the formation here of early nuclei for the growth of continental crust.

Yet another possible mechanism is that the fall of such an asteroid served as the cause of formation of a basin of the Pacific Ocean type. Formation of the Moon was recently related to such a catastrophe (due to the scatter of material beyond the Roche limit). This is actually a new variant of the hypothesis of G. Darwin and Pickering.

It may be assumed in every case that the Late Archaean represented an important stage in the course of formation of the asymmetry of the Earth. At this time (3.0–2.7 b.y. ago) a particularly intense accretion of continental crust and its concentration in the future continental hemisphere, deep metamorphism and granitisation occurred. On the other side of the Earth, intense upward convective currents may have been activated and prevented the origin here of any stable continental mass from that time (according to the models of L.P. Zonenshain for the Palaeozoic, generally adopted in the much later reconstructions). In other words, multicell convection characteristic of much of the Early Proterozoic was replaced by the end of that aeon by large-scale unicell convection, which was probably common for the entire mantle.

PART II
LATE PRECAMBRIAN

5

Early Riphean: New Stage of Destruction and Commencement of Fragmentation of Pangaea I

5.1 REGIONAL REVIEW

5.1.1 North American Craton

After completion of Hudsonian diastrophism, the stabilised territory of North America was essentially uplifted, but in some regions platform cover accumulated, aulacogens developed, formation of felsic volcanoplutonic associations which had commenced right at the end of the Early Proterozoic continued and tectonothermal reworking of the basement took place.

5.1.1.1 *Canadian Shield.* Lower Riphean formations were distributed in isolated basins (Fig. 41) of the type of aulacogens bound by faults with a predominant north-north-westerly strike. Carbonate-terrigenous sediments of the Hornby Bay group were deposited (up to 2.4 km in thickness) in the north-west, in the region of the Arctic coast between Coppermine River and Great Bear Lake on an Aphebian basement (1930–1850 m.y.) with erosion and unconformity. This group is intruded by diabases aged 1400 ± 75 m.y. (Young, 1980). South of Lake Athabasca the same-named ellipsoidal basin is filled with quartzites and conglomerates of the Athabasca group with Carswell carbonate formation overlying them. The total thickness of the sequence is 800 m and the interval of its formation 1500–1430 m.y. (Goodwin, 1991). The main part of the sequence of the basin in the region of Baker and Dubawnt lakes was formed at the end of the Early Proterozoic but the carbonate-terrigenous upper Thelon formation intruded by Mackenzie swarms of dykes evidently pertains to the Lower Riphean. Conglomerates, sandstones and dolomites of Kanuyak, Tinney, Ellis River and Parry Bay formations aged 1760–1200 m.y. are developed on the islands of Bathurst Bay in Coronation Gulf (Goodwin, 1991). Isolated basins filled with quartzites, preceded by Keweenaw-Sioux, Baraboo and Barron basalts are distributed in the northern central states of the USA. Palaeorifts of Lake Superior marked by accumulation of conglomerates and Sibley quartz-sandstones

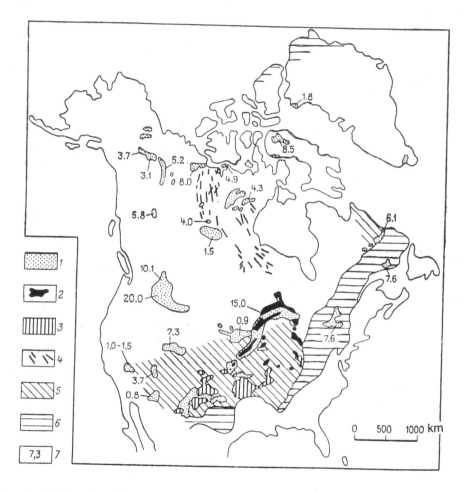

Fig. 41. Distribution of Precambrian rocks in North America with ages ranging from 1700 to 850 m.y. (after Stewart, 1976).

1—sedimentary rocks; 2—basic lavas and intrusive rocks; 3—rhyolitic tuffs, lavas and granulites; 4—diabasic dykes; 5—Elsonian reworking 1.5–1.3 b.y. ago; 6—Grenville orogeny, 1000 m.y.; 7—thickness in kilometres.

(Lower Keweenaw) aged 1537 m.y. and the volcanoterrigenous groups of Bruce River and Lake Leticia in central Labrador resting unconformably on the Aphebian and affected by intrusions aged 1496 ± 37 and 1392 m.y. evidently began forming by the Early Riphean (Goodwin, 1991).

In southern Greenland formation of sandstones and volcanics of the Eriks Fjord formation (1600–1310 m.y.) evidently occurred in the Early Riphean under conditions of the nascent Gardar rift system.

On the south-eastern and eastern fringes of North America, an immense anorogenic volcanoplutonic belt was formed in the Early Riphean. The belt extended from the Labrador peninsula to California for a distance of about 5000 km with a width of 1000 km. Its main constituents are anorthosite, rapakivi granite, granite-rhyolite bodies measuring (on the surface) up to 100 km. Almost undeformed rhyolites, felsites, dacites and their tuffs aged 1.44–1.2 b.y. have been described in the segment from Texas to the western part of Ohio (1700 km) (Anderson et al., 1969). Formation of the bulk of the plutons pertains to the interval 1.49–1.41 b.y. (Silver, 1980).

An extensive zone of Elsonian tectonothermally reworked basement (1400 m.y.) lies immediately north-east of the zone of distribution of the volcanoplutonic association. The reworked zone encompasses the adjoining part of the Midcontinent, the southern Rocky Mountains and the Canadian Shield. Reworking manifested in the intrusion of numerous granitoids aged 1.45–1.35 b.y., and in regional processes of K-Ar and Rb-Sr 'rejuvenation' of mica systems. Similar events occurred in Nain Province in the north-eastern Canadian Shield. Here they were manifest in anorogenic intrusions at the level of about 1.4 b.y. This was preceded by accumulation of metavolcanics of the Petsapiscau group aged 1525 ± 60 m.y. (Rb-Sr method) that experienced folding and metamorphism at around 1490 m.y. (Goodwin, 1991). Intrusion of Michikamau adamellites into the Petsapiscau group 1462 m.y. ago (U-Pb method) should evidently be regarded as the beginning of anorogenic non-metamorphosed intrusions proper. The last outburst of anorogenic plutonism was manifest in the build-up of an anorthosite complex 1388 ± 25 m.y. ago, associated Snegaschuk type granites 1426 m.y. ago and others (Goodwin, 1991).

Outside the Nain province, in the zone of Elsonian reworking, supracrustal extrusive-sedimentary formations of the Lower Riphean are virtually unknown. Discordant intrusions into the ancient basement, consolidated in the Archaean and Early Proterozoic, are noticed everywhere. Thus the Wolf River batholith dated 1485 ± 15 (Rb-Sr) and 1437 ± 34 m.y., Kroner Island intrusive complex dated 1464 m.y. and Busau syenite of the same age are located south of Lake Superior in the Southern Province of the Canadian Shield. Formation of anorogenic granites and anorthosites in the age interval 1.5–1.4 b.y. attained an unprecedented magnitude in North America and encompassed southern and central parts of the Rocky Mountains, adjoining part of the Midcontinent and southern part of the Canadian Shield. Here hundreds and perhaps even thousands of plutons of anorthosites, granites, granite-porphyries, syenites and gabbro with outcrop size of up to 100 km intruded a strip about 6000 km long and 1000 km wide. Intrusion of these bodies was accompanied by regional processes of K-Ar and Rb-Sr rejuvenation of mica systems but manifestation of metamorphism was generally weak. Faults and rift-like structures are extensively distributed in the Elsonian zone of reworking. The entire belt is regarded as riftogenic (Goodwin, 1991).

It is interesting that the zone of plutonism under consideration was spatially associated with the aforesaid zone of formation of volcanoplutonic association extending in a wide belt for 1700 km from Texas (USA) to Ontario (Canada).

5.1.1.2 *Cordillera.* The territory of the outer zones of the present-day Cordilleran belt represented part of the craton in the Early Riphean. Here were formed aulacogens filled with thick carbonate-terrigenous deposits. In the north-west, in the region of the Mackenzie Mountains, such an aulacogen was filled with the terrigenous-carbonate Wernecke supergroup (14 km), the lower part of which is equivalent to the Hornby Bay group described above (Young, 1980), and is affected by intrusions aged 1500 m.y. (Goodwin, 1991) and hence could be regarded as Lower Riphean in age.

In the central part of the Cordillera lies the Belt-Purcell aulacogen (Young, 1979) filled with the same-named supergroup of sandy-clay deposits with dolomites lying on a basement aged 1.7 b.y. Major rhythms of sedimentation are distinguished in the sequence of the supergroup reaching 10–20 km in total thickness. These are: lower turbidites (Prichard-Oldridge), middle quartzites (Ravali-Kreston group), middle carbonates (Helena-Wallace formation) and upper quartzites (Missoula group). The general age range of the supergroup has been determined as 1.8–1.1 b.y. (Hoffman, 1989) and 1700–850 m.y. (Goodwin, 1991). Lower turbidites are intruded by basic sills aged 1.43 b.y. and their weak metamorphism has been dated 1.35 b.y. (Hoffman, 1989). One view holds that the entire sequence of the Belt superseries is not younger than 1.25 b.y. (Elston and McKee, 1982). Belt-Purcell formations are crumpled into gentle folds at the Eastern Kootenay orogeny (900–825 m.y.), were thrust eastward in the Cretaceous period and at present are in an allochthonous position (Goodwin, 1991).

In the southern part of the Cordillera carbonate-terrigenous deposits (1.3 km thick) of the Crystal Spring and Ben Spring formations of Death Valley in eastern California and also terrigenous formations of Ancompagre in south-western Colorado (2450 m) with an age range of 1600–1435 m.y. probably belong to the Lower Riphean.

5.1.2 East European Craton

The Early Riphean is marked by a continuation of cratonisation of the Baltic Shield, its Gothian tectonothermal reworking and the birth of several aulacogens within the territory of the present-day Russian Platform.

5.1.2.1 *Baltic Shield.* Formation of the Trans-Scandinavian granite-porphyry belt (Smäland-Värmland) of granosyenites, monzonites, granodiorites and metavolcanics (1750–1600 m.y.) set off from the west by a major tectonic zone of Protogine (Svecokarelian or Gothian front) from the south-western tectonothermally reworked province of Scandinavia was completed right at the commencement of the Early Riphean. The reworked province is mainly made up of Gothian granitoids (1750–1500 m.y.) represented by tonalite-granodiorite gneisses. Gothian plutons were reworked in the course of the Hollandian (1500–1400 m.y.) and subsequent orogenies. The rocks were metamorphosed predominantly to amphibolite facies while charnockites and granulites are present on the west coast of Sweden. The Subjotnian complex (Dala, Los Homra, Amal etc.) of Sweden synchronously associated with Gothian granites is represented by felsic and mafic metavolcanics (porphyries), quartzites, schists and conglomerates aged about 1650 m.y.

The birth of the major mylonite fracture zone running in a meridional direction for a distance of 450 km between Telemark, Norway and the west coast of Sweden falls in the Early Riphean. Bimodal anorogenic magmatism in the form of intrusions of rapakivi granites accompanied by gabbro and anorthosites (1700–1540 m.y.) is extensively manifest in the central part of the shield and in the southern part of the Svecofennian province (Finland and Sweden).

The Ladoga aulacogen is situated on the south-eastern slope of the shield. In the sequence of this aulacogen, lake region and Salma suites of red sandstones and conglomerates overlain by basalts aged 1350 and 1500 m.y. correspond to the Lower Riphean.

5.1.2.2 *Territory concealed under the cover of the Russian Platform.* The birth of some aulacogens of the East European Craton falls in the Early Riphean (Fig. 42). In the Pachelma aulacogen Lower Riphean forma-tions are represented by red conglomerates and sandstones of the Kaverino series (up to 970 m); in the Pavlov-Posad aulacogen by the Ramenskoye series of red sandstones, shales and siltstones (over 1.5 km); in Orsha by the Rogachev series (240 m); and in Krestets by the lower part of the same-named suite containing an intrusive sheet of gabbro-diabases with an iso-tope age of 1.35 b.y. According to Aksenov et al. (1978), the Lower Riphean is also present within the Kama-Belaya and Abdullino aulacogens.

5.1.3 Preuralides

Works of Ivanov (1979) and Kurbatskaya (1985) demonstrated the rift nature of the trough that existed at the site of the western slope of the central and southern Urals in the Riphean and its ensialic character. In the southern Urals a stratotypical sequence of the Lower Riphean (Burzyan) is seen in the Bashkirian anticlinorium. The Burzyan series is divided into three suites (from

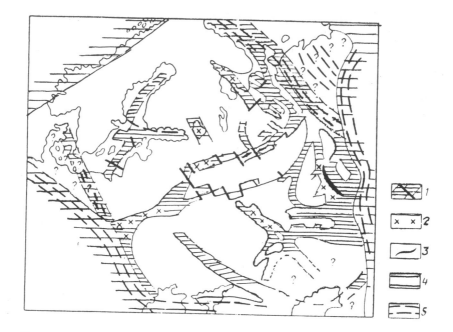

Fig. 42. Riphean aulacogens of the East European Craton (after Milanovsky, 1983).

1—aulacogens; 2—manifestations of Riphean and Early Vendian magmatism; 3—inversion uplifts in aulacogens; 4—Late Proterozoic geosynclinal regions; 5—Baikalian fold deformations.

bottom upwards): Aya (1200–2200 m)—conglomerates, quartz-sandstones, shales and mafic volcanics; Satka[1] (2000–2400 m)—dolomites, carbonaceous shales, limestones and marls; and Bakal (1350 m)—carbonaceous shales and dolomites. This series rests with major unconformity on gneisses of the Taratash complex and granites aged 1625–1590 m.y. (Rb-Sr) intruding them. The Satka suite is intruded by rapakivi granites (Berdyaush complex) dated 1348 ± 13 m.y. (Rb-Sr) and 1355 ± 5 m.y. (U-Pb). This data shows that the age of the Burzyan series falls in the range 1650 ± 50 to 1350 ± 50 m.y. (Keller et al., 1984). In this series the complex of stromatolites and microfossils has been well studied (in: *Riphean Stratotypes. Stratigraphy and Geochronology,* 1983).

The presence of the Lower Riphean in the central Urals is problematic. In the Near-polar and Polar Urals, Man'khoben and Shokur'ia suites (up

---

[1] According to the data of the Syktyvkar Interdepartmental Conference on the Upper Precambrian of the Northern European USSR (1983). In: *Upper Precambrian of the Northern European USSR* (1986).

to 1.6 km) of quartzites, sandstones and amphibolites pertain to the Lower Riphean.

5.1.4 Kazakhstan-Tien Shan Fold Region

Views on the Late Precambrian evolution of the region are quite contradictory. Zaitsev (1984) assigned a significant role to the succession of geosynclinal Riphean troughs from the Early Proterozoic. According to Apollonov and colleagues (in: *Problems of Kazakhstan Tectonics,* 1981), break-up and destruction of the wholly consolidated pre-Riphean supercraton occurred in the Riphean. In any case, the following definitive palaeotectonic environment has been described somewhat tentatively for the Early Riphean: existence of arcuate elongated geosynclinal troughs separated by Early Precambrian consolidated massifs. The position of these troughs and massifs differs in detail in the existing schemes of various authors but a general pattern is discernible. According to Zaitsev (1984), at this time there were major Ulytau-Terskei and Ubogan-eastern Kazakhstan geosynclinal systems and many less larger zones separated by the Balkhash, Kokchetav-Nyaz, southern Turgay, Syr Darya and other massifs (Fig. 43). The Issedon Early-Middle Riphean complex filling these geosynclinal troughs sometimes acquired miogeosynclinal and sometimes eugeosynclinal features. The terrigenous-carbonate sequence extended west of the Kirghiz Range and into the Ichkeletau and Talas ranges in Karatau-Talas zone. In the latter zone it attained a thickness of 4–7 km (Karabura suite, Ichkeletau series etc.).

The eugeosynclinal type of sequence comprising porphyroids, basalts, rhyolites, jaspilites, conglomerates, limestones, tuffs, shales and quartzites, is represented by the Bozdak series of Ulytau and Sarysu-Teniz water divide (3.5 km), Sarybulak Kirghiz-Terskei zone etc. According to determinations made by A.A. Krasnobaev by the α-lead method, clastic zircons from porphyroids of the Bozdak series are aged 1475 ± 150 m.y.

5.1.5 Altay-Sayan Fold Region

As in other regions of Siberia, in the Late Precambrian this territory served as an arena of destruction and disintegration of a vast palaeocraton that formed part of Pangaea I (Fig. 44). Many researchers place this process in the Early Riphean. This might be more definitively said only of the Eastern and Western Sayans, however. The Early Riphean age of the Kuvai and Chatyrly series developed in the Ingyzhei zone along the Main Sayan fault and Dzhebash series of Western Sayan has been demonstrated by the dating of organic remains. Moreover, the Kuvai series falls below the Karagas series, probably belonging to the Middle Riphean, and

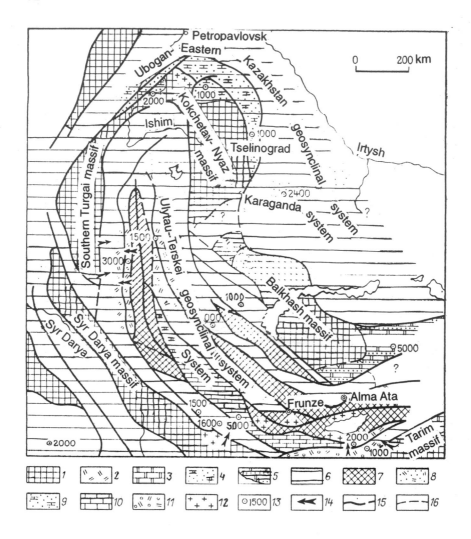

Fig. 43. Palaeotectonic scheme of the Issedon stage [of evolution—N.A.B] of the Kazakhstan-Tien Shan geosyncline (up to 1100 ± 50 m.y.) (after Zaitsev, 1984).

1—residual median massifs; 2 to 5—troughs superposed on median massifs and filled with 2—rhyolite-porphyroid series, 3—carbonate formations, 4—carbonate-terrigenous-schist and 5—quartz-schists and carbonates; 6—geosynclinal troughs; 7—geoanticlinal uplifts; 8 to 11—geosynclinal formations (8—rhyolite-porphyroid-schist, 9—sandstone and carbonate-schist, 10—carbonate and sandstone-carbonate, 11—rhyolite-porphyroid and porphyroid-porphyritoid conglomerates); 12—regions of granitisation; 13—thickness (in m); 14—direction of supply of clastic material; 15—deep faults; 16—assumed boundaries.

is made up of thick (6–10 km) volcanosedimentary strata with gabbroid and alpine-type ultramafic bodies. The trough north of Tunka basin filled with the

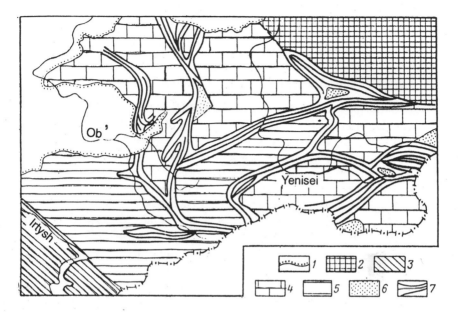

Fig. 44. Scheme showing structure of the Upper Precambrian of Altay-Sayan region (after Borukaev, 1985).

1—boundary of West Siberian Platform cover; 2—Siberian Craton; 3—Ob-Zaisan zone; 4—carbonate strata; 5—clay deposits; 6—outcrops of pre-Riphean formations; 7—ophiolite zones.

Ilchir series and containing ultramafics too is probably of a similar type. At the same time, troughs were formed in which 'miogeosynclinal' type sequences have accumulated. Among them are the Tumanshet trough of Eastern Sayan filled with the same-named series of terrigenous-carbonate rocks dislocated and metamorphosed to amphibolite facies and containing Early Riphean organic remains (Dodin, 1979). The Iya-Urin graben is made up of essentially sedimentary rocks with felsic volcanics in the lower part of the sequence. The Sayan-Altay miogeosynclinal trough is made up of thick (up to 12 km) rhythmically alternated terrigenous series with horizons of lime-stones, quartzites and volcanics (in the eastern part) similar to the Dzhebash and Terekhta series. The rocks have been metamorphosed from greenschist to amphibolite facies, complexly dislocated and containing migmatites and granites. Simultaneous with geosynclinal complexes, the Early Riphean plat-form cover of carbonate sediments with bands of mafic and felsic volcanics was formed in ancient massifs of the Biysk-Barnaul and Khakas type sepa-rating the mobile zones. The Bitu-Dzhida zone of the Khamar-Daban range, trough of Sangilen massif and Khoral zone of Tuva also reveal a similar sequence. Thus within the Altay-Sayan region, ensimatic and ensialic rift structures developed in the Riphean side by side with the platform cover (Fig. 44).

5.1.6 Fold System of Yenisei Range

The origin of this structure is associated with rifting appearing in the meridional zone and manifest in the intrusion of dykes, stocks and sills of the highly alkaline Indygli complex in the sialic basement and formation of bimodal volcanics in association with shales and tuff sandstones in the lower Korda suite. Bodies of the Indygli complex (Fig. 45) are traced as bands up to 110 km long and 5–8 km wide along the boundaries of the outer and inner zones of the Yenisei Range (Postel'nikov, 1980). In the rift trough formed, accumulation of the Korda suite of the Sukhopit series commenced evidently in the Early Riphean (*Resolutions of the All-Union Conference* . . ., 1983). The Sukhopit series is made up of actinolite-biotite microcrystalline schists, quartzites and phyllites to a thickness of 600–2200 m. The rift nature of the Late Precambrian geosyncline of the Yenisei range is demonstrated by the formation of dykes and sills of the highly alkaline Indygli complex in the lower Korda suite under conditions of extension. The Korda suite comprises bimodal volcanics in a band 110 km long and 5–8 km wide (Postel'nikov, 1980). In the present-day structure the western flank of the Yenisei range system is concealed under formations of the West Siberian Platform but according to the data of deep seismic sounding, this flank is well perceived in the form of the Kass salient of the ancient basement, thus confirming the intracratonic nature of the mobile zone under consideration.

Fig. 45. Tectonic scheme of the Yenisei Range (after Postel'nlkov, 1980).

Basement of Riphean geosynclines. *Outcrops of pre-Riphean sialic crust:* 1—Kan series, granite gneisses of Bogunaer complex (AR); 2—Yenisei series (PR); 3—Tarak (1800±100 m.y.) and Girev gneiss-granite complexes; 4—Teya-Abalakov series (PR^2_1—PR_1 ?), possibly even much older formations in the north-west: salients (?) of melanocratic basement; 5—ultramafics (Surnikha complex), gabbroids and ultramafics (Borisikhin complex).

Riphean geosynclinal formations. *Rocks of early geosynclinal stage* (Sukhopit cycloma—PR_{1-2}): 6—orthoamphibolites (Indygli complex); 7—sedimentary; 8—mafic volcanics; 9—granite-gneiss formation (Teya complex), 1 b.y. *Rocks of late geosynclinal stage* (Tunguska cycloma—PR_{2-3}): 10—sedimentary; 11—volcanics and subvolcanic bodies of different compositions (Tokmin complex and others); 12—formations of granite batholiths (Tatar complex, 850 ± 50 m.y.). *Rocks of orogenic stage* (Chingasan and Chapa cycles—PR_{3-4}): 13—sedimentary without subdivisions; 14—olistostromes; 15—trachybasalt formations (including diabases of Vedugin complex); 16—granite-leucogranite formations (Glushikhin complex); 17—granite-granosyenite formation (Noibin complex); 18—alkali-basalt and alkali-syenite formation (central Tatar complex and others); 19—*rocks of platform cover* (Phanerozoic).

Tectonic structures: 20—tectonic sutures; 21—faults; 22—overthrusts.

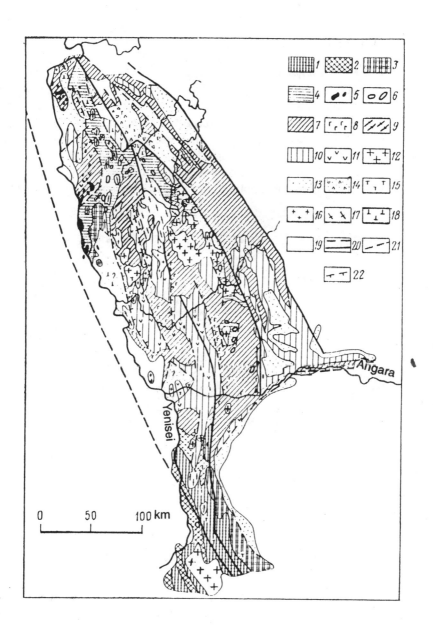

5.1.7 Siberian Craton

Here a generation of aulacogens and, unlike in the East European Craton, the platform cover associated with them in structures of the syneclise type, began forming in the Early Riphean (Fig. 46).

5.1.7.1 *Uchur-Maya region and southern part of the craton.* An example of the syneclise-type structure is the Uchur basin in the south-eastern part of Siberia whose sequence has been well studied as a Siberian hypostratotype of the Riphean (in: *Riphean Stratotypes. Stratigraphy and Geochronology,* 1983). The Uchur series (1190 m), filling a large basin, corresponds to the Lower Riphean. It is made up of shallow-water deposits, i.e., sandstones, siltstones, shales and dolomites, which are often rhythmically interstratified. This series with its content of organic remains and age determined

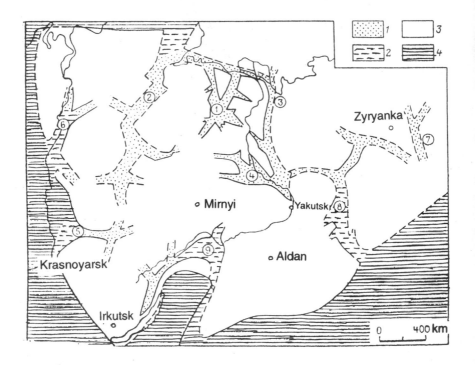

Fig. 46. Riphean aulacogens of the Siberian Craton and Verkhoyansk-Chukchi region (after Shpunt, 1984).

1—aulacogens; 2—aulacogens with considerable thickness of deposits; 3—pre-Riphean basement; 4—geosynclinal regions. Circled numbers—names of aulacogens: 1—Bilir-Udja; 2—western Anabar; 3—western Verkhoyansk; 4—Linda; 5—Irkineevo; 6—Turukhansk; 7—Kolyma region; 8—Yudoma-Maya; 9—Lena-Patom (Urik).

on glauconite (1360 to 1570 m.y.) is comparable to the Burzyan series of stratotypical sequence (Keller et al., 1984; *Riphean Stratotypes. Stratigraphy and Geochronology,* 1983).

Part of the Yudoma-Maya aulacogen east of the Uchur basin, more precisely its northern margin in the region of Gornostakh range, also falls in the Early Riphean. Lower Riphean rocks are not known outside this range. Here the Uchur series is characterised by a very fine composition of terrigenous rocks and considerable thickness (up to 3.2 km). Initially, this basin did not extend to the southern part of the Yudoma-Maya trough and had a north-easterly orientation (in: *Riphean Stratotypes. Stratigraphy and Geochronology,* 1983). At the end of the Early Riphean, structural reorganisation occurred in the region accompanied by folding in the trough and uplift in the Uchur basin.

The Patom-Viluy system of aulacogens connected with the Baikal-Patom zone by the Urin aulacogen began forming also in Early Riphean time in the southern part of the craton.

5.1.7.2 *Northern part of the craton.* Early Riphean rifting has been noticed in the northern part of the Siberian Craton (see Fig. 46). This region was studied in detail by Shpunt (1984) who delineated the major Verkhoyansk, Udja and Maiero-Kheta palaeorift zones with a general north-westerly strike and their associated western and eastern Anabar, Olenek and other regions of distribution of numerous grabens with a north-west and north-easterly strike. In the sequences of aulacogens surrounding the Anabar massif, in the Udja and Olenek uplifts, the Lower Riphean sequences (up to 3 km) lie unconformably and with a hiatus on the ancient basement. The lower parts of these sequences are represented by significantly terrigenous, variegated sandy-clay formations (Mukun and Sagynakhtah suites etc.) and their upper parts by dolomites and limestones containing stromatolites (Ulan-Kurgun and Khutingda suites and their analogues). Sedimentation was not restricted to the territories of aulacogens but extended to the adjoining regions of the frame, i.e., there was a 'spread over' of the cover. A characteristic feature of the Riphean formations of this region, according to the data of Shpunt (1984), is the presence of a considerable amount of volcanic material, mainly in the form of tuffs and tuffites with distinctly manifest potash specialisation.

Dykes of mafic composition and volcanic pipe systems of central and fissure types are widely developed in the south-eastern part of the Anabar uplift.

The Lower Riphean has not been established in the sequences of the central Siberian aulacogens but its presence should not be ignored within the deep Irkineevo aulacogen and other structures concealed under the cover of the Tunguska syneclise. Phyllites and schists of the Ludov suite of Igarka region are tentatively placed in the Lower-Middle Riphean (*Resolutions of the All-Union Conference* . . ., 1983).

5.1.8 Baikal-Vitim Fold Region

Conventionally the regional structure is differentiated into external (mio-geosynclinal) and internal (eugeosynclinal) belts or zones (Salop, 1984). The former zone is developed on the margin of the Siberian Craton, extends in a narrow band along western Peri-Baikal and in the north involves the Patom upland including Mama-Bodaibo region. A system of troughs separated by massifs of Early Precambrian crust arose as a result of the destruction of the Epi-Karelian basement within the territory under consideration at the beginning of the Riphean. The largest of such massifs separating the Baikal and Mongolia-Okhotsk regions were the Malkhan-Yablonovoy zone and also Muya and Gorgan blocks made up of highly metamorphosed and repreatedly reworked Archaean and Lower Proterozoic formations. The Baikal block, delineated in many schemes as a salient of the Archaean basement, is made up of zonal-metamorphosed rocks of the Olkhon series constituting a single complex with formations of the Sarma series of the Primorie range (Bozhko and Demina, 1974) whose age is not wholly clear. Evidently the ancient sialic crust is widely distributed in the Baikal province but concealed by granitoid fields of the Barguzin complex, as also much younger forma-tions. Fold zones located between the basement salients bear intracratonic character. In the composition of formations constituting them, these fold zones have been classified by researchers into 'zones with eugeosyncli-nal and miogeosynclinal regimes of development' (Mitrofanov et al., 1984) or 'structures of eugeosynclinal and miogeosynclinal type' (Bulgatov, 1983). Such a division is justified on the whole but there is much that is not yet clear in the stratigraphic correlation and age of rocks constituting these zones.

Let us first examine the outer zone disposed on the margin of the Siberian Craton and extending through the western Peri-Baikal region and Patom upland (including Mama-Bodaibo region). In this zone the Sharyzhalgai and Chuya salients of the cratonic basement represent the oldest Archaean rocks. A complex of molassoids, quartzites and subaerial volcanics of the Domugda, Malaya-Kosa, Khybelen and Chaya suites of the Akitkan series rests directly on this basement. In the western part of the region volcanics are accompanied by intrusions of rapakivi-type granitoids of the Primorie complex aged 1900 m.y. The distribution of these formations in a northerly direction was recently established by drilling in the north under the cover of the Siberian Craton. This together with the composition of rocks, indicates that the Akitkan complex represents a typical example of volcanoplutonic association, characteristic of the 'cratonisation' epoch manifest at the end of the Early Proterozoic to commencement of the Riphean. Volcanism was maximal in the Peri-Baikal zone. The available values of absolute ages help place these formations in the upper part of the Lower Proterozoic (Salop, 1982) or Lower Riphean (Belichenko, 1977; Gurulev, 1976). Conglomerates, gritstones, sandstones, shales and metadiabases of the Anay suite (1.5–2.0 km), constituting the Anay peak massif, also fall in the Lower Riphean (*Resolutions of the All-Union Conference* ..., 1983). The so-called three-member Baikal complex (Goloustnaya, Uluntui and Kachergat suites) of Upper Riphean age was deposited right on Primorie granites and formations of the Akitkan series (Shenfil', 1991).

The age of the Teptorga series of Patom upland, usually regarded as Lower Riphean (Mitrofanov et al., 1984) or correlated with the porphyric formation of the Akitkan series (Salop, 1982), has not been conclusively resolved. It is relevant here to recall the argument of Gurulev (1976) against such a correlation by pointing out that the volcanics of the Akitkan series did not undergo regional metamorphism as did the Teptorga series and its analogues. Gurulev placed them below the Olokit and Mama series.

According to recent studies (Shenfil', 1991), the Teptorga horizon of the stratigraphic scheme of South Siberia comprising sandstones, conglomerates, shales and siltstones of the Purpol suite (1500 m) lying unconformably on Lower Proterozoic formations, belongs to the Lower Riphean, which indicates probable flooding in the period of the Baikal-Vitim geosyncline disposed south of the interior belt (Fig. 47) and characterised by discontinuous outcrops of supracrustal formations among vast fields of granitoids.

The above zone is made up of regionally metamorphosed (from upper granulite to greenschist facies) volcanosedimentary formations, metamorphic basalt-diabases, basalt-keratophyres and spilite-keratophyre interstratified with tuffs, greywackes, green ortho- and paraschists, quartzites and marbles. Intrusions of gabbro-plagiogranite formations are associated with these rocks. This complex has been traced from Khamar-Daban range in the south-west to the north-western part of the western Peri-Baikal region through Olkhon Island vicinity up to the northern Peri-Baikal region (Muya zone). The interrelationship of these formations with the above-discussed porphyries, quartzites and molassoids of the Akitkan series is not entirely clear although there are many references in the literature to the Akitkan rocks being overlain by formations pertaining to the Sarma series. Tyya and part of the Olokit series of the Olokit trough, the Kilyan and Muya series of the central Vitim region are correlated with the Sarma series (Bukharov et al., 1985).

Apart from obtaining reliable radiometric age data, an understanding of some geological relationships is essential for ascertaining the age of this metamorphic complex. In the central part of the western Peri-Baikal region, Primorie granites are traditionally regarded as much younger relative to rocks

200

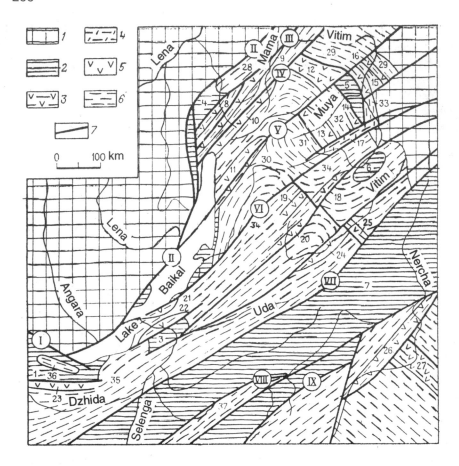

Fig. 47. Scheme showing structural arrangement of the Baikalides at the end of the geosynclinal stage of evolution (after Bulgatov, 1983).

1—Archaean-Early Proterozoic structures in the frame of the Baikalides. B a i k a l   f o l d z o n e: 2—complexes of Archaean-Lower Proterozoic forming blocks and geoanticlinal uplift; 3—eugeosynclines with basalt-rhyolite, ultramafic, gabbro and plagiogranite complex (ophiolite zones); 4—eugeosynclines with rhyolite and plagiogranite complexes; 5—assumed eugeosynclines; 6—miogeosynclines; 7—faults. Arabic numbers in the scheme: B l o c k s a n d   g e o a n t i c l i n a l   u p l i f t: 1—Khamar-Daban; 2—Baikal; 3—Mandrik; 4—Kutim; 5—Muya; 6—Amalat; 7—Stanovoy. E u g e o s y n c l i n a l   z o n e s: 8—Abchad; 9—Tyya-Mama; 10—Kicher-Mama; 11—Tompuda-Svetlina; 12—Kunkuder-Mama; 13—Gorbylok; 14—Kilyan-Irokinda; 15—Yanguda-Kamenka; 16—Karalon; 17—Tsipa-Bambuy; 18—Bagdarin- Usoi; 19—China-Alakar; 20—Vitim-Amalat; 21—Kolok; 22—Burlina; 23—Bitu-Dzhida; 24—Kholoi-Vitim; 25—Yumurchen; 26—Aga-Shilka; 27—Onon. M i o g e o s y n c l i n a l z o n e s; 28—Mama; 29—Delun-Uran; 30—Katera; 31—Uakit; 32—Bambuy; 33—Shaman; 34—Ikat; 35—Zunmurin-Temnik; 36—Utulik; 37—Kunalei.
M a j o r   f a u l t s (Roman numerals): I—Main Sayan; II—Baikal-Viluy; III—Olokit; IV—Mama-Kunkuder; V—Tompuda-Nerpa; VI—Argada-Bambuy; VII—Tungui-Kannging; VIII—Chikoi-Ingoda; IX—Chikokan.

of the Sarma series and its analogues. The same is also true of the volcanosedimentary formations of the Akitkan series regarded as molasse. At
the same time, the volcanoplutonic association of Primorie granites, i.e., of
Akitkan, represents a typical anorogenic association which developed in all
the shields at the end of the Early Proterozoic (see Chapter 4). In the Peri-
Baikal region this association extended to the margin of the ancient Siberian
Craton which was subjected to destruction in the Riphean and two elements
of the Earth's crust with different origin are tectonically joined here. In this
case metavolcanics and metasedimentary formations of the Sarma series
and its analogues formed in a rift structure could well be of Riphean age. The
relationship of these metamorphic series of the Pèri-Baikal region with the
Muya ophiolite complex delineated by Klitin and colleagues (1975) is not
wholly clear. This complex represents a vertical succession of gabbroids
containing relics of ultramafic rocks and amphibolites, amphibolite schists
and metavolcanics. The latter are often combined into the Nurundukan,
Muya and Kilyan series or the Muya complex correlated with the Sarma
series of western Peri-Baikal. Usually the contact of the gabbro-amphibolite
part of the complex with the metavolcanic is tectonic. Plagiogranites of the
Muya plutonic complex are distributed among the aforesaid rocks. These
series constitute narrow zones of the 'eugeosynclinal type' in the contemporary structure, which are controlled by faults (Bulgatov, 1983) and are
located between fragments of the ancient basement like those of the Muya
block, forming together with the latter and zones of the 'miogeosynclinal
type' a Precambrian mosaic-block structural plan of the region (Fig. 47).

The Early Riphean age of the Muya complex is supported by some
radiometric age data of the Tyya-Mama zone of paraschists at 1500 m.y.
(U-Pb) and andesite porphyries of the Yanguda-Kamenka zone at 1448 m.y.
(Rb-Sr) (Bulgatov, 1983). But these ages have not been fully confirmed.

Dobretsov (1982), who studied the gabbro-ultramafic association of the
Muya zone, concluded that it differs from classic ophiolites and placed it
among metamorphosed Lower Riphean analogues of ophiolites of the Red
Sea type.

Zones of the second (miogeosynclinal according to A.N. Bulgatov and
G.L. Mitrofanov) type also represent linear structures isolated among fields
of reworked basement. They are made up of sandy-shale and terrigenous-
carbonate strata with a sharply subordinate amount of volcanics. Rocks are
metamorphosed from greenschist to amphibolite facies and are intruded
by numerous granitoid bodies. A.N. Bulgatov delineated 10 structures of
this type in the Transbaikalian region: Mama, Katera, Delun-Uran etc.
G.L. Mitrofanov placed the Baikal-Patom among them also (in which
terrigenous formations and volcanics of the Medvezhev suite accumulated
in the Early Riphean), as well as the Khamar-Daban and Sludyanka troughs.
The platform complex of shallow-water formations (270–1000 m), the Anay,

Chukchi suites etc., formed synchronously in sequences not affected by rifting (Mitrofanov et al., 1984).

Most interpretations of the tectonic development of the Baikal-Vitim region are based on geosynclinal concepts (Salop, 1984; Bulgatov, 1983) as well as rift trough models (Mitrofanov et al., 1984; Rile, 1991).

Recent studies (Gusev et al., 1992) in the Muya zone of northern Peri-Baikal identified an island-arc complex among the volcanics of the Kilyan suite and also fragments of oceanic crust in the form of ophiolites of the Paramaka massif. This was proved by the investigations of N.A. Bozhko in the Kunkuder-Mama zone. Correspondingly, the evolution of the region is interpreted on the basis of the Wilson cycle. Evidently most zones of the 'eugeosynclinal type' are of island-arc nature while the 'miogeosynclinal type' represents complexes of old back-arc basins.

Based on the presence of diverse units separated by ophiolite sutures in the structure of the inner zone, we proposed an accretionary model of evolution of the Baikal-Vitim region (see Fig. 56). Based on these concepts and also taking into consideration the above discussion, it may be suggested that an oceanic basin existed here already in the Early Riphean and that the formation of island arcs began in it. This opinion is further confirmed by recent determination of the age of the main regional metamorphism and granite emplacement, which also falls in the Early and Middle Palaeozoic.

5.1.9 Sino-Korea Craton

As a result of Luliang folding (2000–1850 m.y.), the basement of the craton was consolidated at the end of the Early Proterozoic. But already in the Early Riphean this basement experienced destruction and rifting. Formation of aulacogens commenced with a predominant north-easterly strike and filled with mafic volcanics, red terrigenous formations and carbonate rocks. It has been suggested (Goodwin, 1991; Qian Xianglin and Chen Yaping, 1985) that two of the largest of these aulacogens, viz., Zhongtiao and Yan-Liao, formerly represented a common palaeorift system intersecting the craton. The Yanshan aulacogen has been best studied in the Jixian region where a classic Upper Precambrian sequence is found. The Lower Riphean in it corresponds to the upper part of the Changcheng system (1500 m). Many geologists delineate it as an independent Nankou system (Keller et al., 1984), comprising two upper suites of the Changcheng system: Dahongyu (400 m) made up of sandstones and felsic volcanics and Gaoyuzhuang (1600 m) represented by carbonate rocks. The lower age limit of the entire Changcheng system has been determined as 1.85 b.y.; the upper boundary is usually dated 1.4 b.y. For rocks of the Gaoyuzhuang suite, the available age values are: 1622–1680 (K-Ar) and 1434 ± 50 m.y. (Pb) (Keller et al., 1984).

Another palaeorift is located in western Henan, which opens in the southwest into the shelf sea joining with the Qinling marine basin. The lower part of the sequence of the western Henan aulacogen comprises the volcanoclastic Xianghe group correlated with the Dahongyu suite of the Changcheng system. In the southern continuation of the aulacogen, the Xianghe volcanics are overlain by the Gaoshanhe terrigenous formation belonging to the upper Changcheng. Farther south, metavolcanics, quartzites and shales of the Kuanping group correspond to the Lower-Middle Riphean formations of western Henan within the Qinling fold belt. These formations represent a complex of the southern ancient margin of the North China Craton. On the northern margin of the Tarim Craton, in the Quruktagh aulacogen, Yangjibulak series of quartzites and phyllites; in central Tien Shan the bottom part of the Kawabulak group (marbles and quartzites); and in the Beishan mountains north of Alxa, metamafics of the Baihu group—all these probably belong to the Lower Riphean. Riphean carbonate strata are widespread on the northern slope of Kunlun.

Cratonic carbonate formations are widespread also in the Yining and Qaidam massifs.

It is quite possible that the Amnokkan and other palaeorifts of the Korean part of the craton also belong to the Lower Riphean.

5.1.10 South China Craton and Indosinian Region

South of the Sino-Korea Craton, within contemporary south-eastern China, Vietnam and Laos, the overall tectonic environment was characterised by the existence of an extensive oceanic basin designated as the Qinling ocean and representing probably a gulf of the Proto-Pacific. Isolated blocks of pre-Riphean consolidation (microcontinents) existed within it. The Proto-Yangtze block, comprising Central Sichuan and the Kam-Yunnan massif, as well as the Kontum, Sinoburman, Vietbak and others were the largest of such blocks.

The Proto-Yangtze block was surrounded from the north and west by passive margins and by a system of island arcs, deepwater troughs along its south-eastern margin. Accumulation of a cratonic cover proceeded in the interior parts of the massif.

Complexes of passive continental margin were widespread in the south-western part of the region east of Lujiang. These are known as the Kunyang group in eastern Yunnan and the Huili group in southern Sichuan. The general sequence of beds at Kunyang town is divided into three parts. The lower four formations are 2500 m in thickness and are represented by schists, dolomites, shales and sandstones. The central four formations are made up of rhythmic terrigenous deposits containing tuffs and pass into carbonate formations. The upper part is represented by carbonate rocks and intermediate-felsic volcanics. Based on determination of U-Pb ages on stromatolites which fall in the range 1760–780 m.y., the age of the Kunyang and Huili groups corresponds to the upper Changcheng, Jixian and Qing-baikou (Yang et al., 1986). Chinese geologists justifiably place these complexes among miogeosynclines. However, these complexes should exclude the upper volcanosedimentary part formed evidently under conditions of an active continental margin.

The lower carbonate-terrigenous subgroup of the cratonic cover of the Shennungjia group formed in western Hubei in the northern part of the Proto-Yangtze block probably falls in the Lower Riphean.

Complexes of ancient island arcs extend in the south-eastern part of China along the Jiannan massif for a distance of about 1500 km from south-west to north-east from northern Guangxi and eastern Guizhou to Zhejiang. In the Fanjingshan Mountains these ancient island arcs include the Fan-jingshan group represented by thick beds of volcanosedimentary rocks, i.e., sandstones, tuffs, spilites, agglomerates and keratophyres, containing ultra-mafics. The Sibao group corresponds to it in the Yuanbao mountains (Yuvan-dashan) of northern Guangxi. The lower Baiyandin formation (4000 m) of the Sibao group is represented by terrigenous flysch while the middle Jiux-iao is sandy-clay with spilites. The upper Wentong formation is made up of metasandstones, spilites, keratophyres, ignimbrites, tuffs and jasperoids. The sequence is topped by a second flysch member (1500 m). The Sibao group is intruded by granites aged 1065 and 1109 m.y. with low $^{87}Sr/^{86}Sr$ ratios of 0.7001 (Yang et al., 1986). At the same time, an isotopic age value of 1401 m.y. is indicated on phyllites of this group (Il'in, 1986). The tectono-magmatic events of Dongan with which the formation of the oldest ophiolites among the Sibao group of rocks in the Yuvandashan mountains is associated also took place at this same level (Zhang et al., 1984).

In the central part of the Jiannan massif covering the northern part of Xiangxi, the lower Shuangqiaoshan subgroup associated with ophiolites and comprising tuffaceous sandstones, shales and volcanics is correlated with the Sibao and Fanjingshan groups. The Rb/Sr method gave an age value of 1410 m.y. for tuffaceous sandstones (Yang et al., 1986).

Based on this data the Sibao and Fanjingshan groups should be regarded as Lower-Middle Riphean.

Ophiolites in the Yuvandashan mountains contain harzburgites, serpen-tinites, gabbro and gabbro-diabases, pillow basalts and deepwater siliceous deposits. The presence of peridotites and pyroxenites in association with spilites and metamafics has been noticed in the Fanjingshan group of east-ern Guizhou and in the lower part of the Shuangqiaoshan group of northern Jiangxi (Yang et al., 1986).

Based on a study of the tectonics of the South China and Indosinian blocks and spatial patterns of distribution of Early-Middle Riphean ophio-lites, Le Zui Bath (1985) concluded a new phase of advance of the Proto-Pacific on the formerly consolidated Early Precambrian blocks involving the latter in geosynclinal development as microcontients—Sinoburman, Kontum, Vietbak etc.

The Khamdyk ophiolite complex of Early-Middle Riphean age was thrust on the northern margin of the Kontum salient in Indochina and overlain by Late Riphean molasse. The complex is represented by an association of cummingtonite amphibolites, gneisses, marbles, serpentinised ultramafics and metagabbro.

The above data bears witness to important events occurring in this south-east Asian region falling between Laurasian and Gondwana rows of ancient cratons. These events point to extensive processes of ophiolite formation associated with the extension of this section of Pangaea I and secondary formation of oceanic crust, followed by involvement of this crust in subduction and formation of active margins, i.e., with manifestation of the Wilson cycle. These events also bear witness to commencement of the opening of the eastern segment of the Proto-Tethys.

5.1.11 South American Craton

Cratonisation of the crust that set in at the end of the Early Proterozoic after completion of Transamazonian diastrophism continued into the beginning of the Early Riphean. This is supported by appropriate (about 1500–1450 m.y.) Rb-Sr ages of the rocks of the Uatuma volcanoplutonic complex. Intrusion of Parguaza rapakivi granites (1545 ± 20 m.y. by U-Pb and 1531 ± 39 m.y. by Rb-Sr), largest in the world (30,000 km^2), occurred in the north-western part of the present-day Guyana Shield (Goodwin, 1991). The volcanoplutonic activity of this period is called the Paraguaçan or Parensian episode in Brazil and was accompanied by isotope rejuvenation of the basement rocks. It was also accompanied by processes of graben formation and accumulation of typical volcanosedimentary cratonic cover that has survived in the form of isolated basins. These include the terrigenous rocks of the Gorotire group of the Central Brazilian Shield overlying the volcanics of the Uatuma complex and intruded by mafic dykes aged 1475 m.y.; Mutun-Parana (1700–1400 m.y.), Beneficiente (1600–1400 m.y.), Caiabis (1400–1200 m.y.) groups and others. Synchronous cover deposits accumulated in the Chapada-Diamantina basin on the Sao Francisco eocraton.

Tectonothermal reworking (TTR) processes manifest in the formation of 'mobile belts' occurred simultaneous with anorogenic tectonic activity. The TTR belt Rio Negro-Juruena with a width of about 500 km extends along the western margin of the Guyana Shield in a submeridional direction. It has the form of a band of ancient reworked and 'rejuvenated' basement that experienced significant mylonitisation, cataclasis, local metamorphism to granulite facies and intrusion of syenites and phonolites in the period 1750–1550 m.y. (Goodwin, 1991; Brito Neves et al., 1990).

Formation of another TTR belt, the Sao Ignacio, took place at the end of the Early Riphean. This belt represents an ancient constituent of the northeastern part of the much larger Rondonian belt, up to 500 km wide, extending along the western margin of the South American Craton from Uruguay to Venezuela. The Sao Ignacio belt comprises reworked, regionally metamorphosed series (granulites to greenschists) and numerous granites aged 1400–1250 m.y. Tectonometamorphic and intrusive processes are associated with Sao Ignacio orogeny (1300 m.y.). Commencement of this TTR has been placed at 1510 m.y. (Brito Neves et al., 1990).

5.1.12 Geosynclinal Zones of Central Brazil

Formation and development of the geosynclinal intracratonic systems of Brazil, Espinhaco and Uruaçu, are placed in the Early Riphean. The Espinhaco fold belt, extending almost parallel to the eastern coast of Brazil from Bahia to Minas Gerais, is made up of the Espinhaco quartzite-phyllite supergroup (up to 8000 m) passing eastwards into the platform cover of the Chapada-Diamantina group (400–900 m). The intensity of fold-thrust deformation in the belt increases from west to east. Commencement of accumulation of the supergroup is placed in the period 1800–1700 m.y. The main phase of deformation and metamorphism has been dated at 1200±100 m.y. (Goodwin, 1991) and 1400–1300 m.y. (Brito Neves et al., 1990).

The Uruaçu belt (Araxaides) of Central Brazil extends south-east for a distance of more than 1000 km from Natividad (Goyas) to Guaxupe (Minas Gerais) and farther east. The lower Araxa group (1850 m) comprising metalutites, quartzites and apovolcanic chlorite schists and the upper Arai group (2240 m) comprising quartzites, metalutites and tuffs take part in the structure of the Uruaçu belt. The main tectonomagmatic events in this belt are usually placed at 1000 m.y. (Goodwin, 1991) but Brito Neves and colleagues date them at 1400–1300 m.y. by distinguishing the tectonic Espinhaco-Uruaçuan cycle in the interval 1750 to 1200 m.y., while diastrophism at the level of 1000 m.y. is regarded as post-tectonic (Brito Neves et al., 1990, 1991). Folds in the Uruaçuan belt have a north-eastern vergence towards the Sao Francisco eocraton. The Rio Grande fold belt, representing a possible continuation of the Araxaides in the south-east, is located north-east of Sao Paolo in which Vasconselos has established a similar succession of events: commencement of sedimentation around 1800 m.y. ago and the main phase of diastrophism around 1400 m.y. ago. Hailbron recorded similar geochronological data for metasediments of the Sao Joao del Rei and Andrelandia groups (Brito Neves et al., 1990). The Tocantins fold system comprising the Baixo-Araguaia supergroup (1.7–1.0 b.y.) consisting of the lower Estrondo group and the upper Tocantins group represents the probable northern continuation of Araxaides. Brito Neves and colleagues also regard the Tocantins belt as a Middle

Proterozoic (1750–1000 m.y.) mobile belt remobilised in the Brazilian cycle (Brito Neves et al., 1990). Other researchers distinguish the Araguaia fold belt comprising the Baixo-Araguaia supergroup consisting of two groups: Estrondo (quartzites and shales) and Tocantins (quartzites, shales and silicites). Mafic and ultramafic massifs are associated with these rocks (Herz et al., 1989).

Thus the Early Riphean of South America is characterised over much of the area by uplift and platform regime (Amazon Craton) and also the birth of rift structures on both sides of the present Sao Francisco eocraton. In these rift structures Espinhaco and Araxa supergroups and their analogues were formed. These events mark the first half of the Uraçuan cycle (1.7–1.3 b.y.).

5.1.13 African Craton

The main territory of Africa was a megashield in the Early Riphean. Formation of subaerial felsic volcanics, conglomerates and arkoses (Eglab and Guelb-el-Hadid series and Tiderijauin complex) in the north-west, in the region of Eglab, and also in western Hoggar was inherited from the end of the Early Proterozoic.

A platform cover was formed in small-size basins in the Zambia and Tanzania massifs in the form of terrigenous formations of the Upper plateau and Kingongolero series.

5.1.14 Intracratonic Mobile Zones of Africa

Reliable data on the existence of geosynclinal troughs of Early Riphean age is available only for Central Africa where the Kibara-Ankolean belt is located (Fig. 48). This belt extends for 1500 km from the upper courses of the Zambezi River (Angola) through Shaba province (Zaire) to the north-eastern fringe of Lake Victoria (Uganda) with a width of 100–500 km. It is made up of formations of equivalent Kibaran, Burundian and Karagwe-Ankolean supergroups lying unconformably on Lower Proterozoic formations. In the stratotypical region (Kibara mountains, Shaba province, Zaire), the Kibara is 10,500 m thick and is represented by four groups: (1) phyllites with subordinate quartzites, felsic and mafic volcanics (1700–4300 m); (2) predominantly quartzites with basal conglomerates topped by thick strata of mafic volcanics (5500 m); (3) sandstones and shales with subordinate quartzites (1500–4000 m) and (4) basal conglomerates, black carbonate shales and carbonate formations with stromatolites. The corresponding Burundian supergroup in the north-eastern belt is 12,000 m thick and is mainly made up of phyllites with subordinate bands of quartzites.

The rocks of the belt are crumpled into folds with a north-north-easterly strike and north-westerly vergence and metamorphosed mainly to green-schist facies. At places in the lower parts of the sequence, the degree of metamorphism attains amphibolite facies.

208

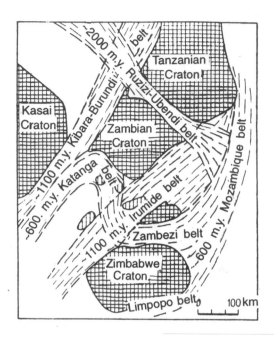

Fig. 48. Intersecting relationship between Proterozoic mobile belts in eastern Central Africa (after Kroner, 1977).

Rhyodacites interstratified with Lower Burundian deposits are dated 1353 ± 46 m.y. (Rb/Sr isochron method) and the main syntectonic granites 1280–1260 m.y. (Rb-Sr) (Goodwin, 1991).

The small systems of narrow north-north-eastern troughs presently located along the eastern coast of Lake Tanganyika, Kigoma, Itiaso and Ukinga evidently belong to the Early Riphean. This is supported by the age of the syntectonic Chimala granites in the Ukinga system.

A narrow trough existed in north Madagascar in which the volcanosedimentary formations of Daraina, Milanoa, Vohemar and Ambohipato series accumulated. Manifestation of Early Riphean non-geosynclinal TTR of the basement is also fixed here (Hottin, 1976).

The troughs described above were riftogenic in their tectonic character. A sialic basement was present in them and they were characterised by a rectilinear structural pattern with blind termination in the body of the craton. These troughs were filled with thick terrigenous complexes with bimodal volcanics, mainly of the mafic type.

The existence of sedimentary basins including those of oceanic type in southern and western Africa and also within the Irumide belt in the Early Riphean may be tentatively conjectured. The available geochronological data reliably points only to the manifestation of diastrophism at the level of

1000 m.y. in these regions. This aspect will be discussed in the next chapter. The same may be said of the Arabia-Nubian region also. Dolerite dykes aged 1.4 b.y. are known here, which according to Kazmin and colleagues (1978) may point to commencement of the opening of the Proto-Red-Sea ocean basin whose existence has been clearly established in the Late Riphean.

5.1.15 Indian Craton

The Early Riphean palaeotectonic environment in India is characterised by a predominance of cratonic conditions. A sedimentary platform cover was formed in the extensive basins of the craton. On the basis of the identical nature of the stromatolite complexes (M.E. Raaben, pers. comm.), it may be suggested that a large syneclise existed in the central part of the Indian peninsula which later became disjointed by erosion into independent Purana basins: Cuddapah, Vindhyan, Kaladgi, Godavari, Bastar and Chattisgarh. The sequences in these basins show considerable similarity. They are filled with quartzites, shales and limestones among which layered intrusions of mafic rocks are present. The thickness of the Cuddapah complex formations ranges from 2–5.7 km. Apart from the identification of stromatolites, the Early Riphean age of these formations is supported by the age of glauconites from the Papagni, Geyar (Cuddapah group) of the Cuddapah basin (1470 ± 60 m.y.), Kaladgi series of the same-named basin (1330 ± 63 m.y.), Semri group of the Vindhyan basin (1400 ± 70 m.y.), Pokhal series of the Godavari basin (1330 ± 63 m.y.) and also lavas from the lower part of the Cuddapah system (1370±60 m.y., Moralév, 1977; 1583±147 m.y., Jackson et al., 1990).

The tectonic nature of the Purana basins has been interpreted by some researchers in relation to the formation of Riphean back-arc basins and passive margins (Kale, 1990).

In the Singhbhum region the Noamundi series was formed at the beginning of the Early Riphean. These rocks are represented by shales, mafic volcanics, tuffs and siliceous rocks with bands of hematite ores and lie unconformably on Dzhanneri lavas (1.6 b.y.). Rocks of the Noamundi series have been dated 1562 m.y. Northward, these formations transit into an intracratonic trough filled with the Singhbhum series (1550–850 m.y.). Its lower part (Porat group) has a significantly terrigenous composition (Banerji, 1977). Collision of the continental margin of the Aravalli belt with the microcontinental arc has been dated to 1.5 b.y. (Deb and Sarkar, 1990).

The Early Riphean was also marked by intense tectonothermal reworking of the basement in the Eastern Ghats charnockite belt, manifest in the formation of granitoids, pegmatite fields, 'isotopic rejuvenation' and origin of schistose zones. Deformation of the Cuddapah basin adjoining the Eastern Ghats belt (Nellore and Kammai belts) is also associated with the movements caused by tectonic reactivation of the Eastern Ghats belt. Khondalites (garnet-sillimanite metalutites), widely developed in the Eastern Ghats, are probably Riphean in age. Similar events manifest in north-western India in the Rajasthan zone of reworking of the basement where ultramafic bodies, apart from granites, formed along deep faults with a north-easterly strike.

5.1.16 Australian Craton

By commencement of the Early Riphean much of the area of the continent represented a craton, its various parts characterised by different extent of consolidation.

Formation of platform cover took place mainly in the northern and north-western parts of the craton and on Sturt plateau, southern slopes of the Pilbara and Yilgarn shields and in MacArthur basin.

The Lower Riphean is most fully represented in the western part of Sturt plateau by the lower part of the Osmond Range complex comprising Mount Parker sandstones (300 m) and Bangle-Bangle dolomites (1 km) containing Lower Riphean stromatolites (Semikhatov, 1974). Dolomites are overlain with an angular unconformity by shales aged 1128 ± 110 m.y. (in: *Geology of Western Austalia,* 1975). In the central part of the region the sequence of the Limbunia group of proximate character (1475 m) is comparable to Mount Parker sandstones and Bangle-Bangle dolomites and the Birrindudu group (6 km) in the region of Birrindudu-Tanami. Isotope age determination of the glauconite sandstones from this group ranged from 1.6 to 1.4 b.y.

In Western Australia significantly terrigenous groups, Bresnahen (4.9 km) and Mount Minnie (780 m) filling the same-named basins on the southern slope of the Pilbara Shield and the Stirling-Barren and Woodline series of similar composition on the southern part of the Yilgarn Shield, belong to the Lower Riphean.

The Lower Riphean has been established most reliably in the MacArthur basin located south of the Gulf of Carpentaria and extending for 1300 km in a sublatitudinal direction. As on Sturt plateau, thick quartzitic sandstones (Tawallah and Katherine River groups) formed initially, were replaced by clay-carbonate sedimentation (MacArthur and Mount Rigg groups) and subsequently by the red-coloured Roper group. The sequence attains a thickness of 9 to 12 km, unusual for platform cover. The Lower Riphean age of these formations was positively demonstrated by identification of stromatolites (Semikhatov, 1974) and also by isotope datings falling in the interval 1.8–1.4 b.y. M. Kralik further determined the age of the MacArthur sedimentation group as 1537 ± 52 and of the Roper group as 1429 ± 31 m.y. (Page et al., 1984).

On Gawler plateau (South Australia) conglomerates, quartzites and dolomites of the Corunna series (700 m) overlain by Gawler Range rhyolites belong to the Lower Riphean. These rhyolites have been dated 1525 ± 14 and 1529 ± 33 m.y. (Goodwin, 1991) while the age of the cement of the

Corunna conglomerates has been put at 1560 m.y. Synchronous volcanics aged 1592 m.y. were deposited in the Williama block on the eastern margin of the Adelaide fold belt.

The mobile zones of Australia in the Early Riphean are represented by aulacogens, Mount Isa intracratonic geosyncline and TTR zones of the basement.

Aulacogens were formed in the MacArthur basin at the site of graben-like subsidence of the basement. Quite a complex network of aulacogens of various magnitude has been established. The largest aulacogen, Batten, is filled mainly with deposits of the MacArthur group (thickness 3.6 km).

The Mount Isa intracratonic geosyncline lies east of MacArthur basin, directly adjoining it, and was mainly developed in the Early Proterozoic. In the Early Riphean this structure with a 'through-going' evolution, uncommon for Early Proterozoic geosynclines, was closed. It was divided by a salient of ancient Archaean basement into two troughs. In the interval 1.67–1.62 b.y., fold deformation and metamorphism occurred after formation of the Pine Creek and Fine Creek volcanics. In the western part of the trough, after intrusion of Sibella granites (1620–1544 m.y.), the terrigenous MacNamara and Mount Isa groups (thickness up to 5 km) were formed on their eroded surface. These two terrigenous groups were separated by the Mount Isa fault. An equivalent group, Mount Albert, was formed in the eastern part of the trough. Sedimentation in the geosyncline ceased with folding and meta-morphism about 1490 m.y. ago and with plutonism about 1425 m.y. ago.[2]

The eastern part of the trough experienced far greater deformation and metamorphism compared to the western. The distribution of thickness and facies was controlled by fault dislocations, some of which were oriented at a large angle to the strike of the geosyncline. Thus arose the narrow (15 to 30 km) Paradise aulacogen entering Mount Isa trough from the west. In the north-west the Mount Isa geosynclinal trough adjoined with the MacArthur basin through the Lawn Hill Platform. The eastern framing of the trough continuing in the north, according to gravimetric data is represented by Georgetown and Koen basement salients and their northern and southern continuations.

Significant regions of the continent were involved in tectonothermal reworking in the Early Riphean. The Georgetown uplift made up of ancient formations (2.49 b.y.) underwent discrete tectonothermal events characterised by metamorphism from amphibolite to granulite facies and generation of compressed isoclinal folds. In the Early Riphean Rb-Sr datings on bulk rock samples fixed the stages of reworking at 1570 ± 20 and

---

[2] Other data is available pointing to much greater antiquity of Sibella granites (1.67 b.y.), regional metamorphism and folding (1.62–1.5 b.y.) and post-tectonic granites (1.5 b.y.) (Page et al., 1984).

1470 ± 20 m.y. (Page et al., 1984). These events preceded formation of the Croydon volcanics and Esmeralda granites aged about 1400 m.y. Three stages of reworking of the ancient crust in the form of deformation and metamorphism in the interval 1660 ± 20 to 1490 ± 40 m.y. have been established in the Brocken Hill block. At the end of the Early Riphean (1.36 b.y. ago), superposed metamorphism to granulite facies manifested in the Musgrave block (central Austalis). Cratonisation was complete at the very beginning of the Riphean, i.e., 1.6–1.55 b.y. ago (Page et al., 1984). The ancient basement of the Albany-Fraser province experienced granulite metamorphism of similar age (1300±12 m.y.). Reactivation of the basement in the form of deformation and low-grade metamorphism (Anmatiera event) was established at the level of 1.4 b.y. in the Arunta block.

5.1.17 Antarctic Craton

The Antarctic territory, excluding the Antarctic peninsula, represented a craton consolidated in the Early Precambrian. In the Shackleton Range (30°W long.) metasedimentary formations aged 1446 ± 60 m.y. were deposited on the basement.

A granulite belt extends from Windmill islands (110°E long.) to Dronning Maud Land (0°). Rb-Sr isochron age determinations on whole rock in this belt fall in the range 1100–1400 m.y. Formations of the belt are intruded by charnockites (1200 m.y.), granites (1100–1400 m.y.), pegmatites and dolerites. Evidently these values reflect TTR processes.

Birth of the Indurance strike-slip fault zone in the central part of the Transantarctic mountains falls in the Early Riphean. The subparallel Late Cambrian continental margin was displaced along the above strike-slip fault zone. It has been suggested that this displacement reflected the presence of oblique subduction of the oceanic platform under the western Antarctic margin of Gondwana in the period from Early Riphean to Cambrian (Goodge et al., 1991).

5.2 TYPES OF STRUCTURES AND TECTONIC CYCLES

The Early Riphean epoch is characterised by the development of some types of tectonic structures that prevailed within the extensive sialic mass of Pangaea. Its existence, as in the case of the Early Proterozoic, has been demonstrated by palaeomagnetic data (similarity of apparent polar wander curves) and existence of transitional palaeostructures. Reconstruction of the Riphean Pangaea is a difficult task. The present authors assume the existence of a stable global structural plan in the evolution of the Earth, ignoring the random movement of continents from epoch to epoch. This premise assumes a general similarity of configuration of supercontinents and continents of different age, which is reflected in the global palaeotectonic schemes for the Late Precambrian in which the hypothetical Pangaea I resembles Palaeozoic Pangaea II (Fig. 49). The main area of Early Riphean Pangaea was uplifted and represented in general a gigantic megashield that was unevenly consolidated at the end of the Early Proterozoic. Platform cover was formed within the most stabilised segments already in the Early Riphean. At the same time, the first manifestation of Pangaea disintegration processes has been detected in south-east Asia: new formation of oceanic basins and the birth of a row of microcontinents on the margins of which calc-alkaline volcanic complexes began to form in the subduction zones. In essence, this is a unique reliable manifestation of active marginal processes in the Early Riphean. The overall palaeotectonic environment in the interior of Pangaea was characterised by development of destructive processes in the ancient crust, which led to the formation of various types of rift structures.

5.2.1 Platform Basins

Lower Riphean platform cover occupies a relatively small area and was formed in basins that preserved some specific features of Early Proterozoic protoplatform structures. Among such features are the small size of sedimentary basins and absence of associated anteclises characteristic of Phanerozoic cover. The Early Riphean platform complex accumulated directly on the ancient basement, bypassing the aulacogen stage. The tectonic regime of formation of this first of the Late Precambrian platform covers was distinguished by a relatively large contrast of vertical movements, as demonstrated by the significant thickness of formations. For example, 6 km of sediments were deposited in the MacArthur basin during the Early Riphean per se. Sandstones, quartzites, shales and siltstones containing bands of limestones and dolomites predominate in the composition of these sediments. Development of flood basalt formations and felsic volcanics is also a characteristic feature.

Platform basins joined with aulacogens and other linear structures at several places, as in Australia where the MacArthur basin joined with the synchronously developing Batten aulacogen and Mount Isa intracratonic geosyncline. Considering that the Early Riphean cover in character was intermediate between the protocover of the Early Precambrian and Phanerozoic platform cover, it would be more appropriate to term the basins in which they were deposited as protosyneclises.

5.2.2 Zones of Epicratonic Volcanoplutonic Associations

Development of these zones was inherited mainly from the end of the Early Proterozoic (Guyana and Reguibat shields and others). A specific feature of the Early Riphean complexes is the increased role of terrigenous, coarse clastic molassoid formations prevailing over volcanics, which may be

214

Fig. 49. Global palaeotectonic scheme for the Early Riphean (after Bozhko, 1988).

1—continental Early Precambrian crust; 2—platform cover; 3—felsic volcano-plutonic associations; 4—aulacogens; 5—ensialic intracratonic geosynclines; 6—ensimatic intracratonic geosynclines; 7—zones of tectonothermal reworking of the ancient basement; 8—passive continental margins; 9—oceanic crust of older age; 10—newly formed oceanic crust; 11—subduction zones; 12—Kibara fold belt; 13—boundary of continental crust.

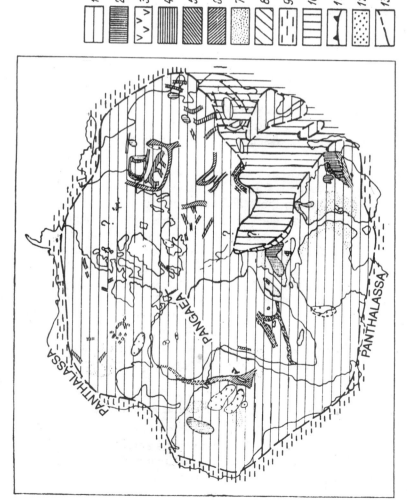

totally absent. Volcanosedimentary formations accumulated in tiny grabens, often branching and merging to form shallow depressions. Mafic and alkaline hypabyssal intrusions affected the sedimentary strata as well as the adjoining shield regions. Rocks were deformed unevenly with a predominance of brachymorphic folds and normal faults. Volcanics, tuffs and ignimbrites of rhyolite-dacite (less often andesite) composition and intrusions comagmatic with them are present in the magmatic products in the zones under consideration.

A distinct intracratonic position of these regions (Rondonia, Eglab and Gawler) is most often established. In this case the separate grabens, volcanic structures and intrusive bodies form structural zones that are generally quite isometric in plan in the background of an extensive, stabilised craton. Here it is difficult to conceive the existence of extended volcanic belts even taking into consideration the role of later erosion causing some disjointing in the distribution of molassoid strata and volcanoplutonic formations. These zones appear like prominent impregnations in the background of the gigantic area of the craton. Other regions are characterised by a more linear structural plan. Among them are a band of felsic extrusions and intrusions into the south-eastern part of the North American Craton extending for not more than 2000 km from Texas (USA) to Ontario (Canada). Zones of the first type (isometric) having died out are not succeeded by other mobile structures while the second type usually transits into much younger Riphean intracratonic geosynclines or zones of reworking.

The tectonic regime of the regions of accumulation of volcanoplutonic associations and molassoid complexes of this type is interpreted variously as orogenic, due to cratonisation, reactivated or marginal-continental. Based on a study of the tectonic status of these Lower Riphean complexes, it could be concluded that most of them are anorogenic. The magmatism of these regions was caused by crustal processes, as supported by strontium isotope determinations on granitoids, volcanics and other rocks. Evidently the formation of felsic volcanoplutonic complexes with an alkaline tendency and the accumulation of continental molassoid strata in association with subaerial volcanics proceeded on a crust less stabilised than in the regions of development of platform cover. It is significant that already in the Early Riphean complexes of this type volcanics vanish and terrigenous, often quartzose-sandy rocks predominate; these rocks accumulated in graben-like depressions. This tendency probably reflects an increase in the maturity of the crust, leading to conditions favourable for the formation of typical platform cover.

5.2.3 Aulacogens (Palaeorifts)

These structures are quite extensively developed in the Early Riphean, especially in the northern row of cratons. They do not differ essentially in inner structure from much younger and older analogues. Yet with respect to some parameters, differences do exist among them in various parts of the Pangaea.

The unusually large thickness of formations (up to 12 km) and their terrigenous-carbonate composition are characteristic of the aulacogens of North America and Australia. The Lower Riphean parts of the sequences of aulacogens of the East European Craton are less thick and are characterised by significant terrigenous composition with extensive development of red-coloured rocks. Early Riphean sedimentation of Siberian aulacogens is transgressive in character. Variegated sediments are replaced in the upper part of the sequence by terrigenous littoral-marine sediments and they, in turn, by siliceous carbonates.

In many aulacogens magmatic activity was manifest in the formation of dykes and sills of diabases. On the Siberian Craton, in the magmatism of aulacogens, which have been thoroughly studied by Shpunt (1984), an intricately differentiated complex of alkaline ultramafics-carbonatites-syenites-trachyliparites and basalts was formed in the Early Riphean.

A characteristic feature of the aulacogens of Siberia, Australia and some aulacogens of North America is their association with formations of the platform cover that synchronously accumulated in the periphery of palaeorifts. Other aulacogens (East European Craton) represent isolated structures.

The Early Riphean represented in general an epoch of the birth of most aulacogens whose evolution continued for almost the entire Late Precambrian.

5.2.4 Intracratonic Geosynclines

The birth of a new generation of structural types, i.e., intracratonic geosynclinal systems—Kibara, Uruaçu, Mauritania-Senegalese, Yenisei and others—falls in the Early Riphean. These are distinctly linear, comparatively narrow zones occupying an intracratonic position. In the nature of formations filling them and their development, these structures are divided into significantly ensialic (i.e., wholly preserving the coherence of the sialic basement) and ensimatic structures in which the coherence of the basement has been disturbed to some extent and mantle material participated in their build-up.

5.2.4.1 *Ensialic intracratonic geosynclines.* These were laid in the form of narrow elongated troughs discordant to the basement structures (see Figs. 48 and 49). Almost everywhere a sharp unconformity is noticed in the attitude of strata filling these troughs in relation to the Archaean or Lower Proterozoic complexes. The superpositioning of Late Precambrian zones on structures of the ancient basement is well expressed, for example at the site of intersection of the northern part of the Kibara-Ankolean system with the Early Proterozoic Ruzizi-Ubendi belt. The salient of the latter divides the zone of the Kibarides throughout its width into two segments—northern and southern. Cases of the inheritance of the Early Precambrian structural plan are more rare (for example, Mayombe or Mount Isa zones).

The systems under consideration have a typical pattern in the plan with parallel fault boundaries and sharp 'blind' attenuation of both ends of the systems in the body of the craton. Such a characteristic structural pattern is also typical of present-day continental rift zones. This pattern is even more evident when the mobile zones under consideration are grouped into branched systems, as in Central Africa where the subparallel Kibara-Ankolean and Irumide belts are connected by a chain of narrow Ukinga-Itiaso zones. In this case individual blocks of the basement (Zambian massif) are separated and the pattern as a whole is highly reminiscent of the present-day East African rift system with its western and eastern segments, isolation of the Dodoma block etc.

Ensialic troughs were filled with very thick, essentially carbonate-terrigenous sediments with volcanics of bimodal composition playing a subordinate role. This is well expressed on the example of the Kibarides sequence (total thickness up to 14 km), mainly made up of quartzites and phyllites with sheets of basalts and bands of graphite schists, limestones and dolomites only in the upper part. In other systems (Mount Isa) the role of volcanics is more significant. Transverse structural-formational zoning is negligible in the Early Riphean ensialic intracratonic geosynclines and the entire system is homogeneous in transverse section, ignoring the salient of the sialic basement in the Mount Isa system separating this geosyncline into two isolated troughs. Development of this system was inherited from the Early Proterozoic and it represents a unique geosyncline that experienced metamorphism, folding, granitisation and total closing in the Early Riphean, i.e., 1490–1425 m.y. ago. Other ensialic intracratonic geosynclines of this generation experienced a corresponding diastrophism at the end of the Middle Riphean and hence it is more convenient to examine these processes and the structures created by them separately (see next chapter).

5.2.4.2 *Ensimatic intracratonic geosynclines*. Evidently the birth of a new type of intracratonic trough, i.e., ensimatic geosynclines, falls in the Early Riphean. This could be said of the Central Brazilian and Mauritania-Senegalese zones, with some reservation due to the absence of reliable age data, and of the Yenisei range. Fold systems arising at the site of these troughs are distinguished from intracontinental systems of the previous type by the presence of ultramafics or rocks of disjointed and usually incomplete ophiolite association. These systems are more extended compared to ensialic systems and are characterised by distinct transverse zoning with the development of eugeosynclinal and miogeosynclinal zones. The ensimatic state of these troughs varies from the presence of separate bodies of alpine type ultramafics to fragments of ophiolite association.

This pattern may not be constant even within a single belt, however. An example is the Baikal-Vitim belt of eastern Siberia. As mentioned above, the complete sequence of ophiolite association has been described in the Muya zone (Gusev et al., 1992). Klitin described such a sequence earlier (Klitin et al., 1975).

Ensimatic intracratonic geosynclines differ in extension. In some belts, as evidently happened in several segments of the Baikal-Vitim belt, destruction already in the early stage may have led to a limited opening of the Red Sea type and paraophiolites in such a case appear at the base of the sequence. In other belts the main development occurred in a regime of continental rifting while ultramafics and paraophiolites appear only in the concluding stage of evolution. Thus the Uruaçu and Yenisei belts developed in the Early Riphean like ensialic zones, having undergone magmatism later.

Having originated as intracontinental rifts, Early Riphean troughs experienced a regime of intracratonic geosynclines and extension, later replaced by compression. During the geosynclinal stage a thick (10 to 15 km) volcanosedimentary complex was formed in ensimatic troughs and a terrigenous-carbonate complex in ensialic troughs. In the Early Riphean only the Mount Isa geosyncline (Australia) fully experienced culmination of an intracratonic geosynclinal regime.

5.2.5 Belts of Tectonothermal Reworking (TTR)

These belts correspond to regions of the pre-Riphean Pangaea affected to some extent by recurrent metamorphism, magmatism, isotope rejuvenation and superposed deformation without significant participation of the Riphean superstructure. In the Early Riphean some such zones occurred in North and South America, Europe, India, Australia and Antarctica. All are linear in form and extend several thousand kilometres. Their substratum is made up of deeply metamorphosed Lower Precambrian formations. Riphean strata corresponding to the stage of reworking are negligible or present only in the form of small patches of volcanics (Nain Province, Canada) or terrigenous rocks (Albany-Fraser, Australia).

Two types of TTR belts have been recorded in the Early Riphean. The first (classic) type is characterised by intense recurrent metamorphism (granulite or amphibolite facies), total rejuvenation of rocks, formation of elongated thrust zones, thrusting, granitisation, formation of pegmatite fields and migmatisation. The Easter Gothian belt, Albany-Fraser zone, experienced such reworking. The other type of TTR is closely associated with formation of felsic volcanoplutonic complexes similar to the volcanoplutonic associations described above. The environment in such belts is considerably less orogenic. Recurrent metamorphism is weak and led in some zones only to isotope rejuvenation while vertical fault tectonics predominate. Among TTR zones of this type are the Mazatzal or Elsonian zone of reworking embracing an immense area in the southern part of the North American continent, Rondonian belt of South America and Gothian belt of reworking of the Baltic Shield. The Rio Negro-Juruena and Sao Ignacio belts are intermediate in character between the two types of TTR zones just described. This classification reflects the characteristic of the Early Riphean as an epoch of the coexistence of two tectonic regimes—'non-geosynclinal' TTR and 'cratonisation', i.e., formation of volcanoplutonic associations, commencement of the first regime and dying out of the second.

Evidently these regimes are somewhat close in their tectonic nature and, in this sense, Bukharov's (1987) use of the term 'protoreactivation' is definitely justified. Both processes regenerate the consolidated sialic basement in a non-geosynclinal environment. Yet distinct differences are apparent in the character of this transformation. Volcanism, with rare exception, is not typical of TTR zones. If, however, volcanics do participate in the superstructure formation, they are of mafic composition. Accumulation of molassoid strata in small grabens and depressions, extensive development of alkaline granitoids and rapakivi granites are not characteristic of these zones. Here normal and calc-alkaline granites, adamellites and anorthosites are developed. On the other hand, metamorphism, more so ultrametamorphism, origin of thick zones of diaphthorites and thrusting, characteristic of TTR zones, are not typical of the formation zones of felsic volcanoplutonic associations. Magmatism here is exclusively crustal while very deep magmatism is distinctly manifest in TTR zones and at places even intrusion of ultramafics has been noticed. What causes these differences in processes and what are the energy and material sources for the formation of these zones? The main factor probably lies in mantle heterogeneity, more precisely in the varying degree of its depletion by heat-producing radioactive elements under different segments of the crust. In this sense, a craton may be represented as a zone of early consolidation of crust, its saturation with granite material and total depletion of the mantle under it.

Formation zones of felsic volcanoplutonic associations may be regarded as sections of depleted mantle but still producing heat sufficient to cause the appearance of fusion centres in the crust with formation of subaerial volcanics and comagmatic intrusions; TTR zones represent zones of continental crust already formed but not yet finally stabilised. In such zones the process of mantle depletion was extended in time, as a result of which the TTR regime was maintained under conditions of its pulsed excitation. It is significant that the TTR regime lasted 500 m.y. longer in the evolution of the Earth compared to the anorogenic volcanoplutonic regime of the end of the Lower Proterozoic-Riphean discussed above.

Another explanation is also possible. The influence of the mantle may not be due to residual active heat sources, but rather to the action of hot spots or convective currents. Having examined the geodynamic environment of the manifestation of the above two regimes, the role of extension in the primary stages of their existence, which causes formation of small grabens and rifts (in TTR zones), must be remembered. Evidently it was replaced in the concluding stages by compression, as supported by deformation of supracrustal strata and also development of overthrusts and strike-slip faults noticed in zones of reworking of the first type. TTR belts of the second type are characterised by a convergence of the characteristics of the two tectonic regimes, i.e., cratonisation and classic tectonothermal reworking, reflecting the specific features of Early Riphean tectonics.

5.2.6 Marginal-cratonic (Marginal-continental) Geosynclines

As pointed out above, these structures were established on the basis of development of ophiolite belts and island-arc complexes within south-east Asia and have recently come to light mainly from the works of Zhang and associates (1984) and Le Zui Bath (1985). Much of the sequence of the eugeosynclinal series of the Indosinian and South China blocks is made up of ophiolite complexes of Early-Middle Riphean age—Khamdyk, Yang-pen and Sibao. The first of these complexes is represented by an association of cummingtonite amphibolites, gneisses and marbles and serpentinised ultramafics and metagabbro. This complex obducted along overthrusts on the northern margin of the Kontum salient. It is unconformably overlain by molasse and intruded by granitoids of Late Riphean age.

The Yangpen ophiolite complex has been established on the western slope of Sikan-Yunnan range (or Kannging uplift) in its central segment where a sequence up to 10 km thick has been described as made up of spilites, pillow lavas, greenschists, siliceous rocks, housing numerous bodies of serpentinites and metagabbro in the lower part, and turbidite-flysch strata in the upper. The intruding gabbros are aged 1315–1210 m.y. and are overlain by Upper Riphean molasse.

The third Lower Riphean ophiolite complex has been detected within the Jiannan uplift in the Sibao group (Yang et al., 1986; Zhang et al., 1984), Fan-jingshan, and is represented by ultramafics, gabbroids and mafic volcanics which are overlain by flysch and contain tonalite bodies aged 1422 m.y.

The Buhang, Kunyang, Huili and Bani complexes pertain to a meso-geosynclinal series. All of them were formed in the marginal zones of Precambrian blocks. In composition and structure of the sequence, the Buhang complex is close to the series of outer non-volcanic island arc with lower terrigenous and upper volcanoterrigenous formations. The Kunyang and Huili complexes were formed in the back troughs of the Sikan-Yunnan Range that arose during extension of the western margin of the Sichuan block. They are somewhat younger than the Yangpen complex. At this same time collision of the microcontinental arc with the continental margin occurred in the Aravalli marginal-continental system of India that developed from the Early Proterozoic.

Thus in the Early Riphean zones with secondary oceanic crust arose in south-east Asia which led to the separation of microcontinents. Already in the Early Riphean, within South China, island arcs appeared and the island-arc series began to form.

So the Early Riphean global environment is characterised by the existence of a cohesive Pangaea which in general was only slightly affected by processes of destruction. Yet processes of destruction and transformation of continental crust were actively manifest at several places. Important among these processes was rifting, which caused the birth of aulacogens, intracratonic geosynclines and grabens. Evidently this process was effective already at the commencement of formation of TTR zones. Thus the geodynamic regime of extension, superposed on crustal segments that stabilised to a different degree, giving rise to development of different types of structures in them, was of great importance. Some of these structures experienced compression at the end of this stage.

All types of tectonic structures characteristic of the Late Precambrian, i.e., shields and platform basins, zones of formation of volcanoplutonic associations, TTR zones of older substratum, aulacogens, all types of intracratonic and marginal-continental geosynclines, were represented in the Early Riphean. Their distribution was uneven however.

6

Middle Riphean: Continuation of Destruction and Break-up of Pangaea

6.1 REGIONAL REVIEW

6.1.1 North American Craton

6.1.1.1 *Canadian Shield and Eastern Cordillera.* The environment prevailing in the preceding stage, i.e., the existence of a vast cratonised region, generally continued to dominate in the Middle Riphean. Aulacogens and zones of reworking of the basement developed on the base of this cratonised region (see Fig. 41).

Platform cover deposition continued in the north-western part of the craton. The Lower Riphean Hornby Bay group was overlain by the Dismal Lakes group of similar composition, which in turn was overlain by mafic volcanics and sandstones of the Coppermine River group. The Rb-Sr age for volcanics of this group is 1.2 b.y.

According to Stewart (1982), formations of the Apache group and Troy quartzites in the southern part of the Rocky Mountains also belong to the platform cover. The Apache group is made up of siltstones, sandstones, conglomerates, dolomites and mafic volcanics while the overlying Troy formation is made up of quartzite-like sandstones. The Apache group lies on quartz monzonites aged 1.42 b.y. while all the other aforesaid formations are intruded by diabases aged 1.15 b.y. (U-Pb method on zircon). Moreover, the Middle Riphean age of the Apache series has been confirmed by finds of stromatolites (Semikhatov, 1974).

In the Wernecke-Mackenzie aulacogen the upper part of the sequence of the Wernecke supergroup and the lower part of the Mackenzie Mountains supergroup are correlated correspondingly with the Dismal Lakes and Coppermine River groups. These formations were crumpled into folds in the epoch of 'Racklan orogeny' (1.2 b.y. ago), equated to Grenvillian diastrophism (Young, 1980).

Accumulation of the upper part of the sequence occurred in the Belt-Purcell aulacogen. The age of the basalt sill at the base of the upper quartzites at 1100 m.y. determines the minimum age limit of the supergroup

(Hoffman, 1989). The birth of the Grand Canyon aulacogen in northern Arizona also pertains to the Middle Riphean. The Grand Canyon supergroup rests here on a basement aged 1.7 b.y. and consists of the Unkar group at the base. This group is represented by shales, quartzites, limestones, sandstones and Cardenas lavas. Sedimentation of this group commenced 1.25–1.2 b.y. ago while the Cardenas lavas are 1.07 b.y. old (Elston and McKee, 1982).

The main stage of development of the narrow arcuate Keweenaw trough (2000 km) occurred in the Middle Riphean and is manifest in extrusions of Middle Keweenaw subaerial volcanics which are predominantly mafic in composition. Their thickness is about 12 km and attains 400,000 km^3. Shallow-water red terrigenous Upper Keweenaw formations (6.5–10 km) contain lavas in subordinate amounts and fall in the age range 1000 to 600 m.y. Middle Keweenaw volcanics are intruded by major Duluth gabbro-anorthosites (1109–1086 m.y.) and also by numerous gabbroid dykes and sills.

The present-day structure of the aulacogen is in the form of an asymmetric graben-syncline with a very steep southern flank complicated by longitudinal faults with an amplitude of up to 3–4 km. A major gravitational and magnetic anomaly corresponds to the axial part of the aulacogen. The various segments of the structure are displaced relative to each other along transform faults.

It has been suggested that mafic intrusions and dykes confined to the axial part of the aulacogen merge at depth and reach the mantle as a single mafic trunk. This unique structure arose as a result of extension of continental lithosphere in all its thickness, possibly under the influence of a hot spot. The theoretical rate of extension has been determined at 60.5 cm per year and the total width of the opening of the lithosphere at 90 km (Goodwin, 1991). According to another viewpoint (Windley, 1989), formation of the Keweenaw rift and synchronous anorogenic magmatism on the margin of the Grenville belt is associated with compression during the intense Grenvillian orogeny. But it is more probable that the influence of this orogeny was confined to the concluding stage of development of the Keweenaw aulacogen.

The zone of distribution of Keweenaw volcanics represents part of an extensive flood basalt province extending in the south under Phanerozoic cover in the form of a wide band from Lake Superior to Kansas (Read and Watson, 1975). Differentiated gabbroids are also associated with this rift. The layered Muskox intrusion (1270 ± 4 m.y., U-Pb method) is associated with the same-aged traps in north-western Canada. Thus this epoch of flood basalt magmatism of North America is distinctly manifest in the interval 1.3–1.05 b.y. The same Middle Riphean epoch is marked by formation of the well-known 'Mackenzie Swarm', a complex of dolerite and basalt dykes

224

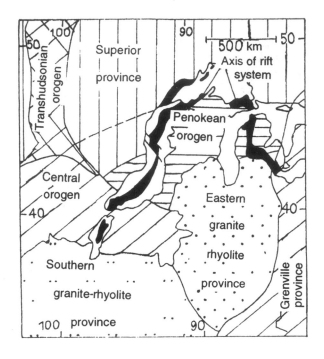

Fig. 50. Midcontinent rift system (after Van Schmus, 1992).

aged 1267 ± 2 m.y. (Goodwin, 1991) intersecting the Canadian Shield from north-west to south-east in a band 2500 km long and 400–1800 km wide.

The Keweenaw aulacogen represents an exposed section of the rift system of the Midcontinent (Van Schmus, 1992). Its south-eastern branch extends from Lake Superior to Michigan and the south-western part up to Oklahoma (Fig. 50). The most active (?) phase of rifting prevailed in the interval 1109–1087 m.y. Cessation of rifting is associated with episodes in the Grenville belt (1030–970 m.y.).

6.1.1.2 *Grenville belt.* The most important event in the Precambrian evolution of the North American continent, viz., formation of the Grenville mobile belt, occurred at the end of the Middle Riphean. This belt extends for a distance of up to 3500 km at an average width of 400 km. In the south-east, along Logan line, it adjoins the Appalachian system running parallel to the Grenville belt from Newfoundland almost up to the Gulf of Mexico. The Oaxacan fold belt, formed about 900 m.y. ago, represents a possible continuation of the Grenville belt in south-eastern Mexico. Traces of Grenville diastrophism have also been established in North-Central Cuba (Renne et al., 1989).

The characteristic features of the Grenville belt are a high degree of meta-morphism of rocks (up to granulite facies), numerous radiometric ages of about 1 b.y. with two peaks at 1100 and 950 m.y. reflecting the culmination of 'Grenville orogeny', wide development of anorthosite intrusions, general north-westerly tectonic vergence, presence of two, i.e., north and north-eastern, structural trends, intense reworking of the ancient substratum, considerable development of overthrusts and absence of preserved molasse. One of the world's largest thrust zones, viz., the Grenville front, represents the north-western boundary of the belt. At present, a three-member tectonic zoning of the Grenville belt prevails (Rivers et al., 1989) from north-west to south-east: north-western zone (para-autochthonous), central zone (allochthonous polycyclic) and south-eastern zone (allochthonous monocyclic).

The north-western zone, i.e., zone of the Grenville front, comprises Early Precambrian rocks of the Central Gneiss Belt that are analogous to the cra-tonic formations in the north but experienced intense tectonothermal rework-ing during the 'Grenville orogeny' with a peak at 1.16–1.03 b.y. Sixty per cent are quartz-feldspathic gneisses. The Malok granite batholith aged 1244 m.y. occurs in the northern Grenville province. This points to the continuation, already into the Middle Riphean, of magmatism responsible for formation of the Midcontinent granitoid belt. Recent data suggests anorogenic mag-matic events in the interval 1450–1250 m.y. in the eastern Grenville province. These events are manifest in the intrusion of anorthosite and granite massifs and mafic and alkaline riftogenic magmatism and continental sedimentation (Gower, 1985). Tectonothermal reworking was manifest in a high degree of metamorphism of rocks under a pressure of 8–10 kbars and their intense plastic deformation. The Grenville front itself represents a cluster of north-western overthrusts intersecting Lower Precambrian foreland formations and thus separating the reworked rocks from the same-aged ones unaffected by this reworking.

The central zone is set off from the preceding by the Allochthonous Marginal Overthrust; it represents a complex allochthon overthrust in a north-westerly direction and includes the Central Granulite Block made up of rocks that have experienced pressures exceeding 11 kbars and contain-ing major anorthosite bodies, and the Baie Comeau segment made up of rocks of similar composition but metamorphosed to amphibolite and not granulite facies. Anorthosites contain xenoliths of supracrustal rocks of the Grenville supergroup. The age of the high metamorphism of the Central zone is correlated with the overthrust formation processes of the south-eastern zone and has been determined as 1.06–1.03 b.y. (Hoffman, 1989). Recent Sm-Nd age determinations of grey gneisses of the Baie Comeau segment (1530 ± 7 m.y.) together with the results of geochemical studies point, according to researchers, to the subduction of the terrane made up of these gneisses and joined later to the margin of the Laurentian continent in the accretionary process (Dickin and Higgins, 1992).

The presence of juvenile Middle Proterozoic crust has also been established in the Adirondack Mountains (Daly and McLelland, 1991). At the same time, at least 30% of the Grenville province here is considered as formed of reworked ancient crust. In the adjoining Canadian territory, reworked crust constitutes 80%.

In the south-eastern zone relics of a protocraton are absent. It has been assumed that its boundary with the Central zone corresponds to the Grenvillian continental margin to which allochthonous terranes were joined in the process of accretion. These terranes now constitute the zone under consideration: Bancroft, Elzevier, Frontenac Mont, Adirondack and Waikham. The first three of these terranes correspond to the 'Central Metasedimentary Belt' comprising the above-mentioned Grenville supergroup (marbles, quartzites, paragneisses, amphibolites and metavolcanics). Based on lead and neodymium isotope ratios, the age of the juvenile crust of these terranes has been put at 1.3–1.2 b.y. (Hoffman, 1989). According to another model (Corriveau, 1990), Elzevier terrane and shoshonite plutons were formed in an environment of subduction (1089–1079 m.y.). Its subsequent coalescence with other terranes occurring after subduction led to formation of the Metasedimentary Belt. Following this, it experienced collision with the Central Polycyclic Zone.

The Grenville belt as a whole is characterised by a reduction of isotope ages south-east of the Grenville front from the reworked Archaean crust of the craton (3 b.y.) through a strip of reworked crust aged 1700–1600, 1500–1400 and 1300–1200 m.y. All the zones experienced superposed Grenville metamorphism 1100–1000 m.y. ago and tectonic transport in a north-westerly direction (Goodwin, 1991).

The above collision model of the Grenville belt is neither the most common nor unique to date. Other models negate the existence of an ocean in the epoch of the Grenville belt formation and its 'closing' and suggest a complex polycyclic evolution of this region accompanied by rifting in the region of the Central Metasedimentary Belt, ending in considerable tectonothermal reworking. Alternative models of collision deserve attention, at least for the reason that no signs whatsoever of oceanic crust of 'Grenville' age have been detected to date. Further integrated studies are required for resolving the acute problems of Grenville belt tectonics (Goodwin, 1991). The Grenville belt probably represents a collage of different types of terranes—island arc and microcontinental.

Milanovsky (1983) regarded the Seal Lake (Nascaupi) graben-syncline extending along the 'Grenville front' and filled with terrigenous formations and basalts (10–12 km) as a fragment of a Middle Riphean graben-like trough sheared by an overthrust from the side of the Grenville belt.

6.1.2 Greenland Shield

The birth and evolution of the Gardar palaeorift zone of southern Greenland fall in the Middle Riphean. Three 'episodes' ranging from 1.33 to 1.16 b.y. are distinguished in the formation of the zone (Upton and Blundell, 1978). Formation of the Gardar terrigenous-basalt complex and of dykes and massifs of alkaline gabbroids, dolerites, granites, syenites and carbonatites occurred in this period.

Metapsammites of the Krummedal series aged about 1200 m.y. developed in the central part of eastern Greenland. The formation of this series immediately preceded Carolinidian orogeny (about 1 b.y.) involving the Eastern Greenland fold belt.

Milanovsky (1983) suggested the existence of a major graben-like basin of the Keweenaw type or an entire palaeorift system in the Riphean at the site of the present-day Labrador and Baffin seas. The palaeorift system consisted of some longitudinal grabens with a row of lateral and diagonal branches penetrating the Greenland and Canadian shields. Some grabens filled with Riphean volcanoterrigenous formations (Thule series etc.) are known in north-western Greenland, in the northern part of Baffin Land and in Lancaster Sound.

Jackson and Yannel link the existence of an extended (3200 km) rift zone along the northern margin of the Canadian Shield with opening of the Poseidon ocean in the interval 1.25 to 1.2 b.y. which closed 1 b.y. thereafter (Goodwin, 1991).

6.1.3 Grenvillides of Scotland, Western Ireland, North-western France and Southern Scandinavia

The Middle Riphean was marked in the North Atlantic region by manifestation of the Grenville tectonomagmatic cycle. Continuation of the Grenville belt of Canada has been traced in Scotland, western Ireland and southern Scandinavia. The initial sedimentation of this cycle is manifest in accumulation of the Glen Finn part of the Moine complex of north-western Scotland, which preceded formation of the Ardur granite-gneisses with U-Pb age of 1020 ± 50 m.y. (Bowes, 1980) and also the Annag gneisses of western Ireland intruded by syntectonic (1070 ± 30 m.y.) and post-tectonic (1000 ± 30 m.y., U-Pb method) granites.

Tectonothermal activity of north-western Great Britain was synchronous with analogous events in southern Scandinavia, as manifest in formation of the Sveconorwegian belt representing a continuation of the North American Grenvillides. In the Armorican massif (France) the crystalline basement (Pentevrian) also experienced tectonomagmatic reworking at the level of 1100–900 m.y., corresponding to Grenvillian events.

The foregoing points to the existence of a single continental platform 1 b.y. ago joining North America, Greenland, northern Great Britain, Ireland,

Scandinavia and north-western France and comprising an elongated linear zone affected by Grenvillian diastrophism (Anderton, 1982; Max, 1979; Young, 1982). The latter is manifest in metamorphism, deformation and granitisation of supracrustal strata formed in the course of the Grenville cycle as well as reworking and 'rejuvenation' of the ancient substratum.

Such reworking is distinctly manifest in the south-western part of Sweden and southern Norway in particular.

The so-called Sveconorwegian (Dalslandian) regeneration (TTR) is manifest most distinctly in a strip of ancient gneisses about 500 km wide west of the Småland-Varmland belt and separated from it by the Protogine fault zone. Intense tectonic movements and formation of pegmatites prevailed in the developing mylonite zone extending meridionally for 450 km between the west coast of Sweden and Telemark in Norway. Dalslandian events—deformation, metamorphism, migmatisation and granite formation—affected the surrounding complexes of Gothides, Svecofennides and development of Sveconorwegian tectonothermal reworking. In southern Norway, the Bamble sector, these processes were studied in detail and correlated with other regions of southern Scandinavia (Starmer, 1991). The following phases have been distinguished: early Sveconorwegian orogenic phase of deformation and granite formation (1250 m.y.), Sveconorwegian anorogenic phase (1175–1100 m.y.), main Sveconorwegian orogenic phase (1100–950 m.y.) and post-tectonic period (915–800 m.y.). Termination of reworking was marked by intrusion of Bohus granites into south-western Sweden, Herefos and Grymstad in Telemark etc.

Accumulation of conglomerates, quartzites and shales of the Dalsland (Dala) formation occurred roughly synchronously in eastern Dalsland in the interval 1030–1080 m.y. These formations are 2000 m thick and metamorphosed to greenschist facies.

Sveconorwegian events of the Baltic Shield are correlated with Grenville tectonics (Gower, 1985). Rotation of Scandinavia relative to North America in the interval 1200–1100 m.y. and the main Sveconorwegian collision at around 1100 m.y. have also been suggested (Gower, 1985; Park et al., 1991).

6.1.4 East European Craton

6.1.4.1 *Baltic Shield.* Formation of the oldest platform cover of the Baltic Shield pertains to the end of the Early to commencement of the Middle Riphean and is represented by Jotnian sedimentary rocks aged 1.4–1.3 b.y. Jotnian molassoid sandstones, shales and conglomerates were deposited almost horizontally in some areas on the eroded Svecofennian formations and evidently represent piedmont plain formations. Their maximum thickness is 800 m in the Dalarna type region. Jotnian sandstones contain sheets of basalts and are intruded by dolerite dykes aged 1220 m.y. (Rb-Sr method) and thus are somewhat older than the Dala formation. The Ovruch series of the Ukrainian Shield is correlated with the Jotnian.

In Finland, Jotnian red arkoses and siltstones accumulated in the depressed blocks of the folded Precambrian complex to a thickness of 1 km. Jotnian rocks (Trucile sandstones) are also known in the easternmost part of Norway. In the eastern part of the Baltic Shield the sedimentary complex of the Terskei coast of the White Sea also corresponds to the Jotnian.

Formation of the Gothian complex of anorogenic volcanic and intrusive rocks and the associated coarse clastic Subjotnian formations of Sweden and Norway concluded in the Middle Riphean.

6.1.4.2 *Russian Platform.* Aulacogens of the central part of the East European Craton, i.e., Pachelma, Soligalich-Yarin, Leshukov-Safonov, Belomorian, Ladoga, Krestets etc. were formed in the Middle Riphean. They were filled with red terrigenous or volcanoterrigenous deposits. The Middle Riphean age of aulacogens adjoining the Timan range has been established from the similarity of their filling with slate formations of the pre-Upper Riphean Chetlas series of middle Timan. In northern Timan the Barmin series corresponds to it. The Middle Riphean is absent in Kama-Belaya and Sergiev-Abdullin aulacogens.

The Rtishchev suite of red sandstones (700 m) in the Pachelma aulacogen, Loginov series of shales and siltstones with bands of sandstones (215 m) in the Pavlov-Posad, the Rudnya suite (300 m) in the Orsha, the Kolomna suite of shales (85 m) and the Tokarevo suite of white sandstones (95 m) in the Krestets and Chukhloma and Kostroma red suites (1500 m) in the Soligalich—all these belong to the Middle Riphean, as do the Solozero and Nenoka red suites and basalts (400 m) in the Onega trough and the Tarkhanov sand-shale series (5000 m) in Kanin. In the Ladoga graben the Middle Riphean is represented by only the lower part of the Kondrat'ev suite of quartz-arkoses with a sheet of volcanics (245 m). In the Roslyatin and Velik-Ustug aulacogens this interval is represented by the Vologda series of shales, sandstones and siltstones (2513 m) dated 1270 m.y. (K-Ar method). In the basement of the Pechora basin, shales revealed in boreholes Seduyakha-58, Malinovka-1 etc. belong to the Middle Riphean (*Upper Precambrian of the Northern European USSR,* 1986).

6.1.5 Basement of European Variscides

The evolution of the Precambrian crust of the Variscides is not yet fully understood. Two complexes, Brioverian and Moldanubicum, are distinguished in the crystalline basement. If the Late Riphean age of the first of these complexes is today regarded as conclusively demonstrated, two versions of the stratigraphic position of the Moldanubicum are considered. According to one version, Moldanubicum rocks belong to the Grenville cycle, i.e., represent Lower (?) to Middle Riphean (Zoubek, 1984). The second version is based on the assumption of facies transition of Moldanubicum into Brioverian and hence the two are assigned the same age. Whatever the case may be, an Early Proterozoic age for Moldanubicum is excluded since the age of clastic zircon from Moldanubicum paragneisses has been established at 2.0–1.8 b.y.

Moldanubicum formations outcrop in the same named zone of Kossmat 750 km long and 200 km wide, running from the south-western part of the Bohemian massif through the Black Forest, Vosges and the French Massif Central to the Armorican massif. The stratigraphic sequence of the Moldanubicum supergroup (Zoubek, 1984) comprises lower and upper groups. The Lower Moldanubicum (6 km) comprises metalutites and metagreywackes, leptynite formation of felsic and mafic volcanics as well as carbonate rocks.

Rocks metamorphosed to almandine-amphibolite facies pass at places into hornblende granulite facies and contain numerous sills of mafic rocks and layered ultramafics.

The Upper Moldanubicum (2 km) is mainly represented by quartzites metamorphosed to epidote-amphibolite facies.

Reverting to the age of the Moldanubicum complex, it must be pointed out that the absence of 'pre-Cadomian' geochronological age data can be explained by the significant reworking of rocks by Cadomian-Baikalian and especially Variscan tectonics. The sharp difference in structure and metamorphism of the Brioverian and Moldanubicum complexes should also be borne in mind. Distinguishing Grenvillian (Dalslandian) diastrophism in the evolution of the basement of the Variscides is therefore justifiable (Zoubek, 1984). It has further been suggested that the Moldanubicum zone inherited a more ancient crustal structure. This zone is set off from the northern rim of the Cadomian-Brioverian belt by the 'Peri-Moldanubicum' fault of ancient origin. Similar tectonic boundaries are known in the southern ('Peri-Pennine' fault) and eastern (Morava zone) frame of this zone.

6.1.6 Mediterranean Belt

6.1.6.1 *Carpathian-Balkan region.* In spite of the difficulty in identifying and correlating Precambrian rocks in young fold belts, manifestation of a tectonic cycle terminating about 850 m.y. ago, close to the Grenvillian (Dalslandian) cycle, could be distinguished in this region. This cycle has been detected by some Rb-Sr age determinations for gneisses and granitoids of the Carpathians and basement of the Pannonian basin at 1180 to 742 m.y. (Rudakov, 1985). The question arises: do these values reflect the overall progressive metamorphism of rocks or the 'rejuvenation' of much older complexes? In the Romanian Carpathians, formations of the Carpian supergroup (Krautner, 1984) metamorphosed to amphibolite facies and unconformably overlain by the greenschist Marsian supergroup are placed in the Grenvillian (Dalslandian) cycle. The total thickness of this supergroup is 18 km; it includes various groups, Aluta and others, whose correlation results in the delineation of six formations (bottom upwards): gneiss-amphibolite, gneiss, leptyte-amphibolite, gneiss-mica schist, carbonate and quartz-mica schist. An analysis of radiometric ages points to the first manifestation of metamorphism at about 850 m.y.

Formations of this pre-Upper Riphean complex (presumably Middle Riphean) are distributed throughout the region although much remains unclear about the Precambrian stratigraphy of various zones.

Proterozoic rocks form the Rhodope supergroup in the Rhodope massif (Kozhukharov, 1984). Based on identification of microfossils, the Asenovgrad group distinguished in the upper part of the supergroup is placed in the Lower-Middle Riphean. It rests conformably, but with a sharp boundary, on the underlying Sitov group formations of Lower Precambrian and is represented by marbles, calcareous shales and mica schists, amphibolites and gneisses to a thickness of about 4000 m.

However, other Bulgarian geologists (Ivanov et al., 1984) are rather sceptical about these age determinations based on microfossils. In their view the Central Rhodope group (generally equivalent to the Rhodope supergroup of Kozhukharov) may be of Precambrian as well as much younger age, Palaeozoic. A significant difference between the schemes of Ivanov and colleagues and Kozhukharov is the denial by the first authors of the presence of two different, Archaean and Proterozoic complexes within the Rhodope massif.

A palaeotectonic environment prevailing in the region before the Late Riphean can only be established tentatively. It may nevertheless be suggested that Grenvillian events occurred on a consolidated pre-Riphean basement. Its relics could be regarded as fragments of the basement of the Pannonian basin, central part of the Rhodope massif and possibly the Serbo-Macedonian massif. Erosion of this basement is supported by clastic zircons identified in the gneiss-schist complex. It may be suggested that formation of this complex proceeded in basins separated by salients of the ancient basement. Diastrophism at the level of 850 m.y. led to progressive metamorphism and folding of strata that filled these troughs as well as to tectonothermal reworking of the ancient basement.

6.1.6.2 *Black Sea region.* Upper Proterozoic formations of the Black Sea region are also characterised by a two-member structure. The lower complex of rocks has metamorphosed to amphibolite facies and is represented by crystalline schists, plagiogneisses, amphibolites and mica schists to a thickness of up to 8–11 km (Stupka, 1986). These are exposed in the Dzirula, Loki and Khrami massifs and in Dobrogea and could be placed in the Lower-Middle Riphean. This is supported by radiometric age determinations on zircons from the Urukh basin schists at 1300 ± 50 m.y., granitoids of the Loki massif at 1200 ± 100 m.y. and porphyric schists of Daut River (Northern Caucasus) at 870 m.y. This complex is set off by an unconformity from the overlying greenschist Baikal complex. Manifestation of an independent Prebaikalian tectonomagmatic stage in southern Crimea has also been established from the radiometric age of granite pebbles from the Upper Jurassic conglomerates of Mt. Demerdzhi (1100 to 956 m.y.). These age values positively reveal the Grenvillian stage in the evolution of the region without, however, resolving the remaining problem of the possible rejuvenation of old formations at this time.

6.1.7 Preuralides

The condition of continental rifting continued to prevail in the territory of the present-day Urals.

In the stratotypical sequence of the Southern Urals zone the Middle Riphean is represented by the Yurmata series unconformably overlying the Lower Riphean formations and divided into a series of suites. The Mashak suite (2 km) is composed of mafic and felsic volcanics, sandstones and carbonaceous shales. The Zigalga suite (up to 1.8 km) is made up of quartzite-like sandstones and phyllitised shales. The Zigaza-Komarov suite (up to 1.5 km) consists of clayey, carbonaceous and micaceous-chloritic shales interstratified with sandstones. The Avzyan suite (1.3 km) is represented by alternation of dolomites, limestones with carbonaceous-clayey shales, siltstones and sandstones. It is also characterised by a corresponding complex of stromatolites as well as microphytolytes of the 'second' complex.

The age of the volcanics of the Mashak suite is 1346 ± 42 m.y. (Pb method on zircon) and 1300 m.y. (K-Ar method) (Keller et al., 1984).

The Yurmata series has not been exposed in the Central Urals and its delineation in the Northern Urals is problematic (in: *Riphean Stratotypes. Stratigraphy and Geochronology,* 1983). In the Near-Polar Urals the Oshiz suite of quartzites and shales (200 m) and Puiva suite of marbles, shales and diabases (2 km) correspond to the Yurmata series.

Commencement of formation of the major Timan structure falls in the Middle Riphean; much of its sequence, represented by the Chetlas series (up to 4.5 km thick), is made up of shales, siltstones and quartzite-sandstones. Based on a geological survey of the bed of the Barents Sea and also a study of the Riphean strata of the Rybachyi and Srednii peninsulas and Kildin Island and the geological data for the Precambrian of northern Norway, a common extended Timan-Varanger linear zone has been established. Red sandstones of the Terskei suite of the Middle Riphean play a significant role in the structure of the aforesaid linear zone in the coastal part of the Murmansk shelf (Stepkin and Samoilovich, 1984).

6.1.8 West Siberian Platform

The presence of Lower-Middle Riphean formations in the pre-Jurassic basement of the platform can only be tentatively conjectured. Based on the latest geological and geophysical data, extensive development of pre-Riphean metamorphic complexes in the form of blocks framed by the Baikal fold system has been suggested. This points to processes of Riphean destruction in this part of Pangaea I. At the same time the long duration of Baikalian tectonics makes for age differences among the various fold systems of this stage of development. Here the presence of pre-Upper Riphean (in any case, Middle Riphean) fold complexes cannot be disputed. This is indicated by submergence of the Yenisei range structures under the cover of the West Siberian Platform. The Kass ancient massif evidently represents the western framing of this structure. The age of Riphean schists and granite-gneisses reached by boreholes is 960–1260 m.y. The horizon with boundary velocity 6.2–6.4 km/s is identified with the surface of the Baikal geosynclinal complex (Zhero et al., 1979). Rocks of this age are extensively distributed in the cores of the Hercynian anticlinoria—Gort, Polui and Sartynya, where such rocks were reached by drilling, and also within all the Mesozoic uplifts—Pudin, Surgut, Nizhnevartov etc. It may be suggested that Grenvillian orogeny, manifest in Kazakhstan and Siberia, left its traces also within some fold systems of the vast territory of the basement of the West Siberian Platform.

6.1.9 Yenisei Range Fold System

The rift stage was succeeded in the Middle Riphean by establishment of a geosynclinal regime. The trough at this time was characterised by a distinct lateral zoning. In the outer eastern zone sedimentary and volcanosedimentary rocks of the Upper Sukhopit series represent a black shale formation; in the western zone a complex of greywackes, schists, tuffs, volcanics and mafic dykes formed. Together with numerous bodies of ultramafics of the Surnikha complex, these Isakovka beds form an ophiolite association (Postel'nikov, 1980). It was recently described in greater detail by Volobuyev (1993) and its age is probably Upper Riphean (see below).

More to the north, within the Turukhansk uplift, siltstones and sandstones of the Besymen (650 m) and limestones of the Linok (310 m) and Sukhotungusik (800 m) suites are placed in the Middle Riphean sequence (its upper part) (in: *Riphean Stratotypes. Stratigraphy and Geochronology,* 1983). These formations continue in the Near-Yenisei part of the West Siberian Platform where they fill the grabens. Formation of plagiogranites of the Teya complex aged 950 ± 50 m.y. is associated with this period of evolution of the Yenisei geosyncline.

6.1.10 Siberian Craton

The active development of aulacogens and the basins associated with them continued in the Siberian Craton in the Middle Riphean. In the hypostratotypical region, after structural reorganisation at the end of the Early Riphean, sedimentation extended to the southern part of the Yudoma-Maya trough and to the Maya basin but is absent in the Uchur basin. The major transgressive cycle of the Aimchan series (up to 2.2 km) consisting of terrigenous rocks in the lower part (Talyn suite) and carbonate in the upper (Svetlin suite) was formed. Glauconites in the lower suite recorded an age of 1.23–1.21 b.y. while a complex of Middle Riphean stromatolites is characteristic of the upper suite. The overlying Kerpyl series also belongs to the Middle Riphean, as has been established on the basis of microfossils and stromatolites contained in its upper Cipanda suite and the diminishing age values of glauconites (1170–970 m.y.) up along the sequence. Yet other researchers place the Kerpyl series already in the Upper Riphean (Shenfii', 1991). The Kerpyl series (thickness 2.3 km) is shaly-silty in the lower part and carbonate in the upper. This transgressive complex has been traced without significant changes in other regions of Central and Eastern Siberia and forms the base of the platform cover of the Okhotsk and Omolon massifs, Kolyma uplift region and also the lower part of the sequence of the Turukhansk-Norilsk aulacogen (in: *Riphean Stratotypes. Stratigraphy and Geochronology,* 1983).

Middle Riphean formations comparable with the Aimchan suite along the western margin of the Siberian Craton are almost not exposed. Phyllites, shales and limestones of the Ludov suite are tentatively placed at this level. East of the Yenisei range, in the Chadobets uplift, the Semenov, Dalchikov and Chuktukon suites are generally regarded as analogues to the upper parts of the Sukhopit suite of the Yenisei range.

In the aulacogens surrounding the Anabar massif in Northern Siberia dolomites of the Billyakh series accumulated in the Middle Riphean (up to 1025 m.y.); tuffosiltstones, basalts, tuffoagglomerates, dolomites and siltstones of the Unguokhtakh suite (600 m) in the Udja aulacogen; and trachybasalts, ash tuffs, variegated sandstones, siltstones and dolomites of the Arymas (380 m) and Debendy (460 m) suites in the Olenek dome while formation of dolerite sills proceeded. Formation of the Kharaulakh aulacogen began concomitantly with formation of tuffs, tuffosiltstones and dolomites of the Ukta (up to 200 m) and Eselekh (up to 400 m) suites. Sedimentation proceeded and a platform cover of rather small thickness formed on the limbs of the aforesaid linear zones in intrablock depressions (Shpunt, 1984).

6.1.11 Kazakhstan-Tien Shan Fold Zone

Geosynclinal troughs laid in the Early Riphean were affected by Issedonian folding (1100 ± 50 m.y. ago) corresponding to Grenvillian. Extensive manifestation of granitisation in the Kokchetav massif (1200 ± 70 m.y.), the Aktau-Mointy anticlinorium (1100±33 m.y.) and in Northern and Central Tien Shan (1070–1270 m.y.) occurred in this epoch. This is supported by data on the Early (?)-Middle Riphean age of geosynclinal deposits based on stromatolites and also the existence of Issedonian unconformity at the base of quartzites of the Upper Riphean of Central Kazakhstan. Thus the Issedonian tectonic cycle corresponding to the Early and Middle Riphean is identified. Its lower age limit is not wholly clear in the sense that Lower Riphean formations are often delineated only tentatively in the scope of the Middle Riphean Sarybulak, Ortotau, Kenkol and Ichkeletau series. Manifestation of Issedonian diastrophism within the ancient massifs separating individual troughs is reflected in the isotope 'rejuvenation' of rocks (Zaitsev, 1984).

6.1.12 Altay-Sayan Fold Zone

Development of the structural plan created by destruction which commenced right in the preceding stage continued in the Middle Riphean. There are more possibilities for dating the strata filling different rift troughs separating the ancient stable blocks.

In the Urik-Iya graben the Zunteya series including the Ermosokh and Ingasha suites made up of shales, sandstones and conglomerates evidently corresponds to the Middle Riphean. The age of the Karagas series of the Sayan region made up of terrigenous-carbonate rocks forming three cycles of 1.5 km total thickness remains controversial. Based on the age of diabases intruding it, it is 1194 m.y. (K-Ar) while its age based on phytolites falls in the Middle Riphean (Altukhov, 1986). Some researchers place this series in the Upper Riphean (Dodin, 1979; *Riphean Stratotypes. Stratigraphy and Geochronology,* 1983).

In the central part of Eastern Sayan volcanoterrigenous formations of the Kuvai series are conformably overlain by carbonate and carbonate-terrigenous rocks of the Chatygos suite containing Middle Riphean phytolites. Sedimentation continued in the Tumanshet trough which contains a carbonate-volcanoterrigenous formation (4–5 km) overlain by the Karagas series of rocks.

Some researchers place the Naryn and Chakhyrtoi suites of Sangilen, the Kadain suite of the Argun basin and the Okhem suite of north-eastern Tuva in the Middle Riphean (Altukhov, 1986). Accumulation of significantly carbonate cover with bands of mafic and felsic subalkaline volcanics occurred synchronously within the stable ancient massifs (Biysk-Barnaul, Khakas, Arzybei, Derbin, Dzhugnym etc.) (Dodin, 1979).

At the verge of the Middle and Late Riphean, the Siberian territory under consideration underwent diastrophism and reorganisation of the strucural plan, as established by the major regional unconformity at the base of

Upper Riphean strata. The Lower-Middle Riphean complex of the intracratonic troughs of the Sayan region, Altay and Tuva was crumpled into linear folds, metamorphosed predominantly to greenschist facies and underwent granitisation.

6.1.13 Baikal-Vitim Fold Region

In the Patom upland the significantly terrigenous Ballaganakh series (1400 m) accumulated in the Middle Riphean; this series is comparable with the Aimchan series of hypostratotype (in: *Riphean Stratotypes. Stratigraphy and Geochronology*, 1983). The upper part of the Middle Riphean corresponds to the carbonate-terrigenous Dzhemkukan, Moldoun and Barakun suites of 1700–2700 m total thickness and also the siltstone-calcareous Valukhta series of about 1000 m thickness. According to other stratigraphic schemes (Shenfil', 1991), only sandstones, siltstones, shales with horizons of the Medvezhev suite of metadiabases lying on the eroded surface of the Purpol suite belong to the Middle Riphean. Correspondingly, the Ballaganakh series belongs to the Upper Riphean.

Based on stromatolite complexes, some researchers regard the lower half of the Baikal three-member complex (Goloustnaya and lower part of the Uluntui suites) as belonging to the Middle Riphean but there are more grounds to regard it as Upper Riphean (see Chapter 7).

In the inner zone, formation of island-arc complexes that probably commenced already in the Early Riphean continued. In Abat, Tyya-Mama, Kicher-Mama, Tompuda-Svetlin, Holoi-Vitim, Aga-Shilka and Onon zones, a basalt-rhyolite formation is delineated in the lower part although mafic volcanics sharply predominate generally in the sequence. In the Kunkuder-Mama, Gorbylok, Kilyan-Irokinda, Yanguda-Kamenka and Karalon zones, basalts, rhyolites and tuffs of intermediate composition are delineated. Volcanic strata contain greywackes, sandstones, siliceous-terrigenous schists, carbonate-terrigenous rocks and jasperoids. Volcanosedimentary rocks (up to 8 km) are intruded by gabbroids and granitoids and contain ophiolite thrust sheets (Gusev et al., 1992).

A significant extent of the Baikal montane region is occupied by structures of miogeosynclinal type which are also mainly made up of Middle Riphean formations. Sandy-shale and terrigenous-carbonate rocks attaining a thickness of 8–10 km play a decisive role in the sequences of these zones. The bulk of the Transbaikalian granite batholiths is confined to these rocks. In many schemes (Bulgatov, 1983; Mitrofanov et al., 1984) Patom, Katera, Mama, Delun-Uran, Ikat and other zones have been assigned to the type under consideration. It is possible that these complexes may reveal traits of similarity to syngeosynclinal platform cover but others could be formations of ancient marginal seas.

Among the radiometric age determinations falling in the interval under consideration, the age of tholeiite-metabasalts of the Nurundukan suite at 1.05 b.y. (Sm-Nd isochron on bulk samples) is particularly important (Rytsk, 1991).

Thus the palaeotectonic setting of the Middle Riphean of the inner zone is represented by an oceanic basin which housed island arcs of different ages, marginal sea and microcontinents. Later (Late Riphean), their accretion occurred and, in the present structure, they are 'fused' with ophiolite sutures forming a complex assembly of terranes (microcontinents) of different genesis. At present, nine such terranes have been established but further research would undoubtedly establish the more complex structure of the region under study.

6.1.14 Northern Mongolia

South of the Malkhan zone of Transbaikalia, within Mongolia, the palaeotectonic environment of Early-Middle Riphean is analogous to that of the Baikal-Vitim belt. A complex has been delineated in Mongolia corresponding to the time interval 1.7–0.9 (0.8) b.y., which is set off from the over- and underlying strata by a hiatus, folding and granitisation (Zaitsev, 1984). Troughs arising as a result of the break-up of the pre-Riphean basement developed during this period. As in the Baikal-Vitim belt, two types of intracratonic troughs can be distinguished here. These are the significantly ensialic (filled with schists, metasandstones, quartzites and marbles) and those filled with volcanic formations. Rocks of Early-Middle Riphean age have been metamorphosed to greenschist facies and are grouped under the name greenschists. Migmatites saturating the greenschist complex of the Kerulen River basin are aged 1058–950 m.y. (Rb-Sr). The age of granite pebbles from basal conglomerates overlying the Near-Khubsugul greenschist complex is 828 m.y. (K-Ar method) and of pegmatites from Gobi metamorphic rocks 1100, 1050 and 970 m.y. (Zaitsev, 1984).

The above evidence justifies the conclusion that folding was manifest in Mongolia at the end of the middle Riphean.

6.1.15 Sino-Korea and South China Cratons and Indosinian Region

In the stratotypical Qisyan (Yanshan) aulacogen the Jixian (Qisyan) system corresponds to the Middle Riphean. This system lies with an erosion on the Changcheng system and is over 4 km thick. The system comprises four suites made up of carbonate-terrigenous deposits containing a complex of stromatolites, Conophyton and Jacutophyton. The radiometric ages of these rocks vary from 1243–1019 m.y. (Keller et al., 1984).

Sedimentation continued in the western Henan aulacogen where the significantly terrigenous Ruyang group was formed with the Fenjivan carbonate formation in its southern continuation. In eastern Qinling, the Taowan group comprising marbles and schists, in Quruktagh aulacogen, the Arjigan carbonate group, in Central Tien Shan the central part of the Kawabulak group, in Beishan, the Pingtoushan carbonate group and in central Qinian the Huashishan group—all these correspond to the Jixian system. The northern Qinian aulacogen is filled with turbidites, tholeiites and andesite lavas.

Most of the Lower-Middle Riphean aulacogens of China experienced folding around 1000 m.y. ago, following which a platform regime was established (Qian and Chen, 1985). The Amnokan, Okchon, Phennam and Hesan River series deposited in the continental palaeorifts of Korea are probably Middle Riphean also.

Systems of tholeiitic dykes with a north-western strike intersecting the more alkaline sublatitudinal dykes are widespread in the Sino-Korea Craton. The age of both systems is 1200–1300 m.y. (Chen, Qian, 1985).

Formation of continental crust continued in the Middle Riphean, at the site of the future South China Craton in an environment of island arcs and marginal seas.

Folding (diastrophism) of the Sibao group manifested at the end of the Middle Riphean (1050 m.y.) in the Jiannan massif as fixed by Bendon granites aged 1065 and 1109 m.y. (Rb-Sr isochrons on whole rock) (Yangefae, 1986). Recent data (Chen et al., 1991) shows that the collision of island-arc complexes and the Yangtze block occurred only quite recently in the eastern part of the region, i.e., southern Anhui and north-eastern Jianxi. Ophiolites of Zhanshuduin and Fuchnan occurring among rocks of the Banxi group (Shanxi) were dated by the Sm-Nd method at 1034 ± 24 and 935 ± 10 m.y. respectively. The Banxi group (10,000 m) unconformably overlies the Sibao group and is represented by flysch and spilite-keratophyre formation. Granites intruding this group are aged 900–950 m.y. A similar Rb-Sr dating (950 m.y.) was obtained for andesites of the Banxi group (Yang et al., 1986). South and east of the Jiannan massif, Presinian strata of western Fujian and south-western Jianxi, i.e., the Dingwuling formations (3000 m), comprising terrigenous and siliceous rocks and volcanics belong to the island-arc complexes described above.

In the south-western part of the region, east of the Lujiang fault, beds of passive continental margin continue to accumulate. Known as the Kunyang group in eastern Yunnan and the Huili group in southern Sichuan, these beds are represented by carbonate-terrigenous formations with horizons of volcanics. Uranium-lead age determinations fall in the range 1760–780 m.y. An age of 1200 m.y. was recorded for carbonate rocks from the upper part of the middle subgroup. The basin corresponding to this passive margin deepened south and westward. On the west bank of the Yalujiang River, sandy-clay formations with a large amount of metamafics of the Yanbian, Huangshuihe and Obnan groups correspond to this basin. Rocks analogous to the Kunyang group known in western Yunnan constituted a common passive margin with the Kunyang group but were separated from it in the Palaeozoic, forming a new microcontinent.

In western Hubei of southern China unmetamorphosed carbonate-terrigenous formations of the Shennungjia group (6000 m) rest on a pre-Riphean basement. The central part of this group recorded a U-Pb age of 1332 m.y. and the diabase intrusive 950 m.y. (K-Ar method) (Yang et al., 1986). In the Himalaya migmatised schists, gneisses, carbonate and siliceous rocks of the Jomolungma group (20,000 m) developed along the border of China and Nepal are comparable with the pre-Sinian formations of China.

The radiometric ages of about 1.2 b.y. recorded in the Himalaya and Nyanchen-Tangla and the miogeosynclinal character of the Middle Riphean formations in the Himalaya evidently suggest that the Tibet-Himalayan block in the period under consideration represented the northern margin of Eastern Gondwana.

Lower and Middle Riphean formations of the Sinoburman block are represented in north-eastern Burma (Myanmar) in the form of thick flysch-like formations of the Chaung-Magi group (3 km) lying unconformably on the Mogok rock series and overlain with an angular unconformity by the Upper Cambrian Molohein group. Three types of turbidites are noticed in the Chaung-Magi group, namely greywackes, feldspathic greywackes and silty shales, and shales. The accumulation of rocks occurred in a rapidly subsiding ensialic trough. Flyschoid-terrigenous Chaung-Magi formations underwent intense folding and weak metamorphism and were intruded by dolerites, diorites and biotite-microcline granites. Radiometric determinations (K-Ar method) for dolerite and diorite gave age values of 834 ± 15 and 982 ± 20 m.y. Chaung-Magi rocks are comparable to the Maching formation of north-western Malaysia whose lower part is represented by greywackes, siltstones and shales (1100 m) and upper part by variegated formations (900 m). Comparable formations in South China are the Kunyang-Sinan series of the Yunnan range, in Thailand, the Changmay series and in Vietnam, the Phu Hoat series. The Phu Hoat series could be much older and is represented by binary mica schists and gneisses with garnet, staurolite and cordierite with subordinate amphibolites and marbles. These were primarily terrigenous-carbonate deposits metamorphosed with the formation of dome-like zoning. High-temperature metamorphics and migmatites are found in the dome cores and mica schists along their periphery. Two strata are distinguished in the Phu Hoat series (Fan Chyong Thi, 1981). The Shong Chay formation in the central part of Shonglo massif is 2 km thick and is represented by crystalline schists, gneisses and migmatites. Zircons from gneisses gave an age of 1 b.y. (Pb). The formations under consideration are closely associated with Shongtai granites in northern Vietnam aged 1376–1000 m.y. The Buhang formation is developed in the northern-central part of Vietnam in Phu Hoat and the Buhang Mountains. It forms a major dome-like structure in the core of which biotite plagiogneisses and migmatites are exposed as are binary mica schists in the flanks. Similar rocks in the Hatin region have been dated 1.9–1.3 b.y. In formational composition the Buhang complex is close to the series of outer non-volcanic island arcs with lower subaerial volcanic and upper volcano-terrigenous formations which were metamorphosed to amphibolite facies within a typical thermal dome of concentric zoning. A similar zoning has been traced in Changmay (Thailand) region.

The period under consideration also covers manifestation of the intraplatform alkaline magmatism of Vietnam in the form of granites of the Red River valley aged 1386 m.y., syenite-dolerites and granosyenites of Deo Mai, alkaline granites of Myong Hum in the axial part of Fansipan aulacogen, Shong Chay massif and others.

6.1.16 South American Craton

6.1.16.1 *Guyana Shield.* The Middle Riphean tectonic environment in this craton was similar to that of the Early Riphean. Here the formation of volcanoplutonic complexes and molassoid strata continued to form in impulses in graben-like depressions and local 'rejuvenation' of basement rocks and reworking of the basement by disjunctive tectonics occurred. The volcanosedimentary formations formed had already experienced fold deformation. In the Middle Riphean the first such impulse in Brazil was called the Madeira and dated 1.4–1.25 b.y. (de Almeida et al., 1976).

Its commencement in the Carajas subprovince was marked by the intrusion of granites (1.4 b.y.) into the Lower Riphean terrigenous complex and also alkaline granites of Ticki, Teles-Pires in Rio Negro, Madeira, Xingu and other subprovinces. The volcanosedimentary formations of Praina, Akari and Paresis were formed.

The concluding Rondonian impulse (event) of Amazon Craton reactivation falls at the end of the Middle Riphean (1050–900 m.y.). In the Madeira subprovince of northern Brazil graben-like structures were formed, bound by faults with latitudinal and western-north-western strike filled with volcanosedimentary subaerial formations. The latter are represented by Nova Floresta volcanics in the base and by Panas Novos and Palmeiral terrigenous formations in the roof. Aguapei, Cubencranquem and Dardanelos formations were deposited in the other parts of the Amazon Craton.

Felsic magmatism in the form of anorogenic, ring-like subvolcanic Rondonian alkaline granitoids and also Caripunas and Costa Marces volcanics and Piraparana formations in Colombia etc. were extensively manifest in this period. Basaltic magmatism was detected in Roraima (Kacherasena, Rio Pardo formations) and Madeira (Sirikun, Arinos, etc. complexes) subprovinces. In the Cashimbo and Dardanelos basins in the southern Amazon basin, mafic and alkaline magmatism was intensely manifest in the form of Katamba, Supunduri, Guariba etc. intrusions. The Orinoquense episode in Venezuela, the Nikerie episode in Surinam and the Kmudku episode in Guyana were nearly synchronous with the Rondonian and partly Madeira episodes. These events are characterised by reactivation of tectonic processes and are fixed by isotope ages of 'rejuvenation', development of zones of mylonitisation and manifestation of mafic and alkaline magmatism. Dyke complexes aged 1100–1000 m.y. are distributed in the Sao Francisco eocraton (Renne et al., 1990).

The eastern zone of the Andes evidently represented a part of the Amazon Craton in the Middle Riphean, which also experienced reworking at several places. This is supported by the ages of gneisses (1050±100 m.y.) from borehole cores which reached the basement at Altiplano (Bolivia). Proximate ages coinciding with the 'Rondonian' stage of reworking of the basement, i.e., 1250 m.y., are known in Venezuela.

Studies in the Garson massif (northern Andes) pointed to the presence of an ancient gneiss basement (1600 m.y.) and also supracrustal strata metamorphosed to granulite facies aged 1200 m.y. Orogenic processes of this period are associated with a continental collision and correlate with Grenvillian orogeny of North America (Priem et al., 1989).

6.1.16.2 *Central Brazilides.* The Middle Riphean was a period of concluding events in the systems of central Brazil. In the ensimatic Uruaçuan system, rocks of the Araxa, Canastra, Serra das Messa, Arai, Natividad, Estrondo and Tocantins groups experienced metamorphism from greenschist to amphibolite facies, folding, intrusion of numerous granites and pegmatites. The build-up of ultramafic bodies also possibly pertains to this period. Diastrophism is reflected in K-Ar and Rb-Sr ages of metamorphic and intrusive rocks, i.e., 1050 m.y. (metamorphism of Estrondo group), 1170 ± 24 (metamorphism of Arai group) and numerous dates in the interval 1000–900 m.y. reflecting intrusive magmatism. This diastrophism is synchronous with events in the Espinhaco system. This is supported by the age of metamorphism of volcanics of the Santo Onofre group (1000 ± 100 m.y.), Chapada-Diamantina group (1250 ± 51 and 822 ± 20 m.y.) etc.

The deformation history of the Araguaia fold belt (Herz et al., 1989) consists of four regional phases comprising successively the formation of westward inverted folds, repeated folding in a meridional direction, intense folding and concluding thrusts from east to west. Rocks experienced medium- or high-grade metamorphism with the formation of rare granite-gneiss domes with isograds of staurolite, kyanite and fibrolite. According to the age of metamorphism, these rocks pertain to Uruaçuan diastrophism at the level of 1100 m.y. (Herz et al., 1989).

Conclusion of the main events of the Uruaçuan cycle and formation of the Araxaides and Espinhaco fold systems coincide with the ending of the Middle Riphean.

6.1.17 African Craton

As in the Early Riphean, much of the area of the African continent represented a megashield. However, a platform cover formed only locally in small basins in the Tanzania massif. It was represented mainly by terrigenous formations with traps of the Kisii (1200 m.y.) and Aberkorn (1100–990 m.y.) series.

Mobile zones appearing in the Early Riphean and experiencing folding and intrusion of syntectonic granitoids at the end of that period were active in this stage also.

6.1.17.1 *Kibarides*. Magmatism that commenced 1350 m.y. ago in the Kibara belt, including intrusion of bimodal granites, continued up to 1260 m.y. and was accompanied by folding and metamorphism. The general north-easterly strike of the folds was disturbed close to the granite-gneiss domes that formed in the early stages of deformation. Intrusion of the younger granite plutons associated with the formation of open folds has been dated 1185 ± 59 m.y. and intrusion of the alkaline granites along strike-slip faults with a north-easterly strike 1125–1068 m.y. (Rb-Sr) ago; these did not significantly influence the earlier developed structural plan. Post-tectonic granites of the Kibarides are aged 990 to 970 m.y. (Goodwin, 1991). Whatever be the case, the end of the Middle Riphean was marked in this fold system by the reactivation of an orogenic regime manifest in the intrusion of granitoids and the general uplift of the mountain system, which was accompanied by molassoid formation in the Bushimaye trough, Kasai-Lomami region. The Bushimaye system lies unconformably on Kibara rocks and is represented by quartzitic sandstones, gritstones, limestones, shales and basalts. Its age is 1130–940 m.y. (Cahen et al., 1978).

The Irumide belt developed mainly in the Middle Riphean. This belt, extending for a distance of 1000 km from Lake Malawi in the north-east to Lake Kariba in the south-west, is made up of quartzites and lutites of the Muva supergroup (up to 10,000 m) and is divided into two zones. The outer, north-western zone is characterised by a fold-overthrust structure with a general structural vergence in a westerly direction. Mpokorozo basin in the north, filled with formations of the Muva supergroup (5 km), represents the probable foreland of the outer zone. The inner, south-eastern zone is filled with very thick deepwater formations and is characterised by complex divergence. Irumide structures transit into gneiss-granulites of the Mozambique belt, representing reworked Lower Proterozoic rocks (2300 m.y.) and forming the basement relative to metasedimentary rocks of the Muva supergroup. These are intruded by syntectonic granites aged 1100 m.y. (Rb-Sr). Unreworked granulites of Irumidian age and also Irumide klippen are encountered farther south-east. The continuity and similarity of structures and the common Rb-Sr age, suggest a common evolution of Irumide and Mozambique belts during diastrophism 1100 m.y. ago. In the light of this, the Irumides are regarded as the western intracratonic marginal zone of the Mozambique belt, thereby helping in the study of the southern part of this belt in the period of Kibaran tectonics. Several zones are delineated here which are interpreted as possible ophiolite sutures based on the presence of serpentinites and mafic lavas to some or the other extent. Among them are the Lurio belt with an east-north-easterly strike containing amphibolites, gneisses, metavolcanics with associated serpentinites and also the Namama belt with the same direction 200 km away towards the south. The main rocks here are associated with granulites of Irumidian age. The existence of possible sutures and granulites of the same age as the Irumide formations justifies assuming an accretionary nature of this part of the Mozambique belt and Middle Riphean collision tectonics on the margin of the African Craton. Nevertheless, the tectonic nature of the Lurio and Namama zones remains controversial. It is quite possible that they represent intracratonic ensimatic geosynclines. Northward, within the Kenyan part of the Mozambique belt, Mukogodo migmatites aged 1200 m.y. are related to its basement (Key et al., 1989). Even more northwards, in Ethiopia, volcanosedimentary rocks with ophiolites of the Adola fold-overthrust belt have been dated 1030 ± 40 m.y., pointing to the existence of an island arc at the end of the Middle Riphean (Berak et al., 1989).

According to some researchers (Teixeira et al., 1989), completion of development of the Mayombe belt (Gabon-Angola) is placed in this period but the question of the age of its constituent formations is controversial and will be examined in the next chapter.

Closing of the other branches of the Kibara-Ankolean belt (Ukinga, Itiaso and Kigoma) also occurred at the end of the Middle Riphean.

Some Middle Riphean rift basins were situated in northern Namaqualand-Rehobolt belt of South Africa. These were formed in the period 1300–1000 m.y. and filled with volcanosedimentary rocks. They form an arc around the western and northern margins of the Kalahari Craton. A chain of such basins extends for 900 km from Koras to Sinclair. Am east-northeast rift system more northwards runs at a distance of 1200 km from Klein Aub through Ghanzi to the Shikamba hills in Namibia. These continental palaeorifts were filled with red beds and bimodal volcanics up to a thickness of 8000–15,000 m (Goodwin, 1991).

The Namaqualand-Natal belt extends for 2000 km through the southern margin of the Kaapvaal Craton from the Atlantic coast in southern Namibia to Natal and is divided into the respective provinces. Namaqualand province is made up of gneisses, metasedimentary rocks, gabbro and Lower Proterozoic granites that experienced tectonothermal reworking at the level of 1200 m.y. In the northern part of the province, in Gordonia belt, on its eastern margin at the boundary with Kheis belt, a greenstone association with serpentinites of small thickness aged 1350 m.y. occurs in a narrow zone of thrusts. This association is thrust onto ancient quartzites of the Kheis system. The Jannelsepan volcanoplutonic complex located more westwards is represented by calc-alkaline amphibolites and borders a strike-slip fault aged 1100 m.y. with Kakamas terrane made up of highly metamorphosed rocks intruded by granites and charnockites. The rest of the Gordonia belt is characterised by the development of plutons aged 1200–1100 m.y. According to isotope studies, the Jannelsepan island arc ensemble and the Uppington greenstone complex reflect crust-forming events in the interval 1300–1200 m.y. There is a view that Kakamas terrane houses a concealed geosuture along which the Kaapvaal Craton joins Namaqualand province. The western margin of the province further represented in part an ancient microcontinent joined to the Jannelsepan back 'arc' 1300 m.y. ago and partly a collision suture of two continents during Namaqua tectonics (Goodwin, 1991).

Natal province is made up of highly metamorphosed granite-gneisses with an east-north-easterly strike. In the tectonic melange below the Tugela nappe composed of amphibolite, tuffs, lavas and metasediments aged 1200 m.y. are seen. The available age values fall in the interval 1200–900 m.y., which is regarded as the period of main tectonothermal reworking. According to some authors, Natal zone represents a region of accretion of isotopically juvenile material with an obducted ophiolite complex. The age interval is 1118 ± 35 to 990 m.y. and the zone is not considered one of reworking of the ancient crust.

6.1.17.2 *Arabia-Nubian Shield.* The oldest reliable Late Precambrian age values within this shield pertain to the interval 900–950 m.y. These belong to metavolcanics of the Jidda group and granitoid plutons intruding the Baish Bahah complex (Fleck et al., 1980). The values of radiometric age of metamafics from the south-western part of the Arabian Shield, i.e., 1165 ± 110 m.y., cited earlier, proved inaccurate (Kröner et al., 1989). In the light of this observation, the main tectonomagmatic events in the region under consideration associated with the development of island arcs, are assumed to have commenced after 1000 m.y. (Kröner et al., 1989; Johnson et al., 1987; Goodwin, 1991). These events, however, reflect the closing of the ocean basin whose origin could have entirely occurred at the end of the Middle Riphean as a result of rifting that affected this part of Pangaea I.

6.1.17.3 *West Africa.* In the Mayombe system in the western part of Central Africa, monzonite granites aged 1027 ± 56 m.y. intruded, at the end of the Middle Riphean, evidently after accumulation of the Sikila formation (Cahen et al., 1978).

Formation of the comparatively narrow intracratonic meridional zone of central Hoggar made up of metasedimentary rocks of the Egere and Alexod series that experienced folding and granitisation at the level of 1270 ± 110 to 910 m.y. (Bertrand et al., 1978) and also of Maru trough in Nigeria, for phyllites from which an age of 1060 ± 65 m.y. was recorded (Holt et al., 1978), pertains to this period.

The existence of an extended Mauritania-Senegalese mobile belt in this period can be conjectured. In this belt Middle Riphean formations constitute the Central Zone ('Mauritanian Axis', Serpentinite Belt, Mauritanian Range, etc.) extending between the Western and Eastern zones symmetrically disposed on both sides. These zones are made up of Upper Riphean and Vendian formations. Volcanosedimentary groups (series) Gadel and Auija and their equivalents as also numerous serpentinite bodies take part in the structure of this belt. Together, they represent a geosynclinal complex with a distinct ophiolite character. Serpentinites form outcrops extending along the axis of the zone under consideration and constitute the 'Serpentinite Belt' of the Mauritanides. Their structural position has not been conclusively understood although special investigations on this subject were carried out by Chiron (1974). According to him, ultramafic rocks intruded into the lower horizons of the mica-shale complex of the Gadel series and are localised there. It has further been suggested that the formation of ultramafics based on unconformable relations with host rocks was of an intrusive character. Also suggested are the presence of xenoliths of the host rocks in the ultramafics. Small massifs of gabbro are detected along or adjacent to the band of serpentinites. Chiron emphasises the association serpentinites-jaspilites-carbonates in the composition of the Gadel series and spatial independence of ultramafics from mafic volcanics of the overlying Auija group unconformably resting on the Gadel series and comprising predominantly mica schists. This feature does not suggest the presence of one single complete sequence of ophiolite association. At the same time, basalts of the Auija series surround the Serpentinite Axis from the east although the spatial relation between them is not clear and is usually tectonic.

Thus in the Central Zone of the Mauritanides, almost all component units of ophiolite association are present but in a scattered manner.

The much younger andesite formation of the Rabra series is also localised along the Mauritanian Axis but already west of it. Thus the Serpentinite Belt is surrounded on one side by an andesite series and on the other a basalt series. Blocks of the basement outcropping in the form of individual windows play a significant role in the structure of the Mauritanian Axis. The width of the ensimatic zone here is considerably reduced. The Auija series is reduced to a narrow strip at sites of sharper pinching (for example, north of Moudjeria region). This is also true of ultramafic rocks. These relations demonstrate the primary conditions of uneven extension under which the belt originated. Rocks constituting the Central Zone are crumpled into isoclinal folds with a mean dip of 45° westwards. Tectonic slices are encountered everywhere and two major strike-slip faults along which the Mauritanian Axis experienced displacement westwards have been described.

The structure of the Central Zone of the Mauritanian belt changes its form somewhat towards the south. In eastern Senegal the fold belt is divided into two parts, western and eastern, separated by the Yukunkun syncline enlarging southwards and transiting into the Palaeozoic Bove syneclise. This splitting of the belt into two branches is evidently associated with the deflection caused by a major block of the basement. The western limb made up of Kuluntou shales represents a continuation of the western Mauritanides. The eastern zone is filled with Bassari schists. South of the Bove basin, the Bassari zone is traced in the Rokellides fold belt.

There is little information about the southward spread of ultramafics and mafics of the Central Zone. One can only refer with certainty to a general reduction of ultramafic and mafic magmatism in this direction. Mafic magmatism is noticed at the base of the Kuluntou and Bassari series, however. Basalts and andesites are encountered in the Rokell series in the upper part of the terrigenous sequence. The general uneven distribution of ophiolites in the belt and the disappearance of abundant ophiolites could reflect the uneven extension at the commencement of its formation. Further, the disappearance of massive ophiolites is accompanied by the absence of andesite volcanism. At the same time, outcrops of alpine-type serpentinised ultramafics are traced in the basement among Lower Proterozoic rocks along the western margin of the Rokellides and continue south of the line of the Serpentinite Belt of Mauritania. Their formation is associated evidently with the zone of faults arising during general extension at the time of formation of the entire Mauritanian complex, attenuating on the southern extremity of the belt.

In South America the Guyana Serpentinite Belt falls roughly in the continuation of the above zone. Unfortunately, radiometric data is scant for rocks of the Mauritania-Senegalese system. Its lower age limit has been fixed by their bedding on Birrimian granitoids aged 1.8 b.y. It has been suggested that the axial part of the system experienced folding and metamorphism at the level of 1 b.y. (Manev et al., 1976).

6.1.18 Indian Craton

A sharp contraction of the area of platform sedimentation occurred in the Middle Riphean. It continued into Cuddapah basin where the predominantly terrigenous Kollamalai group accumulated but mainly into the Vindhyan basin where the Kaimur group of beds (2 km) was formed. Its sequence is represented (bottom upwards) by quartzites, siliceous shales, quartzites, shales and sandstones. The rocks of the Kaimur series contain Lower-Middle Riphean stromatolites (identified by M.E. Raaben) while the kimberlites intruding them have been dated 1140 ± 12 m.y. (Moralev, 1977).

In Singhbhum, accumulation of terrigenous formations and the Noa-mundi series of haematite ores continued in the southern part of the region in the first half of the Middle Riphean. In the second half the Noamundi series was replaced by the Kholkhan series of carbonate-terrigenous composition; terrigenous sediments of the Porat group continued to accumulate in the northern rift trough. The Noamundi-Porat orogenic cycle (1100–950 m.y.) led to folding of rocks along a latitudinal fault zone of compression and build-up of a batholith complex formed over a prolonged period and in several phases in the interval 1140–900 m.y. Radiometric determinations of lutites of the Kholkhan series gave an age value of 988 m.y.

An active continental margin continued to exist in the Aravalli zone. Against a background of westward subduction of oceanic crust, a marginal sea arose at the back of the island arc due to rifting at the site of the present-day South Delhi belt between 1400 and 1100 m.y. ago. The mafic crust of this marginal sea later began to subduct in an easterly direction under the ancient island arc. At the level of 1000 m.y., this led to collision and obduction of the Fulad ophiolite complex of the South Delhi belt (Deb and Sarkar, 1990). Fulad ophiolites represent regionally metamorphosed mafic rocks in the Delhi supergroup of north-western India. In geochemistry these ophiolites correspond to basalts of mid-oceanic ranges while the Ranakpur metabasalts associated with them correspond to island-arc tholeiites (Volpe and MacDougall, 1990).

TTR zones of the basement in the Middle Riphean largely inherited their Early Riphean position. The Narmada-Son lineament zone was reacti-vated. Reworking of the basement in Sri Lanka was marked by 'rejuvenation' (1150 m.y.) and intrusion of Tonigala granites (970 m.y.).

6.1.19 Australian Craton

The structural plan of the Australian craton evolved essentially in the Middle Riphean. Major shields (Yilgarn, Pilbara, Gawler, North Australian etc.) have been identified in the western and central parts of the continent. The space between these shields was occupied either by zones of non-geosynclinal reworking or depressions in which the accumulation of platform cover proceeded.

South-east of the Pilbara Shield lies the elongated Bangemoll basin filled with the same-named group of sediments (9620 m) and lying on Lower Riphean formations of the Bresnahen and Mount Minnie groups. The basin sequence comprises dolomites, sandstones, greywackes, shales and felsic volcanics; it also contains sills of dolerites. Dating of black shales provided an Rb-Sr isochron age of 1057 ± 80 m.y. and of rhyolites 1075 ± 42 m.y. (Page et al., 1984). Semikhatov (1974) places the Bangemoll series in the upper Middle to the lower Upper Riphean. In the north-eastern part of the

Bangemoll basin, the degree of rock deformation increases with the formation of isoclinal folds with a north-westerly strike, symmetrical and slightly inclined towards the south-west, as well as overthrusts.

In eastern Kimberley, on the margin of the Sturt plateau, the upper, predominantly terrigenous part of the sequence of the Osmond range (965 m) corresponds to the Middle Riphean. Shales recorded an age of 1128 ± 110 m.y. The composition and age of the Glidden group (560 m) (1031 ± 23 m.y.) in the Kimberley basin, are close to those of the former.

In Northern Australia the Victoria River basin stands isolated. Accumulation of terrigenous formations with subordinate dolomites proceeded here in the Middle Riphean. These formations (1431 ± 440 m.y.) correspond to the Watti, Bulita and Tolmer groups and the Wandon Hill and Stubb formations to the lower parts of the Overn group. The total thickness of the sequence is 3500 m. Glauconites from the lower parts of clastic formations have an Rb-Sr isochron age of 1165 ± 30 m.y. and K-Ar age of 1090 ± 14 m.y. It is assumed that deposition in the Victoria River basin occurred between 1.3 and 1.0 b.y. ago (Page et al., 1984).

Intrusion of numerous dolerite sills and dykes into the Roper series of the MacArthur syneclise aged 1150–1128 m.y. evidently occurred by the Middle Riphean. The mafic Rupena volcanics underlying the Sturt Platform (shelf) sequence in south-eastern Australia with an Rb-Sr age of 1317 ± 30 m.y. (Page et al., 1984) and 1370 ± 31 m.y. (Goodwin, 1991) also fall in this period. At this time mobile zones were represented only by numerous aulacogens and fairly extensive zones of reworking of the basement.

Halls Creek aulacogen inherited the same-named Early Proterozoic intracratonic geosyncline and is filled with the Carr Boyd terrigenous group (9 km) of the same age and similar in composition to formations filling the Victoria River basin adjoining from the east. The Fitzmorris aulacogen situated at the boundary of Western and Northern Australia and made up of the same-named group, is also significantly terrigenous in composition and attains a thickness of 6 km.

TTR zones continued to actively develop in the Middle Riphean. In the Albany-Fraser province reworking concluded in superposed granulite metamorphism (1300 ± 12 m.y. ago) as did the intrusion of granitoids in the interval 1300–900 m.y.; in Paterson province, east of the Pilbara Shield, intrusion of adamellites (1080 m.y. ago); in the Musgrave block intrusion of granitoids, superposed metamorphism and deformation at the level of 1.2 b.y., build-up of the mafic-ultramafic Jilles intrusive complex 1100–1200 m.y. ago and formation of latitudinal overthrusts; in the Arunta block intense migmatisation and reactivation in the course of Ormiston orogeny (1050 ± 50 m.y. ago); and in Georgetown massif intrusion of granites 976 ± 28 m.y. ago.

Total consolidation of the basement of the Australian Craton by and large set in with the cessation of Middle Riphean tectonothermal reworking.

6.1.20 Antarctic Craton

Formations much older than the Upper Riphean have been found to date only at some places in the Antarctic region. Thus, based on the available isotope age determinations, part of the Patuxent formation (up to 1 km) outcropping in the Pensacola Mountains should be placed in the Middle Riphean. Its sequence comprises rhythmically alternating greywackes and dark grey clay-mica shales with bands of conglomerates. Extrusion of basalts (pillow lavas) and rhyolites occurred synchronously with sedimentation and they are most widespread in the form of sills in the western part of the Neptune Mountains. Eastin (1970) puts the age of the Gorecki rhyolites of this region at 1210 ± 76 m.y. by the Rb-Sr isochron method and the age of basalt lavas of the Patuxent formation at 1267–778 m.y. The Patuxent formation is overlain by a Middle Cambrian formation with sharp unconformity. Small bodies of granitoids intruding the formation are aged about 555 m.y. (Elliot, 1975). Sedimentation in the Pensacola Mountains commenced before 1.2 b.y. but, at the same time, as justifiably pointed out by Elliot, there is no doubt that the Patuxent formation was deposited simultaneous with the Transantarctic mountain formations which, evidently, are Late Riphean in age.

Lavas similar to felsic rocks of the Patuxent formation were formed 500 km east-north-east of the Pensacola Mountains in the Tachdaun Hills (nunataks of Littlewood and Bertrab) on the margin of the East Antarctic Craton. Rhyolites of the Littlewood nunatak are about 1 b.y. old. According to Eastin and others, the Rb-Sr isochron ages of these rhyolites of the Littlewood nunatak fall in the range 1044–985 m.y. and of rhyolites of the Bertrab nunatak 999 ± 19 m.y. (Eastin, 1970). Rudyachenok (1974) concluded the affinity of felsic and mafic metavolcanics formed in the range 1250–750 m.y. to a definite facies zone extending from the Tachdaun Hills in the north to the western periphery of the Neptune Mountains in the Pensacola Mountains.

Intense reworking of the margin of the East Antarctic Craton continued roughly within the same range. Its development was recently established in the region of Molodezhnaya station (western part of Enderby Land) where granulites and charnockites aged 2.12 b.y. have been found. At the level of 1 b.y., these rocks experienced granitoid plutonism, metamorphism and tectonic deformation (Elliot, 1975).

Recent studies (Goodge and Dallmayer 1992) contradict the prevailing view regarding the so-called 'Nimrod reworking' at the level of 1100 m.y. in the Transantarctic mountains. According to these studies the Lower Proterozoic Nimrod group did not experience 'Pre-Ross' tectonothermal events. Such processes extended up to the region falling between the Pensacola and Ellsworth mountains where the age of Hak nunatak gneisses was determined by Rb-Sr method as 1 b.y. and also towards the Falkland (Malvinas) Islands, the age of rocks here being 1031–991 m.y. (Craddock and Campbell, 1980).

6.2 TYPES OF STRUCTURES AND TECTONIC REGIME

The progression of marginal-continental processes from south-east Asia to the Arabia-Nubian region points to an increase in the rate of ocean formation and to the continuing break-up of Pangaea I. But, in general, an intraplate tectonic regime predominated and all types of structures characteristic of the Early Riphean existed, albeit their importance and role differed.

6.2.1 Platform Basins

Platform cover continued to form in an active tectonic environment of rapidly subsiding basins of relatively small size. This is suggested by the thickness of formations (thousands of metres), sharp facies variations within the sediments and the presence not only of mafic but also felsic volcanics.

The largest basins have been preserved in India and Australia although their size has relatively decreased. Australian basins (Victoria River and MacArthur) are characterised first of all by the great thickness of carbonate-terrigenous deposits. The newly formed Bangemoll basin represents an interesting structure of this type. It is slightly extended in a south-easterly direction from the Pilbara Shield but is generally isometric and filled with carbonate-terrigenous beds of great thickness (about 10 km) containing mafic and felsic volcanics and crumpled into folds. In characteristics this structure falls between a syneclise and an aulacogen.

Moderate thickness has been recorded for the Cuddapah and Vindhyan syneclises of India separated by the differentiation of a single Early Riphean platform.

The small (less than 100 km in cross-section) isolated basins of Aberkorn and Kingongolero are characteristic of the Tanzanian block of Africa. Formation of platform cover on the limbs of aulacogens continues (Siberian Craton and Lensua in Brazil).

6.2.2 Regions of Formation of Epicratonic Volcanoplutonic Associations

A reduction in scale of formation of volcanoplutonic associations, noticed already in the Early Riphean, continued in the Middle Riphean and ultimately led to a global extinction of this typical regime that continued for about 1 b.y. In principle, only the formation of rhyolites, dacites and their tuffs and clastic rocks of the San Lorenzo-Palmeiral complex and intrusion of anorogenic alkaline tin-bearing Rondonian granites into the basement occurred over a comparatively small territory in the Amazon basin in north-western Brazil. Formation of a similar Gothian complex was completed in Sweden and Norway by the intrusion of Late Gothian granites. Felsic volcanics of the Tachdaun were formed in the Antarctic region. All these regions are similar in characteristics to the Early Riphean regions of this type. A proportionate increase in intrusions compared to volcanics and sedimentary rocks is discernible.

6.2.3 Aulacogens

These structures continued to play a significant role in the structural plan of the Middle Riphean. Along with the development of aulacogens formed in the Early Riphean, a new generation arose: Grand Canyon, Keweenaw, Pachelma, Krestets etc. The character of sedimentation noticed in the Early Riphean persisted: terrigenous-carbonate in the aulacogens of Cordillera and north-western North America, significantly terrigenous in the European and volcanosedimentary in the Siberian aulacogens. Everywhere, Middle Riphean strata correspond to a new cycle of sedimentation.

Keweenaw and Gardar aulacogens, newly formed in the Middle Riphean, are located in North America and Greenland respectively. These structures have been well studied.

The Keweenaw (Midcontinent) aulacogen is unique in the characteristics of its internal structure and the analogy of its gravitational and magnetic anomalies with the corresponding features along the margin of the Atlantic Ocean can justifiably be regarded as reflecting the prespreading stage of rifting, as an 'arrested' rift of the Red Sea type.

The Gardar palaeorift zone situated in south-western Greenland is at present evidently represented only by the roots of a once large graben of which a fragment about 25 km long filled with coarse clastic rocks and basalts is preserved. This zone is represented by dykes of alkaline gabbroids, granites, syenites, trachydolerites and also ring-shaped alkaline complexes. Rocks here have experienced only block movements.

Rearrangement of the structural plan occurred in some aulacogens. Thus in the Wernecke-Mackenzie aulacogen this rearrangement was accompanied by folding and metamorphism at the level of 1.2 b.y. (Racklan orogeny) and led to uplift and erosion in the Belt aulacogen. In many aulacogens of the East European Craton, tectonic activity at the end of the Middle Riphean was fixed in the sequence by intrusion of diabase sills. Accumulation of rocks in aulacogens of the Siberian Craton continued in a specific regime. Shpunt (1984) identified two stages of reactivation of magmatic processes in the Middle Riphean. One took place about 1050–1000 m.y. ago when extrusion of tephra of silicic-alkaline composition occurred in the region of the Kharaulakh and Olenek uplifts. The second stage occurred about 1000–950 m.y. ago when the entire northern part of the Siberian Craton was affected by processes of alkali-mafic magmatism.

6.2.4 Intracratonic Geosynclines

Evolution of Early Riphean intracratonic geosynclines of both subtypes continued in the Middle Riphean.

6.2.4.1 *Ensialic intracratonic geosynclines.* As a result of diastrophism that commenced about 1.3 b.y. ago, some intracratonic geosynclines were transformed into fold systems. This is primarily true of the Kibara-Ankolean system. The folding style in the Kibarides is quite simple. In the well-studied southern part of the system, some subparallel synclinoria and anticlinoria with a north-easterly strike and upturned north-west were noticed. The synclines are usually strongly compressed and complicated by minor isoclinal folding, anticlinorial structures are simpler. In the axial part of the Irumides rocks were crumpled into isoclinal vertical folds which, along the periphery of the system became upturned and complicated by numerous thrusts. Regional metamorphism of the Kibara-Ankolean system is weak. Only the lower parts of the sequence attained the stage of binary mica schists. Granites of two types are developed in the Kibara system: synkinematic biotite granites with elongated bodies confined to anticlinoria and post-kinematic leucocratic tin-bearing granites. The main phase of granitoid magmatism in the southern part of the belt occurred in the interval 1330–1280 m.y. and in the north-eastern part 1250 m.y. ago. Tin-bearing granites intruded in the interval 1000–950 m.y. Many ensialic fold systems—Alexod in West Africa, some 'Issedon' troughs of Kazakhstan and Tien Shan, Altay-Sayan region and Transbaikalia—reveal similar features of structure and magmatism.

Middle Riphean data aids the study of the regime of proper geosynclinal evolution of ensialic intracratonic geosynclines and, in particular, the processes of folding, metamorphism and granitisation occurring in them, following which many were transformed into fold systems. In the Kibara-Ankolean system a classic example of structures of this type, i.e., deformation and granitisation of beds, proceeded in several stages. Two or three phases of deformation are distinguishable at different places in the system. Thus in the north, in Burundi, the first phase was associated with formation of near-surface overthrusts and intrusion of granitoids. The second phase was manifest in the formation of open folds oriented in a north-easterly direction. The third phase was manifest locally in the form of narrow strike-slip fault deformation. The pattern of some younging of deformation from north to south from the Kibarides through the Irumides towards the Lurio belt in Mozambique is noteworthy. Folding is manifest unevenly not only within the same system, but also in different systems within even the same region. The same could also be said with regard to granitisation. The phase of intrusion of syntectonic granites in the southern part of the Kibarides is somewhat older (1.3 b.y.) than in the northern (1.25 b.y.). A weak metamorphism, not beyond greenschist facies, is typical of ensialic geosynclines as a whole.

6.2.4.2 *Ensimatic intracratonic geosynclines.* This structural subtype was represented in the Middle Riphean by the Mauritania-Senegalese and

Central Brazilian belts, Yenisei range system, and ensimatic zones of Kazakhstan, Mongolia and the Baikal-Vitim belt. As in the Early Riphean, newly formed belts are characterised by uneven development of dismembered and usually incomplete ophiolite association along the strike of structures (Mauritanides) or intrusion of alpine-type ultramafics in the filling of the rift troughs (Tocantins). Interestingly, the subsequent development of segments of belts enriched with 'paraophiolites' was accompanied to some extent or the other by the extinction of calc-alkaline volcanism (Mauritanides) concomitant with the 'degeneration' of ensimatic conditions along the strike (Rokell).

The belt of alpine-type ultramafics continues into the zone of the Tocantins and Araxa of South America where, too, it is confined to the axial part of the geosynclinal belt characterised by a zonal structure. Sedimentary and metamorphic formations comprising the various zones of Paraguay-Araguaia, Araxa-Estrondo and Brazil, are generally equivalent to Mauritania-Senegalese formations. Uneven distribution of mafic-ultramafic material is also noticed here. It is more fully developed in the Araxa-Estrondo zone which, in this respect, is equivalent to the 'Mauritanian Axis'. A sequence is present here which resembles the ophiolite association (Pilar de Goias group), serpentinite melange and numerous bodies of alpine-type ultramafics. Like the African branch, the amount and nature of ophiolite material decreases towards the Atlantic but in a northerly direction.

A significant difference of the South American belt is the total absence of andesite volcanism and extremely insignificant development of basalt volcanism. Also absent are serpentinite-jasperoid-carbonate associations noticed in the Gadel series. The amount of ultramafic rocks and gabbro, as in the African branch, is uneven. A similar intrusion of them into the salients of the ancient 'frame' (e.g. in Goiania massif of Brazil at the joint with the Araxaides zone) is noticed. Basement blocks are seen within the Araxaides as well as within the Mauritanides. At the same time, a common line of development of ultramafics for the two continents, similarity of sedimentary filling, common zoning, synchronicity of tectonic events and their uniformity positively point to the existence of a common geosynclinal system of Western Gondwana extending from Mauritania to Central Brazil. A distinctive feature of this system is the manifestation of ultramafic magmatism all along its strike. It is uneven in character and is accompanied at places (Mauritanides and Araxaides) by the formation of a complex similar to ophiolite association and at other places only by the formation of alpine-type ultramafics. Available data does not provide a clear clue to the nature of this formation. As pointed out above, more data favours the intrusive character of ultramafics. At the same time, these could be protrusive bodies or melange. The intracratonic character of the system under study is obvious. The sialic basement is present or assumed everywhere along both its flanks and very often has the form of salients and bridges within the system itself, which is significantly ensialic in many sections. Such an internal structure, characterised by extremely uneven and irregular distribution of magmatic material, may reflect only an uneven break-up of crust against a background of tensile stresses of varying intensity. In some sections this has led to the formation of a deepwater intercratonic marine basin with oceanic characteristics and in others only to the origin of riftogenic seas on a continental basement and intrusion of some ultramafic rock massifs into the basement and the geosynclinal complex.

An interesting feature of the development of ensimatic intrageosynclines through inversion has been established within the Mauritania-Senegalese and Central Brazilian belts. Central uplifts ('Serpentinite Belt' and 'Mauritanian Axis') arose almost synchronously throughout the length of the aforesaid belts. Symmetrical troughs developed already in the Late Riphean formed on both sides of these two central uplifts. Such an inversion provides indirect proof of the intracratonic character of these systems.

The Lurio belt, recently discovered in Mozambique, evidently belongs to the subtype of ensimatic intrageosynclines. This belt extends from Lake Malawi to the Indian Ocean. The Namama belt situated more to the south and almost of the same age as the Lurio also belongs to this subtype (Sacchi, 1984). The Lurio belt is made up of a 'paraophiolite' complex and is characterised by the development of deeply metamorphosed formations and systems of major gentle overthrusts reaching the lower crust by their 'roots'. These overthrust slices are mainly made up of granulites. The general vergence of the belt is south-easterly. Similar cover structures in the Namama belt have an eastern trend.

Discreteness of deformation and granitisation is also characteristic of ensimatic troughs but the degree of deformation and metamorphism in them is higher than in ensialic troughs, pointing to the relative intensification of the endogenic regime in the former. Thus in the Araxaides and in some zones of the Mauritanides metamorphism goes up to amphibolite facies and in the Lurio belt to granulite facies.

A common feature of all intracratonic geosynclines of the Middle Riphean, as in the Early Riphean, is the predominance of extension in the initial stage of their evolution. Further, in ensimatic geosynclines the formation series fixes the stage of continental rifting and the impulse of opening of the trough and build-up of 'paraophiolites', i.e., complexes proximate to the ophiolite association but concomitantly differing somewhat from it. The uneven nature of destruction of the crust of the Middle Riphean geosynclines is well expressed along the strike of the Mauritania-Brazilian belt. Here sections with development of nearly complete ophiolite association are replaced by zones in which the intrusion of only alpine-type ultramafics has occurred.

The concluding stage of evolution of the intracratonic geosynclines proceeded against a background of compression of the system.

255

6.2.5 Belts of Tectonothermal Reworking

These geostructures actively developed in North America (Grenville belt), in the Transantarctic mountains (Nimrod reworking), on Madagascar, in South Africa (Natal-Namaqualand belt), in India and in Australia. The type of reworking closely associated with cratonisation, discussed in the previous chapter, continued to develop in the Rondonian belt of South America but the classic type of tectonothermal reworking accompanied by granitisation, superposed high-grade metamorphism, total isotope rejuvenation, intrusion of anorthosites and mafic intrusions in general is sharply prevalent. The problem now is to design a reliable tectonic model of zones of tectonothermal reworking. Models associated with evolution conforming to the Wilson cycle and ending by plate collision dominate the literature at present. Thus, the collision hypothesis (Baer, 1976) assumes formation of the Grenville belt during the collision of two cratons separated originally by an ocean basin. This collision (Himalayan type) concluded the process of subduction and led to thickening of the crust as a result of the sliding of one plate over another. The ophiolite suture is further buried and transformed into a cryptosuture. In the thickened crust, temperature and pressure rise, partial fusion occurs and rocks undergo metamorphism and are intruded by plutons. Reactivation in this process diminishes in a north-westerly direction, away from the collision zone, as seen in the Grenville belt. Although this interpretation explains many structural features, in our view the model of Himalayan type collision cannot be accepted unreservedly for several reasons:

1. Data on palaeomagnetism excludes major oceanic openings up to 1 b.y. and runs in favour of the existence of a single supercontinent (Piper, 1982 and 1983) (Fig. 51).

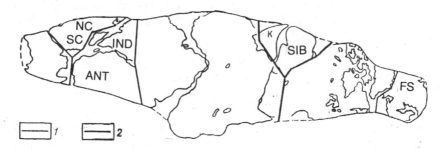

Fig. 51. Proterozoic supercontinent reconstructed mainly on the basis of palaeomagnetic data (after Piper, 1982):

1—rifts aged 1100 m.y.; 2—rifts aged 560 m.y.; NC—North China; SC—South China; K—Kazakhstan; IND—India; SIB—Siberia; FS—Fennoscandinavia; ANT—Antarctic region. Palaeomagnetic data is not available for North and South China and Kazakhstan. Their position in the reconstruction is tentative.

2. Continuation of ancient Archaean and Lower Proterozoic structures from the cratons within TTR zones. Thus structures of the Labrador trough are traced south of the Grenville front, granulite belt of the Tanganyika massif within the Mozambique belt etc.
3. Absence of thick molasse accompanying Himalayan type collision.
4. Rarity or absence of traces of ophiolites, i.e. relics of ancient oceanic crust.
5. Synchronism of main events in the TTR zones with corresponding ones in the adjoining intracratonic geosynclines and the spatial relationship of these structures noticed in many cases.
6. The affinity of TTR zones to the ancient granulite belts established in several cases.

For all these reasons I (N.A.B.) cannot fully support the proponents of a Himalayan type collision model and totally oppose the hypothesis of formation of TTR zones as a result of thinning of the crust due to diffuse spreading and subsequent compression without complete destruction of the coherence of the lithosphere. In this context, the old hypothesis of Wynne-Edwards (1976) continues to be of interest. It propounds the elastic spreading in a heated crust fixed by lines of anorthosites and leading not to break-up, but to slow displacement of the lithosphere over a thermal source, leading to its subsequent subsidence. This process thus terminates not in subduction but compression.

In view of the prevailing controversy over the tectonic nature of TTR zones, it would be appropriate to regard them as special structural units of the Precambrian.

Commencement of tectonothermal reworking is evidently associated with the process of continental rifting, i.e., proceeds against a background of extension. Thus Sveconorwegian reworking has been marked by the formation of a complex of mafic dykes aged 1180 m.y. (Solyom et al., 1992).

6.2.6 Regional Dyke Complexes

Dyke complexes were very widely developed in the Middle Proterozoic and can be regarded as an independent structural type. These formations have not been well studied to date, however. Mention has already been made of a dyke complex in Gardar zone (southern Greenland). The famous 'Mackenzie Swarm', an even more grandiose structure of this type, represents a complex of dolerite and basalt dykes aged 1.3–1.0 b.y. intersecting the Canadian Shield from north-west to south-east in a belt 2500 km long and 400 km wide. It fixes a powerful extension of the Earth's crust.

6.2.7 Marginal-cratonic Geosynclines

Marginal-plate tectonics continued to develop in a pure form in southeast Asia at the site of the present-day Yangtze Craton and adjoining territories and also in the Aravalli system of India. Development of the Sibao

Fig. 52 Global palaeotectonic reconstruction. Middle Riphean (N.A. Bozhko, 1988):

1—continental Early Precambrian crust; 2—platform cover; 3—felsic volcano-plutonic associations; 4—aulacogens; 5—ensialic intracratonic geosynclines; 6—ensimatic intracratonic geosynclines; 7—zones of tectonothermal reworking of the basement; 8—newly formed fold systems; 9—accretionary complexes; 10—passive margins; 11—oceanic crust of older age; 12—newly formed oceanic crust; 13—subduction zones; 14—boundary of continental crust.

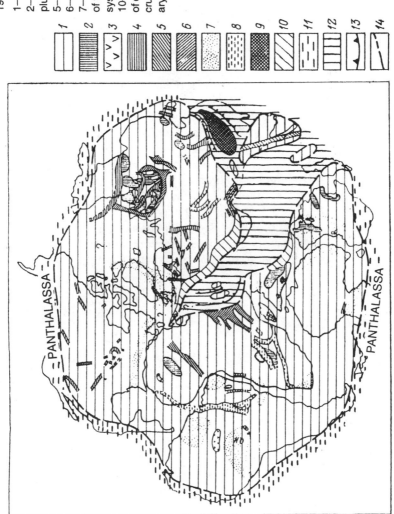

island arc and its migration occurred in a regime of similar environment in the Phanerozoic. The same could also be said of processes in the Aravalli system. Beds of the Kunyang group in the western part of the yangtze craton are entirely comparable to the complexes of passive margins.

As pointed out before, there is positive evidence that opening of an ocean basin in the region of the present-day Red Sea and Saudi Arabia began at least at the end of the period under consideration.

It must also be noted that there are some Middle Riphean belts for which no reliable tectonic models can be designed at present. Some researchers interpret the evolution of these zones according to the Wilson cycle but such an interpretation is far from undisputable and calls for further proof. Among such belts are the Grenville of North America, the Namaqualand-Natal of Africa etc.

At the same time, the advance of marginal-plate processes westwards and a general increase in the number of intracratonic geosynclines that developed Red-Sea-type rifts point to global development of the mechanism of plate tectonics and intensification of destruction of Pangaea.

The global Middle Riphean palaeotectonic environment (Fig. 52) bears many features in common with that of the Early Riphean: a significant portion of Pangaea remains uplifted; formation of epicratonic volcanoplutonic complexes still continues at some places in it; cover accumulation is more extensive; and zones of tectonothermal reworking and intracratonic rift structures are formed. The process of break-up of Pangaea evidently extends farther westwards, reaching the present-day Red Sea region where the build-up of an accretionary complex with ophiolites and island arc volcanics had begun already at the end of the Middle Riphean at the level of about 1 b.y. The simultaneous development of intracontinental and marginal-continental processes remains an important feature of tectonic activity.

7

Late Riphean: Disintegration of Pangaea into Gondwana and Laurasia. Origin of Mobile Belts against the Background of Development of Plate Tectonics

7.1 REGIONAL REVIEW

7.1.1 North American Craton

At the beginning of the Late Riphean, sedimentation continued in most of the already existing structures which had experienced diastrophism to a varying extent simultaneous with Grenville events. In the north-western Canadian Shield, in the region of accumulation of platform cover, red sandstones and siltstones of the Rae group containing sills aged 718–605 m.y. overlie the Coppermine River group with small angular unconformity. This group correlates with the Shaler group (Victoria Island), total thickness 4000 m, and is significantly made up of evaporites and unconformably overlain by Natkusiak basalt formations aged about 700 m.y. In the Mackenzie aulacogen the upper part of the Mackenzie Mountains supergroup corresponds to these formations. These have been combined into the B formation (Young, 1980) or the Little Dala group (2000 m) containing stromatolites which correspond to the age interval 1100–800 m.y. (Goodwin, 1991). Clastic material in this region came from an uplift formed in the south-east as a result of Grenville orogeny.

In the eastern part of central Alaska the Lower Tindir carbonate-terrigenous group corresponds to this level (Young, 1984); corresponding formations in the Grand Canyon aulacogen are the Chuar group shales and siltstones (up to the Sixtymile formation; Elston and McKee, 1982); and in the Keweenaw trough, sandstones, siltstones and shales of the upper Keweenaw group falling in the Upper Riphean (Semikhatov, 1974).

The structural plan of North America underwent reorganisation in the middle of the Late Riphean (850–750 m.y.) as a result of tectonic movements along faults and also a general uplift and folding in Cordilleran aulacogens. This event has been called the 'Eastern Kootenay Disturbance' or 'Heyhook Orogeny' (Young, 1980) or 'Grand Canyon-Mackenzie Mountains Disturbance' (Elston and McKee, 1982). The age of this event in the Grand Canyon and Mackenzie Mountains was more accurately determined by Elston and McKee at 823 m.y. based on isotope analysis of Cardenas lavas. According to Stewart (1982), reorganisation of the tectonic plan of North America occurred roughly 850 m.y. ago. It concluded in the general uplift of the central part and transition from isolated basins and aulacogens to a continuous passive margin and miogeosynclines, encircling almost the entire continent (Fig. 53).

In Northern Cordillera, following the above event, accumulation of thick formations of the Eqvi supergroup and the Rapitan group occurred, their

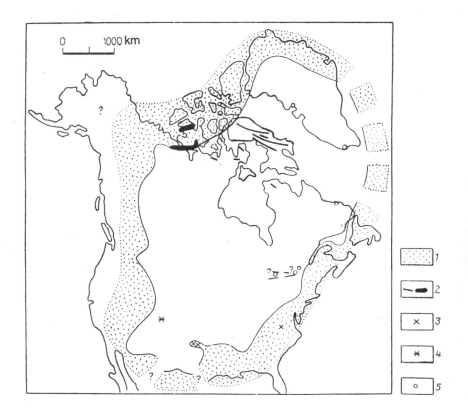

Fig. 53. Precambrian and Lower Cambrian rocks of North America aged 850-540 m.y. (after Stewart, 1976).

1—miogeosynclines; 2—diabase dykes and sills (about 675 m.y.) and probably basalts associated with them; 3—granites; 4—alkaline igneous complexes; 5—carbonatites and alkaline intrusives associated with them.

sequences terminating respectively in the Wernecke and Mackenzie Mountains. These subdivisions have been grouped into the C formation (Young, 1980). In the central part of the Rocky Mountains, the Windermere superseries and, in Alaska, the upper part of the Tindir group are correlated with the C formation. In carbonate formations of the group, upper Tindir microfossils of the Upper Riphean correspond to the interval 620–780 m.y. (Kaufman et al., 1992) and are also correlated with the C formation. Formations of this group are represented by tilloids, mafic lavas and turbidites. The thickness of formations and the content of terrigenous material increases from east to west. The formation of these complexes occurred in the conditions of miogeosyncline newly formed as a result of rifting (passive margin).

The littoral-marine formations of the Windermere supergroup and its equivalents, formed along the western margin of the North American Craton in the interval 800–570 m.y., are presently exposed in the form of an interrupted belt about 4000 km long, extending from Alaska to California and northern Mexico. The thickness of the formations increases in a westerly direction from zero to 10,000 m within a distance of 300–350 km (Goodwin, 1991).

Accumulation of the Windermere supergroups preceded the Eastern Kootenay orogeny of the central part of the Rocky Mountains (850–800 m.y. ago) and was manifest in the deformation of the Belt formation and remobilisation of the basement.

Age determination by lead isotope method for granite-gneisses from British Columbia province (Canada) at 1850 m.y. (gneisses of the Sifton range) and 728 m.y. (gneisses of the Desert range) fix the age of the basement and the lower age limit of the Windermere supergroup of the Riphean Rocky Mountains miogeosyncline (Evenchik et al., 1984). It has been further established that the latter supergroup may have been deposited directly on the basement without the intermediate Purcell series. Tobi tilloids (up to 1.8 km) were usually deposited at the base of the Windermere sequence (but not everywhere).

Considering the maximum age limit of the Windermere supergroup, this basal bed alone or at least part of it should probably be placed in the Upper Riphean.

A similar picture has been suggested for the eastern margin of North America. Thick prisms of sediments were formed here in the Late Riphean along the rift margin of the embryonic Iapetus (Proto-Iapetus)—an ocean that existed at the site of the North Atlantic in the Early Palaeozoic.

Volcanosedimentary deposits of the Upper Riphean and Vendian were deposited to a thickness of up to 8000 m within the entire Atlantic belt for a distance of over 3500 km from Alabama (USA) to Newfoundland and overlie a basement of Grenville age. In the western Appalachians of Canada the Humber zone represents the old western margin of Iapetus bound in the east by the Palaeozoic Baie-Verte-Brompton ophiolite suture. The thick terrigenous series overlying the Grenville basement and increasing in thickness eastwards was formed in the interval 800–600 m.y. The old eastern margin of the Iapetus ocean corresponds in Newfoundland to the Gander zone located east of the suture. The pre-Ordovician Gander group (3 km) is made up of greywackes and is close to sequences of the Avalon zone. Some researchers regard the Upper Riphean Avalon zone formations as African-European terranes (Goodwin, 1991). These are represented by almost unmetamorphosed and undeformed formations. In Newfoundland they are 6000 m thick and made up of subaerial volcanics (Harbor Main group), greywackes, tillites (Conception group), arkoses and conglomerates (Signal Hill group). Ediacaran fauna described in the roof of the Conception group (Goodwin, 1991) justifies placing a significant part of this sequence in the Upper Riphean.

In the territory of the USA the old western margin of the Iapetus has been reconstructed within the Appalachians based on turbidites of the Ocoee supergroup (8000 m) extending on the western flank of the Blue Ridge-Green Mtns-Long Range metamorphic belt resting on a basement aged 1250–1000 m.y.

Thus in the middle of the Late Riphean there was a highly uplifted craton in North America encircled by miogeosynclinal zones arising as a result of massive rifting involving the Laurasian supercontinent.

7.1.2 North Atlantic Region

Following relative stabilisation, intense destruction of the old substratum 'shaken up' before by Grenville events occurred. The territories of Newfoundland and the Avalon blocks, Eastern Greenland, Scotland, Scandinavia and France, which lay close to each other at that time in the Pangaea, were affected by rifting. The break-up process was accompanied at places by dyke formation and volcanism. As a result the Proto-Iapetus palaeostructure was formed, comprising a system of basins and straits extending from the Southern Appalachians to Spitsbergen. At this time the Eleanore Bay (13 km) group accumulated in the central part of Eastern Greenland, represented by lutites, greywackes and quartzites, replaced by mafic lavas towards the west. Shallow carbonate-terrigenous formations of the Hagen Fjord group (5 km) and a dyke complex aged 988 m.y. intruding the Thule group were formed in north-eastern Greenland. The middle Hecla-Hoeck greywacke-quartzite series was formed on Spitsbergen in the Late Riphean. Thus the Eleanore-Hecla ensialic trough lay between the Greenland and Barents Sea cratons in the Late Riphean and a geosynclinal series of the type continental slope and shelf accumulated in this trough (Roberts and Gale, 1978). A thinned riftogenic crust served as the basement of this trough.

Late Riphean sedimentation in Scandinavia proceeded in the graben-like Mjosa Osterdalen trough in south-eastern Norway and eastern Finnmarken.

It was preceded by the formation of a dyke complex aged about 1 b.y. In northern Norway the sequence is represented by terrigenous formations of the Barents Sea group (9 km), the Bagko group (800 m) and the Tana Fjord group (1–3 km) overlain by Vendian tillites. Lutites of the Bagko group recorded an Rb-Sr age of 825 m.y. A deepening of the basin in a north-easterly direction in the east and a north-westerly direction in the west is noticed. Unlike in Greenland, the sequence of northern Norway does not contain volcanics. This is explained by its disposition close to the interior of the craton.

Upper Riphean Scandinavian complexes extend into south-east Great Britain where a series of rift troughs of north-easterly trend are situated between the two salients of the basement (Roberts and Gale, 1978).

In Scotland formations of the Torridonian supergroup represented pre-dominantly by clastics resting on the gneisses of the Lewisian complex and filling the palaeorifts pertain to the Upper Riphean. The lower Stoer group (2 km) is made up of conglomerates, sandstones and shales dated 970 ± 25 m.y. (Rb-Sr). The upper Torridonian group (7 km) rests uncon-formably on the Stoer group and is represented by red sandstones and shales dated 790 ± 20 m.y. (Rb-Sr) (Goodwin et al., 1991).

In the zone of British metamorphic Caledonides the Moine complex comprising meta-arkoses, mica schists, gneisses and migmatites pertains to the Upper Riphean. Schists dated 720 ± 120 and 700 ± 10 m.y. (Rb-Sr) and pegmatites dated 815 ± 30, 780 ± 10 and 740 ± 30 m.y. reflect Morarian orogeny (750 m.y.).

In the zone of British unmetamorphosed Caledonides and Midland, volcanosedimentary rocks correlated with the Moine complex are represented by quartzites, turbidites and basic lavas of the Mona complex. In south-eastern Ireland the Kullestone formation corresponds to the Moine complex. The rocks of this formation are crumpled into folds with a north-easterly strike. According to Cogné and Wright (1980), the Celtic Sea (Proto-Iapetus) separating southern Scotland and Wales underwent reconstruction in the Late Riphean (1000 m.y.). Formation of the Moine complex concluded by the Morarian orogeny and occurred in the north-western margin of this ocean while rocks of the Mona complex associated with glaucophane schists, serpentinites, gabbro and olistostromes accumulated in Anglesey and south-eastern Greenland in the south-eastern margin of this ocean.

7.1.3 East European Craton

Riftogenic structures continued to evolve in the Late Riphean. In the Pachelma aulacogen variegated arkoses, shales and siltstones accumulated in the Tsna (400 m), Irgiz (205 m), Vorona (up to 360 m) and Red Lake (up to 170 m) suites; in the Pavlov-Posad aulacogen in the Pavlov-Posad (626 m) and noginsk (167 m) series; and in Orsha in the Orsha (600 m), Lapich and

Blonsk suites. In the Krestets aulacogen the red-coloured Polotsk (up to 600 m) and Dvoretsk suites correspond to the Karatavian of Bashkinia.

Formation of the Central Russian aulacogen is marked by deposition of the Obnorsk suite (red beds, 350 m). A similar sequence of Upper Riphean with a predominance of feldspathic-quartzitic sandstones, shales and silt-stones is present in the Onega, Leshukov-Safonov and Yaren aulacogens. Development of sedimentation along the entire Timan Range from Kanin Peninsula to Poludov Kamen pertains to the Late Riphean. Accumulation of sandstones, shales, limestones and dolomites formed the Bystrin series (up to 2.6 km) and sandstones, shales and siltstones the Kisloruchei series (up to 1.7 km) and their analogues. At the end of the Late Riphean (Kudash), mafic, intermediate and felsic volcanics as well as shales were formed in the Pechora-Kolva aulacogen and Khoreiver basin.

7.1.4 Basement of European Variscides

As pointed out already, the Brioverian supergroup, characterised by a two-member structure, is distinguished in the basement of the European Variscides (Dupret, 1984) (Fig. 54) apart from the Moldanubicum and Pentevrian. In the Armorican massif the Lower Brioverian evidently pertains to the Upper Riphean. Its sequence from the bottom upwards is represented by (1) spilite lavas and felsic pyroclastic rocks, (2) siltstones, black siliceous schists, dolomites and sandstones and (3) alternation of siltstones and tuff-sandstones. In geochemical characteristics the lower spilite-keratophyre complex pertains to island arc tholeiites (Dupret, 1984). The nature of the pre-Brioverian basement on which supracrustal Brioverian rocks were deposited is not yet fully understood because of its sporadic exposure. Pentevrian gneisses (see Chapter 3) characterised by old age values possibly represent fragments of Gondwana set off from the West African Craton (Cahen et al., 1984). According to Schafelbotom, Strachan and Roach, the other part of the Pentevrian may represent rocks of an Early Cadomian volcanic arc on which the Brioverian rests unconformably (Nance et al., 1991). Lower Brioverian formations are intruded by tonalites aged 670 m.y. (Dupret, 1984).

The plate-tectonics subduction model of Auvrlay (Goodwin, 1991) assumes formation of the Brioverian in the La Manche ocean, south of which was located the Pentevrian continent. This ocean corresponds to the earlier mentioned Celtic Sea separating the British Isles, parts of which similarly moved away from North Africa as well as Armorica.

In the Massif Central the upper part of the Precambrian sequence of 'the Auvergne core' has been correlated with the Brioverian of Armorica (Goodwin, 1991). This sequence is represented by 15-km zones of thick metalutites, metapsammites, mafic metavolcanics, marbles and greywackes as well as conglomerates, quartzites and mafic lavas of Limousin.

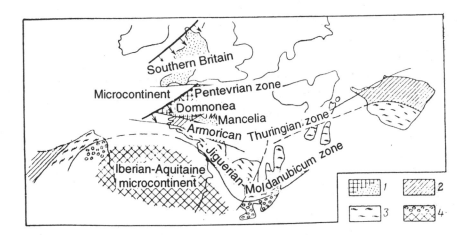

Fig. 54. Palaeogeotectonic zones of the Upper Proterozoic in the Variscan massifs of Central and Western Europe (after Cogné and Wright, 1980):

1—zones of development of the Lower Proterozoic basement (2000 m.y.) and Lower-Upper Brioverian represented by a complex of active margin in the zone of subduction (Domnonea Cordillera, Cadomian folding I + II); 2—zones of development of Upper Brioverian represented by a complex of marginal sea (Cadomian folding II); 3—zones of development of eugeosynclinal complexes from Brioverian to Lower Palaeozoic which underwent Variscan metamorphism and deformation (Ligerian Cordillera of Devonian age); 4—zones of development of Lower Proterozoic basement and Upper Proterozoic shale-greywacke strata and porphyroids.

In the largest Variscan massif, the Bohemian, Upper Precambrian rocks of the Brioverian cycle rest on the old remobilised Moldanubicum Craton and are characterised by a three-member sequence: (1) carbonate schists, (2) psammites, spilite lavas and tuffs and (3) upper black schists with diamictites. Rocks are metamorphosed to amphibolite facies.

The tectonic nature of the Armorican-Thuringian (Armorican-Barrandian) geosynclinal zone is not well understood. Its formation is assumed to be the result of ocean opening as well as of the thinning of continental crust (Zoubek, 1984). Considering the general palaeotectonic environment, it may be assumed that this zone represented a branch of a more extensive Proto-Tethys ocean that arose some 800 m.y. ago between Laurasia and Gondwana containing various microcontinents.

Ophiolite sutures arising on the closing of this ocean are traced through the southern part of the Armorican massif, northern margins of the Central and Bohemian massifs and small outcrops of ophiolites in the Saxothuringian zone. Upper Riphean ophiolites overlain by Vendian formations have been described in the frame of the Saxonian granulite massif (Werner, 1985). In the Sudeten Mountains small exposures of Sleza, Grohova-Brashovice and Skliary surround the Pre-Riphean Sowie Gory block and are associated with the Baikalian island arc complex.

7.1.5 Mediterranean Belt

7.1.5.1 *Alps and Western Mediterranean region.* The presence of the Precambrian in the Pre-Hercynian basement of the Alps is demonstrated by the radiometric ages of the gneisses of Detztal nappe (540 ± 25 m.y.) and the Gottard massif (560 m.y.) pointing to the age of rock metamorphism. But it is possible that these values reflect only 'rejuvenation' while the gneiss complex of the Alps was formed in the Grenville cycle or earlier. The greenschist complex evidently pertains to the Baikalian cycle. In this case it may be assumed that the Alps were formed on a basement in common with Hercynian Europe (Khain, 1984). Even more difficult to solve is the question of relating the greenschist complex (primarily volcanosedimentary) to the Upper Riphean or Vendian. Their comparison with the Brioverian of the Armorican massif runs predominantly in favour of the Late Riphean, in any case to the pre-Late Vendian age of these formations. The folding and metamorphism of the complex could be assigned to the Cadomian epoch which is known to have occurred at the level of about 640 m.y. It must be added that the absence of Cambrian and Lower Ordovician in the Alpine sequence points to the later uplift of the territory. Evidently the same is true of a good part of the Vendian sequence. Sedimentation in the eastern Alps commenced towards 800 m.y. (Schmidt and Sollner, 1983). The age of the ophiolites of the Berisal complex in the Pennine Alps (1020 m.y.; Stille and Tatsumoto, 1985) may serve as an indicator of the destruction of the continental 'bridge' between the East European and Gondwana cratons (Khain and Rudakov, 1991) and the birth of the Proto-Tethys.

The ages of metamorphic rocks· outcropping in the various parts of the Pyrenees, Provence, Atlas, Cordillera Betica and Corsica by various methods fall in the range 600–535 m.y. These values point to the manifestation here of Cadomian folding and metamorphism.

Ophiolites of the Iberian peninsula as also those of Anti-Atlas are evidently of Late Precambrian age. Thus the entire western Mediterranean Sea region can justifiably be regarded as formed during extension of the Early Precambrian continental crust that originally united Europe and Africa at the beginning of the Late Riphean. Thereby the process of opening up of the Proto-Tethys advanced farther westwards from the Arabian region. By the commencement of the Palaeozoic, this Proto-Tethyan geosyncline experienced compression, metamorphism and granitisation (Khain, 1984).

7.1.5.2 *Carpathian-Balkan-Dinarides region.* Separation of the Upper Riphean from the Vendian in the sequences of this region is even more difficult than in the western Mediterranean since Salairian and not Cadomian folding was manifest at the end of the Cambrian. The greenschist complex resting on the Dalslandian one is younger than 850 m.y., i.e., the age of metamorphism of Carpian strata and their analogues (Fig. 55). Its upper

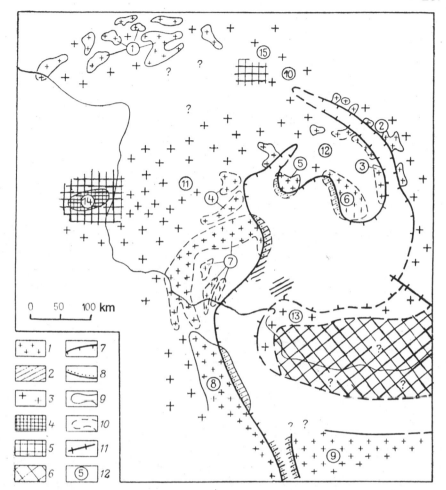

Fig. 55. Suggested relative emplacement of the oldest complexes of the Carpathian-Balkan region towards commencement of the Salairian tectonic stage (after Rudakov, 1985).

1—Dalslandian complexes of the oldest consolidated blocks of the Carpathian-Balkan region; 2—same, at the base of Salairian troughs; 3—same, at the base of Alpine basins; 4—Predalslandian complexes on the surface; 5—same, at the base of Alpine basins; 6—Moesian basin; 7 to 10—boundaries: 7—Salairian troughs (bergstrichs turned within the troughs), 8—transgressive Salairian cover on Dalslandian complexes, 9—para-autochthonous outliers of Dalslandian and much older blocks, 10—possible original position of allochthons; 11—Ferbintsi-Vaklino deep fault; 12—Dalslandian and much older blocks. Circled numbers: 1—Tatro-Veporides of the Western Carpathians; 2—Belopotok-Bretila and 3—Divnaya-Rareu unit of the Eastern Carpathians; 4—Baya-de-Aries and 5—Bihor units in the northern Apuseni Mts; 6—eastern and 7—western parts of Getikum and Suprageticum of the Southern Carpathians; 8—Serbo-Macedonian massif; 9—Balkan Sredna Gora and Rhodopian massif; 10—basement of Transcarpathian trough; 11—basement of Pannonian basin; 12—basement of Transylvanian basin; 13—basement of the northern periphery of Moesian basin; 14—zone of Mecsek hills and environs; 15—Vii-Vitany zone.

boundary is determined by the conformably resting Cambrian floor, however. The Marsian supergroup of Romanian Carpathians (Krautner, 1984) containing Vendian acritarchs represents a good sequence of this complex. Its lower parts are compared by Romanian geologists with the upper part of the Riphean and Vendian. Possibly, ophiolites (gabbro and associated serpentinites) of the Plavica as well as Corbu series contacting with Plavica ophiolites in the lower volcanosedimentary formation of the combined sequence of the Marsian supergroup are Late Riphean. However, rocks synchronous with this formation do not contain mafic volcanics in all the zones. Such groups as the Leota, Bihor and others have a shaly composition.

The northern margin of Epicadomian Gondwana is marked by the southern line of ophiolite sutures (Khain and Rudakov, 1991). Apart from the above-mentioned ophiolites of the Eastern and Pennine Alps, ophiolites of the northern Dinarides around the Fruška Gora mountains and in the Medvednice mountains of Yugoslavia also fall in this category. These latter ophiolites recorded ages of 1010 and 750 m.y. Small bodies of serpentinites in gneisses aged 592 m.y. occur between the above outcrops of Yugoslavian ophiolites in the northern part of the Serbo-Macedonian and Rhodope massifs in the Mecsek Mountains of Hungary.

The northern line of ophiolites formed in the Salairian epoch is represented by ophiolites of Western Sudetes, west Carpathian Veporides, south Carpathian Danubicum and Stara-Planina. In spite of the fact that the age of emplacement of ophiolites does not agree with the formation period of rocks constituting the ophiolite association, the oldest recorded ophiolite ages of about 1000 m.y. are confined to the southern, i.e. Cadomian, line.

7.1.5.3 *Anatolian region.* The gneiss complex of the Pelagonian and Rhodope massifs and also in the tectonic windows in the nappes for which some age values of about 1 b.y. have been recorded, may represent either formations of the Lower-Middle Riphean or an older basement 'rejuvenated' in the Grenville tectonic epoch. The salients of this complex may be regarded as fragments of a pre-Late Riphean 'bridge' connecting Europe, Africa and Arabia (Khain, 1984). Predominantly greenschist rocks, granitised at the level of about 600 m.y., pertain to the Upper Riphean-Cambrian. These include the Vlasina complex of the Serbo-Macedonian massif and the lower part of the sequence of the Durmitor zone (Inner Dinarides) on the periphery of the Pelagonian massif. These metamorphites arose after terrigenous and carbonate-terrigenous formations containing at places mafic volcanics.

Metamorphic formations of central Anatolia comprising the Menderes, Kirshehir and Bitlis massifs are represented by gneisses, mica schists, marbles, quartzites and amphibolites. These rocks are unconformably overlain by Cambrian formations and the lower parts of these sequences may pertain to the Late Riphean. This is suggested by the Vendian age of the complex overlying the southern Anatolian metamorphites. C. Şengör and his colleagues emphasise the manifestation of Pan-African tectonics in the Menderes massif. Its weakly metamorphosed molassoid cover (Kavakdere group) unconformably overlying the Chine group and experiencing Baikalian diastrophism was discovered comparatively recently. In view of the similarity of rock complexes with the Central Iranian massif, central Anatolia may be interpreted as its western margin (Khain and Rudakov, 1991). The formation of a platform cover in Anatolia began in the Vendian-Cambrian, as in the adjoining Arabia-Nubian Shield.

7.1.5.4 *Iranian region.* Precambrian rocks were detected here in small outcrops, most extensively developed in Central Iran. As in southern Anatolia, Upper Proterozoic formations (evidently, Upper Riphean to a large extent) are overlain unconformably by red beds, dolomites and salts of the Upper Vendian. Among pre-Vendian formations of central Iran, two complexes are distinguished: gneiss-marble-amphibolite and schistose. Both complexes are intruded by granitoids aged 850–600 m.y. and overlain unconformably by Upper Vendian rhyolites and carbonate platform sediments. Very few age determinations are available for these rocks: 1307 m.y. (Rb-Sr) recorded by R. Crawford for mica schists of the Sarkhuh complex of the eastern part of Central Iran and 1075 m.y. for the Boneh-Shizrou complex. In the Alborz and Gorgan massifs pre-Vendian formations are represented by metavolcanics, sericite-chlorite schists, phyllites and quartzites.

The Central Iranian massif is regarded as a Peri-Gondwana unit which could be a direct continuation of the Arabia-Nubian Shield (Kale, 1990). The northernmost part of the massif falls in Caucasus Minor in the Miskhana and Megri massifs of the Zangezur Mountains. Metamorphic complexes of Caucasus Major are interpreted as belonging to the Peri-Fennosarmatian part of the Proto-Tethys (Khain and Rudakov, 1991).

7.1.5.5 *Afghan-Baluchistan-Pamirs region.* This segment of the Alpine-Himalayan belt is complex in structure and characterised by Lower Precambrian blocks between which bands of Riphean greenschists overlain unconformably by Vendian strata have formed. This complex may include Middle and Upper or only Upper Riphean formations. It is set off everywhere from the Lower Precambrian formations by faults. This complex is known within the Kabul and Hazara massifs, in central Afghanistan (Chaman, Barmanai and Rudi Gaz series), in the northern part of Afghan Badakhshan, in the Pamirs etc. According to Khain (1984) and Karapetov (1979), the complex could have formed in the Baikalian eugeosynclinal 'ocean hiatuses' arising as a result of break-up and extension of the Central Asian section of the Pangaea megacontinent. Further, the presence of melanocratic magmatites, possibly pertaining to metaophiolites, must be taken into consideration. The initial width of the Baikalian eugeosynclines was evidently much more than that of the present-day belts of development of the Baikalian complex.

This complex evidently corresponds to the 'third Precambrian complex' of Perfil'ev and Moralev (1975) and is represented by metamorphosed sandy-clay (greywacke-phyllite) formations, quartzites, limestones and mafic volcanics. To it belong the Chinozar (Kabul block), Attaka and Hazara (Hazara massif), Chaman and Barmanai (central Afghanistan), Vanch and southern Alichur series (Afghan Badakhshan and Pamirs) etc. In the Argandab block of the Central Afghan median massif, the Baikalian complex is made up of volcano-terrigenous formations of mafic composition and attains a thickness of 16 km.

7.1.5.6 *Himalayan region.* The Himalaya and the northern part of the Indian Shield before the break-up of the Gondwana continent in the Early Triassic are characterised by common evolution. Gansser (1964) regards the Lesser Himalaya as the northern continuation of the shield, a complex platform region (passive margin in the modern sense) bordering the shallow Tethys ocean. The gradual facies transition from shield to northern Tethyan facies in the contemporary structure has been disturbed by the Main Central Thrust. The Central Crystalline Complex and its nappes within the Lesser Himalaya were displaced into the present position in the Tertiary period.

The Middle Riphean shelf carbonate sedimentation in the Lesser Himalaya and the peninsular region led to formation of the Shali-Deoban group correlated with the Lower Vindhyan (Semri group) of the shield. In the Late Riphean this regime was replaced by the Simla-Jaunsar miogeosynclinal cyclome (Virdi, 1988) and formations of predominantly terrigenous Simla and Jaunsar groups corresponding to the peninsular Upper Vindhyan (Kaimur, Rewa and Bander groups).

7.1.6 Preuralides

In the Peri-Polar Urals (Lyapin anticlinorium) the Upper Riphean is represented by quartzites, schists, marbles of the Khobein (1.1 km) and Maroin (up to 1.5 km) suites as also by the overlying mafic and felsic volcanics of the Sablegor (up to 2 km) suite. The differentiated volcanic complex and schists of the Bedamel series (1 km) and also sandstones and mudstones of the Enganepei suite (up to 700 m) correspond to this age as well.

In the Riphean stratotypical region, on the western flank of the Bashkirian anticlinorium, the Upper Riphean sequence (Karatavian) commences with terrigenous, predominantly coarse clastic rocks of the Zilmerdak suite (400–3300 m), which is overlain by the Katav carbonate formation (300 m), Subinder clay-carbonate formation (120 to 350 m) and Inder sandy-silt-clay, glauconite-bearing rocks (up to 1150 m). The sequence terminates in dolomites and limestones of the Minyar suite (700 m). Terminal Riphean (Kudash) is structurally closely associated with the Karatavian series. The lower terrigenous-carbonate Uk suite (600 m) and upper sandy-silt Krivoluk suite (560 m) are distinguished in the composition of the Kudash.

Thus one could suggest that island-arc conditions existed in the Late Riphean in the Peri-Polar Urals while the southern and central Urals continued to develop in a continental rift environment.

7.1.7 Kazakhstan-Tien Shan Fold Region

At the commencement of the Late Riphean, a single platform region existed over the vast territory of Central Kazakhstan and Tien Shan. Within this region, from Kokchetav to southern Junggar, the Epi-Issedon (Epi-Grenville) platform cover was formed (Zaitsev, 1984). Its lower horizon (Kiselev et al., 1992a) or the Dzheltysni complex (Kiselev et al., 1992b) is represented by weakly metamorphosed quartzite-sandstones and shales of the Kokchetav, Andreev and Svyatogor series (800–1500 m) of northern Kazakhstan, Ushtobin suite of Ulytau, quartzites of the Buzkhan suite (1 km) of Junggar Alatau and Dzheltysni suite (500 m) of Tien Shan. The lower age limit of these formations was determined by the age of clastic zircons (1100–1000 m.y.) and the upper limit by the superposition of volcanics (850–800 m.y.).

The second half of the Late Riphean corresponds to the embryonic geosynclinal stage of the Palaeozoic evolution of Kazakhstan and Tien Shan (Zaitsev, 1984), i.e., destruction of continental crust under geodynamic conditions of extension. At this time a system of arcuate structures[1] arose i.e., Beleuti trough (Ulytau), Kalmykkul and Baikonur synclinoria, Great Karatau and Chatkal-Naryn zone, Karatau-Talas trough, Kirghiz-Terskei, Sarytum, Dzhalair-Naiman etc. These were separated by salients of the old basement and rhyolite, rhyolite-dacite and basalt-rhyolite series accumulated in them. These troughs filled with rhyolites and bimodal Late Riphean volcanic formations and conglomerates adjoined geosynclinal troughs in which synchronous basalt, rhyolite-basalt and volcano terrigenous complexes were formed.

According to Kiselev et al. (1992b), two horizons were formed in this period. The middle volcanic horizon (2500 m) corresponds to the Terskei complex of bimodal volcanics (Kiselev et al., 1992a) and includes the Kainar series of Great Karatau, Koksui series of Ulytau, Baiepshin suite of Atasu-Mointy zone, Terskei series of Kirghiz-Terskei zone etc. Volcanics of this level have given age values of 785–880 m.y. The upper horizon comprises the Karagin terrigenous-carbonate-flyschoid complex (3000 m) of the Talas-Karatau zone of Northern Tien Shan and its analogues and is intruded by granitoids aged 753–692 m.y.

7.1.8 Altay-Sayan Fold Region

A large number of local stratigraphic subdivisions have been proposed for the Upper Precambrian of this region but as yet have not been well

---

[1] This arcuate pattern is regarded by other investigators as a secondary one.

correlated. Upper Riphean formations lie here with a sharp unconformity on the underlying formations and together with Vendian-Cambrian strata, constitute the Salairian tectonic complex disposed at its base. The Salairian and Caledonian troughs of the Altay-Sayan fold region represent extended zones bound by faults and filled with a rock assemblage of the eugeosynclinal type with ophiolites. The main epoch of formation of ophiolites and thus of the maximum destruction of continental crust in the region was the Vendian-Cambrian (Mossakovsky and Dergunov, 1983).

What were conditions like in the Late Riphean? Possibly, this period was characterised by a continental-riftogenic regime. Evidently, conditions stabilised in the Altay-Sayan territory after diastrophism, manifest at the end of the Middle to the beginning of the Late Riphean. These conditions were marked by the accumulation of formations of a complex of significantly 'calcareous' Mongoshin level (Man'kovsky and Poroshin, 1981). The lithological composition of this complex includes dark grey limestones, silicilites, carbonaceous phyllites and lenses of mafic volcanics, especially in the upper part. V.K. Man'kovsky and E.E. Poroshin relate the following suites to the Mongoshin level: Mongoshin, Mana, Bakta, Sarlyk, Chatyrly etc. (Eastern Sayan); Kabyrzin, Pasechnaya, Poludennaya, Prokopiev, Malorastay etc. (Kuznetsk Alatau); Bledzhyn, Poludennaya, Turin etc. (Batenev range); a significant part of Baratal, Sagalak, Arydjan etc. (southern Gornyi Altay); lower part of Ailych (north-eastern Tuva), Chekyrtoi, Naryn (Sangilen) etc. The western Siberian level corresponding to much of the Upper Riphean and lower part of the Vendian is traced above the Mongoshin throughout the Altay-Sayan territory. It includes laterally persistent, significantly dolomitic phosphate-bearing formations with subordinate volcanics of exceptionally mafic composition. The following suites have been placed in the western Siberian level: western Siberian, Bagzass, Ashar, Tartul, Sukharin etc. (Kuznetsk Alatau, Gornaya Shoriya, Batenev range); much of Ovsyankov and Gorlyk (Eastern Sayan); Ulan-Arshin, midportion of Ailych and Naryn (Tuva); Eskonchin and part of Manzherok (Gornyi Altay); and lower part of Chyngin (Western Sayan). According to the above authors, the territory was covered in the Late Precambrian by a nearly continuous 'blanket' of basaltoid-siliceous-carbonate complex of persistent thickness and uniform structure over a vast expanse. This complex lies transgressively on formations of the Lower Precambrian and older Riphean strata. The absence of contrasting movements, distinctly manifest uplifts and troughs, shallow deposits of epicontinental sea and absence of manifestation of magmatism other than basaltoid are emphasised by Man'kovsky and Poroshin.

The above generalisation is of considerable importance. It brings some order to the contradictory, debatable local schemes of correlation of numerous subdivisions. The conclusions of Man'kovsky and Poroshin (1981) largely coincide with the concept of Yu.A. Zaitsev about the Epi-Issedonian platform cover of Central Kazakhstan. But this is only one of the several contradictory stratigraphic schemes of the region. The presence of ensialic rift troughs which evidently existed already in the Late Riphean should not be overlooked. This is supported by the increased thickness of rocks and sharp increase in the volume of volcanics in some zones. In any case, the concept of a common 'blanket' should not be extended to the next level, i.e. Vendian-Cambrian Belkin-Sornin, characterised by sharp facies variation of its constituent basaltoid-siliceous-carbonate deposits. Evidently the region was again affected at the end of the Late Precambrian by destruction, rift formation occurred and it was later drawn into a regime of intracratonic geosynclines. Such troughs were the eastern Tuva, Mana, Agul, eastern Salair, northern Sayan, central-western Sayan, Katun, Tebel, Boruss, Shuya etc.

Some of these troughs closed in the Salairian (Early Caledonian) epoch of folding and others in the Caledonian epoch.

7.1.9 Yenisei Range

Some researchers commence the Upper Riphean in the Yenisei range, as in the case of other regions of Siberia, with formations that are correlated with the Kerpyl series (Shenfié, 1991), of the Uchur-Maya region and others with the Lakhanda series of Eastern Siberia (in: *Riphean Stratotypes. Stratigraphy and Geochronology*, 1983). According to the first viewpoint, it consists of the Pogorui, Kartochka and Aladyin suites of the Sukhopit series, Tungusik, Oslyan and Taseev (Chingasan and lower parts of Chapa) series. According to the second, the Upper Riphean sequence commences only with the Tungusik series.

In the intracratonic geosyncline of the Yenisei range, according to E.S. Postel'nikov, the first half of the Late Riphean corresponds to the late geosynclinal stage (see Fig. 45). At this time, after formation of the Teya granite-gneiss complex, accumulation of the Tungusik cyclome proceeded. These rocks were terrigenous carbonate formations and volcanics of the Potoskui (1.9 km), Greben (1.5 km) and Kyrgytei (up to 4 km) suites. T. Ya. Kornev divided the volcanics and subvolcanic bodies of the Tungusik cyclome into bimodal rhyolite-basalt (Potoskui suite) and differentiated rhyolite-andesite-basalt (Greben and Kyrgytei suites) formations. The metarhyolite-andesite-basalt association of the Lower Surnikha series corresponds to the Tungusik series in the internal (western) zone of the geosyncline (Postel'nikov, 1980). At the level of 850 m.y., significant events occurred in the Yenisei range: folding, metamorphism and formation of large multiphase granitoid batholiths of the calc-alkaline series (Tatar, Chirimba and Kalamin massifs). The overlying unmetamorphosed formations of the Late Riphean constitute large isolated troughs and basins filled with terrigenous and terrigenous-carbonate rocks of the Oslyan, Vorogovka and Chingasan cyclomes (Postel'nikov, 1980). These formations correspond to the orogenic stage of development of the range geosyncline corresponding to the Baikal horizon of the scheme of Novosibirsk geologists (Shenfil', 1991). The Oslyan and Taseev series of the Angara-Pit trough are represented by sandstones, shales, siltstones and subordinate limestones exceeding 5000 m in total thickness. The Chingasan and Chapa (except for the Nemchanka suite) series of the Teya-Chapa trough are made up of conglomerates and sandstones with subordinate dolomites exceeding 4000 m in total thickness. The Chingasan and Oslyan series are characterised by age values of 730–750 m.y. Complexes of microphytolites and stromatolites have been described in these formations (Shenfil', 1991).

7.1.10 Siberian Craton

In the hypostratotype of the Riphean of the Uchur-Maya region, the Lakhanda and Ui series correspond to the Upper Riphean. But some researchers commence the Upper Riphean with the base of the Kerpyl series (Semikhatov et al., 1991). The Lakhanda series lies with a hiatus on Cipanda dolomites and consists of the lower Neyuren suite (200–700 m) of carbonate rocks and shales and the upper Ignikan suite (200–300 m) of limestones and dolomites. Age determinations on glauconites (K-Ar) fall in the interval 1000–780 m.y. and correspond to the Inzeria stromatolite complex. The Ui series gradually replaces the Lakhanda and consists of the lower quartzite-sandstone-shale Kandyn suite (200–4000 m) and upper Ust-Kirbin suite (200–1500 m) mainly made up of siltstones and shales (in: *Riphean Stratotypes. Stratigraphy and Geochronology*, 1983).

The Upper Proterozoic formations of the Magan pericratonic trough of the eastern Stanovoy range represented by the terrigenous-carbonate Neldin (1000–1265 m) and volcanocarbonate Nemerikan (1.5 km) suites correlate with the Ui series. Evidently terrigenous and terrigenous-carbonate formations of the Bogauktin and Bulzhinei series constituting part of the platform cover of the Aldan Shield correspond to the Upper Riphean. These suites are overlain by formations of the Porokhtakh (Yudoma) suite. Riphean carbonate formations (Kamov series), according to drilling data, were developed in the form of a thick, almost continuous cover on a vast territory in the south-eastern part of the Siberian Craton. The top of the Kamov series corresponds to the Lakhanda horizon. Their replacement by clay-silt formations of the miogeosynclinal type is noticed only in the south-western part of the craton on transition to the eastern zone of the Yenisei range and also in the north-west in the Igarka-Norilsk region and Taimyr peninsula. The uppermost subdivision of the Upper Riphean (Baikal horizon) is noticed in the much thicker borehole sections in the southern part of the craton predominantly as analogues of the Uluntui and Kachergat suites of siltstones, shales and limestones.

Farther north, deep within the craton, the thickness of the analogues of the Baikal series decreases sharply. More westward, in the regions adjoining the Sayan and Yenisei ranges,these deposits are present in Bel'sk, Tyretsk and other boreholes. Dark grey siltstones and sandstones of the lower sub-suite of the Uluntui suite are gradually replaced by cherts of the lower part of the Oselkov series. Farther north-west, in the southern part of the Yenisei range, the Taseev series belongs to the Baikal horizon.

The sequence of Middle-Upper Riphean formations of the western margin of the craton (Irkineevo salient, Chadobets uplift and Kuiumba-Taiga region) is characterised by the alternation of terrigenous and carbonate rocks. The total thickness of these formations is 3 km.

North-east of the Peri-Baikal trough in the inner regions of the Siberian Craton adjoining the Patom trough, Baikal formations have been reached by Ust-Biruk, Olekma and many other boreholes. These are represented by dolomites, limestones and shales, similar to the Kalanchev, Biruk and Tor-gin suites of the Patom trough. West of Ust-Biruk, pre-Vendian formations have been reached in the Parshina borehole and are represented by con-glomerates, variegated sandstones and clay-carbonate rocks which could be placed among the analogues of Baikal formations of the Patom trough (Shenfil', 1991).

Direct geological data is not available on the presence of Riphean forma-tions north-east of the boreholes described above. They probably continue under formations of the Viluy syneclise.

However, the following formations correspond to the Upper Riphean: in the Anabar massif (see Fig. 51) accumulation of the Riphean Billyakh carbonate suite continued; in the Udja uplift the dolomite-siltstone Khapchanyr (400 m) and red-coloured Udja (200 m) suites; in the Kharaulakh uplift the Sietachan suite (380 m) of limestones, dolomites and tuffites; and in the Olenek uplift, siltstones, sandstones and dolomites of the Khaipazh suite. As in preceding stages, the spatial distribution of Riphean formation in the northern part of the Siberian Craton was uneven and disconnected (Shpunt, 1984).

In north-eastern Russia (Okhotsk, Omolon, Peri-Kolyma and Taigonos massifs), formations similar to the Lakhanda series are represented by terrigenous-carbonate rocks of the upper parts of the Doribin, Zarosshin and Chebakulakh suites. Analogues of the Ui series of the hypostratotype closely associated with them are composed of terrigenous formations of the Sibegan, Oldyan and Spiridonov suites (in: *Riphean Stratotypes. Stratigraphy and Geochronology*, 1983).

Manifestation of magmatic activity was evidenced in the craton at the end of the Late Riphean epoch. In the interval 780–760 m.y. mafic sills and in the interval 710–680 m.y. rocks of alkaline-ultramafic composition were formed (Shpunt, 1984).

7.1.11 Verkhoyansk-Chukchi Fold Region

The most complete sequence of the Riphean of the north-eastern fold rim of the Siberian Craton is known in the Sette-Daban range. A characteristic feature of its Upper Riphean part is the presence of mafic volcanics. In the north, in the Kharaulakh (Tuora-Sis) salient, the Upper Riphean is made up of predominantly carbonate formations containing a large number of tephroid bands. As in the Yudoma-Maya trough, sequences are marked by a sharp increase in thickness east of the Siberian Craton.

In the Peri-Kolyma anticlinorium the sequence of Middle-Upper Riphean formations is similar to that of Yudoma-Maya. Felsic and mafic volcanics play a significant role here in the roof of the Upper Riphean. In the Omolon and Okhotsk massifs similar formations (Doribin series etc.) are characterised by a significantly smaller thickness (in: *Riphean Stratotypes. Stratigraphy and Geochronology*, 1983). In the Polousnyi anticlinorium the Middle-Upper Riphean Tommot series (2.5 km) is made up of carbonate rocks and phyllites.

An analysis of the structure of Riphean formations of the region, i.e., a general increase in thickness in the east and north-east, accompanied by a predominance of sandy-shale strata and the disappearance of red rocks and hiatuses from the sequence, long ago led some researchers (Yu.A. Kosygin and L.M. Parfenov) to the conclusion of the existence of a passive continental margin in north-eastern Asia in the Riphean. To these arguments should be added only the possible presence of Riphean ophiolites within the Polousnyi block, Uyanda and Moma zones. Although the existence of a Riphean eugeosynclinal zone has long been discussed in the literature (Smirnov, 1976; Yan Zhin-shin, 1983), it cannot be regarded as proven even today. The position of ultramafics of the Bilakchan zone has not been clearly understood. These have been placed in the Lower Proterozoic on the basis of only indirect data while similar rocks of the Uyanda zone have been assigned an age of about 700 m.y. (K-Ar). Some researchers place the Yudoma-Maya trough among aulacogens (Gusev, 1973; *Riphean Stratotypes. Stratigraphy and Geochronology*, 1983).

The concepts discussed above suggest isolation of the Siberian Craton within its present boundaries at least from the Late Riphean (manifestation of mafic volcanics) and the existence of a passive continental margin from Sette-Daban to Kharaulakh due to intense rifting and separation of a block of continental crust with the formation of an oceanic basin. This block (microcontinent) included the Omolon, Okhotsk, Peri-Kolyma and Chukchi massifs, which today are separated.

7.1.12 Baikal-Vitim Fold Region

Upper Riphean formations attain maximum thickness in the two major palaeotroughs of the outer zone: the Peri-Baikal and Patom.

The most complete sequence is developed within the Patom trough. Here, as everywhere in Siberia, the volume of the Upper Riphean is interpreted differently. Some geologists commence it with the Ballaganakh series (Shenfil', 1991), some with the upper part of Kadalikan (Semikhatov, 1974; *Riphean Stratotypes. Stratigraphy and Geochronology*, 1983) while some others hold that the volume of the entire Upper Riphean in Patom upland ends with the Zhuia series (Dol'nik and Vorontsova, 1974). In the unified schemes the age of the Ballaganakh series is dated as Middle Riphean (*Resolutions of the All-Union Stratigraphic Conference*, 1983). An unconformity is seen within the terrigenous-carbonate Kadalikan series (2000 m) consisting of Dzhemkukan, Barakun and Valukhta suites along the eastern margin of the Patom upland. The upper transgressive part of this sequence (Sen suite) shears the underlying deposits and transits to the basement. The Zhuia series (1000 m), conformably resting on the Kadalikan series, comprises clay-carbonate strata of the Nikol'sk and Chenchyn suites.

South-west of the Patom upland, Upper Precambrian formations extend in an almost uninterrupted band through thin sequences of northern Baikal upland into western Peri-Baikal forming a homocline dipping westwards and complicated by folds and overthrusts.

Siberian geologists (Shenfil', 1991) united the formations of the Upper Riphean Peri-Baikal trough into a large interregional subdivision named the Baikal series (Baikal horizon). Thus Peri-Baikal represents a stratotypical region. Formations of the Baikal series rest unconformably on underlying rocks of various age and consist of three suites; hence the series is often called a three-member complex. The lower Goloustnaya suite (up to 500 m) is represented by dolomites and sandstones; the middle Uluntui (800–900 m) by shales, black limestones, siltstones and sandstones; and the top Kachergat (1400 m) by rhythmically alternating terrigenous rocks. The last suite is reliably compared with the Zhuia series of the Patom upland. The Goloustnaya suite, consisting mainly of tillite-like conglomerates, may be compared on the basis of this feature with the Dhemkukan series (Shenfil', 1991). According to its complex of microfossils and stromatolites, the Baikal series pertains to the Upper Riphean, more precisely to the upper subdivision of the Upper Riphean (Baikalian) whose lower limit is about 800–820 m.y. old, based on correlation with the sequence of the Yenisei range (Shenfil', 1991). In some older schemes the stratigraphic volume of the formations under consideration was considerably increased and these formations were partly placed already in the Middle Riphean (Dol'nik and Vorontsova, 1974).

In the inner zone of the Baikal-Vitim region, evidently in the middle of the Late Riphean, accretion of terranes of different genesis (volcanic arcs, microcontinents and marginal seas) occurred and merged into a common neocratonic structure of ophiolite sutures (Fig. 56). The presence of ophiolites has been demonstrated in the works of Klitin et al. (1975), Bozhko (1994),

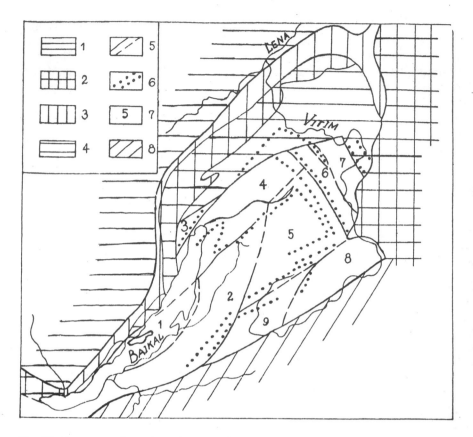

Fig. 56. Terranes and post-accretionary complex of the Baikal mountain region (after Bozhko, 1994).

Pre-Riphean crystalline formations outside and within the Baikal mountain region: 1—overlain by platform cover; 2—not overlain by cover; 3—Baikal-Patom marginal trough; 4—Mama-Bodaibo outer zone; 5—ophiolite sutures (a—established and b—assumed); 6—tectonostratigraphic terranes; 7—contours of post-accretionary troughs; 8—Udino-Vitim Palaeozoic fold system.
Terranes: 1—Baikal, 2—Barguzin, 3—Upper Angara, 4—Olokit, 5—Central, 6—Muya-Mamakan, 7—Karalon, 8—Amalat, 9—Upper Vitim.

Gusev et al. (1992) and Grudinin (1992). The duration of accretion has been established indirectly on the basis of age determinations of post-accretionary complexes which were already formed in the second half of the Late Riphean in inherited troughs as well as in newly formed rift basins. Among them are the Tsipa-Vitim, Zhanok, Tuluya, Padra, Levomama, Ondoka, Bambuy and other troughs filled with volcanosedimentary but predominantly sedimentary formations. Among the available age determinations, the most significant are 765 and 715 m.y. (Rb-Sr) characterising the age of the extrusive Padra series (Bulgatov, 1983; Mitrofanov et al., 1984) and also the age of 700 m.y. (Sm-Nd, Rb-Sr and Pb-Pb methods) pointing to the age of the Dovyren layered pluton and felsic metavolcanics in the cover of the Olokit series (Olokit region) (Rytsk, 1991). This data suggests a period of accretion at 1050–765 m.y., roughly at the level of 850 m.y., taking into consideration the manifestation at this time of diastrophism in the Yenisei range and formation of the Yangtze neocraton in south-eastern China.

7.1.13 Sino-Korea Craton, Yangtze Craton and Indosinian Region

In the type section (Yanshan aulacogen) terrigenous formations (600 m) of the Qingbaikou system resting with erosion traces on the Tanin suite of the Jixian system correspond to the lower part of the Upper Riphean. The strata contain the stromatolites Inzeria, Jurisania etc. Age values of about 850 m.y. for illite and glauconite establish the age of pre-Sinian unconformity.

Formation of the intracratonic Korean troughs Amnokkan, Hesan-Revon, Phennam and Okchon occurred as a result of rifting in the Late Riphean, or probably earlier.

In the Phennam trough Riphean formations are represented by the Sanwon system divided into three series. The lower Chikyon series rests unconformably on Early Precambrian formations and is represented by chlorite-muscovite schists, phyllites, quartzites and marbles (300–2000 m). The middle Sadangu series comprises dolomites and limestones with Collenia (1600–1800 m). The upper Kuhyon series consists of tillites and sandstones and is compared with the Nantuo tillites of the Sinian system of China. The age of glauconite from quartzites of the Sanwon system has been put at 853 m.y. (Geology of Korea, 1987).

In the Okchon trough the lower part of the Okchon group unconformably overlain by the Choson group of Cambrian age evidently corresponds to the Upper Riphean. The Choson group consists of pebbly phyllites, graphite-bearing phyllites and chlorite schists.

The Hesan-Revon trough is made up of dolomites, quartzites, mica schists of the Hesan-Revon and Hugan groups belonging to the Neoproterozoic (1000–650 m.y.) (Geology of Korea, 1987).

Within the future Yangtze craton, active processes commenced before the middle of the Late Riphean. In the Kam-Yunnan region the Middle Riphean passive margin was transformed into an active margin of the Andean type, as reflected in the formation of a granitoid belt and synchronous volcanosedimentary formations. The latter correspond to the upper part of the Kunyang group (Junshao formation in Imen, Liubatang in Jinnin, Niutoushan in Liulyan and Jegun in Yuanmu) represented by intermediate and felsic calc-alkaline volcanics, sandstones and carbonate formations. Identification of the Upper Riphean stromatolites Inzeria and Katavia in the Liubatang formation helped place it in the Qingbankou system (Yang et al., 1986). Granitoids (granodiorites and normal granites) aged 867–808 m.y. are distributed along the Kam-Yunnan axis in central Yunnan and western Sichuan in the form of a continuous meridional zone.

These events correspond to the Jinning orogenic cycle. During the first phase of deformation (900 m.y.) fold structures with a latitudinal strike arose. The last phase (800 m.y.) caused formation of a meridional fold belt and the basement of the Yangtze block.

A fairly distinct palaeotectonic environment arose in the Jiannan after Sibao diastrophism. In the west, in Bendon region, an Andean type margin obviously originated on the margin of the Sibao neocraton. The corresponding complexes are represented here by intermediate volcanics and turbidites of the Banxi group (Danzhou) developed in the Xuefengshan and Leigong mountains. The major part of the Banxi group, however, accumulated in a back-arc basin and island-arc environment. It is represented by greywacke flysch with relatively high content of Al_2O_3 and SiO_2. In the type sequence of Banxi, Yanmin, the Banxi group is represented by the lower Madiyi formation (sandstones, andesites and dolomites) and the upper Wuqiangxi (rhythmic alternating green sandstones and shales with gradational bedding). In the Yuntjyan region of south-eastern Guizhou the Banxi group is 10,000 m thick. In northern Guangxi, the Banxi (Danzhou) group lies unconformably on the Sibao group and is intruded by diabases aged 837 m.y. Andesites in the midportion of the Madiyi formation are 950 m.y. old (Rb-Sr).

Late Riphean was a period of consolidation of the Yangtze and Tarim cratons following the Yangtze or Jinning orogeny (850–800 m.y.). This diastrophism led to the final closing of interarc basins separating the Sibao and Fanjingshan island-arc systems and collision of the margin of the Sichuan microcontinent with the island arcs. Jinning orogeny was manifest in folding, intrusion of granitoids, formation of migmatisation zones in Anhui, north-western Jiangxi and north-western Guangxi. According to U-Pb and K-Ar datings, the age of these intrusions falls in the interval 840–900 m.y. Formation of molassoid collision complexes also falls in this epoch. The Luokedong formation (236 m) in northern Jiangxi, between the upper Shuangjiaoshan and lower Sinian formations in the Bunin region is represented by red molasse.

Mica schists, quartzites and marbles of the Gaoligonshan group of western Yunnan extend into Burma where they are described as the Chaungmagi group pertaining to the Qingbaikou system. They are cut by migmatites aged 806 m.y. (Rb-Sr). In formational affinity, they obviously belong to a rift complex. Evidently synchronous with Yangtze diastrophism within Kontum salient, intrusion of the palingenetic Chulai granitoids overlain by Sinian formations occurred.

Following the conclusion of Jinning-Yangtze diastrophism (850 m.y.), the newly formed Yangtze Craton represented a heterogeneous continental block overlain on a significant part by the shallow epicontinental Yangtze sea. After Jinning orogeny, the region entered the Sinian epoch (800–590 m.y.) corresponding to the younger part of the Late Riphean and the Vendian of our Precambrian scale. In the Sinian stratotypical sequence on the Yangtze Craton, two series and four formations of 1000 m in total thickness are delineated. The lower Sinian is represented from the bottom upwards by the Liantuo formation (160 m) of red sandstones and arkoses and Nantuo tillites and sandstones (60 m). The upper Sinian consists of the Dougshantuo formation (200 m) of siliceous carbonates and the Danying formation (700 m) also of carbonate composition. For rocks of the Dougshantuo formation, Rb-Sr ages of 693 m.y. are known and for the upper part of the Liantuo 740 m.y. (U-Pb on zircon). The boundary between the lower and upper Sinian lies at the level of 700 m.y.

Within south-western Sichuan, local extension occurred in the course of the Late Riphean collision in a regional field of compression. This extension resulted in development of the Susong aulacogen (Wu Geyao, 1986) filled with volcanosedimentary lower Sinian formations. In eastern Yunnan the lower part of the Chenjang formation is also correlated with lower Sinian formations. Basalts of the Chenjang formation gave an Rb-Sr isochron of 885 m.y., which slightly depresses the lower boundary of Sinian and Jinning movements in this region (Sun Jiacong, 1985).

Evidently the metamorphosed carbonate-terrigenous Poko formation of the Kontum salient filling a palaeorift and overlying Chulai granites belongs to the considered interval. Following Jinning diastrophism, much of the Yangtze Craton was relatively uplifted but rifting preceding formation of future passive margins was already manifest in the Early Sinian along the northern part of the craton facing the Qinling ocean and along the south-eastern margin adjoining the Cathaysian basin.

In western Hunan and northern Guangxi the lower Sinian is represented by three formations of the Jiangkou group: diamictites of Chang'an, mafic volcanics of Fulu and marine glacial formations of Nantuo.

Sinian formations are extensively developed within the Tarim block. In the Wuxi region the lower Sinian is made up of tillites and turbidites of the Ermenaik and Janbulak formations (Yianping et al., 1991).

7.1.14 South American Craton

7.1.14.1 *Shields.* Processes of destruction of the old basement manifest in the Early-Middle Riphean in the territory of present-day Brazil in the form of the Araxaides and Espinhaco rift structures intensified even more in the Late Riphean. In this epoch the structural plan of the Precambrian persisting to the present was essentially laid down, namely, zones of stabilised continental crust (cratons) surrounded by linear mobile zones (Fig. 57).

After conclusion of the cratonisation regime marked by felsic and alkaline magmatism, the largest Amazon (Guapore) Craton including the Guyana Shield, western part of the Central Brazilian Shield and Apa and Missiones massifs, was definitely separated. A probable continuation of the Amazon Craton is the so-called Rio de la Plata Craton including the territory of Buenos Aires and south-eastern Uruguay provinces.

The second largest craton (Sao Francisco) occupied the territory corresponding to the present-day basin of the same-named river in Minas Gerais and Bahia states. Commencement of formation of terrigenous-carbonate rocks of the Sao Francisco supergroup (Macaubas and Bambuy groups) comprising a platform cover of the craton extending over an area of 300,000 km^2 and passing into formations of adjoining fold zones can be dated as Late Riphean. The Macaubas group (1250 m) extends for 500 km north of Minas Gerais. It is made up of basal tillites overlain by a thick suite of quartzites, arkoses, siltstones and greywackes. Similar tillites and tilloids have been established in Minas Gerais, Bahia and Mato Grosso. On the eastern margin of the Sao Francisco craton, the Macaubas group experienced fold deformation and metamorphism from greenschist to amphibolite facies and constituted the Aracuai belt of north-north-easterly strike. Marine glacial rocks of this belt are correlated with Jequitai tillites within the craton. Among the other rocks filling the belts are metalutites and metapsammites, limestones and mafic volcanics. The age of metamorphism has been evaluated at 600 m.y. (Rb-Sr) (de Almeida et al., 1981).

The overlying Bambuy group (820 m) has been divided into six formations: basal conglomerates (Jequitai tillites), dolomites and limestones, siltstones with lenses of limestones, alternation of carbonate rocks and siltstones, siltstones with lenses of arkoses, and calcareous siltstones. The Una and Estancia groups represent the stratigraphic equivalents of the Bambuy group in the north-east.

An age value of 812 ± 22 m.y. was recorded for the Bebeduro formation from the lower part of the supergroup in the northern part of Chapada-Diamantina in the Lencois Craton set off from the Sao Francisco Craton, and correlated with the Jequitai.

A small cratonic region, Sao Luis, existed in the north-eastern part of Brazil in the Late Riphean. Evidently the territory now occupied by the Parana basin also represented a craton.

7.1.14.2 *Brazilides.* An extensive mobile belt continued to develop east of the Guapore Craton in the Late Riphean. A meridional fold zone with a serpentinite belt in the axial portion formed here after Uruaçuan diastrophism (1 b.y.) and inversion. On both sides of the zone, in the troughs disposed respectively along the margins of the Guapore and Sao Francisco cratons, terrigenous-carbonate formations accumulated. The western Paraguay-Araguaia trough is wheel-shaped and extends for 2500 km, disappearing under the Phanerozoic Parana and Paraiba basins. Formations falling in the age range 1000–570 m.y. (de Almeida et al., 1981; de Almeida, 1978) are crumpled into folds with a distinct westward vergence towards the Guapore Craton. Rocks of the Baish-Araguaia supergroup (3000–5000 m) participate in the structure of the northern segment of the Paraguay-Araguaia belt (north of the Goyas massif). The lower Estrondo group is represented bottom upwards by quartzites, metalutites, biotite-muscovite schists, metagreywackes, mafic and ultramafic lavas and biotite-plagioclase schists. The overlying Tocantins group is made up of phyllites, ferruginous quartzites, talc-actinolite schists and metamafic tuffs. It is unconformably overlain by Cambrian-Ordovician formations.

The lower Ciciaba group (1000–650 m.y.) and upper Corumba group (650–570 m.y.) participate in the southern structure of the belt. The Ciciaba group is represented by conglomerates, quartzites, arkoses and metasiltstones.

Westward, formations of the southern segment of the Paraguay-Araguaia belt pass into the same-aged (1000–650 m.y.) rocks of the cover of the Guapore Craton. They are represented by sandstones, conglomerates, mafic lavas, Boqui and Jacadigo groups of limestones (total thickness ranging from 500 to 2000 m) developed in Bolivia, Brazil and Paraguay. Limestones and dolomites with local tillites of the Corumba group and the Araras in Brazil, Murcielago and Tucavaca in Bolivia and the Itapocumi groups in Paraguay occupy an analogous position.

The eastern Brazilian fold belt extends along the western margin of the Sao Francisco craton for 1100 km at a width of up to 300 km. Rhythmically alternating quartzites, phyllites and dolomites of the Paranoa group (2000 m) and overlying folded analogues of the Bambuy group with tillites at the base participate in the structure of this fold belt. The rocks of the belt are intensely crumpled with a general eastward vergence, towards the Sao Francisco Craton. The Rio Preto fold system made up of phyllites, schists and gneisses falls in the northern continuation of the Brazilian belt. Phyllites gave an Rb-Sr age of 762 m.y. The Rio Preto belt was transformed later into the complexly structured Borborema province in north-eastern Brazil (Kariri belt). A possible version of the structure of the initially joined Brazilian and Kariri belts is shown in Fig. 57. The structural plan of the Borborema province or Kariri belt is characterised by the presence of major faults separating the unreworked zone of gneiss-migmatite Early Precambrian basement ('median massifs' of de Almeida) from the zone of Riphean fold complex that was formerly known as the Ceara group. The following fold zones have been delineated south-westerly from the margin of the Sao

Fig. 57. Late Riphean palaeotectonic scheme of South America (after N.A. Bozhko, 1988):

1—uplifts of the basement of cratons; 2—platform cover complexes; 3—felsic volcanoplutonic complexes of cratons; 4—aulacogen complexes (palaeorifts); 5—complexes of intracratonic ensialic geosynclines (a—volcanosedimentary and b—predominantly sedimentary);
6—complexes of ensimatic intracratonic geosynclines; 7—complexes of outer zones of intracratonic geosynclines; 8—intracratonic fold systems; 9—orogenic complexes of intracratonic
geosynclines; 10 to 13—marginal-cratonic complexes: 10 and 11—miogeosynclinal passive
margins (10—outer zones—shelf and 11—inner zones—continental slope); 12—accretional fold
complexes; 13—volcanoplutonic and greywacke complexes of island arc type; 14—oceanic
complexes; 15 to 26—formations: 15—marine terrigenous (a—arkose, b—greywacke,
c—lutite), 16—carbonate, 17—turbidite, 18—tilloid, 19—spilite, 20—trap, 21—andesite,
22—rhyolite, 23—granitoid, 24—ophiolite (tentative), 25—stromatolites, 26—tectonothermally
reworked zones of the basement; 27—thermal reworking (isotope rejuvenation); 28—folding;
29—isotope age in m.y. (a—Rb-Sr method and b—K-Ar method); 30—faults; 31—zones of
cataclastic diaphthorites; and 32—palaeostructures. Circled numbers: 1—Proto-Andean belt,
inner zone, 2—the same, outer zone, 3—Amazon craton, 4—Paraguay-Araguaia system,
5—Araxa-Estrondo fold system, 6—Brazilian system, 7—Gurupi system, 8—Sao Luis Craton,
9—Medio Coreau system; 10—Santa Quiteria massif, 11—Jaguaribe and Curu-Independencia
systems, 12—Rio Piranhas massif, 13—Serido system, 14—Nova Floresta massif, 15—Pianco-
Alto Brigida system, 16—Sergipe system, 17—Bambuy basin, 18—Espinhaco fold belt,
19—Chapada-Diamantina basin and system, 20—Macaubas basin structure of Ribeira belt,
21—Apiai system, 22—Joinville massif, 23—Tijucas system, 24—Pelotas massif, 25—eastern
Uruguay system).

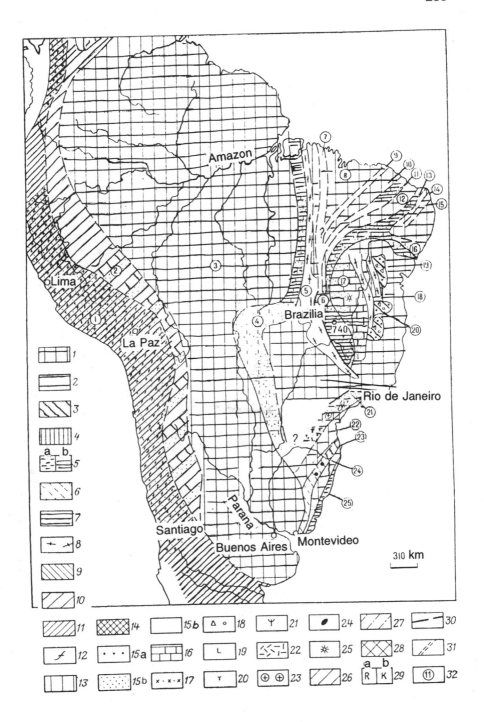

Luis Craton: Medio Coreau, Curu Independencia, Serido, Jaguaribe, Pianco-Alto Brigida, Pajeu-Paraiba, Riacho do Pontal and Sergipe. All these zones are separated by reworked basement blocks: Pernambuco-Alagoas, Coldas Brando, Rio Piranhas, Troy and Santa Quiteria (Brito Neves, 1986). The largest faults are of the strike-slip type separating and intersecting the fold systems and basement salients with a sublatitudinal strike: Sobral, Senador Pompideu, Pernambuco and Patos.

Some difference is noticed in the structure of peripheral and inner belts of the Borborema province (de Almeida et al., 1981; Brito Neves, 1986; Davison and Santos, 1989; Goodwin, 1991). Two marginal belts (Sergipe and Medio Coreau) respectively adjoining the Sao Francisco and Sao Luis cratons are made up of carbonate-terrigenous formations of the miogeosynclinal type metamorphosed to greenschist and amphibolite facies with a thickness of a few thousand metres and crumpled into isoclinal folds vergent towards the aforesaid cratons. In general, for formations forming these belts, a four-member stratigraphic division has been established (bottom upwards): 1) conglomerates and quartzites with subordinate volcanics in the Sergipe belt; 2) carbonate formations with subordinate shales; 3) quartzites, lutites and conglomerates with subordinate limestones; and 4) molasse unconformably resting on the preceding formations. Geosynclinal development of peripheral belts commenced 1100 ± 100 m.y. ago. The upper age limit as determined on syntectonic granitoids is 650 m.y. and the age of molasse 550 ± 30 m.y.

The six inner belts are predominantly formed of terrigenous rocks with subordinate volcanics metamorphosed to amphibolite facies, crumpled into isoclinal folds vergent towards the adjoining basement massifs. A general two-member stratigraphic division of the sequence has been established. Unconformably lying molasse has been noticed only in two belts, the Jaguaribe and the Pianco-Alto Brigida. The origin of inner belts is assigned to Early Middle Riphean, even Early Proterozoic (Goodwin, 1991). Two phases of deformation of the Brasiliano cycle (700–500 m.y.) have been recorded within these belts.

The structure of the south-eastern orogenic belt (Ribeira, Paraiba and Atlantic) is close to that of the Kariri belt and extends north-easterly along the Atlantic margin of the continent from the southern part of the Brazilian state Bahia to the Uruguayan boundary and consists of alternation of old reworked massifs of the basement and Riphean fold systems. From west to east the Apiai fold system, Joinville massif, Tijucas system, Pelotas massif and eastern Uruguayan system are usually delineated (de Almeida et al., 1976). These systems are respectively made up of the Porongos, Acungui, Brusque and Lavalleja groups of similar sequence. The lower part of the belt is predominantly made up of schists, the middle part is carbonate-terrigenous and the upper part terrigenous. Mafic volcanics are developed almost everywhere. Rocks are repeatedly crumpled into folds and meta-morphosed to greenschist and amphibolite facies. The age of the Acungui group has been evaluated at 900 m.y. based on stromatolites (Wernick et al., 1978). Migmatites of this series gave an age of 783 m.y. The above for-mations are overlain at places by Vendian molasse and intruded everywhere by granitoids of the same age. The massifs separating the fold systems were formed during the Transamazonian geotectonic cycle (Early Proterozoic) and reworked by Riphean tectonic activity.

Three fold systems, the Tijucas, Apiai and Sao Roque, separated by two massifs, have been distinguished in recent schemes (de Almeida et al., 1981; Brito Neves, 1986; Goodwin, 1991). A marginal salient of the Rio de la Plata Craton is exposed in the south-western extremity of the belt. The major part of this craton is concealed under Phanerozoic formations. The north-eastern continuation of this salient, according to the reconstruction of Brito Neves and Cordani (Brito Neves et al., 1991), is represented by the Luis Alves Craton together with the marginal Curitiba massif made up of an old block between the Apiai and Tijucas systems. The latter system contin-ues in the south-west up to the salient of Rio de la Plata Craton in Uruguay and constitutes the Don Feliciano pericratonic belt. The Pelotas massif sit-uated immediately to the east represents a complex batholith which, in its petrochemical characteristics, corresponds to a calc-alkaline granitoid mas-sif of subduction zones (Brito Neves, 1986). The Don Feliciano belt is further regarded as a marginal (miogeosynclinal) zone of the Adamastor Neopro-terozoic ocean; its opposite African margin corresponds to the Gariep zone (Brito Neves et al., 1991).

7.1.14.3. *Proto-Andean belt.*

Formation of the Proto-Andean belt on the western margin of the continent evidently pertains to the Late Riphean. It was formed almost everywhere on the old continental crust whose existence has been confirmed by radiometric age determinations of granulites from the Coastal Cordillera of Peru (Arequipa massif). Only in the northern and southern Andes is the crystalline Early Precambrian basement absent and an oceanic crust possibly formed here in the Riphean. The Early Precambrian gneiss-migmatite complex is overlain by volcanosedimentary, flyschoid, chlorite schists, phyllites and greywackes with subordinate bands of amphibolites. This greenschist complex is developed in the Eastern Cordillera (Colombia), Sierra de Merida (Venezuela), Central Cordillera (Colombia) and its southern continuation, i.e. Cordillera Real (Ecuador). Formations of the Sierra Pampas in Argentina, Puno region and Eastern Cordillera of Argentina (Puncoviscana formation) and Eastern Cordillera of Peru (Maranao formation) belong to this same complex.

In the eastern periplatform part of the belt these formations of flyschoid form pass into the miogeosynclinal carbonate strata of Borello (1969). It could be concluded from this that the Proto-Andean belt arose as a passive continental margin on the western margin of the South American Craton.

Within this margin one may distinguish somewhat tentatively an outer eastern zone corresponding to the shelf and an inner western zone with turbidite sedimentation corresponding to the continental slope. The large extension of these zones suggests formation of a continental margin of the Atlantic type at the site of the contemporary Andes.

The available radiometric age determinations for this complex are scant and reflect mainly the age of metamorphism of rocks (680–500 m.y.). At the same time, values of 1400–1200 m.y. for highly metamorphosed Colombian rocks (lower complex) point to processes of reworking and rejuvenation of the basement which were widely manifest in the Early to Middle Riphean on the Guyana Shield. This is also supported by the age of 1050 ± 100 m.y. recorded for gneisses of Altiplano and Colombia. This data on the age of reworking of the Andean basement indirectly supports Borello's view that formation of flyschoid Andean beds occurred in the interval 1100–580 m.y., i.e., Late Riphean and Vendian.

7.1.15 African Craton

The Late Riphean was marked here by formation of a platform cover in vast basins, syneclises of Western and Central Africa, and also by the origin and development of a new generation of various types of mobile zones (Fig. 58).

7.1.15.1 *West African Craton.* This structure was distinctly separated in the Late Riphean as a result of the formation of the Taoudenni Trans-Sahara belt. The zone of formation of the platform cover embraced the territory of the present-day Taoudenni and Volta syneclises joined into a single basin. In the northern part of the Taoudenni syneclise are seen the well-studied Upper Riphean sequences combined by Trompette (*West African Orogens* ..., 1991) into supergroup I. They have a three-member structure and are characterised by a thickness reaching 2 km. In the stratotypical locality (Mauritanian Adrar), the lower Char group is mainly made up of quartz sandstones and schists. Clauer (West African Orogens ..., 1991) obtained age values of 998 ± 32 and 1058 m.y. on clay minerals of the group by the Rb-Sr method and 990 m.y. on glauconite from the correlative series in Mali (Bozhko, 1984). The middle Atar group is made up of shales, limestones and dolomites and is characterised by Rb-Sr ages of clay minerals ranging from 890 ± 35 to 775 ± 52 m.y. Bertrand-Sarfati identified the complex of Upper Riphean stromatolites. The upper Assabet el Hassaine group, represented by sandstones, shales and conglomerates, correlates with the volcanosedimentary Hoggar beds for which age values in the interval 836–635 m.y. are known (*West African Orogens* ..., 1991).

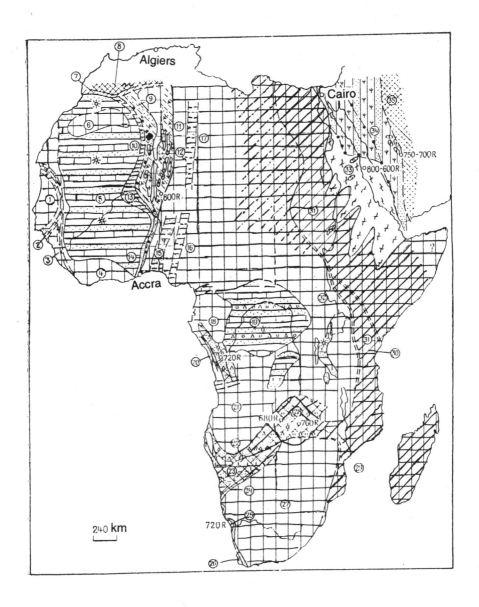

Algiers

Cairo

Accra

240 km

290

In the south-western and southern margins of the syneclise the Madina-Kuta series (Guinea and Senegal), Sotuba, Sikasso and Kuluba (Mali and Burkina Faso) and in the south-eastern part (Gourma trough), Irma, Hombori or formations 1, 2, 3, 4 and 5 attaining a thickness of up to 8 km correspond to the age level considered. The nature of the Gourma trough is not clear and it is interpreted as an aulacogen, foredeep or miogeosyncline.

A similar three-member Upper Riphean sequence has been established in the Volta basin where it is represented by the Morago series and partly by the Tamale series in Ghana (Bozhko, 1984) corresponding to the Bombouka supergroup of Affaton; clay minerals from this supergroup recorded an age of 993 ± 65 m.y. (*West African Orogens* ..., 1991).

7.1.15.2 *Central African craton.* Upper Riphean formations of the major Congo syneclise are exposed along its periphery in north-eastern Zaire, the Central African Republic, western Zaire, Gabon and Angola and are concealed under Phanerozoic formations in the central part.

In north-eastern Zaire, Riphean formations are combined into the Linda supergroup including the Ituri (140–190 m), Lokoma (400–850 m) and Aruwimi (1650–1750 m) carbonate-terrigenous groups. The Alolo shales of the Aruwimi group are dated 700 ± 9 m.y. (Rb-Sr). The Akwokwo tillites of the Ituri group are aged 950 ± 50 m.y. Thus, with the exception of the concluding Banalia sequence of arkoses, the Linda supergroup is older than 700 m.y.

In the Central African Republic the sequence commences with tillites correlated with the Akwokwo tillites above which lies the Bugbala group, an analogue of the Lokoma group, Nacondo quartzites and Bakuma and Jalanga formations. The latter is correlated with the Alolo shales of Zaire (Goodwin, 1991).

The south-western flank of the Congo syneclise is made up of platform analogues of the Western Congolides folded complex.

Fig. 58. Palaeotectonic scheme of Late Riphean of Africa (1000–700 m.y.) (after Bozhko, 1988)..

Legend same as in Fig. 55. Palaeostructures: 1—Mauritanian system; 2—Kuluntou system; 3—Rokell system; 4—Leon-Liberian Shield; 5—Taoudenni syneclise; 6—Reguibat Shield; 7—Anti-Atlas, outer zone; 8—Anti-Atlas, inner zone; 9—Ougarta system; 10—western Ahaggar system; 11—In-Ouzzal block; 12—central Ahaggar; 13—Gourma zone; 14—Atakora zone; 15—Western Nigerian trough (Anka); 16—Eastern Nigerian trough (Maru); 17—Tiririne system; 18—Chaillu massif; 19—Congo syneclise; 20—Western Congolese system; 21—Angola massif; 22—Damara system, northern outer zone; 23—Damara system, inner zone; 24—southern outer zone; 25—Gariep system; 26—Malmsberry system; 27—Kalahari Craton; 28—Katanga system; 29—zone of Manika cataclasites; 30—zone of Bubu cataclasites; 31—tectonothermally reworked Mozambique belt; 32—Asva zone; 33—Arabia-Nubian neo craton; 34—Arabian island arc belt; and 35—Arabian oceanic basin.

In their tour folded deposits of the Katangides pass into platform analogues in the Shaba province forming 'Katanga Bay'. Accumulation of the
Bukoba supergroup continued in the same-named trough of the Tanzania Craton. Here the upper part (dolomites) of the Busondo group and
the Uha carbonate-terrigenous-volcanic group containing Gagwe lava aged
815 ± 14 m.y. evidently pertain to the Upper Riphean.

The Late Riphean was marked in Africa by the formation of a new generation of intracratonic mobile zones.

7.1.15.3 *Mauritania-Senegalese belt.* According to Chiron (1974), a
median geosynclinal zone ('Serpentinite Belt' or 'Mauritanian Axis') was
formed in the Mauritania-Senegalese belt on both sides of which troughs
were disposed symmetrically. In the western trough the Quesh-Quesh,
Akjoujt (Mauritania) and Bakal-Kuluntou (Senegal) groups were deposited
and in the eastern trough the Sangarafa and Bassari series. Some of these
series (Quesh-Quesh and Akjoujt) contain significant amounts of andesite-
basalts, andesites and rhyolites.

New data has been given in recent works on the geological structure
and geodynamic evolution of the Mauritania-Senegalese belt and its south-
eastern continuation, the Rokellides system. In the light of this information,
let us consider the different segments of the belt from north to south.

Mauritanides (after Lecorche et al., 1991). The outer zone adjoining the
West African Craton is comparable with the foreland and does not exhibit
the increase in thickness of formations that is typical of miogeosynclines.
Windows of the Birrimian basement are present in it. In the north, in the
region of Akjoujt, the Regs complex is analogous to the Adrar supergroup I.
Conglomerates and mixtites of Jonaba and Mbanu, comparable with tillites
of the base of supergroup II, have been respectively detected in central and
southern Mauritania but formations overlying supergroup II have not been
detected.

In the present-day structure of the inner zone, the Upper Precambrian is
exposed in the so-called zone of infracrustal allochthons of the central and
southern Mauritanides but are poorly exposed in the north, especially in the
region of Akjoujt and western Sahara cover. The ophiolite complexes of El
Auja and Qued Amur in the central Mauritanides and Hassilay and Sene-
galese Gabon in the southern Mauritanides made up of metamafic volcanics,
metagabbro and serpentinites have been distinguished in the composition
of this zone.

The Farkaka complex, made up of metasedimentary rocks, quartzites
and kyanite-staurolite schists as well as amphibolites formed after low titanium continental tholeiites, represents rift formations of the continental margin. Calc-alkaline magmatic complexes are represented in the western part
of the belt as Kelbe granitoids and El Neikat differentiated metavolcanics.

In the southern continuation Guidimaka granodiorite and Mbout volcanics in association with Oua-Oua metasedimentary strata and the Niocolo Koba complex even more south in the Bassari zone correspond to the above magmatic complexes.

The geodynamic evolution of the Mauritanides is depicted as follows:

1100–700 m.y.—accumulation of sedimentary strata of supergroup I on the base of the West African Craton;

700–680 m.y.—continental rifting and subsequent formation of oceanic crust of indeterminate but limited width. The rifting phase was marked by intrusion of Bou Naga syenites aged 680 ± 10 m.y. (Pb), extension by dykes of Farkaka amphibolites and opening of ocean basin by ophiolites; and

680–640 m.y.—partial fusion of crust and formation of calc-alkaline complexes in an environment of an ensialic island arc. The Niocolo Koba granite massif formed in this cycle is dated 680–685 m.y. (Pb).

Bassarides (after Villeneuve et al., 1991). The Bassarides represent a segment of the Mauritania-Senegalese belt within eastern Senegal and northern Guinea consisting of two branches separated by the Phanerozoic Yukunkun and Bove basins. The north-north-eastern branch (Bassari) is made up of two parallel and fault-separated complexes, the Termess and Guingan groups, represented by basalts, tuffs, breccia, quartzites, jaspers and shales. Serpentinites have been described in the southern part of Senegal.

The north-eastern branch (Kuluntou) is made up of the Niocolo Koba (mafic and felsic lavas, tuffs, jaspers and conglomerates) and Kuluntou (dacites, rhyolites, tuffs, shales and quartzites) groups separated by faults but an unconformity is assumed between them. The Niocolo Koba granites, as already mentioned, are dated 685 ± 15 m.y. The age of calc-alkaline Kuluntou magmatism is also put at 680 m.y. (Rb-Sr) as determined on Linki-Kuntu granites. Tholeiite volcanism of Bassari has not been dated but is tentatively estimated at 700 m.y.

A westward inclined suture has been established between the Bassari system and the West African Craton based on geophysical data.

The Late Riphean geodynamic model of the Bassarides assumes an initial extension of crust and formation of the Madina-Kuta basin at the level of 1050 m.y., destruction of the West African Craton with formation of a continental rift and possible limited opening of a Red Sea type basin between 800 and 700 m.y., and formation of the Termess and Guingan groups. In the period 700–675 m.y., the rift closed with supposed subduction under its western margin and formation of the calc-alkaline complexes Niocolo Koba and Kuluntou on the western flank of the rift.

Rokellides. The Rokellides belt extends for 600 km from western Guinea through Sierra Leone into Liberia. The presence of Upper Riphean formations within it has not been established but could be supposed from the stratigraphic considerations of various authors. According to the latest scheme

(Culver et al., 1991), the Upper Precambrian-Lower Cambrian Rokell group pertains to the interval corresponding to the Vendian and hence the problems concerning this belt will be studied in the next chapter.

7.1.15.4 *Trans-Sahara belt* (Hoggar-Atakoran, Pharusian-Dahomea and Libya-Nigerian). This belt represents the largest of the complex African Precambrian structures formed in the Late Riphean east of the West African Craton and thus determining the corresponding boundary of the latter. In the east the Trans-Sahara belt adjoins the intensely reworked margin of the East Saharan Craton. The Mali-Nigerian syneclise divides the belt into two parts corresponding in the present-day structure to the Tuareg (Ahaggar) Shield in the north and the Akwapim-Nigerian Shield in the south.

Tuareg (Ahaggar) Shield. This segment of the belt has recently been well studied by Caby, Bertrand, Boullier and other (Fig. 59). It is divided into four zones.

1) Miogeosynclinal zone of the eastern margin of the West African Craton made up of terrigenous-carbonate rocks. Its fragments have been established in Gourma where the thickness of formations goes up to 8 km (formations 1 to 5). In characteristics, this outer zone represents a remnant of the flank of a major rift structure arising during formation of the belt. In the present-day structure this zone is characterised by a nappe structure.

2) Eugeosynclinal zone. The Pharusian belt has been divided in half by the Archaean In-Ouzzal block. Taking the data for Adrar Iforas into consideration, Boullier (1991) divided the Pharusian belt into the east Pharusian belt, west Pharusian belt, central Iforas batholith and Tilemsi island-arc region.

Within the west Pharusian belt, the old basement is made up of Archaean granulite and a Lower Proterozoic granulite-free complex. Riphean formations are represented by carbonate 'stromatolite' series of the platform type resting on a crystalline basement and equivalent to formations of the margin of the West African Craton, in particular the Atar group (Goodwin, 1991). In Adrar-Iforas the Kidal complex is correlated with the aforesaid formations. The latter are overlain by basalts and intruded by numerous ultramafic and mafic sills, stocks and laccoliths covering a total area of 2000 km^2. The oldest mafic pretectonic sills intruding the stromatolite series are dated 793 $\pm$ 32 m.y. (Boullier, 1991). A distinct lithological similarity between the stromatolite series and the Taoudenni syneclise formations as well as the facies transitions between them suggest the existence of a single West African-Tuareg Craton, at least in the interval 1035–820 m.y., following which rifting commenced and separated the West African and Tuareg Cratons (Caby et al., 1981).

According to Boullier (Boullier, 1991), ultramafic-mafic intrusions and lavas were formed under conditions of back-arc spreading during subduction. The overlying Série Verte is almost unmetamorphosed and made up of greywacke flysch in association with andesite lavas and pyroclastic rocks of

294

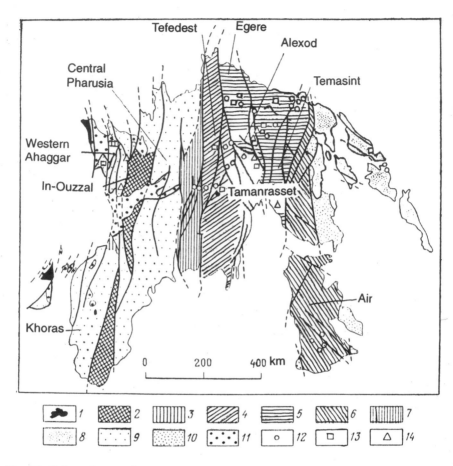

Fig. 59. Schematic map of structural and geochronological regions of Ahaggar (after Bertrand and Lassere, 1976):

1—salients of the West African Craton; 2—In-Ouzzal granulite block (2.1–3 b.y.); 3—basement of west Ahaggar aged 2 b.y.; 4—basement reworked during Pan-African orogeny; 5—basement reworked during Kibaran orogeny; 6—Air basement of undefined age; 7—Pharusian; 8—Pretiririne basement of undefined age; 9—Pharusian formations that experienced only Pan-African orogeny; 10—Tiririne formations; 11—molasse of Pan-African orogeny; 12—present-day volcanoes; 13—site of In-Ouzzal granulite facies; 14—eclogites.

typical calc-alkaline series. The Série Verte adjoins the stromatolite series through a tectonic contact and unconformably overlies quartz-diorites aged 696 m.y. and is intruded by rhyolite dykes dated 634 m.y. (Boullier, 1991).

In the southern part of western Hoggar, in the Tilemsi valley and along the western margin of the Adrar-Iforas massif, a volcanosedimentary complex of greywackes, andesites, basalts and felsic rocks evidently equivalent to Série Verte was formed. The Tilemsi magmatic island arc (Boullier, 1991) with a width of 100 km represents a bimodal series of tholeiite metabasalts and rhyodacites overlain by turbidites. The rocks are affected by numerous intrusions of gabbro-norites and diorites with U-Pb datings of 710 ± 6, 726 and 635 m.y. The evolutionary model of this zone (Caby et al., 1981) includes formation of the Tilemsi intraoceanic island arc 730 m.y. ago; high temperature metamorphism at the level of 710 m.y.; and accumulation of a volcanoclastic complex caused by erosion of andesites at 695 m.y.

In the east Pharusian belt (central Pharusian trough; Goodwin, 1991) the stromatolite series corresponds to carbonate platform formations of the Timesselarsine series (Pharusian I) resting on an Eburnean basement with calc-alkaline granites and quartz diorites (Taklem batholith) for which uranium-lead datings are 868 ± 8 and 839 ± 4 m.y. The Timesselarsine series is crumpled into isoclinal folds. This data points to rifting succeeded by subduction (Boullier, 1991) and orogenic events in this zone at the level of 850 m.y.

The Pharusian II complex rests unconformably on Pharusian I and is represented by the terrigenous Amded series and overlying terrigenous-volcanic calc-alkaline Irrelouchem series for which an Rb/Sr isochron of 680 ± 36 m.y. has been recorded. The rocks are intruded by granitoids that are geochemically close to volcanics. The Pharusian II complex corresponds to an old island-arc environment or active continental margin (Boullier, 1991).

3) Polycyclic central Hoggar represents an extensive region of development of the Eburnean basement (Suggarian) reworked in the Pan-African epoch.

4) Eastern Hoggar is related to the assumed 'East Saharan Craton' (Bertrand et al., 1978). Late Precambrian is present here in the Tiririne belt extending for a distance of 750 km along the meridian 8°30′ E at a width of 100 km. It is filled with Tiririne carbonate-terrigenous formations (8000 m) resting on a gneiss basement intersected by a batholith aged 729 ± 8 m.y. (U-Pb), metamorphosed and intruded by gabbro, diorites and granodiorites 660 ± 5 m.y. (U-Pb) ago.

Akwapim-Nigerian Shield. A lateral zoning from west to east is noticed within this shield too:

1) Miogeosynclinal zone made up of carbonate-terrigenous formations of the Oti trough on the eastern margin of the Volta basin passing westwards into the platform cover of this basin.

2) Akwapim (Atakora) zone extending for about 900 km located in the Akwapim (Atakora) mountains in Ghana, Togo and Benin, is characterised by an allochthonous nappe series made up of folded and metamorphosed quartz-schist rocks of the Togo series (Atakora or Akwapim) in the eastern part of the belt and tillites, sandstones, shales, dolomites, basalts and

rhyolites of the Buem series in the west. The Togo series has been compared with the Morago and Bassa series (lower Volta) of the Volta basin and the Buem series with tillites and supratillites of Oti and Tamale series (central Volta) of this same basin (Bozhko, 1984). This concept was followed in later studies (Goodwin, 1991). Recent mapping carried out by Jones (1985) adopted the structural succession of Buem-Togo as a stratigraphic standard and the entire sequence is compared with the central Volta of the Volta basin (Goodwin, 1991).

In such an interpretation the Atakora belt would essentially belong to the Vendian. If, however, the first correlation is adopted, the Atakora series should pertain to the Upper Riphean.

Small bodies of serpentinites are traced within the Atakorides forming a chain parallel to the main fault zone regarded as a possible suture zone inclined eastwards (Black, 1980; Goodwin, 1991).

3) Polycyclic Nigerian complex. The geological structure of this region has not yet been adequately studied. Vast expanses of Nigeria are made up mainly of gneisses, migmatites and highly metamorphosed supracrustal rocks (old metasediments) of Lower Precambrian reworked in the Pan-African epoch.

Supracrustal weakly metamorphosed formations (juvenile metasediments) are represented by mica schists, phyllites and quartzites. They fill the narrow rift zones Maru, Anka and others (Holt et al., 1978) laid evidently in the Late Riphean and intruded by granites aged 700 to 600 m.y. In structure, this Nigerian zone mostly resembles north-eastern Brazil.

Given the above data, the general structural features of the Trans-Sahara belt can be recognised:
1) Disposition between two pre-Riphean megablocks—the West African Craton and the reworked margin of the East Saharan Craton.
2) Sharp wedging out of a Late Precambrian geosynclinal-fold zone from north to south (Pharusian-Atakora belt).
3) Reduction in volume of alpine-type ultramafics, gabbro and calc-alkaline volcanism in the same direction (until their total disappearance).
4) Sharply manifest asymmetric zoning with western miogeosynclinal and eastern eugeosynclinal zones, extensive development of nappe structures and meridional strike-slip faults.

There is no doubt that formation of the Trans-Sahara belt was associated with continental rifting. This is supported by dyke complexes extending along the margin of the West African Craton and also intruding the stromatolitic Hoggar series and In-Ouzzal block. The first complex is noticed in Tanezrouft and is manifest as a system of extended parallel dykes trending meridionally for a distance of about 500 km. Dykes are represented by rhyolites, syenites and intrusions into an earlier mafic dyke complex. Caby and colleagues demonstrated the alkaline and subalkaline nature of these magmatites and

297

also their origin from the upper mantle. A similar type of dyke complex is noticed at a distance of 120 km along the western tectonic boundary of the In-Ouzzal block. Its composition is also distinctly alkaline and it intruded into a much earlier mafic dyke complex. This data suggests that formation of this mobile zone occurred in an environment of extension close to continental rifting conditions.

Further history of the evolution of the Trans-Saharan belt is variously interpreted. According to the more widely accepted approach, evolution of the belt corresponded to the Wilson cycle and concluded with a collision and suturing between the West African Craton and Pan-African mobile belt (Black, 1980; Caby et al., 1981; Boullier, 1991). Other interpretations assume a general intracratonic character of the Trans-Saharan belt (Fig. 60). Evaluation of the width of opening of the primary basin is of vital importance.

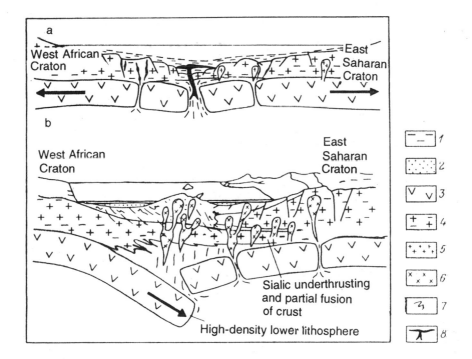

Fig. 60. Sketch showing evolution of the Pharusian belt between the West African and East Saharan cratons (after A. Kröner and Clauer, 1979):

a—opening of Pharusian ocean and b—closing of Pharusian ocean; 1—sediments; 2—volcanic flysch; 3—lower lithosphere; 4—lower crust; 5—calc-alkaline granitoids; 6—calc-alkaline and alkaline granitoids; 7—island-arc complex; 8—mafics and ultramafics.

Bertrand and Caby (1978) assume 'an opening of 100–150 km wide ocean basin' in the course of subsequent 'basification'. This width corresponds to the opening of the present-day Red Sea. It is difficult to check the reliability of these values which generally correspond to a limited character of destruction and opening of an ocean basin. In our view, this is the most appropriate interpretation, however.

The model of an extensive ocean and development conforming to the Wilson cycle cannot be accepted unreservedly for the following reasons:

1) A full sequence of ophiolite association is nowhere encountered within the Trans-Saharan belt. The bulk of the ultramafic and mafic rocks is intruded in the basement of the craton and formations of the platform cover. Thus the question of formation of a true oceanic crust remains unsolved.

2) A large opening should be excluded on the basis of data of palaeomagmatism pointing to the similarity of Precambrian polar wander curves of Gondwana cratons (Fig. 61), which led to the conclusion regarding the existence of a single Proterozoic supercontinent (Piper, 1982, 1983).

3) The period of opening of the main oceanic basin still remains unclear. Destruction of continental crust in the western Pharusian belt, well dated at the level of about 800 m.y., took place after manifestation of calc-alkaline magmatism and folding in the eastern Pharusian belt or coincided with it in time. Boullier's model (Boullier, 1991) of marginal back-arc basin does not solve all the problems of the Trans-Saharan belt.

4) 'Degradation' of the eugeosynclinal zone in terms of width as well as nature of formations and magmatism occurred in a southerly direction. The plate-tectonics model for Atakorides encounters far more problems than for Hoggar due to the greater 'ensialicity' of the former. This tendency finds logical confirmation in continuation of the Akwapim-Nigerian Shield in Borborema province of Brazil where Riphean belts are essentially ensialic.

Evidently the vast extension embracing West Africa in the middle of the Late Riphean was uneven and weakened southerly and easterly. The Trans-Saharan belt should evidently be regarded as a rifting 'apophysis' that intruded wedge-like deep into the continent from the sublatitudinal Anti-Atlas region, part of the Proto-Tethys.

7.1.15.5 *Western Congolides.* In the Late Riphean, a complexly branched major mobile belt arose in the territory of Southern and Central Africa. The eastern part of the Proto-South-Atlantic belt that existed in the form of a single zone before the disintegration of Gondwana and is now divided into South American and African halves represents the Peri-Atlantic segment of this mobile belt. The African half is made up of fold systems of the Western Congolides, submeridional coastal branch of the Damarides and the Gariep and Saldania (Malmsberry) systems.

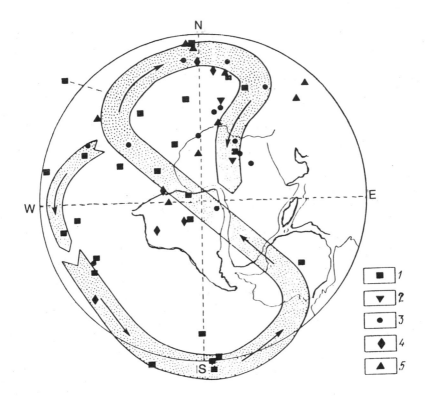

Fig. 61. Curve showing polar displacements for Gondwana continents (shaded) in the interval 1000–400 m.y. (after McElhinny and Embleton, 1976, modified).

Position of poles determined from rocks of different ages (from 1200 m.y. to Early Palaeozoic): 1—from Africa; 2—from Antarctica; 3—from Australia; 4—from India; 5—from South America.

The Western Congolides belt extending parallel to the coast of the Atlantic Ocean through Gabon, Congo, Zaire and Angola is essentially ensialic and is made up of formations of the Western Congo supergroup resting on Lower-Middle Riphean Mayombe supergroup formations. Quartzites, sandstones and shales of the Sansikwa group (up to 1.5 km) lie at the base of the sequence. On top is a bed of Lower Tillite (400 m) alternating with limestones, shales and sandstones of the Upper Shiloango group (up to 1 km). The Upper Tillite (up to 150 m) alternating with a 'shale-calcareous group' up to 1.2 km thick lies higher up in the sequence with a hiatus. These formations are overlain by the Mpioka quartzite-shale group (up to 5 km). The sequence terminates with the Inkisi continental group. It is admitted that the Western Congo supergroup rests unconformably on Mativa granites intruding the Mayombe group whose age is 1027 ± 56 m.y. (Cahen et al., 1984). At the same time, the reliability of this value is rather dubious.

This helps enlarge, according to some researchers, the stratigraphic volume of the Western Congo supergroup by including in it the volcanosedimentary formations of the Zadinian and Mayombe groups. The main phase of folding (pre-Inkisi) of the Western Congolides is dated as the age of the Noqui granites at 734 ± 10 m.y. (Rb-Sr and U-Pb). The second phase (post-Inkisi) is estimated in the interval 734–625 (600) m.y. Basalts of the Sansikwa group are dated at 970 m.y. (Goodwin, 1991). Based on stromatolites, the shale-calcareous group is compared with the Mauritanian supergroup I, which is considered younger than 777 ± 19 m.y. Evidently the Western Congo supergroup accumulated in the interval 1050 to 625 m.y. (Goodwin, 1991) and corresponds to the Upper Riphean except for the Inkisi formation.

Formations of the supergroup are crumpled into folds, verging easterly and flattening out in the same direction, and pass into the horizontally situated rocks of the Congo Craton.

The suggested geodynamic models assume the opening of an oceanic basin in the inner zone of the belt (Zadinian-Mayombe), with subsequent subduction and collision, or a model of continental rifting with extension, thinning of crust, volcanosedimentary filling and subsequent folding (Goodwin, 1991).

A submeridional coastal branch of the Damara fold system made up of carbonate-terrigenous formations, tilloids and volcanics of the Damara supergroup is situated in the southern continuation of the Western Congolides, within Namibia.

7.1.15.6 *Gariep system, Saldania system, Damara-Katanga belt.* At the boundary of the Republic of South Africa and Namibia, in the region of the Orange River estuary lies the Gariep system of the African part of the Proto-Atlantic belt. It is filled with the same-named group correlated in age with the Damara supergroup. The structure of the fold system comprises an eastern miogeosynclinal zone made up of sandstones, mixtites, dolomites, quartzites of Stinkfontein formation (3000 m), quartzites, gritstones, siltstones and dolomites of Hilda formation (up to 2000 m) and tillites and dolomites of Numees formation; The western eugeosynclinal zone is filled with basalts and andesites of Grootderm formation (4000 m), dolomites, greywackes, quartzites of Oranjemund formation and greywackes and shales of Holgat formation (up to 600 m) (Kröner, 1974).

The subvolcanic granitoid complex comagmatic with the volcanic Gariep group is dated 920 ± 10 m.y. (Rb-Sr), granite intruding Stinkfontein formation 780 ± 10 m.y. (U-Pb), and deformation and metamorphism 700 m.y. (Kröner, 1974). The group was deposited in the interval 950–700 m.y.

Rocks of the Gariep group are crumpled into folds and disturbed by major overthrusts. The presence of volcanic rocks with characteristics of ophiolites in the overthrust slices and the development of glaucophane schists provide justification for a plate-tectonics model of evolution of the Gariep system.

The Gariep system continues into the south-western Cape province in the form of the Malmsberry fold system. Within it are similarly distinguished a western eugeosynclinal zone made up of flyschoid and Tygerberg amygdaloidal lava formations and an eastern miogeosynclinal zone, in which the mainly terrigenous Frenchhook and Moorreesburg formations and the Porteville formation equivalent to them take part (Hartnady et al., 1974).

Thus some change in conditions from the almost amagmatic West Congolese system to the Gariep and Malmsberry systems with distinctly manifest volcanic eugeosynclinal zones is noticed from north to south in the Riphean mobile belt stretching along the Atlantic coast of Africa from Gabon to Cape province of the Republic of South Africa. This mobile belt is characterised by the presence of branches setting off from its eastern boundary into the north-eastern and sublatitudinal directions and forming triple junctions. Such a triple junction is distinctly noticed in south-western Africa between the two branches of the Damarides: submeridional coastal and eastern intracontinental.

The main Damara-Katanga branch is set off from the main submeridional belt of Namibia and extends north-easterly up to Shaba province (Zaire) terminating arcuately in the body of the basement. This fold system has now been well studied (Fig. 62) and comprises the Damara supergroup of formations.

Formation of the Damara belt and deposition of the Nosib group began 1000–900 m.y. ago in three isolated rifts 70–50 km wide, by a thick clastic accumulation accompanied by felsic and alkaline volcanism, intrusion of carbonatites and syenites aged 840 ± 12 m.y. (Kröner, 1980). Subsequently sedimentation covered the entire present-day territory of the Damarides but a very deepwater (eugeosynclinal) trough was formed in the central and southern parts while accumulation of shallow-water carbonate-terrigenous formations (miogeosynclinal zone) occurred in the north under a shelf condition. The latter formations are represented by carbonate-terrigenous rocks of the Otavi group (3000 m) overlain by the Mulden group of molasse aged 660–570 m.y. Thus the period of accumulation of the Otavi group occurred in the interval 840–660 m.y.

In the eugeosynclinal zone, in Khomas trough, the filling of the Nosib rifts was overlain by metaturbidites and mafic metavolcanics of the Swakop group. Flysch was overlain by a thick formation of mica schists which includes, in Windhoek region, a narrow and elongated (350 km) zone of Matchless amphibolites made up of metamorphosed tholeiite-basalts, gabbro and mafic and felsic pyroclastic rocks. These rocks correspond to a complex of limited oceanic rifting and were formed in the interval 830–760 m.y. (Kröner, 1980).

302

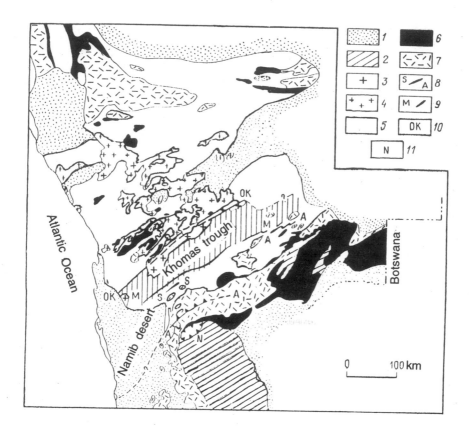

Fig. 62. Sketch showing tectonic zoning of the Damara belt (after Kasch, 1983):

1—Karroo and much younger formations; 2—Namama group; 3—Donkerhok granites; 4—Salem type granites; 5—Otavi and Swakop groups; 6—Nosib group; 7—Predamaran basement; 8—Shlezen-Amerugen ultramafics; 9—Matchless amphibolites; 10—Okajandja fault; 12—Naukluft nappe.

The littoral branches of the Damarides represent minor modifications of the intracontinental zone. The northern branch (Kaoko belt) is made up of thick (5000 m) Nosib quartzites and Swakop carbonate rocks in the east; in the latter, metamorphism and deformation increased westwards along with an increase in the amount of terrigenous material. Similar features are noticed in the structure of the southern branch.

Tectonic models of the Damara system are associated with the effect of hot spots (mantle plumes) causing triple junctions. Later versions of the evolution of the system vary from models of intracontinental ensialic rifting to the development of an oceanic basin of the Red Sea type or even a normal ocean (Goodwin, 1991). The most reliable from our point of view is the development of an intracratonic Damara geosyncline (eastern branch) as a branch of the Proto-South Atlantic palaeo-oceanic basin identified in Gondwana reconstructions at the junction of South America and Africa.

The Katanga fold system, arcuate (Lufilian arc) in its northern extremity within Shaba province, represents the north-eastern continuation of the Damarides within Zaire and Zambia. It can justifiably be regarded as part of a single Damara-Katanga geosynclinal trough which existed in the Late Riphean. The Katangides are made up of Katanga supergroup formations (Cahen et al., 1984; Goodwin, 1991). The lower Roan group (about 2 km) is composed of quartz sandstones, shales and dolomites and ends in conglomerates, arkoses and phyllites of the Mwashya formation (600–800 m). The Great Conglomerate (300 m) replacing the Mwashya rocks upwards along the sequence, is represented by tillites with lenses of shales and sandstones. Mafic lavas are associated with them at places. The overlying Lower Kundelungu group (up to 2 km) is composed of sandstones and conglomerates passing on top into stromatolitic Kakontwe limestones. The Small Conglomerate (80 m) following it is represented by tilloids of marine-glacial origin. The Upper Kundelungu (3 km), next in the sequence, has a three-member structure and is made up of carbonate-terrigenous formations with a characteristic red-coloured Plateau suite (1500–500 m) ending the Katanga sequence.

The similarity of sequences of the Katanga and Western Congo systems is striking. A good correlation is noticed also between these sequences and the Damara supergroup sequence. At the same time there are no analogues of the Great Conglomerate and the Lower Tillite in the Damara system. These were possibly eroded before the accumulation of the Otavi and Swakop groups.

The lower Katanga age limit is determined by its bedding on Nchanga granites (1200–1100 m.y.). The age of granite fragments from the Great Conglomerate is 976 ± 10 m.y. and of associated lavas 948 ± 20 m.y. Lufilian orogeny affecting the Upper Kundelungu has been dated in the interval 656–602 m.y. The age of the Roan group is thus 1100–950 m.y., of the Lower Kundelungu 950–850 m.y. and of the Upper Kundelungu 850–602 m.y. (Cahen et al., 1984; and Goodwin, 1991).

Folded and weakly metamorphosed Katanga rocks of the Lufilian arc pass towards the north-east into horizontally lying platform deposits of the 'Katanga Bay' reaching Lake Tanganyika. The isolated Bushimaye basin located in western Shaba and eastern Kasai comprises conglomerates, sandstones, dolomites and mafic lavas of the Mbuji Mayi (Bushimaye) supergroup and is correlated with the Roan group of Katanga based on lava datings of 948 ± 20 m.y. and pegmatites 976 ± 10 m.y. (Goodwin, 1991).

From the northern margin of Lake Tanganyika to Mobutu Lake, the Itombwe belt extends for a distance of 800 km. This belt comprises the same-named supergroup resting on Kibaran formations and has a two-member structure. The lower terrigenous Nyakasiba group (1000–1500 m) is replaced on top by mixtites (2500 m) passing upwards along the sequence into the sandy-clay Tshibangu group (2000 m). The rocks are crumpled into open folds and intruded in their lower part by alkaline granites and syenites aged 774 ± 44 m.y. The entire group is intruded by granitoids dated 660 ± 10 m.y. (Cahen et al., 1984).

The Zambezi belt represents a narrow arcuate zone of highly metamorphosed, intensely deformed formations including gneisses of the basement, metasedimentary rocks and granitoid intrusives situated between the northern margin of Zimbabwe Craton in the south and the Karroo basin in the north. The boundary with the craton is manifest by overthrusts along which gneisses and metasedimentary rocks are thrust in a southerly direction onto the craton. The Urungwe klippe is seen 40 km away from the overthrust front. The Zambezi belt comprises complexes of different ages. The old basement (Ar-Pr$_1$), intensely reworked by Pan-African metamorphism (575–430 m.y.), is unconformably overlain by the much younger Macuti group correlated with the Katanga. All the rocks are intruded by the pretectonic Mpande batholith aged 1100 m.y. and the syntectonic Ngoma batholith aged 820 ± 7 m.y. The age of Zambezi deformation preceding Pan-African reworking has been determined as 950–850 m.y. (Goodwin, 1991).

The southern latitudinal branch (eastern Saldania system) of the main meridional trunk of the Proto-South Atlantic belt is traced along the southern tip of Africa in three isolated outcrops among Phanerozoic formations, i.e., Kango, Gamtus and Kazimans (George). The carbonate-terrigenous formations of Kango and Gamtus accumulated in a riftogenic ensialic trough and are intruded by mafic dykes aged 782 m.y. (Hartnady et al., 1974), with which the flyschoid clay formation of Kazimans outcrop beside George is correlated.

7.1.15.7 *Mozambique belt.* In the complex polycyclic structure of the Mozambique belt following Irumidian (Kibaran) events in the Namama, Lurio and Adola zones, the palaeotectonic environment continued to be characterised by the existence of ensimatic intracratonic geosynclines with limited spreading of oceanic crust. In the southern part of the Mozambique belt collision of the Nampula and Namarroi blocks and formation of the Namama imbricate-overthrust structure took place evidently in the period 850–800 m.y. ago, with renewed tectonic activity, thrusting in a southward direction and dextral displacement along the Lurio zone. Tectonothermal events in Kenya at the level of 820 m.y. (Samburuan-Sabachian orogenies) were manifest in metamorphism to granulite-amphibolite facies, formation of recumbent folds that involved basement slices and volcanosedimentary rocks including ophiolites (Goodwin, 1991). These processes represented a consequence of collision between the Archaean Tanzania Craton and the assumed Kibara Craton in the east.

Processes of rapprochement and collision of blocks of the Earth's crust continued in the interval 800–650 m.y. Collision of the Ethiopian block with the volcanoplutonic calc-alkaline complex occurred in the period 750–700 m.y. (Kazmin et al., 1978), which resulted in the imbricate structure of the Western Ethiopia zone. Collision of the Ethiopian block with the Sidamo block occurred probably in the concluding phase of Katanga tectonics in the period 680–650 m.y.

At present, however, the prevalent plate-tectonics model of the Mozambique belt assumes the collision of two or more continental fragments. This is supported by palaeomagnetic data which reveals differences in apparent polar wander curves for Western and Eastern Gondwana.

The question of relation between the Mozambique belt and structures of the Arabia-Nubian region (see below) continues to remain controversial. According to one viewpoint, the 'tectonic disappearance' of island-arc complexes of Arabia, Sudan and Egypt within the Mozambique belt was caused by transcurrent displacement along the eastern continental margin. This model admits the spread of the Riphean ocean throughout the Mozambique belt and Arabia-Nubian Region, followed by a continental collision between Eastern and West Gondwana. The sharp difference in the extent of metamorphism between rocks of the Mozambique belt and the Arabia-Nubian Shield is explained by a more intense erosion in the south which, in turn, was caused by a more significant increase in the thickness of crust during the collision.

Another model is based on wedging out of the ancient Arabia-Nubian oceanic basin southwards and the transformation of its 'apophyses' into ensimatic intracratonic zones. Predominance in the Mozambique belt of an ancient reworked basement and its subordinate development (only in the extreme north-east) within the Arabia-Nubian region as also the differing evolution of these structures favour the second approach. The Mozambique belt contains traces of Kibaran orogeny while formation of the ancient island-arc complexes of the Arabia-Nubian region commenced only in the Late Riphean.

Evidently Madagascar and the Seychelles developed in a manner similar to the Mozambique belt. The ancient Early Precambrian and Lower-Middle Riphean substratum was involved in Late Riphean reworking as manifest in the presence of intrusions aged 740 ± 40 m.y. and 710 ± 9 m.y. and also the development of metamorphism at the level of 860 ± 35 m.y. and 740 ± 40 m.y. (Goodwin, 1991).

7.1.15.8 *Arabia-Nubian Shield and Anti-Atlas.* The Arabia-Nubian Shield comprises Precambrian formations of Arabia and the north-eastern part of Africa and is now one of the most well-studied regions of Precambrian development of the Earth. The Arabia-Nubian Shield represents the most convincing example of the effect of the Wilson cycle in the Precambrian, i.e., evolution conforming to the plate-tectonics model.

The Arabian part of the shield, located east of the Red Sea, is made up of volcanosedimentary and intrusive formations of Late Precambrian. Lower Precambrian rocks are present only in the north-west, in the Afif and Ar Rayn massifs. A few tectonostratigraphic terranes (microplates) have been iden-tified in the present-day Arabian structure. Their number varies in different schemes from three to ten (Johnson et al., 1987). These terranes have been separated by sutures with ophiolites (Fig. 72, Chapter 8). The volcanosedi-mentary formations are divided into three complexes—A, B and C (Goodwin, 1991). The oldest complex, C, exposed in the south-western Peri-Red Sea part (Asir terrane), is represented by metabasalts, meta-andesites, turbidites, siliceous rocks, marbles of the Baish, Bahah and Jidda groups (Bidah-Hali and Abha-Bishah complexes; Hsü et al., 1990) intruded by quartz diorites aged 910 ± 22 m.y. Metavolcanics of the Asir terrane gave an Rb-Sr age of 912 m.y. (Johnson et al., 1987). Volcanics of Birak, Al Lith and Nukra respectively from Hijaz, At Taif and Afif terranes probably pertain to complex C (Johnson et al., 1987).

Complex B of the Asir terrane is represented by terrigenous formations and calc-alkaline Ablah volcanics resting unconformably on complex C and overlain by the Halaban group (10,000 m) made up of basalts, andesites, dacites, pyroclastic and terrigenous rocks. Sharp facies transitions are typi-cal of the Halaban group and its correlatives (Hulayfah; and others). All vol-canics have distinct island-arc petrochemical characteristics (Johnson et al., 1987). Rocks of complex B are intruded by granite batholiths aged 810 ± 4 and 743 ± 11 m.y. and underwent regional metamorphism in the interval 800–714 m.y. (Goodwin, 1991). The main part of the volcanosedimentary complexes Zaam and Bayda (Midyan terrane), Al Ays, Farri (Hijaz), Mala-hah (Malahah-Nadjran), Hulayfah (Afir) and Ha II and Ad Dawadimi of the same-named terranes evidently pertain to complex B (Johnson et al., 1987).

Complex A lies unconformably on much older formations and is made up of clastic (Murdama group) as well as volcanoclastic formations (Shammar group) aged 678–650 m.y. In the western part of the shield, Fatima volcanics have been dated about 700 m.y. The uppermost part (Jubaylah group) rests with a break on the preceding formations. Metasedimentary rocks of the Abt trough (Dawadimi terrane) and Tathlith of the same-named terrane and Wassat (Malahah terrane) evidently pertain to complex A.

Ophiolites concentrated among rocks of complex B are of vital impor-tance in the make-up of the Arabian part of the shield.

The well-preserved ophiolite association of Jabal Ess (26°20' N lat. and 37°30' E long.) forms an allochthonous slice crumpled into folds. The age of ophiolites has been estimated at present in the interval 870–700 m.y. and of synchronous island-arc volcanics at 840–650 m.y. (Goodwin, 1991).

The Nubian part of the Arabia-Nubian Shield encompasses regions of Egypt, Sudan and Ethiopia situated essentially between the Red Sea and the Nile River.

The question of the presence of ancient (950 m.y.) crust in the Eastern Desert of Egypt has so far remained unresolved although proximity to such crust has been demonstrated in the west by the age of clastic zircons (2060 m.y.) from metaquartzites. Some geologists (El Gabi and others) suggest an extensive development up to Pan-African gneisses in the region.

The metasediments of Abu Sweil, their upper age limit estimated at 800 m.y., are correlated with the Arabian Baish-Bahah complex (C). The much younger volcanosedimentary complex petrochemically corresponds to Riphean island-arc associations of Arabia and is represented by basalts, andesites, rhyolites, their tuffs, greywackes and conglomerates. It corresponds to the Nafirdeib and Awat groups in Sudan and Shadli and Dokhan in Egypt. The Nafirdeib volcanics have been dated 723 ± 4 m.y. (Rb-Sr isochron). The rocks are intruded by gabbro and granite massifs forming large complex I type batholiths, dated 700–500 m.y. As in Arabia, Al Ghadir, Onib and other ophiolites are present among volcanic formations.

The first full ophiolite sequence was described in 1976 in the Jabal Al Wask region (Bakor et al., 1976). The lower part of the sequence is represented by serpentinised ultramafics conformably succeeded by gabbro and on top by spilite-keratophyre rocks. The main portion of magmatites is intruded by mafic dykes. Other complete sequences of the ophiolite association—Wadi Ghadir, Al Amar etc.—have since been established.

The Al Ghadir ophiolites are represented by dunites and harzburgites, layered gabbro with a small amount of trondhjemites, distinctly manifest parallel dykes, tholeiitic pillow lavas and siliceous formation. These rocks are associated with extensive zones of tectonic melange. In spite of the complex structure of the region, four or five subparallel belts separated essentially by calc-alkaline magmatites have generally been established. A definite pattern has been observed in the increase from south-west to north-east of potassium content in the volcanics, age of rocks and $^{87}Sr/^{86}Sr$ ratios (generally low), pointing to their mantle origin. The petrochemical characteristics of basalts and andesite-basalts generally correspond to an island-arc series. Further, a systematic change is noticed in the composition of volcanics from more tholeiitic in the ancient complexes to considerably calc-alkaline in the younger ones, which is accompanied by a significant increase in potassium content. The Onib ophiolites of northern Sudan are aged 712 ± 58 m.y.

Taking into consideration the above data, the evolution of the Nubian part of the Arabia-Nubian belt is divided into two stages.

In the first stage the region developed in an environment of passive continental margin, evidently of a marginal sea, since island arcs developed simultaneously in the east in Arabia..Transformation of the Nubian passive margin into an active one occurred roughly at the level of 750 m.y. At the same time, it is quite possible that the assumed ancient crust in the Eastern Desert of Egypt developed in the form of individual massifs (microcontinents at some time attached to the ancient African margin) (Kröner et al., 1989).

The width of the opening of the Arabia-Nubian ocean in the Riphean is not quite clear due mainly to lack of palaeomagnetic data. It has been estimated at 1000 to 6000 km (Kröner, 1977). The upper limit was arrived at by allowing for a rate of subduction of 1 km per annum during the period 1100–500 m.y., which corresponds to the period of closure of this ocean. Church (1979) assumes a comparatively smaller opening of the ocean basin. In spite of the complex structure preventing construction of a real geodynamic model of the evolution of the Arabia-Nubian region in the Riphean (direction of the slope of Benioff zones, the problem of the number of island arcs etc.), the presence of all constituents pertaining to the Wilson cycle in the rock complexes of the region helps reconstruct its evolution as a lateral accretion in an environment of island arcs and marginal seas. Accretion led to the build-up of several hundred kilometres of new continental crust added to the margin of the African Craton, mainly during the Late Riphean. This process proceeded in impulses (episodes), as reflected in the corresponding unconformities, formation of allochthonous ophiolite nappes and melange and also calc-alkaline volcanics and granitoid batholiths, and thick beds of greywacke flysch. Over time, this process shifted from south-west to north-east, reflecting the migration of island arcs, as pointed out by the direction of change in the corresponding parameters mentioned above. At the same time, details of the course of this process of continental accretion are not yet clearly established. Herein lies the contradiction evidenced in the models of various investigators: some admit the existence of Benioff zones inclined in different directions, others the presence of a Benioff zone inclined north-east, and still others a Benioff zone inclined south-west. According to Gass (1977), formation of the calc-alkaline complexes of Saudi Arabia and Egypt occurred in a system of simultaneously existing island arcs. It is also quite possible that conditions of a margin of Andean type prevailed along the margin of the continental crust of Africa, which led to a combined model of development of Andean and island-arc type (Bozhko, 1984). Recent data on the presence of ancient continental crust in the south-eastern part of the Arabian Shield points to the need for taking into consideration the existence of microcontinents when constructing models of the evolution of the region under consideration (see Fig. 72, Chapter 8). Closure of ocean 'wedges' intruding into the continent in the Ethiopian and Sudanese territories occurred in the same period, possibly earlier, but the internal structure of these zones compels the suggestion of limited spreading during their formation. The latter begins to manifest northwards when the above 'wedges' enlarge and merge with the ocean.

In spite of the complexity of the structure and geodynamic evolution, there is no doubt that the Arabia-Nubian Shield represents an accretionary complex of ensimatic island arcs, fragments of ancient continental crust and ophiolite belts. The oceanic basin opened evidently during the Late Riphean. Formation of island arcs, marginal seas and their passive margins and also the accretion of various terranes followed by the formation of a neocraton occurred in the interval 950–670 (640) m.y. The collision and post-collision stage took place between 670 and 550 m.y. (Fig. 63).

Marginal-continental processes extended in the Late Riphean to the north-western part of Africa in the region of Anti-Atlas. Here, in the erosion window of Bou-Azzer quartzites and limestones deposited on Lower Proterozoic granitoids are comparable to the Upper Riphean 'stromatolithic' series of Ahaggar and transit in a northerly direction along the Main Anti-Atlas fault into a very deepwater formation of the Tachdamt-Bleida series.

At the base of it lies a suite of mafic volcanics (500 m) represented by tholeiites with an alkaline tendency. Above them lies a volcanosedimentary suite (500–1000 m) made up of rhythmic alternating shales and siltstones with bands of quartzites and mafic lavas and tuffs of keratophyres. This complex and the margin of the craton are tectonically overlain by an

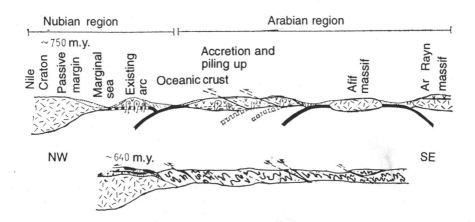

Fig. 63. Hypothetical profiles through the southern branch of the Eastern Desert of Egypt (Nubian region) and the Arabian Shield (Arabian region), demonstrating the assumed evolution of the region at the level of 750 (upper) and 640 (lower) m.y. (after A. Kröner, 1982). Ophiolites and melange zones are depicted in black in the lower profile.

allochthonous slice of ophiolites, especially well developed in the Bou-Azzer-El Graara boutonniere (Fig. 64). The age of ophiolites has been tentatively determined as 800 m.y. from dating on gabbro (788 ± 8 m.y.) intruding them along the margin of the West African Craton (Goodwin, 1991).

Synthesising the data of numerous recent publications, the tectonic structure of Anti-Atlas may be represented in the following manner. The south-western segment, as already mentioned, is located on the northern margin of the West African Craton, consolidated in the Eburnean epoch (2 b.y.). On this basement lies a platform cover of quartzite-calcareous series overlain by a volcanosedimentary series developed in the north-east. The passive margin of the West African Craton was formed as a result of rifting. This is supported by numerous intrusions of gabbro-diabases traced along the margin of the craton parallel to the Anti-Atlas zone. North of the passive margin, beyond the line of the present-day Main Anti-Atlas fault, was located an oceanic basin whose existence could be justifiably assumed on the base of allochthonous superposition on the margin of the craton of a nappe of ancient oceanic crust whose sequence was described by M. Leblanc (Fig. 64). The period of opening of this ocean has been tentatively dated about 800 m.y. (Leblanc and Lancelot, 1977). Evidently a similar passive margin has been fixed on the northern side of the ocean by the mio-geosynclinal Upper Riphean prisms fringing the North American continent (Stewart, 1976). The rift stage was succeeded by obduction of oceanic crust and overthrusting of the ophiolite complex onto the margin of the craton, which proceeded from north to south. Deformation was followed by epizonal metamorphism and accompanied by syntectonic intrusion of quartz diorites and post-tectonic intrusion of granodiorites. The age of sericite schists of the Tachdamt-Bleida series (685 ± 15 m.y.) corresponds to the main manifestation of this stage and the age of granodiorites (615 ± 12 m.y.) to its conclusion (Leblanc and Lancelot, 1977). Thus an active marginal regime of the Andean type has been reliably established in the Precambrian for the first time. Development of the Anti-Atlas in the Late Riphean represents yet one more example of the complete manifestation of the Wilson cycle in the Precambrian.

Saquaque and colleagues (1989) have proposed a significantly different model by considering the ophiolite complex of Bou-Azzer as a piling up of terranes. In their opinion, ophiolites per se form only the central part of the complex. The southern part of the ophiolite complex represents an accretionary serpentinite melange. This interpretation, the authors claim, is confirmed by petrochemical data.

Thus data for the western Mediterranean, south-eastern margin of the North American Craton and Anti-Atlas point to the opening of the western part of the Proto-Tethys and to completion of thorough dismembering of

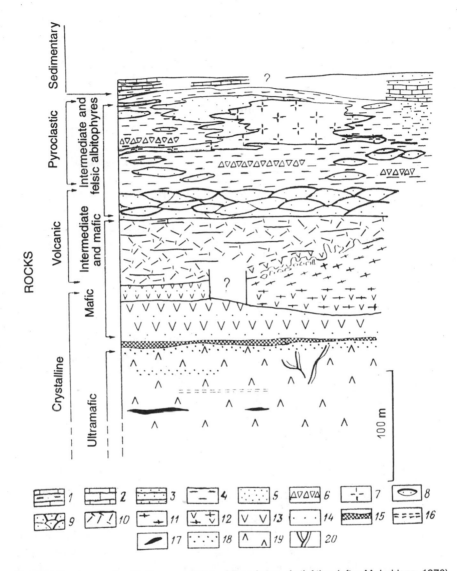

Fig. 64. Sequence of ophiolite association of Bou-Azzer, Anti-Atlas (after M. Leblanc, 1976).

1—shale-limestone formation; 2—limestones; 3—quartzites; 4—tuffs; 5—greywackes; 6—pyroclastic breccia; 7—keratophyres and rhyolites; 8—spilites; 9—pillow lavas and jaspers; 10—diabases; 11—oriented quartz diorites; 12—gabbro-diorites; 13—leucocratic gabbro; 14—melanocratic gabbro; 15—pyroxenites; 16—magnetite lenses; 17—chromite lenses; 18—lenses of pyroxenes; 19—serpentinites; 20—rodingites.

Pangaea I in the Late Riphean. These processes, which developed in the Early Riphean, moved in a westward direction.

312

7.1.16 Indian Craton

The Late Riphean of India (Fig. 65) was marked by expansion of the regions of platform sedimentation compared to the preceding epoch. Upper Riphean beds with a monotypical terrigenous-carbonate type of sequence have been established in all the Late Precambrian basins. The commonality of stromatolitic complexes described by M.E. Raaben suggests, as in the Early Riphean, a common sedimentation basin that comprised all the basins which are now separated.

The Upper Riphean of the Cuddapah syneclise is represented by the Kurnool series unconformably overlying the Krishna series. It is represented by alternation of sandstones and limestones which contain stromatolites of Late Riphean age (M.E. Raaben's determinations). Formations of the Bhima series of the same-named basin correspond to the Kurnool series.

The Chandrapur series (800 m) deposited at the base of the sequence of Chattisgarh basin is made up of conglomerates, shales, quartzites and sandstones with bands of limestones. It is overlain by the Raipur series (450 m) of limestones, shales and sandstones.

The Upper Riphean of the Godavari basin is represented by the terrigenous-carbonate Penganga series and overlying terrigenous Sullavai series. The Rewa group (250–3300 m) overlying the Middle Riphean Kaimur series corresponds to the Upper Riphean in the Vindhyan basin. The Rewa group is made up of sandstones, shales and conglomerates.

In the north-western Indian peninsula, in the Kirana hills, a volcanoplutonic complex with rhyolites and hypabyssal granitoids was established. Volcanics of this complex are dated 870 ± 40 m.y. An approximate age was obtained for Malani rhyolites (745 ± 10 m.y.) forming a volcanoplutonic complex in western Rajasthan together with hypabyssal granitoids (Moralev, 1977). These volcanoplutonic associations are exposed in a comparatively small area but their presence at the level of the Upper Riphean is extremely interesting. They are probably the youngest representatives of similar complexes that developed largely at the end of the Early Proterozoic main 'cratonization' epoch) which considerably diminished in the Early Riphean and almost disappeared in the Middle Proterozoic. Formation of the volcanoplutonic Malani complex concluded the plate-tectonics processes of the Rajasthan-Delhi region.

At the commencement of the Late Riphean in Singhbhum, accumulation of the carbonate-terrigenous Kholkhan series still continued in the southern part of the 'craton' and the Gangpur series in the northern trough. Sedimentation ceased as a result of the orogenic Gangpur-Kholkhan cycle at the level of 850 m.y. Further, rocks of the Gangpur series were crumpled into linear folds while the Kholkhan series experienced only local deformation as a result of mafic intrusions.

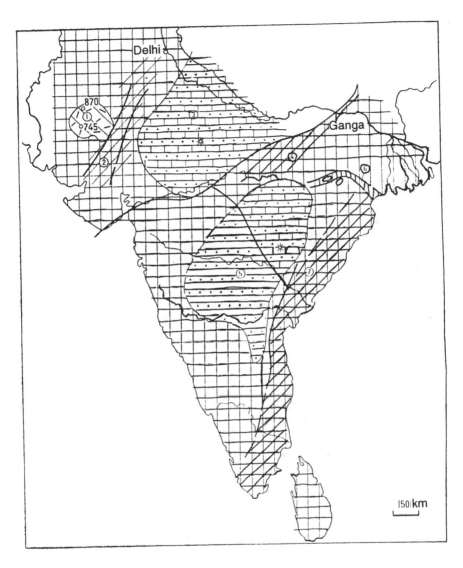

Fig. 65. Late Riphean palaeotectonic scheme of India (1000–700 m.y.) (after Bozhko, 1988).

Legend same as in Fig. 55. Palaeostructures: 1—Malani basin; 2—TTR Rajasthan basin; 3—Vindhyan syneclise; 4—TTR Satpura zone; 5—Central Indian syneclise; 6—Singhbhum aulacogen; 7—TTR Eastern Ghats belt.

The existence of three main zones of non-geosynclinal tectonothermal reworking of the basement is noticed:

1. E a s t e r n G h a t s, represented by alkaline granites and syenites aged 790 and 726 m.y.
2. S a t p u r a, separating the Vindhyan basin from the southern platform basin and extending from the Satpura mountains in the west to the Bidar pegmatite field in the east. Fixed by the age of pegmatite minerals at 850–950 m.y.
3. R a j a s t h a n, with extensive development of pegmatites, granites and alkaline rocks.

7.1.17 Australian Craton

The Late Riphean stage is characterised by predominance of a platform regime almost throughout the territory of Australia except for the eastern sector (Fig. 66).

The formation of platform cover continued in aulacogens in north-western and northern Australia, i.e., upper parts of Bangemoll, Carr Boyd, Glidden and Fitzmorris-Guvergne. Over much of the area of eastern Kimberley and Sturt plateau, these formations are overlain by equivalent sandy-shale Kuniandi and Duerdin groups containing tillites at their base and midportion. The age of the lower tillite horizon (Fargu) has been tentatively evaluated at 723 ± 3 m.y. and of the upper horizon (tillites of Moonlight Valley) 680 or 660 m.y. (Goodwin, 1991).

In the region of the present-day coast of the Arafura Sea, accumulation of the sandy-siltstone Gassel group proceeded; glauconites from this group are dated 805 and 790 m.y. (Semikhatov, 1974). The sublatitudinal aulaco-gens Amadeus, Ngalia and Georgina, filled with thick sedimentary rocks of a shallow-marine and continental origin, have been delineated in the central part of Australia. Sedimentation commenced around 900 m.y. ago as sup-ported by the occurrence of these beds on dolerites with an isochron age of 897 ± 9 m.y. (Page et al., 1984).

Based on biostratigraphic and isotope data and correlations, a part of the basin sequence (about 3 km) including the lower quartzites of Heavytree, predominantly carbonate Bitter Springs formation with evaporites, terrigenous Areyonga formation with tillites and siliceous-carbonate Aralka overlain by Vendian formations pertain to the Upper Riphean (Semikhatov, 1974). Sequences of the Ngalia and Georgina basins are well correlated with the sequence of the Amadeus basin (Burek et al., 1979). The same correlation is applicable to the formations filling Officer Basin situated south of the Musgrave block.

7.1.18 Adelaide Fold Region

The Adelaide zone in south-eastern Australia was formed in the stage considered here. This trough joins in the west along the Torrens fault with the horizontally lying Sturt 'shelf' formations.

315

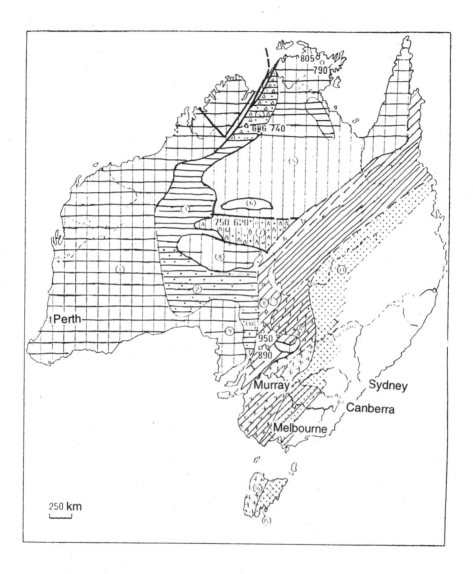

Fig. 66. Late Riphean palaeotectonic scheme of Australia (1000–700 m.y.) (after Bozhko, 1988).

Legend same as in Fig. 57. Palaeostructures: 1—West Australian Shield; 2—Officer basin; 3—Gibson basin; 4—Victoria River basin; 5—North Australian Shield; 6—Ngalia basin; 7—Amadeus aulacogen; 8—Musgrave Shield; 9—Gawler Shield; 10—Sturt Platform; 11—Adelaide zone and outer part (shelf) of East Australian passive margin; 12—Kanmantoo zone and inner part (continental slope) of passive margin; 13—east Australian part of Palaeo-Pacific; 14—Rocky Cape massif, Dundas trough (small oceanic basin); 15—Eastern Tasman zone.

The initial stage of Adelaide trough evolution coincides with the accumulation of thick (6 km), predominantly terrigenous (quartzites, sandstones and phyllites) 'Calanna' beds lying sharply unconformably on an old metamorphic complex and containing a suite of rhyolites, andesites and basalts of Wultana (600 m). The Burra group (2 km) lying on top of Wultana with a break is represented by quartzites containing a band of dolomites (350 m). The overlying Umberatana group consists of three subgroups: quartzite-shale with Udnamutana tillites (5 km), Forina silt-shale (3 km) and Willokra sandstones with tillites at the base comparable to Egan tillites of the Kimberley basin. The age of these tillites is 680–660 m.y. (Chumakov, 1978) and corresponds to the lower boundary of the Vendian. Thus the Upper Riphean part of the sequence is bound on top by the Willokra tillites. Data on the period of accumulation of the lower part of the Adelaide sequence is contradictory. A reference point is the correlation of the Wultana volcanics with those of Beda on the Sturt plateau which is generally accepted at present. Beda volcanics recorded an Rb-Sr isochron age of 1076 m.y. (Webb and Horr, 1979).

Dacites from Upper Calanna are aged 802 ± 10 m.y. Isochrons of 750 ± 53 and 724 ± 40 m.y. have been obtained on rocks of the Umberatana group (Goodwin, 1991). Thus sedimentation in the Adelaide trough commenced about 1000 m.y. ago and the main part of the Adelaide sequence below the Willokra tillites pertains to the Upper Riphean.

The tectonic nature of the Adelaide fold system remains an unresolved question. The solution largely depends on which type of crust extended eastward in the Adelaide zone. According to one point of view, continental crust extended in the Riphean under the entire contemporary Palaeozoic Tasman region. In this case this system could evidently be an aulacogen or an intracratonic geosyncline. It must be pointed out that the recent Sm-Nd investigations of bulk samples of rocks confirmed to some extent the spread of Middle Proterozoic crust under much of the Lachlan belt (Page et al., 1984). According to another point of view, the eastern part of Australia experienced rifting and drift, with the Adelaide region in the Riphean representing a passive continental margin broken into blocks. Development of corresponding facies favours the existence of a much deeper eastern zone. In the neighbourhood of Brocken Hill Block, an analogue of Adelaide is the Torreaunji group made up overwhelmingly of shales with bands of volcanics. The 'orthotectonic' Penguin zone of Tasmania, according to Rutland (1976), represents an interior zone relative to the 'paratectonic' Delamere belt of the Adelaide system and should have extended to northern Tasmania island. Metamorphic rocks detected in the cores of Queensland anticlinoria may also possibly fall in this same interior zone. It is interesting that Adelaide formations experienced Delamerian diastrophism simultaneous with the Lower-Middle Cambrian Kanmantoo series containing ophiolites. The

Kanmantoo series (10 km) is distributed directly east of the Adelaide system. Its lower part pertains to the Late Precambrian. All of this direct and indirect data points to the existence of an oceanic basin in the Riphean east of the Adelaide belt that represented a passive margin. At the same time, it is difficult to judge the width of the opening. It is possible that this structure was proximate to the Red Sea type. In any case, the presence of large blocks of continental crust should be assumed at this time eastward in the future Tasman belt (Williama massif); this is supported by the aforesaid isotope studies. These blocks may be compared to the detached massifs of the Arequipa massif type on the margin of South America, Mary Byrd Land microcontinent of the Antarctic etc.

In the west the Adelaide geosyncline is connected to the Amadeus aulacogen in the type of a triple junction, as supported by drilling data in the Permian Pedrika basin and also gravimetric data.

On Tasmania island the Precambrian is exposed in the west. Tyennan massif ('core') is set off from the Rocky Cape anticlinorium situated northwest by the Early Palaeozoic Dundas trough. Both regions are made up of terrigenous metasedimentary rocks with bands of dolomites and mafic volcanics. The minimum age of formations has been estimated at about 1.1 b.y. and that of the main tectonothermal events about 800 and 580 ± 40 m.y. Some volcanic activity has been noticed at the level of 700 m.y. All these values are difficult to assess due to superposed Palaeozoic reworking (Page et al., 1984). Considering that in the ophiolite Palaeozoic Dundas trough, the base of the sequence is made up of Precambrian, the existence of a narrow oceanic trough can be assumed at the end of the Late Riphean-Vendian in western Tasmania.

7.1.19 Antarctic Craton

This stage is marked by formation and development of the Transantarctic (Ross) belt and distinct separation of the East Antarctic Craton within its contemporary frame. The main territory of the craton represented a shield. Supracrustal complexes formed comparatively small sequences of platform cover and filled the aulacogens.

In Shackleton range the platform cover is represented by a thick carbonate-terrigenous series (Turnpike group) containing Riphean stromatolites. It rests on granites aged 1446 ± 60 m.y. and is tentatively placed in the Middle-Upper Riphean. The Spann formation in the Argentinian range is compared to the Turnpike group.

Formation of the Penck aulacogen in the western part of Dronning Maud Land and MacRobertson, filled mainly with Vendian formations, in the East Antarctic region, probably pertain to the Late Riphean. In contemporary structure, they are represented by graben-synclinoria. The lower age limit of the Friendship series forming the MacRobertson aulacogen is 1.1 b.y.

(based on age determinations of mafic intrusions overlain by formations of the aforesaid series).

7.1.20 Transantarctic (Beardmore) Belt (Fig. 67)

Formation of this belt represented a major event of this stage. In the region of the contemporary Transantarctic range a thick shale-greywacke formation was laid from Horlick region to the north-eastern part of Queen Victoria Land. The most complete sequence of this formation, several kilometres thick is fixed in the south. In the Queen Maud Mountains these formations are combined into the Beardmore group (Stump, 1976) including the equivalent formations of Goldie, La Gorce and Duncan.

In the region of Beardmore, Ramsay and Nimrod glaciers the Goldie formation (6 km) is made up of rhythmically alternating metagreywackes, metasiltstones and grey and dark grey shales. Gradational bedding is seen in the rocks. The Duncan formation (4.6 km), developed in the same-named mountains, and the La Gorce formation (3 km) of the Wisconsin range are similar in composition, i.e., flyschoid alternation of metagreywacke sandstones and metalutites with bands of quartz sandstones. Metamorphism of the La Gorce formation has been dated 728 ± 27 m.y. (Rb-Sr) (Rudyachenok, 1974).

The Berg group (1.4 km), developed on the northern margin of Queen Victoria Land in the Leo Berg Mountains, comprises quartz-sericite-biotite-chlorite schists, marmorised limestones and quartzitic sandstones. Rocks of similar composition are distributed in the upper parts of Rennick Glacier (Rennick group) in the Seqwens Mountains. Quartz-biotite schists from the Seqwens Mountains recorded an Rb-Sr age of 770 ± 20 m.y. (Faure and Gair, 1970). Riphean microfossils have been detected in rocks of the Berg group (Rudyachenok, 1974).

The Priestley formation, representing the equivalent of the Berg group in the central part of Queen Victoria Land is made up of dark grey schists, metashales and metasiltstones with subordinate bands of limestones. The Skelton group (3 km), developed in MacMerdo region and made up of sandstones, shales and limestones metamorphosed to greenschist facies, is comparable to the Priestly formation.

The Robertson Bay group (5 km), developed in north-eastern Queen Victoria Land in the Victoria, Admiralty and other mountains, is made up of rhythmically alternated greywackes, argillites and shales. Lutite constitutes 60–80% of the volume of the sequence. V.M. Rudyachenok identified Upper Riphean microfossils in the shales of the group.

The West series (5 km), developed on Saunders Coast (Mary Byrd Land), is characterised by highly similar composition. An Upper Riphean acritarch complex has been detected in it. The Swenson formation (4.3 km) in the Ford Ranges (Mary Byrd Land) is correlated with the West series. The Minaret group and the lower part of the Heritage group of the Ellsworth

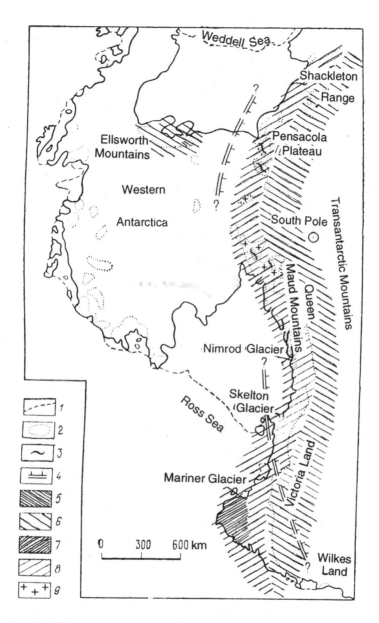

Fig. 67. Late Precambrian Beardmore orogen (after Elliot, 1975).

1—margin of shelf ice; 2—boundaries of exposed territory; 3—structural directions; 4—boundaries of Late Precambrian deformation and metamorphism; 5—exposures of Upper Precambrian shallow-water formations; 6—assumed distribution of shallow-water formations; 7—exposures of Upper Precambrian shale-greywacke beds and volcanics; 8—assumed development of shale-greywacke beds; 9—Upper Precambrian intrusions.

Mountains as well as the upper part of the Patuxent group of the Pensacola Mountains evidently represent formations equivalent to those of the Transantarctic range (Rudyachenok, 1974).

The Minaret group (1 km) is mainly made up of basal conglomerates, which are followed by limestones upward along the sequence. The lower part of the Heritage group (1.3 km) is represented by metashales, schists, phyllites and conglomerates and the midportion by basalts while Cambrian fauna has been identified in the upper, carbonate part.

In the Pensacola Mountains basalts of the upper part of the Patuxent group are aged 778 m.y. (Elliot, 1975). This group is correlated with the basal part of the Beardmore group of the Transantarctic range. Thus the sequence of the Transantarctic range reveals shale-greywacke flysch and less deep sandy-carbonate formations. The latter are characteristic of the lower part of the Beardmore group and its analogues. At the same time, sequences with a greater role of carbonate deposits are generally confined to the eastern pericratonic part of the zone under consideration, which could imply their formation under shelf conditions more or less synchronous with flysch rocks.

The Explanatory Note to the Tectonic Map of the Antarctic Region on Scale 1:10,000,000 (1978) suggests the possible existence of two structural-facies zones within the Ross fold belt: It is quite possible that the western (pericratonic) part of this belt comprises not so much the typical Ross complexes as their miogeosynclinal equivalents laterally replacing the true Rossides, and supracrustal formations of the intracratonic (intraplatform) aulacogens.

The region from Pensacola to the Tachdaun mountains is characterised by a different type of deposits. It differs from the same-aged shale-greywacke formations of the central part of the Transantarctic range in the development of abundant volcanics of mafic and felsic composition, great thickness of formations and far less manifestation of subsequent granitoid magmatism. To these features must be added its meridional strike transverse to that of the Rossides. All this points to the different tectonic nature of the Ross and Pensacola zones. The latter was intracratonic and lay between the western margin of the East Antarctic Shield and cratonic region in the west (Hak Craton). Data on tectonothermal reworking of the latter at the level of 1 b.y. is available. The Ross zone, however, represented a passive margin in the Riphean, as supported by development of a turbidite shale-greywacke formation to which volcanism was not characteristic at this level, especially the bimodal (graben) type noticed in the Pensacola zone.

It may be assumed that the Transantarctic range structures continue in the region of Ellsworth Mountains in which Ross complexes are also present. This is supported by the general character of the disposition of major positive forms of topography forming a subglacial continuation of the Ellsworth

Mountains towards the south-west. Such a combination of the Ellsworth with the Ross fold system forms a single continental margin extending from the Ellsworth Mountains to Victoria Land, from which the intracratonic rift type Pensacola geosyncline was set off.

The carbonate-terrigenous formations exposed in the Ellsworth Mountains comprise the outer shelf part of this passive margin while its assumed inner zone corresponding to flyschoid formations of the Transantarctic range is probably concealed under ice. At the same time, this interpretation cannot be straightaway confirmed at present since a vast expanse extending north-east from the Transantarctic range to Mary Byrd Land is also concealed under ice. Here, in our view, an analysis of the characteristics of the geological structure of Victoria Land is of great importance. In its extreme north (Pennel and Borchgrevink coasts) the north-eastern zone is distinguished by the total absence of syntectonic and late tectonic Ross granitoids; the Robertson Bay group sedimentary complex (less metamorphosed!) is fully analogous with the Beardmore group of the Transantarctic range. It is extremely significant that this zone is set off from other parts of Victoria Land by the Bowers Early Palaeozoic trough. In the contemporary structure it represents a graben-synclinorium extending north-west for 350 km at a width of about 50 km and transiting later along the strike into a major fault zone.

In the anticlinal bends of the flanks of the synclinorium Vendian formations are exposed to a thickness of not less than 3.5 km, represented by the Sledgers group. The latter is divided into the lower volcanic Glasgow formation and terrigenous Molnar formation. The former is represented essentially by ungraded basalt breccia with sheets of pillow lavas. Vendian acritarchs have been identified in the Molnar formation. Palaeontological data is scant on the age of the lower part of the sequence of the Sledgers group, which is not fully exposed, and probably relate it to the Riphean and not to the Vendian. Continuation of the Bowers trough is found in Tasmania, most probably in the Dundas (Elliot, 1975) or Adamsfield trough.

Taking into consideration the structure of the Bowers synclinorium, the above differences in mode of evolution of the Victoria Land blocks separated by it and also its continuation in Tasmania into the Dundas ophiolite trough in Gondwana reconstructions, the existence in the Riphean of an intracratonic marine basin with signs of oceanic crust can be assumed along the north-eastern margin of the East Antarctic Shield. Relative to the latter, the region of the Transantarctic range developed as a continental margin.

The Mary Byrd Land Craton (microcontinent ?) lay north-east of the trough and turbidite formations of the Robertson Bay group accumulated on its north-eastern margin. Such an interpretation can explain the sharp difference in the subsequent evolution of these two regions of Victoria Land.

In the western part of Mary Byrd Land very thick (up to 8–10 km) formations of the West series lie on Archaean metamorphic formations (Fosdic series). The West series correlates well with the above-described Robertson Bay group in composition, degree of metamorphism and nature of fold structure.

7.2 TYPES OF STRUCTURES AND TECTONIC REGIME

The Late Riphean is characterised by distinct intensification of secondary oceanic formation processes and development of marginal-cratonic (orthogeosynclinal) tectonic regimes. These processes led to complete disintegration of Pangaea I due to oceanic formation in the western region of the Proto-Tethys between Africa on the one hand, and North America and Europe on the other. As a result of this, two first-order structural elements of the Earth, i.e., Gondwana and Laurasia supercontinents, were isolated for the first time (see Fig. 67).

Within the supercontinents this epoch was associated with the formation of a new generation of tectonic structures. It included all the types that existed in the first half of the Late Precambrian except for regions of volcanoplutonic associations, which had almost ceased to exist by this time.

7.2.1 Platform Basins (Syneclises)

The largest structures of this type have been formed in Africa, South America and India. The Taoudenni syneclise in the western part of Africa is one of the most representative platform basins laid in the Late Riphean and probably the first of such size in the history of the Earth.

The Taoudenni syneclise is presently almost rectangular (1000 × 1800 km^2) but in the Riphean was joined with the Volta basin and formed an even larger flat structure. Upper Riphean shallow-water formations directly rest on the basement forming the lower of three platform complexes separated by unconformities and laterally shifted relative to each other. The rocks are monotypical and are represented predominantly by quartz sandstones with bands and layers of shales that reach a few hundred metres in thickness. The sequence is complicated by the development of marine carbonate formations (Adrar). A characteristic feature is the transition of the platform cover of the syneclise into the same-aged and much thicker formations of adjoining geosynclinal troughs of the Mauritania-Senegalese, Ahaggar-Atakora and Anti-Atlas systems. Evidently transgressions from these troughs made for the vast interior expanse disposed between them to be covered by an epicontinental sea. The degree of deformation of syneclise formations also changes correspondingly. In the central part they lie horizontally, having been disturbed only by some faults. On transition to zones of geosynclinal systems, dips increase and folding is gradually manifest. The Congo, Sao Francisco, Vindhyan and Central Indian syneclises are characterised also by significant dimension and carbonate-terrigenous filling. Further, the Upper Riphean sequence of these basins, unlike the Lower and Middle Riphean, is characterised by moderate thickness. Except for the Malani region in India, volcanoplutonic associations together with molassoids are not known in the Late Riphean.

Thus the Late Riphean is characterised by the commencement of changes in the style of formation of platform covers, viz., formation of the most extensive platform basins of the Taoudenni syneclise type. It may be pointed out that such platform basins began to form on Gondwana continents directly on the old basement of cratons, bypassing the aulacogen stage. As shall be seen later, this tendency continued to develop in the Vendian too. Flatter extensive structures with a far smaller amplitude of subsidence than in the Middle Riphean, suggest an even more mature nature of substratum.

7.2.2 Aulacogens

Some aulacogens continued to develop in the Late Riphean (most East European and Siberian) while others ceased to exist (North American, Orsha, Peri-Ladoga, Viatka and others in Eastern Europe). A new generation also arose: Amadeus, Ngalia, Officer, Itombwe and others. At the site of some closed aulacogens, intracratonic fold structures (Keweenaw and Sill Lake) arose in the form of graben-syneclises complicated by much smaller folds, overthrust faults, or zones of moderate linear folding (Belt and Fitzmorris). Most Siberian and European aulacogens did not experience fold deformation. Inversion was manifest in them as uplift.

Four major aulacogens were formed in central Australia in the Late Riphean (from south to north): Officer, Amadeus, Ngalia and Georgina. Major negative Bouguer anomalies are characteristic of these structures and corresponding positive anomalies of the salients of the basement separating them. The structure of these aulacogens represented linear, slightly asymmetric grabens, which is a common characteristic of this stage of development. In the course of the Late Riphean considerable subsidence and accumulation (up to 10 km) of thick carbonate-terrigenous formations occurred in them.

An interesting group of aulacogens arose at the commencement of the Late Riphean in the territory of eastern Zaire—Niangara, Zemio, Irumu, Bunoro, Itombwe and Bushimaye. These comparatively small (length ranging from 100 to 400 km and width from 20 to 50 km) structures filled with carbonate-terrigenous formations and tillites and rocks similar (but much thicker) to those of adjoining platform cover are remarkable for the brief duration of their evolution. Having originated at the level of about 1 b.y., some (Bushimaye) had already closed by 950 m.y. and the rest by 800–750 m.y. The inversion of these aulacogens was accompanied by folding. These structures are confined to ancient deep crustal faults (Villeneuve, 1983).

The same-aged Bukoba trough of Tanzania too evidently pertains to this group.

The Western Congo trough extending for 1800 km from Gabon to Angola generally parallel to the coast of the Atlantic Ocean represents a peculiar structure of the Late Riphean. This trough inherited to some extent the Mayombe intracratonic geosynclinal system, having been located immediately east of it. This geosynclinal system closed at the end of the Middle Riphean. The formational series of the Late Riphean corresponds to the rift stage and stage of post-rift subsidence and is represented in the lower part by significantly terrigenous strata and in the upper by terrigenous-carbonate with tillites. It fully corresponds to the formational series of synchronous intracratonic geosynclines (Damarides and Katangides) but granitisation and metamorphism have not been encountered in the Western Congo trough. This trough in the nature of its structure and evolution occupies an intermediate position between an intracratonic geosyncline and aulacogen. Evidently the Peri-Baikal trough filled with a three-member complex pertains to this same type.

7.2.3 Intracratonic Geosynclines

Data for the Late Riphean provides a base for studying these structural forms at different stages of their development since the closing of some of these troughs as well as the formation of their new generation occurred in this epoch.

The Late Riphean is remarkable for the formation of Pan-African mobile belts concentrated in West Gondwana. They include ensialic as well as ensimatic intracratonic systems (see preceding chapters) but, unlike the disconnected geosynclines of the Early-Middle Riphean, the Late Riphean belts are more extended and form branched systems in which the various belts are spatially interconnected and isolate the cratons and massifs of unreworked Early Precambrian crust enclosed between them.

Two such systems deserve special attention. One encircles the West African Craton (Mauritanides, Bassarides, Rokellides, Trans-Saharan belt and Atakorides) while the other falls in the central and southern Afro-American part of the continent and separates the Congo, Kalahari, Tanganyika and Sao Francisco cratons and much smaller blocks (Proto-South Atlantic belt, Damarides, Katangides and Western Congolides). These systems were probably associated with the Late Riphean palaeo-oceans, Proto-Tethys and Proto-Pacific, and represented their apophyses within the Gondwana continent. This has been established more definitely in the case of the Trans-Saharan belt in western Africa where the western Hoggar (Pharusian belt) structures join the Anti-Atlas orthogeosyncline through the Ougarta zone. Evidently joining of the Mauritanides with the Proto-Tethys region also occurred.

A similar branch, but already going from the south, from the side of the Proto-Pacific Ocean margin of Gondwana could be outlined in the form of a Proto-South Atlantic belt; its fragments are now found in a disjointed form in Africa (Saldania belt, Gariep and the coastal branch of Damarides) and South America (Ribeira).

The proposed tectonic models of these geosynclines vary from intraplate to marginal plate, with the latter models clearly predominating.

Thus the evolution of the Trans-Saharan belt is represented as an oceanic development conforming to the Wilson cycle that began about 800 m.y. ago and ended in collision and formation of a suture between the West African Craton and the Pan-African mobile belt (Black 1980; Caby et al., 1981). This interpretation is based on the presence of ophiolites, island-arc complexes, volcanoclastites of marginal basins, calc-alkaline plutons and paired metamorphic belts. Yet there are certain difficulties in adopting this model. Foremost among them is determination of the width of the ocean that arose as a result of rifting along the eastern margin of the West African Craton. There are some grounds for assuming this oceanic basin to have been relatively small in size. The identity of the stromatolitic Hoggar series with the Atar group of the West African Craton unambiguously points to the existence of a single supercontinent of West Africa roughly up to 800 m.y. At the same time, even at the level of 730 m.y., the Tilemsi volcanic arc, manifestation of high-temperature metamorphism (710 m.y.) and folding (696 m.y.) occurred. Formation of an ultramafic-mafic complex intruding into the basement and the stromatolitic series after which the Série Verte volcanics accumulated (696–634 m.y.) has been assumed under conditions of back-arc spreading (Boullier, 1991) while Bertrand and Caby (Bertrand et al., 1978) believe that basification led to the 'opening of the ocean basin to a width of 100 to 150 km' only. It must be pointed out that palaeomagnetic data indicating a similarity of APWP (apparent polar wandering curve) for Gondwana cratons in the Precambrian (see Fig. 61), excludes a significant opening of intracratonic troughs.

It has justifiably been pointed out (Boullier, 1991) that before formation of Série Verte, Pharusian II and Adrar-Iforas greywackes, the Hoggar Shield was a single structure. Considering the foregoing and also the condition that a true and complete ophiolite association has not been described so far within the Hoggar, it would be more logical to assume a limited opening of an oceanic basin of the Red Sea type branching in the form of apophyses from the triple junction with Anti-Atlas.

The Pharusian belt wedges out sharply and transits into the Atakora system of same meridional strike south of the Mali-Nigerian syneclise. A correlation has been established between the fold series of Togo (Atakora) and Buem with formations of the western limb of the Volta basin. In the Late Riphean the latter formations form a common cover of the West African

Craton together with those of the Taoudenni syneclise. Concomitantly there is a significant reduction in volume of magmatic rocks within the Atakorides compared to western Hoggar. They are represented here only by scattered small bodies of alpine-type serpentinites and basalt-rhyolite lavas in the cover of the Buem series. In our view this data does not permit assuming development of the Atakora zone in strict conformity with the Wilson cycle. It would be more appropriate to assume 'degeneration' of the ensimatic Hoggar trough in a southerly direction and its transformation into a narrow, almost ensialic intracratonic rift structure.

A nearly similar picture is observed in the western part of the craton in the Mauritania-Senegalese belt. As pointed out above, in the northern part of the belt, within the Mauritanides and Bassarides, present data permits the design of tectonic models of development based on limited oceanic rifting, followed by the formation of island-arc complexes in the subduction zones. Yet a primary opening of the oceanic basin and the presence of cryptosutures have not been confirmed for the southern continuation of the belt, represented by the Rokellides. Unlike the Mauritanides, the Rokellides developed most probably at the site of an intracratonic trough with a thinned continental crust (Culver et al., 1991).

Thus the Hoggar-Atakora and Mauritania-Senegalese systems should evidently be regarded as meridional riftogenic apophyses, diverging wedge-like from the latitudinal palaeo-oceanic region of the Proto-Tethys deep into the continent with the sialic proportion of their crust increasing towards the south. This is confirmed furthermore at the junction of Africa and South America in Gondwana reconstruction wherein significantly ensialic Gurupi and Medio Coreau zones are found at the continuation of the Rokellides and Atakorides respectively.

Development of such wedge-like zones is characterised by the interlacing of intracratonic (intraplate) and marginal-cratonic (plate) tectonic styles. The former is represented by a small width of opening, intrusive character of formation of mafics and ultramafics, absence of subduction or its insignificant scale, general degeneration of magmatism and attenuation of the belt in the body of the craton; the latter is manifest in the presence of mantle material against a background of overall extension and subsequent profuse calc-alkaline magmatism.

A similar overall picture is discernible in the extension of the Proto-South Atlantic belt wherein a gradual reduction of the ensimatic part and wedging out of the belt are apparent from the Gariep system to the Western Congolides. This belt represents a branch of the Proto-Pacific wedging deep into Gondwana but already in a northerly direction.

The Western Congo and Damara-Katanga systems, formed in the Late Riphean, represent the ensialic north and north-eastern branches of the Proto-South Atlantic belt respectively. The Ribeira zone forms a constituent of the western part of the latter. These structures also arose as a result of rifting. Three (Martin and Porada, 1977) or two (Kasch, 1983) grabens were initially formed in the Damara zone. This rift stage in the sequences of these geosynclines is manifest as thick clastites (Nosib, Sansikwa and Roan) with mafic (Western Congolides and Katangides) or felsic volcanics and alkaline intrusions (Damarides). Later a very deep trough or narrow oceanic basin (Red Sea type ?) was formed in the southern part of the Damara trough, evidently tapering off towards the north-east. Its existence is marked by the presence of flysch, amphibolites and lenses of serpentinites along the Matchless line in the Damarides while in the rest of the Damara geosyncline, i.e., north and south of this trough, in the miogeosynclinal zones of the Damarides as well as in the Western Congo and Katangan zones, carbonate-terrigenous formations with tillites of the Western Congo 'system', Otavi and Swakop groups and Kundelungu 'system' were formed. At the end of the Late Riphean, the geosynclines under consideration experienced multiphase diastropism, which led everywhere to formation of complex fold systems and was accompanied by granitisation and metamorphism in the Katangides and Damarides. The general structural plan of the Western Congolides is asymmetric. In sections adjoining the Chaillu massif, at places where the geosynclines formations pass into rocks of the platform cover, only a gentle undulation is noticed in beds, but intensity of folding rises sharply in a south-westerly direction and folds acquire a distinct vergence towards the north-east. Narrow folds verging north-eastwards, complicated by overthrusts, occur close to the axial part of the Mayombe range.

The internal structure of the Katangides is quite complex. A characteristic feature of the Lufilian arc is the presence of a 'chain of granite salients', i.e. granite-gneiss domes made up of ancient formations. Linear folds with undulating axes and also domes and synclines represent the predominant Riphean structures. Close to the foreland, folds are overturned to the north and north-east and cut by thrusts. Metamorphism of Katanga rocks is manifest unevenly. Its extent increases west and south-west of the Copper Belt but usually does not exceed the chlorite-sericite stage of the epizone. Metamorphism in the Damarides is also uneven. It attains the maximum degree (amphibolite stage) in the central part of the system where granite bodies are concentrated and is almost lacking in the north in the miogeosynclinal zone. Folding of the Damarides is extremely intense and isoclinal. Typical nappes are developed in the south. Vergence in the northern part is northwards towards the Congo Craton. Thrusting of the frame onto adjoining cratons is a general structural feature of intracratonic fold systems.

Difference in opinion as to the nature of the Hoggar is also noticed in the geodynamic interpretation of the Damara geosyncline. The plate-tectonics model assumes the opening of an ocean between the Congo and Kalahari cratons and subsequent subduction ending in continental collision. The

328

intracratonic model is based on limited rifting succeeded by 'continental subduction', the latter caused by delamination of the lithosphere (Fig. 68). In spite of the controversy over the suggested mechanism of compression and piling up in the second model, it enjoys preference in view of the intracratonic character of the Damara geosyncline, absence of fragments of true ophiolite association, complex of calc-alkaline volcanics, sharp reduction and disappearance of 'ensimatic proportion' towards the Katangides and palaeomagnetic data. This also concerns the evolution of the Gariep-Malmsberry zone and others (Bozhko, 1984).

The nature of the Baikal-Vitim belt is not so certainly defined, given the diverse views on the stratigraphy of the region. As in some Gondwana belts, here too, after diastrophism, contiguous troughs were formed in the Early-Middle Riphean time. Some researchers regard these contiguous

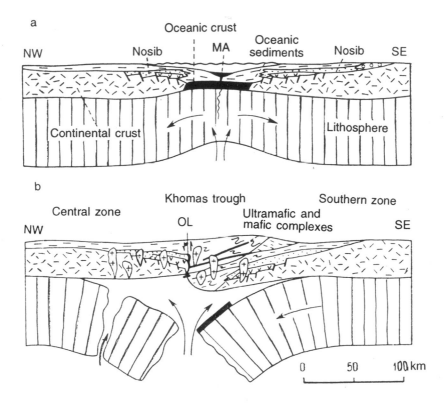

Fig. 68. Intracratonic model of development of the Damara belt including limited spreading along the southern graben followed by thickening of the crust (a) and subduction (b) (after Kasch, 1983):

OL—Okahandja lineament; MA—Matchless amphibolites.

troughs as pertaining to the orogenic stage of the Baikalides (Bulgatov, 1983). Foremost is the Peri-Baikal trough which in structure and tectonic position greatly resembles structures of the type found in the Western Congo system, troughs of Paraguay-Araguaia, Brazil and Faleme, that arose after inversion and folding of corresponding Early-Middle Riphean zones and filled with non-metamorphosed and non-granitised flyschoids and carbonate-terrigenous Upper Riphean formations and Vendian molassoid rocks, moderately deformed and passing into horizontally lying strata of adjoining syneclises. Angara-Pit, Teya-Chapa and Vorogovka troughs of the Yenisei range that appeared in the second half of the Late Riphean after inversion and folding at the level of 850 m.y. are similar in character.

Many aspects of the geodynamics of formation of the Atakora orogen are still not quite clear. It can only be confidently said that it proceeded in an environment of compression encompassing an extensive region east of the West African Craton. Further, compression forces were largely directed from outside and were generally global in character, taking into consideration the simultaneous manifestation of Pan-African orogeny in the various belts of Africa and West Gondwana as a whole. This compression within the system was compensated by folding, nappe structures and strike-slip faults. Manifestation of calc-alkaline magmatism can be explained as due to the mixing of crust and mantle material during separation of a block of heavy lithospheric mantle and lower crust, their subsidence into the asthenosphere and partial melting. This argument of density could be promoted by massive intrusion of ultramafics. It is significant that calc-alkaline magmatism and volcanism manifested only at such places (Hoggar) where basification had previously proceeded intensely. The lithosphere thus transformed very rapidly into eclogite in its lower part and could have separated in the form of independent blocks or slices submerged in the asthenosphere. Many authors have recently indicated the possibility of production of andesites outside the zones of subduction. In this context attempts to reconstruct a zone of subduction based only on establishing the calc-alkaline nature of magmatism without taking into consideration the petrochemical polarity of volcanics are not convincing.

The tectonic regimes of the Late Riphean do not differ on the whole from the corresponding regimes of the Early and Middle Riphean characterised in the early stages of development by a distinct riftogenic style and in the concluding stages by orogeny accompanied by metamorphism, usually to greenschist facies, fold deformation and intrusion of granitoids. As before, several phases of deformation and plutonism have been delineated.

In concluding this description of the intracratonic geosynclines of the Late Riphean, attention may be drawn to a significant complication in their tectonic structure. In the ensimatic troughs that continued to develop, inversion caused formation of marginal troughs. The newly formed zones have a far more complex structural plan, with not only transverse but longitudinal zoning, manifest particularly in the transition of ensimatic into ensialic zones along the strike.

7.2.4 Belts of Tectonothermal Reworking (TTR)

The first impulses of the so-called Pan-African orogeny, i.e., manifestation of tectonomagmatic activity encompassing almost the whole of the southern continents in the Vendian and Early Cambrian, evidently were manifest towards the end of the Late Riphean. Along with formation of fold belts, an important aspect of this process was the development of non-geosynclinal reworking of the ancient substratum on extensive expanses of Gondwana. Against the background of effects of much later, more intense Vendian-Cambrian processes, it is difficult to isolate the effect of the Late Riphean reworking. The consequences of the latter can be seen in the relict values of isotope ages of 700, 750 and 800 m.y. within the Mozambique, Libya-Nigerian and other TTR belts.

Unlike the intracratonic geosynclines that are interrelated and linked with the Proto-Tethys and Proto-Pacific by triple junctions, the structural plan of Late Riphean troughs within linear zones of tectonothermal reworking follows the general plan that was typical of the Middle-Late Riphean, i.e., comparatively narrow subparallel ensialic rift troughs separated by very broad salients of reworked basement.

Late Riphean reworking in regions of Middle Riphean consolidation is very clearly delineated. This process has been well studied in the northern Kibarides in Burundi (Tack, 1984). Reworking proceeded for almost 30 m.y. after folding of the belt and was manifest in the formation of major strike-slip faults (1100 ± 38 m.y.), formation of tin-bearing granites (980 m.y.), formation of local strike-slip fault zones with a meridional strike, intrusion of an undifferentiated alkaline complex (775–740 m.y.), intrusion of syenites and carbonatites along meridional faults (739 ± 7 m.y.), and renovation of movements along faults and isotope rejuvenation (707–699 m.y.).

7.2.5 Marginal-cratonic Geosynclines

In the Late Riphean these structures were in different stages of evolution of the Wilson cycle.

Formation of accretionary complexes in an environment of island arcs and marginal seas occurred in South-east Asia (Yangtze Craton and Indosinian complex) and the Arabia-Nubian region.

As mentioned before, although these processes commenced as early as the Middle Riphean (Sibao orogeny) in the Jiannan massif, they did not lead to the final consolidation of the neocraton. The latter occurred during Jinning orogeny (850 m.y.), as determined by formation of collision granitoids and molasse. Thus in South-east Asia development conforming to the Wilson cycle proceeded with some lag compared to the other marginal-cratonic geosynclines. Here subduction processes in active continental margins around the Yangtze block were transformed in the middle of the Late Riphean into collision processes and ended with formation of the Yangtze neocraton.

In the Arabia-Nubian region formation of island-arc magmatic complexes commenced and continued throughout the Late Riphean. The intraoceanic island-arc environment is demonstrated by the low $^{86}Sr/^{87}Sr$ ratios in granites and volcanics of the Ablah and Halaban groups and their analogues and also the composition of the fragments in the greywacke flysch. At the same time, microcontinents Afif and Ar Rayn were involved in the formation of an accretionary complex in the north-east.

Formation of a secondary oceanic crust (see Fig. 64) and passive continental margin in the region of Anti-Atlas led to completion of the opening of the Proto-Tethys.

Data on the Riphean formations of south-eastern China and the Arabia-Nubian Shield makes a positive contribution to the study of the evolution of these regions in the light of modern plate-tectonics models.

The environment of an active margin of the Andean type evidently arose at the end of the Late Riphean in Anti-Atlas on the margin of a newly formed ocean when the formation of quartz diorites commenced within the former miogeosynclines. Similar transformation of a passive margin into an active one occurred in the western part of the Yangtze block in the Sikan-Yunnan region at the end of the Middle and commencement of the Late Riphean.

A different tectonic environment prevailed within the present-day Circum-Pacific belt. In the Late Riphean there appears conclusive evidence of the formation of this largest of the mobile belts on the Earth. At this time there was an almost uninterrupted strip of miogeosynclines (passive margins) girdling the Proto-Pacific ocean. These structures are fixed in the contemporary sequences of the Riphean carbonate-terrigenous and turbidite series of Windermere, Tindir and their analogues in the Cordillera; El Allabique, Maranao and Puncoviscana in the Andes; Beardmore series of the Transantarctic ranges; Adelaide series of Australia; and Lakhanda and Ui suites of Siberia. Further, a significant feature has been established in the contemporary structure: these miogeosynclinal sequences together with the much younger geosynclinal complexes adjoining them were blocked in many cases and set off from the Pacific Ocean by pre-Riphean massifs of the type Okhotsk, Chukchi, Arequipa, Mary Byrd and others in which a Riphean platform cover of the same age developed. Corresponding eugeosynclinal zones have not been detected in most such cases. They are either concealed under much younger rocks in the form of fragments, as in the Uyanda zone of Verkhoyansk-Chukchi region, or sheared by faults, as evidently occurred in the Bilakchan zone separating the Okhotsk massif from the Yudoma-Maya trough. Ophiolite complexes are preserved in the Bowers and Dundas troughs in Antarctica and Tasmania. A sutural zone has been established between the Yangtze and Henan blocks (Hsu et al., 1990). In most cases the Riphean miogeosynclinal prisms under consideration experienced major deformation not in the Precambrian, but in the Phanerozoic, as for example the Yudoma-Maya trough in the Mesozoic, Transantarctic range in the Early Palaeozoic Ross epoch, and the Adelaide series in the Delamere (Salairian) epoch. All this provides justification for assuming a chain of newly formed oceanic basins as a result of rifting in the region of the Late Riphean Pacific Ocean margin of Pangaea, more correctly, of Gondwana and Laurasia. These oceanic basins were set off from Panthalassa by continental massifs that were moved aside and could be compared to the borderlands of Ch. Schuchert. In the subsequent complex Phanerozoic history these massifs disintegrated into smaller ones and, as a result of the closing of the aforesaid basins, attached themselves to the continents as individual blocks or. exotic lands (Saleeby, 1983). This resulted in the evolution of the contemporary structural plan and enables us to depict the continental crust or its fragments which were separated by rifting from Pangaea I in the course of formation of the Circum-Pacific belt and formations of a strip of its Riphean passive margins. Without this, it is difficult to reconstruct the course of such rifting. It follows from this scheme that formation of the Circum- Pacific geosynclinal belt occurred on secondary oceanic crust without significant influence of the then already inert zone of the Palaeo-Pacific or Panthalassa proper. The belt itself represented a global network of riftogenic oceanic basins along the margin of Pangaea I. The formation process of such basins could be compared to formation of the Tasman Sea in the Cretaceous period by the separation of New Zealand and Lord Howe continental blocks from Australia.

It must be pointed out that most publications interpret the nature of these continental fragments from the viewpoint of terrane tectonics. Further, an extremely remote primary disposition and subsequent transport of such blocks regarded as exotic terranes cannot be ruled out. This interpretation contradicts the similarity of the geological structure of these blocks to the basement of adjoining cratonic margins that was detected in detailed investigations (Milov, 1990; Ross and Bowring, 1990). This helps suggest the relative proximity of these above blocks to the Riphean passive margins of the Peri-Pacific ring. They evidently represent fragments of much larger blocks that broke apart in the course of Phanerozoic tectonics. Thus the character of the Circum-Pacific belt is essentially similar to the other secondary riftogenic oceans of Pangaea.

In the Late Riphean the future Early Paleozoic ocean Iapetus survived its embryonic rift stage. A major continental rift structure (Proto-Iapetus) and possibly a whole system of subparallel rifts arose along the large zone

Fig. 69 Late Riphean global palaeotectonic reconstruction (after Bozhko, 1988).

1—Early Precambrian continental crust; 2—platform cover; 3—aulacogens; 4—ensialic intracratonic geosynclines; 5—ensimatic intracratonic geosynclines; 6—zones of non-geosynclinal tectonothermal reworking of the basement; 7—newly formed fold systems; 8—accretionary complexes; 9—passive margins; 10—pre-Riphean oceanic crust of Panthalassa; 11—oceanic crust newly formed in the Riphean; 12—Avalonian island arc; 13—zones of subduction; 14—ophiolite belts in accretionary complexes; 15—boundaries of continental crust.

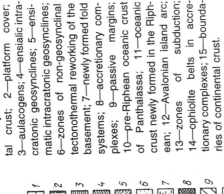

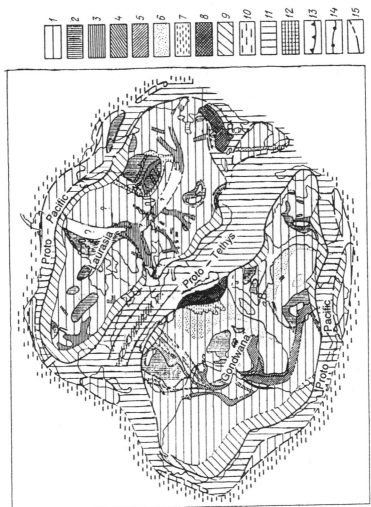

from the Avalonian Platform to Spitsbergen. Fragments of this single Late Riphean rift system are now disjointed and exposed as thick prisms of clastic Upper Riphean formations of the type seen in the Eleanore Bay group (eastern Greenland), Hecla Hoek group (Spitsbergen), eastern Finnmarken supergroup (Norway), Torridonian group (Great Britain) and Gander group (Newfoundland). The thickness of these wedges increases eastwards along the western margin of the main Proto-Iapetus structure and westwards along its eastern margin. The development of this rift system led to the opening of the Iapetus ocean proper in the Vendian-Cambrian.

Thus in the Late Riphean conditions of all the stages of the Wilson cycle, from continental rifts to collision zones, manifested simultaneously in the different marginal-cratonic geosynclines.

Palaeotectonic analysis of the Early, Middle and Late Riphean points to the existence of structures with a very prolonged polycyclic development, such as those seen in the Yenisei range, Proto-Urals and Mauritanides that developed throughout the Late Precambrian and mono- and dicyclic structures whose life is measured in terms of one or two epochs, such as Hoggar-Atakorides, Kibara-Ankolides, Damara-Katangides and others.

The general complexity of the Late Riphean global structural plan (Fig. 69), i.e., formation of basins with secondary oceanic crust in the form of ensimatic intracratonic troughs—apophyses of larger oceans, joining in turn with ensialic troughs and forming triple junctions, points primarily to intensification of rifting. This represented a precondition for further development of marginal-cratonic processes in an environment of growing active margins. At the same time, intracratonic structures continued to be active and the mode of their evolution became more complicated (inversion of ensimatic troughs and formation of marginal troughs). All this leads to the conclusion that in the Late Riphean a typical global environment set in and was almost equivalent to a combination of intraplate and plate-marginal-tectonic regimes. An outstanding event of the Late Riphean was complete opening of the Proto-Tethys and corresponding disintegration of Pangaea I into two supercontinents—Gondwana and Laurasia—as well as formation of the Circum-Pacific mobile Belt.

8

Vendian: Consolidation of Gondwana and Formation of Mobile Belts of Laurasia against the Background of Its Disintegration

8.1 REGIONAL REVIEW

8.1.1 North American Craton

The major part of this craton was uplifted in the Vendian and represented land surrounded on all sides by oceanic basins.

In the west, along the present-day Cordillera belt, formation of a passive margin continued in the form of the upper part of the Windermere super-group and its analogues conformably preceded by Cambrian formations. In the Rocky Mountains the Tobi tilloids of the Windermere supergroup are replaced towards the top of the sequence by mafic Ayrin volcanics alter-nating with conglomerates and sandstones (up to 2.5 km) and higher up by the Horstiff Creek suite of dark shales, siltstones and sandstones (up to 2.5 km). The thickest Vendian sequences have been established west of the suture of the Rocky Mountains Trench. Here Cambrian formations con-formably replace Vendian ones, participating together with Late Precambrian ones in the formation of lower complexes of the Cordilleran geosyncline.

Similar processes developed in the east where the western passive mar-gin of the Iapetus Ocean was formed within the Appalachian geosyncline in the interval 800–600 m.y.

The miogeosynclinal Humber zone in the Canadian Appalachians and a band west of the Blue Ridge-Green-Long uplift in the USA Appalachians correspond to this western passive margin of the Iapetus Ocean (Goodwin, 1991). The Ocoee supergroup and its analogues accumulated along the future Blue Ridge-Green-Long uplift. The Great Smoky group (4500 m) and a part of the overlying Walden Creek group made up of greywackes, siltstones and shales characterised by graded bedding evidently correspond to the Vendian part of the Ocoee sequence.

336

The eastern zones of the Appalachian belt (Avalonian zone of Canada and eastern flank of the Blue Ridge-Green-Long uplift (USA) are made up of calc-alkaline volcanics, siltstones and shales of the Caroline Slate Belt and their analogues in Newfoundland. Felsites from the roof of the lower part of the sequence of the Slate Belt have been dated 584 m.y. (U-Pb) and 554 m.y. (Rb-Sr). Calc-alkaline granitoids aged 630 ± 15 m.y. (U-Pb) to 516 ± 13 m.y. (Rb-Sr) are spread in New England (Goodwin, 1991).

As pointed out above, formations of the Avalonian zone and other eastern zones of the Precambrian Appalachians are regarded as Afro-European, primary Gondwanian terranes. This view is supported by the age of the Dartmouth alkaline pluton of New England, determined as 595 ± 5 m.y. (Don Hermes and Zartman, 1992). The presence of Vendian post-collisional alkaline intrusions following the calc-alkaline plutons with a small interval is typical of Pan-African belts.

Formation of the Southern Oklahoma and Western Texas aulacogens entering the craton in the form of passive arms of triple junctions is also associated with these events.

The Vendian part of the sequence of the Southern Oklahoma aulacogen is represented by a thick suite (5 km) of terrigenous rocks and mafic lavas. Much of the North American Craton was uplifted in the Vendian and represented land surrounded by oceanic basins on the east, west and south.

8.1.2 North Atlantic Region

The characteristic features of the Vendian tectonic evolution of the region are associated with its main event, i.e., opening of the Late Proterozoic North Atlantic Iapetus Ocean (Fig. 70).

Accumulation of the Eleanore Bay group (13 km) continued in the central part of eastern Greenland and was succeeded by the tillite-bearing Vendian Merkeberg formation (1 km). Mafic lavas with some amount of ultramafics in the western sequences correspond to this formation.

A similar complex, Hecla Hoek (Spitsbergen), also corresponding in volume to Riphean-Vendian, is overlain by a complex of Varanger tillites.

In the region of Mjosa Osterdalen in south-eastern Norway, formation of a graben-shaped sparagmite basin in a rift pertains to this period; this rift diverges in a south-easterly direction from the main continental margin. The sparagmite Hedmark group (3 km) contains the Storskawern, Rendalen and Littlesberger continental formations in the eastern part of the trough, changing into marine sands (Atna formation) and turbidites and greywacke sandstones (Brettum formation). Higher in the sequence are seen Biri limestones; Osdalen, Biscopas and Ring conglomerates and sandstones; Muelv tillites; Acre shales and the Vangas suite of quartzite-sandstone in the upper part already falling in the Cambrian (Nystuen, 1982). Dolerite dykes are found towards the north-eastern part of this region in the basement and

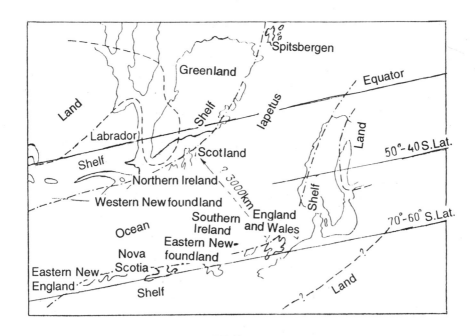

Fig. 70. Palaeogeographic disposition of the Iapetus Ocean (after Toghill and Chell, 1984).

allochthonous sparagmite formation of Sweden. The Trondheim dykes in Norway are analogous. They are associated with the early phase of formation of the Iapetus (Roberts and Gale, 1978). Upper Riphean formations in the extreme northern part of Norway, in eastern Finnmarken, are overlain by the Vendian-Cambrian Westertan group with tillites aged 668 ± 7 m.y.

In the region of the British Isles development of a series of rifts with a north-easterly orientation, deepening towards the south-west away from the framing, was evidently replaced by a continental shelf where the Dalradian supergroup was formed. The lower Appin group is represented by alternation of shales, quartzites and limestones. The basement of the overlying Argyll group is made up of tillites aged about 650 m.y. replaced in the upper part by quartzites and turbidites with dolerite sills. Thus the comparatively shallow-marine shelf deposits are replaced up the sequence by deep-water turbidites (Anderton, 1982), pointing to the formation of a passive continental margin. The basement of southern Great Britain was formed within the southern flank of the Iapetus Ocean. According to palaeomagnetic data, this region was set off from Laurentia at the end of the Precambrian and represented a microplate connected to the northern part of Great Britain as a result of Caledonian subduction (Thorpe et al., 1984). Gabbro, tonalites and albitic granites of Wales, Stunner-Hanber and Malvernia

338

aged 700–640 m.y. are regarded as the earliest island-arc formations in the Iapetus Ocean. Thus there is no crust older than 700 m.y. south of the Iapetus suture (Solvay-Shannon line). In central and northern England, Rushton schists aged 667 ± 20 m.y. (Rb-Sr) are unconformably overlain by Cambrian quartzites, which come into contact along a fault, with Uriconian tholeiite-basalts, somewhat older than Longmindian (600 m.y.). Formation of the first island-arc complex was probably synchronous with accumulation of the Mona complex that experienced metamorphism and granitisation in the interval 600–550 m.y. Thus the basement of the Caledonides in the southern part of Great Britain was formed in the interval 700–950 m.y. and represents an island-arc accretionary subduction complex.

The southern passive margin of Laurentia developed in the interval 700–530 m.y. while a tectonothermal event at the level of 590 m.y. occurred in the Dalradian block, enabling it to be regarded as a fragment of Gondwana (Bluck and Dempster, 1991). Evolution of the northern Armorican massif mostly resembled that of the Avalonian and Dalradian blocks. Brioverian turbidites and calc-alkaline volcanics and intrusions comagmatic with them reflect an environment of ancient island arc and marginal sea. The interval of manifestation of Cadomian magmatism has presently been determined as 700–540 m.y.

8.1.3 East European Craton

Formation of the lower cover complex in structures of the syneclise type and gently sloping troughs commenced in the Vendian. Upper Vendian formations (Valday series) overlie almost the entire territory of the craton in the form of a continuous cover. Tilloids and grey and red terrigenous and carbonate-terrigenous formations are extensively developed in the Vendian formations.

In the Pachelma aulacogen analogues of the Vilchan and Volhyn series are represented by tillites, sandstones and siltstones (125 m), in the Pavlov-Posad aulacogen by the Rakhmanov suite (144 m) and in the Orsha aulacogen by red sandstones, siltstones and tuffs (550 m). The platform cover complex in the Moscow syneclise is made up of the Valday series, the Redkin suite (up to 500 m) of shales, siltstones, sandstones, ash tuffs and clays, and the Lubim (up to 570 m) and Rashma (up to 400 m) suites of variegated shales, siltstones and sandstones.

The Baltic syneclise represents a stratotypical locality for the Upper Vendian. The Valday series here is divided into the Redkin (up to 192 m), Gdov (60 m) and Kotlyn suites. The lithological composition of the Redkin suite is close to that of its sequence in the Moscow syneclise, the Gdov suite is represented by clays, gritstones, sandstones and siltstones, and the Kotlyn suite by grey siltstones, laminarite clays and sandstones. Vendian formations have been discovered by drilling along the south-western margin of the

Vetrenny Poyas range within Pomorie lowland. Here they consist of coarse clastic rocks, shales and sandstones up to 220 m in thickness in the Onega River region.

In the Mezen syneclise the Valday platform cover complex comprises the Ust-Pinega suite of shales with bands of tuffs and sandstones (up to 400 m), Mezen suite and Zimneberezhnaya (Padun) suite of variegated terrigenous rocks (up to 358 m). Tillites, sandstones and shales of the Churoch suite (450 m) and Upper Vendian Potchur series (4.5 km) of sandstones, siltstones and shales accumulated in Timan in the Vendian.

In the Sergiev-Abdullin aulacogen the Vendian is represented by the Bishbulyak series (157 m) of sandstones, siltstones and shales (Aksenov et al., 1984). Red sandstones, shales, siltstones of the Gotan and Shtandisk suites, gritstones, conglomerates, sandstones of the Nizhnekairov, shales and siltstones of the Verkhnekairov and Shkapov suites are developed in the Kama-Belaya aulacogen (I.E. Postnikova).

In the south-western part of the East European Craton, lagoonal-continental formations of the Gorbashev suite accumulated at the commencement of the Volhyn epoch followed by manifestation of active volcanic activity. Subaerial basalt sheets, red shales, tuffs and tuffites of the Berestovetsk suite are extensively developed in the Volhyno-Podol area and attain a thickness of 465 m. During the Valday epoch a large epicontinental marine basin originated and covered the entire south-western part of the craton. Formations of the lower Gdov horizon (up to 180 m) are represented by quartz-arkose sandstones, siltstones and shales with phosphorite concretions. The Kotlyn horizon is represented by sandy-clay formations (up to 200 m). The south-western margin of the craton experienced uplift at the end of the Kotlyn epoch (Chit and Rizun, 1984).

8.1.4 Mediterranean Belt

8.1.4.1 *Western Mediterranean region.* Following formation of an oceanic basin in the Late Riphean between Gondwana and Laurasia, the region under consideration experienced a complex tectonic evolution during the Vendian to the Early Palaeozoic. Opening of the Iapetus led to formation of an extensive ocean expanse between Laurentia, Africa and the east European part of Eurasia within which there were microcontinents and island arcs. A significant event was manifestation of Cadomian diastrophism (640–600 m.y.) which affected Northern Africa, the territory of contemporary Western Europe and Anatolia. Precambrian ophiolites are present in the Cadomian fold belt in north-western Spain. Low (less than 0.705) $^{87}Sr/^{86}Sr$ ratios are characteristic of all Cadomian plutons of France and northern Spain. This data suggests that the Cadomian belt was formed on oceanic crust and that the continental crust in Western Europe (with the exception of small fragments of the Pentevrian type) are not older than 700 m.y.

(Rast and Skehan, 1983). Iberian, Armorican, Central France and Bohemian massifs were involved in Cadomian folding. It has been suggested that already in the Early Cambrian, as a result of new rifting, a significant part of the newly formed Cadomian continental crust together with the Avalonian Platform formed a new microcontinent, viz. Armorica or Avalonia, located between the Palaeo-Tethys and the Iapetus.

An island-arc environment has been recognised in the Eastern Alps in the Vendian (Neubauer, 1991). Metavolcanics (boninites, basalts and andesites) and metasedimentary rocks of the Chabak group aged 650–500 m.y., ophiolites of the Stubak complex of the Pennine region, paragneisses and amphibolites of the Koriden complex (716 ± 30 and 693 ± 39 m.y.) and gneiss-amphibolite association after calc-alkaline and tholeiite-basalts falling in the age interval 600–500 m.y. correspond to the above island-arc environment. Similar island-arc complexes are noticed in the Balkan region between the Alps and Turkey. Two phases of Cadomian folding, about 670 and 570 m.y., noticed in the Eastern Alps correspond to episodes of Pan-African orogeny manifest in metamorphism, deformation and magmatism and provide justification to assume the origin of these complexes along the margin of northern Africa (Neubauer, 1991). At the same time, ophiolites aged about 500 m.y. are present in the Western Alps.

Vendian formations in the Iberian massif are represented by the upper parts of calc-alkaline complexes of the Ossa-Morena zone and flysch extending north of this zone. Here, too, two phases of Cadomian folding leading to collision of independent terranes have been noticed (Quesada et al., 1991).

8.1.4.2 *Dinar-Carpathian-Balkan region.* Active accumulation of greenschist complex formations occurred in different zones of the region during the Vendian. In the summarised stratigraphic sequence of the Marsian supergroup of the Romanian Carpathians, schists and quartzites of lower blastoclastic, graphitic schists and limestones of g r a p h i t i c, volcanics and terrigenous rocks of r h y o l i t e v o l c a n o s e d i m e n t a r y, quartzites and phyllites of q u a r t z-p h y l l i t e, quartzites, schists and metavolcanics of upper b l a s t o c l a s t i c f o r m a t i o n s rest above mafic volcanics and schists of v o l c a n o s e d i m e n t a r y f o r m a t i o n. The Vendian age of the Marsian supergroup of 5 km in total thickness has been confirmed by numerous identifications of acritarchs and also U-Pb age determinations of felsic metavolcanics (560–600 m.y.). Metamorphism and granitisation of the rocks of this supergroup took place at the end of the Cambrian (Krautner, 1984). Thus the region under consideration remained unaffected by Cadomian (640–609 m.y.) diastrophism and, in the Vendian, represented a residual basin of the Proto-Tethys which could be traced in the direction of Greater Caucasus through Dobrogea and the buried Novotsaritsin-Simferopol' uplift of the Crimean plain where a greenschist flysch complex deformed in the Salairian epoch was also recognised.

At the same time, manifestation of Cadomian diastrophism was noticed in the Rhodope, Pelagonian and Serbo-Macedonian massifs, which probably belonged already to the Gondwana margin.

8.1.4.3 *Anatolian region.* The Baikalian tectonic stage evidently concluded over much of Anatolia by the beginning of the Vendian and was succeeded by Salairian rejuvenation. Manifestation of Pan-African reworking has been noticed in the Menderes massif. Although a conclusively Vendian cover has not been detected in this region, there is indirect evidence (supply of clastic material from the north in the Ordovician) that suggests the existence here of a stable shallow shelf on the margin of the Precambrian Gondwana Platform commencing from the Cambrian or the Vendian. A weakly metamorphosed molassoid cover (Kavakdere group) overlying the Baikalian complex was detected recently in western Anatolia.

8.1.4.4 *Iranian region.* The basement of the Central Iranian massif is made up of gneisses and crystalline schists containing serpentinites. These rocks underwent metamorphism not later than 660 m.y. ago and are overlain by a cover of Upper Vendian-Lower Cambrian represented by terrigenous molassoid rocks, evaporites, limestones and subalkaline felsic volcanics (2660 m).

Manifestation of Cadomian orogeny in the form of granitoids aged 620 m.y. has been noticed in Zangezur, Armenia (Miskhana and Megri massifs). Here itself, as well as in north-western Iran (Takab) and different parts of Alborz, is developed a cover with Vendian formations at its base.

Formation of the Kerman and Hormuz salt-bearing basins pertains to the Vendian (Stöcklin, 1968). Red sandstones, dolomites, cherts, quartzites, salt and gypsum were deposited in these basins. Formation of these rocks was associated with volcanic activity, as supported by the presence of andesites, quartz porphyries and tuffs in the sequence.

8.1.4.5 *Afghan-Baluchistan-Pamirs region.* In this territory an Epi-Baikalian Craton arose towards the Vendian and at some time (before commencement of the Palaeozoic) again united the northern and southern continental blocks. Vendian formations unconformably overlie the Riphean fold complex. They include conglomerates, dolomites of the Argandab River basin in central Afghanistan and the Zarabat suite of the central Pamirs.

8.1.4.6 *Himalayan region.* In the Krol nappe of the Lesser Himalaya variegated terrigenous-carbonate beds of the Blaini and Krol series with Ediacaran fauna belong to the Vendian. They extend in an almost continuous band along the marginal mountain ranges of the Lesser Himalaya resting unconformably on underlying formations and are formationally close to the same-aged sedimentary cover of the more western regions of the Proto-Tethys and the Vindhyan rocks of India. In the eastern Himalaya, in Bhutan, the Buxa series of carbonate-terrigenous composition and the Garbiang formation in the Greater Himalaya pertain to the Vendian.

In the Salhan range of Pakistan the Punjab salt series of about 600 m in thickness also belongs to the Vendian. The magmatic activity of this period was manifest in the intrusion of tonalites and granodiorites in Chail, Jutto and Almora regions aged 500–600 m.y. (Goodwin, 1991).

Thus, based on the age of ophiolites and volcanosedimentary rocks and their metamorphism and granitisation, rock complexes that experienced Cadomian (Pan-African) and much younger Salairian (500 m.y.) diastrophism can be identified within the Mediterranean belt. Correspondingly, these complexes constituted the northern (Peri-Fennosarmatian) and southern (Peri-Gondwanian) margins of the Proto-Tethys (Khain and Rudakov, 1991).

The details of palaeotectonic evolution are not wholly clear. One version assumes complete closure of the Proto-Tethys before the Vendian and collision of Gondwana and Eurasia. Another version suggests the existence of a residual basin in the north in the Vendian simultaneous with Cadomian (Pan-African) orogeny in the south. According to a third version, in the western part of the Proto-Tethys, including Central and Western Europe, complete closure of the ocean occurred in the Vendian while conditions in the east, from the Western Sudeten Mountains and the Carpathians to the Asian part, corresponded to the second version (Khain and Rudakov, 1991).

8.1.5 Preuralides

As already pointed out, a terminal Riphean (Kudash) is delineated in the stratotype of the Riphean in the southern Urals comprising terrigenous-carbonate rocks of the Uk suite (400–600 m) and terrigenous rocks of the Krivoluk suite (560 m). Vendian formations represent analogues of the Valday series of the Moscow syneclise made up of the thick complex of sandstones, conglomerates, shales and siltstones of the Asha series.

The Kudash-Vendian association in the central and northern Urals is represented by terrigenous and volcanic rocks of the Baseg series (2.5 km) and sandstones, tillites, mafic volcanics and carbonate rocks of the Serebryanka series and terrigenous formations of the Sylvytsa series.

8.1.6 West Siberian Platform

Riphean-Vendian formations take part in the structure of the platform cover of the Tunguska syneclise of the old craton and are also developed in the Peri-Yenisei belt of the West Siberian Platform where dolomite complexes to a thickness of 2–3 km are known to be palaeontologically Vendian.

This complex has been reached by boreholes; its thickness ranges us to 1 km (Rudkevich and Latypova, 1979).

Many investigators confidently distinguish a Salairian tectonic complex in the composition of the pre-Jurassic basement of the south-eastern part of the platform (Zhero et al., 1979; Surkov and Zhero, 1981) developed on the northern subsidence of Kuznetsk-Alatau. Formation of this complex had evidently commenced already at the end of the Late Riphean, continued into the Vendian and was completed in the Early to Middle Cambrian. Its composition comprises carbonate-volcanic formations, volcanics of mafic composition, algal dolomites and limestones, mica schists and metasandstones. The presence of mafic and ultramafic formations and granitoids is a characteristic feature. In the structure of the pre-Jurassic basement all these rocks form horsts: Middle Chulym, Yulu-Yula, Vezdekhod, Izhmor and others. These horsts are distinctly discernible in the magnetic and gravity fields.

It is important to find out the north-westerly extension of the Salairian (Early Caledonian) complex. This system, whose western boundary runs along Beloyarsk fault while the eastern boundary is tentatively shown along the high gradient zone of gravity anomalies (Surkov and Zhero, 1981), is not extended very far within the platform in many schemes. At the same time, attention must be paid to the reconstruction of Peive and Savel'ev (1982) who admitted the junction of this system with the early Caledonides of the Polar Urals. It is based on the similarity of the latter with the Salairides and their disposition along a common strike, taking into consideration the sharp incongruities of the structural plan of the Preuralides and Uralides.

8.1.7 Yenisei Range and Taimyr Peninsula

According to comparison with the Riphean hypostratotype, the Ostrovnaya suite (270 m) in the southern part of the Yenisei range and Upper Nemchanka subsuite in the northern part (*Riphean Stratotypes. Stratigraphy and Geochronology,* 1983; Khomentovsky, 1976) belong to the Yudomian (Vendian) period.

In Igarka region limestones of the Chernorechenka suite (800 m) and Izluchina red beds (675 m) overlain by the Gravilsk suite, pertaining already to the Cambrian, evidently fall in the Vendian.

In Taimyr peninsula formation of a platform cover proceeded in the Vendian. Bituminous carbonate rocks and conglomerates of the Sovkin suite (550 m) accumulated in a flat trough (Shrenk River basin). Before the Cambrian these rocks experienced deformation and formed a large gently sloping syncline (Zakharov and Zabiyaka, 1983). Away from the craton, carbonate-terrigenous formations are replaced by significantly sandy-clay and molassoid formations; flyschoid ones are also manifest. In the north-western regions of Taimyr a close association between Vendian and Riphean formations has been suggested while an unconformity is noticed in the south-east at the base of the Vendian formations.

8.1.8 Siberian Craton

Commencing from the Vendian, there was a single extensive platform region at the site of the present-day Siberian Craton. All Riphean aulaco-gens before the Vendian experienced significant reorganisation and were transformed into syneclises (Berzin, 1981; Khomentovsky, 1976).

Pre-Vendian reorganisation was more intensely manifest in the south. Vendian formations were deposited here on different Riphean horizons and also on the pre-Riphean basement. In the northern part of the platform Vendian strata are closely associated with the underlying Riphean formations. At the same time, as pointed out by Khomentovsky (1976), the base of the Ven-dian formations reveals no unconformity and there is nothing in the aulaco-gens to suggest conclusive folding. The maximum activity of pre-Vendian tectonic movements occurred not in the aulacogens, but on the margins of ancient uplifts surrounding them. All traces of Vendian structural reorgani-sation disappear on the inner slopes of uplifts (Khomentovsky, 1976).

The Yudoma series of Uchur-Maya region represents the Siberian ana-logue of the Vendian complex. This series is spread out within the eastern, central and northern slopes of the Aldan Shield where Vendian formations form a thin platform cover gently subsiding in the north-west under the Cam-brian formations. Simultaneously, in the opposite direction, these formations rest unconformably on all the older rocks including the basement in the Aldan Shield.

The lower, Aill suite is made up of sandstones, shales and dolomites up to 400 m in thickness while the upper, Ust-Yudoma suite is predominantly made up of dolomites up to 450 m in thickness. The formations of the series contain a fourth Yudoma complex of microphytolites and a stable associ-ation of stromatolites (*Riphean Stratotypes. Stratigraphy and Geochronol-ogy,* 1983). Radiochronological age determinations place formation of the Yudoma series in the interval 650 to 570 m.y. (*Riphean Stratotypes. Stratig-raphy and Geochronology,* 1983; Khomentovsky, 1976).

In Birusinka region, Eastern Sayan north-eastern slope, conglomerates, sandstones, shales, siltstones and dolomites of the Ust-Tagul suite (240 m) pertain to the Vendian.

The Early Vendian epoch in the northern part of the craton was char-acterised by close association between terrigenous sedimentation and out-bursts of volcanism. Extrusion of lavas and eruptions of tephra of mafic and alkaline-ultramafic composition occurred. In the Late Vendian the north-ern part of the Siberian Craton was involved in transgression extending from east to west from the Kharaulakh uplift to the western slope of Anabar massif (Shpunt, 1984). Carbonate sedimentation (Stara Rechka, Khorbusuan and

Kharanstek suites) predominated. Predominantly carbonate Vendian strata accumulated also in a cratonic environment on the Okhotsk and Omolon massifs and Peri-Kolyma uplift areas (Marevka, Troitsk, Vinkem, Korkodon and Kirpichnikov suites).

Vendian formations of the internal regions of the Siberian Craton were reached by drilling. Here, analogues of all three regional Vendian horizons—Ushakovka, Kurtun and Irkutsk—were established in the sequences of the frame. They are correlated on the basis of identifying characteristic key lithological horizons from the data of field geophysics. Shenfil' (1991) has done a detailed review of this data.

8.1.9 Kazakhstan-Tien Shan Fold Region

Destruction of the previously consolidated metamorphic basement and tectonic differentiation that commenced at the end of the Late Riphean continued in the Vendian. Vendian formations take part in the composition of the Palaeozoic structural complex corresponding to its lower continental-riftogenic part (Kiselev et al., 1992a). A sparagmite-like formation with bimodal volcanics and tillites accumulated in several troughs. According to Zaitsev (1984), these troughs appeared like narrow furrows extending for hundreds or several thousands of kilometres with a width of 60–100 km. The most extended was the Baykonur trough including the Kalmykkul as well as troughs of the Greater Karatau and Chatkal-Naryn systems. Accumulation of chert-basalt series occurred in some troughs. The appearance of ophiolite formation in the region is evidently associated with this time interval (possibly also with the Late Riphean). In the contemporary structure ophiolites outcrop in narrow zones and reflect the mosaic-fold structure of the region. Their composition and structure differ from typical ophiolite associations, primarily in abundance of terrigenous rocks in the sequence, absence of normal sequences (triads) and other features. In this sense the conclusion that the ophiolites of Kazakhstan should be regarded as 'predominantly relics of riftogenic structures with limited spreading' could be justified.

Having completed the review of Late Precambrian tectonics of the Kazakhstan-Tien Shan region, let us briefly mention the model of evolution of the crust in this region given by Makarychev and colleagues (1983), which differs sharply from the models reviewed above. It is based on the concepts of zonal age variation of a granite-metamorphic layer that arose during transformation of an initially melanocratic basement. According to these authors the Makbal-Burkhan zone was formed about 2 b.y. ago, Aktyuz-Peri-Issyk-Kul at 1230–700 m.y.; and Talas zone of Tien Shan at 970–700 m.y. The Kenkol-Eastern Terskei and Sonkul-Ashuturukh systems were formed in the Late Riphean-Cambrian. Formation of Ulytau continental crust lasted in this manner almost throughout the Riphean.

8.1.10 Eastern Part of Central Kazakhstan, Western Sayan and Western Mongolia

This region[1] is characterised by a continuous sequence of Vendian-Early Palaeozoic spilite-diabase and deepwater siliceous deposits in association with other fragments of ophiolite association. This makes for the palaeotectonic environment of the Vendian-Early Cambrian, viz., the existence of the Kazakh-Siberian basin in the form of deep troughs-gulfs (1600–2000 km in cross-section) relative to the Central Asian palaeo-ocean (Mossakovsky and Dergunov, 1983) (Fig. 71). Recent studies (T.N. Kheraskova, N.N. Kheraskov, R.M. Antonyuk, A.B. Dergunov, et al.) have pointed out that the spilite-diabase and siliceous formations lie on a melanocratic basement, which is usually represented by a serpentinite melange. The Vendian-Early Cambrian age of these formations has been established in the Lake zone of Western Mongolia, in Mongolia-Altay and Gorny (High) Altay, and in Western Sayan, and is conventionally assumed for other places in the region also.

The western and north-western boundaries of this oceanic basin with the western Kazakhstan continental margin is arcuate and is characterised by the development of terrigenous complexes including turbidites along it. This helps reconstruct a passive continental margin of the Atlantic type. The eastern boundary is irregular and characterised by a subduction type of jointing of oceanic and continental crust, i.e., represents an active margin. All along its extent from the eastern slope of Salair, along the northern and southern margins of Western Sayan, parallel to the margin of the Dzabkhan zone of Mongolia, spilite-keratophyre and andesite series are developed.

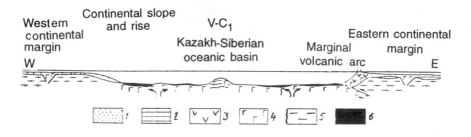

Fig. 71. Kazakh-Siberian oceanic basin in the Vendian-Early Cambrian (after Mossakovsky and Dergunov, 1983).

1—terrigenous formations; 2—carbonate formations; 3—andesites and andesite-basalts; 4—marine tholeiite-basalts; 5—pre-Vendian sialic basement; 6—pelagic sediments.

---

[1] Mainly after Mossakovsky and Dergunov (1983).

Metabasalts of these series correspond in petrochemical features to the contemporary island-arc volcanics. Thus it has been established that extension conditions predominated and a large oceanic basin existed in the Vendian-Early Cambrian. Continental slope and rise deposits were formed on the western margin of this basin while a system of island arcs, i.e., an active continental margin, arose on its eastern margin.

8.1.11 Eastern Tuva, Eastern Sayan, Kuznetsk Alatau, Gornaya Shoria and Dzabkhan Zone of Western Mongolia

In the Vendian-Cambrian this region underwent intense rifting with formation of a system of variously oriented narrow intracratonic geosynclinal troughs separated by salients of ancient, pre-Vendian crust. Such stable massifs were Khakas, Byia-Katun, Ondum-Buren, Hamsara among others. The troughs separating them were ensimatic with characteristics of ophiolite association with thick spilite-keratophyre series and siliceous and terrigenous-carbonate rocks as well as the ensialic, mainly carbonate Khubsugul series of northern Mongolia with flyschoid and carbonate filling of the type of Sarkhoy-Bokson suite of Eastern Sayan. The Kalmykkul, Telbessa, Lerass, Chernuss, Kizir-Kazyr, Boruss, Hamsara, Dzhida and other troughs pertain to the first type while the Mirichun, Bokson-Khubsugul-Sarkhoy and others represent the second type of troughs.

In the Bokson-Khubsugul trough the Vendian of the Sarkhoy series rests with angular unconformity on underlying formations and is made up of variegated conglomerates, sandstones and volcanics of basalt-rhyolite (bimodal) series to a total thickness of 4000 m. The overlying Bokson series (Vendian-Middle Cambrian) consists of limestones and dolomites (4.5 km). According to Berzin (1981), a predominance of zones formed on a sialic crust in the peripheral part of the region adjoining the Siberian Craton while areas of destruction enlarged in the interior of Altay-Sayan and troughs with oceanic crust were formed. On the whole this coincides with the interpretation of the history of the region as given by Mossakovsky and Dergunov (1983) who placed it in the eastern Vendian-Early Palaeozoic margin of the Kazakh-Siberian oceanic basin.

At the same time, Mossakovsky and Dergunov drew the right conclusion that the continental margins of the Kazakh-Siberian oceanic basin (western Kazakhstan as well as Altay-Sayan) experienced intense extension to a varying degree in the Vendian-Early Cambrian, causing numerous zones of rifting and tectonic hiatuses as a result of which a mosaic block structure was formed.

Extension causing trough formation was limited and did not result in the formation of extensive oceanic basins. In the Palaeozoic the folded intracratonic Salairides and Caledonides systems arose at the site of these troughs.

Concomitantly, carbonate and siliceous-carbonate beds overlying similar Riphean formations accumulated on the massifs, forming their cover.

8.1.12 Baikal-Vitim Fold Region

Vendian formations are closely associated here with those of the Lower Cambrian. On the other hand, this complex was formed within the same linear structures which existed at the end of the Riphean. Further, no unconformity, either stratigraphic or angular, has been established in some troughs while such features have been traced in others. These troughs developed synchronously with the Dzhida and Udino-Vitim eugeosynclinal zones. Molassoid, sandy-shale and terrigenous-volcanic formations mainly developed in the Baikal-Vitim region. Thus a more general spatial pattern is observed. The Baikal-Vitim area was characterised by a significantly ensialic type of trough structure while Vendian-Cambrian troughs with ophiolites were typical of the Altay-Sayan region situated more westwards and the Udino-Vitim zone towards the south-east.

Let us examine the composition of the Vendian formations in some specific zones.

The Zherba, Tynnov and part of the Nokhtui suite pertain to the Vendian in the Patom trough. The Zherba suite (300–500 m) is made up of quartz sandstones with sandy dolomites and dolomites in the lower part and dark grey siltstones and shales with bands of quartz sandstones on top. Throughout the extent of the Patom arc the Zherba suite rests on Chengin limestones with visible conformity but begins to progressively cut the underlying formations east of the Chara basin. Age values of 561 to 620 m.y. (K-Ar) have been determined on glauconite from the Zherba suite. The Tynnov suite is mainly made up of limestones and dolomites (about 400 m). The Nokhtui suite is subdivided into two subsuites. The lower (150 m) is made up of marls, dolomites, sandstones, shales and siltstones. Formations of the subsuite reveal stromatolites characteristic of the Yudoma (Vendian) complex of Siberia (Semikhatov et al., 1991).

Vendian formations in the Peri-Baikal trough, include the Ushakovka, Kurtun and Ayankan (Irkutsk) suites. Siberian geologists regard them as a Vendian regional stratotype and delineate them as corresponding to horizons of the regional scale.

The Ushakovka suite rests conformably on the Kochergat Upper Riphean formation and is represented by polymictic sandstones, siltstones, shales and conglomerates (360–1400 m). G.A. Vorontsova (1975) recorded an age value of 644 m.y. for siltstones of this suite.

The overlying Kurtun suite conformably replaces the Ushakovka and is made up of sandstones, shales and siltstones with subordinate dolomites in the upper part of the sequence (263 m). The thickness of the suite decreases to 110 m south-east of the Kurtun River basin towards the craton but rises again to 320 m towards the Irkutsk region of the Peri-Sayan zone at village Moty where the suite corresponds to the red beds of the Moty suite. Microphytolites and stromatolites encountered in the Kurtun suite confirm its Vendian age. In Irkutsk per se the Irkutsk suite (220 m) is made up of alternating dolomites and variegated shales in the lower part and interleaved dolomites and clayey dolomites in the upper part. Formations of the suite contain stromatolites and microphytolites of the Yudoma complex. An age of 575 (609) m.y. has been determined for glauconite from the upper part of the suite (Semikhatov et al., 1991).

The Bagdarin trough of the Vitim upland is filled with continental terrigenous formations reaching 3.2 to 3.5 km in total thickness. Other troughs of Transbaikalia—Birama-Namakan, Adyan, Tataurov, Yanguda etc.—contain a similar terrigenous-carbonate Vendian-Cambrian complex. Thus differential movements intensified in the Vendian and a new generation of troughs formed, with their plan essentially inherited from the Riphean, which closed in the Caledonian era. These troughs experienced a complex evolution. Rocks were deformed into linear and brachymorphic folds, intruded by gabbroids and granites.

The Vendian-Cambrian complex of the Olokit zone comprises the weakly altered carbonate-terrigenous formations of the Anamikat series including basal conglomerates at the base. Formations of the complex fill superposed troughs and are intruded by Upper Palaeozoic granites and alkaline rocks. Yet zonal metamorphism of the rocks of the Olokit trough has been determined as occurring 0.6–0.5 b.y. ago (*Resolutions of the All-Union Stratigraphic Conference . . .*, 1983).

The important role of Palaeozoic tectonics and magmatism in the Baikal-Vitim region is currently being recognised. Mel'nikova (1992) has shown that the Mama-Oron and Kunkuder-Mama granite complexes belong to a single cycle.

Various earlier interpretations of this stage of evolution of the region found in the literature were as follows. Bulgatov (1983) considered it deuteric, i.e., repeated orogenic; E.N. Altukhov (1986) cratonic; and K.A. Klitin et al. (1975) orogenic. V.G. Belichenko (1977) termed this stage geosynclinal and recognised in the Vendian-Cambrian a 'calcareous geosyncline' with manifestation of intense granitoid magmatism.

We consider this development as post-accretionary riftogenic; it commenced at the end of the Riphean and continued during the Vendian (see Fig. 56). According to this accretionary model, tectonic evolution of the Baikal-Vitim region in the Precambrian-Palaeozoic would have been quite similar to development of the south-eastern China (Yangtze craton) and Arabia-Nubian Shield.

The Vendian-Cambrian tectonic events of the region under consideration were generally synchronous with events in the adjoining Altay-Sayan region.

They correspond to a resurgence of rifting, maximally manifested in the Czernaya (Lake)-Kuznetsk zone of the Salairides, which gradually lost its intensity on both sides of this zone in the direction of the Kazakh-Baikal region.

8.1.13 Yangtze Craton, Indosinian Region

Following completion of Jinning diastrophism, the newly formed Yangtze Craton represented a continental block covered for the most part by epicontinental sea. The craton was surrounded on all sides by oceanic basins of the Proto-Pacific in the south and south-east, Qiankingian ocean in the west and Qinling ocean in the north.

In the stratotypical Sinian sequence, in Yangtze gorge, two series and four formations of 1000 m total thickness were delineated. The Lower Sinian is represented bottom-upwards by the Liantuo formation (160 m) comprising red sandstones and arkoses and the Nantuo formation (60 m) comprising tillites and sandstones. The Upper Sinian is represented by the Dougshantuo formation (200 m) comprising siliceous carbonates and the Dengying formation comprising dolomites and limestones with Ediacaran fauna. Rocks of the Dougshantuo formation have been dated 693 m.y. (Rb-Sr) and rocks of the upper part of the Liantuo 740 m.y. The verge between the Lower and Upper Sinian is marked at the level of 700 m.y. Thus the Vendian corresponds only to the upper part of the Sinian system.

Within south-western Sichuan and eastern Yunnan, local extension manifested in the process of Late Riphean collision in a regional field of compression and led to formation of the Susong aulacogen (Wu Geyao, 1986). Filling of this palaeorift is represented by the Susong, Kaijianqiao and Lieguliu formations of the Lower Sinian, which contain continental basalts and tuffs and are overlain by Upper Sinian dolomites. The lower part of the Chenjian formation in eastern Yunnan correlates with them. An Rb-Sr isochron of 885 m.y. recorded for basalts of the Chenjian formation slightly depresses the lower boundary of the Sinian and Jinning movements in this region (Sun Jiacong, 1985).

The metamorphosed Poko formation of the Kontum massif may be tentatively placed among the rift complexes. It rests with unconformity on the Lower Proterozoic formations and Chulai granitoids and is represented by coarsely stratified quartzites interbedded with shales and dolomites. Rocks are folded and metamorphosed in greenschist facies. Their riftogenic nature is proven by the distinct linearity of many of their outcrops and recollection of the fact that tectonothermal reworking preceded their accumulation. Still the possibility of their belonging to formations of a platform cover cannot be totally ruled out.

Formation of a passive margin around the Yangtze Craton concluded in the Late Sinian. Throughout the Early Sinian much of the craton was uplifted.

Only its northern and south-eastern margins, turned towards the Qinling and Cathaysia seas respectively, experienced subsidence.

In western Hunan and northern Guangxi the Lower Sinian is represented by three formations of the Jiangkou group: Chang'an diamictites, mafic volcanics of Fulu and marine-glacial formations of Nantuo. The Upper Sinian is made up of black carbonate chert formations of Liaobao and its analogues.

The Late Sinian south-eastern margin corresponded to the deep Jiannan sea situated between the epicontinental Yangtze sea in the north-west and the Cathaysian ocean in the south-east and was separated by a chain of islands from the latter. The Jiannan sea comprised the outer shelf and continental slope zones (Il'in, 1986). Black coaly shales with concretions of phosphorite accumulated within this basin. Among these black shales silty, siliceous and illite shales are distinguishable. They are replaced by flyschoid formations in the extreme south-east. The Jiannan sea has been reconstructed as a typical stagnant basin set off from the main Proto-Pacific water body by a narrow zone of uplifts (Il'in, 1986). The lithological-facies features of Sinian formations of this region, their geographic position between carbonate formations of epicontinental sea and deepwater Cathaysian flysch enable reliable classification of them as complexes of passive continental margins. The line of transition of this complex into platform dolomites in the north-west coincides roughly with the Hunan-Guizhou boundary and is manifest by a band of terrigenous-carbonate formations with phosphorites.

The western passive margin of the Yangtze continent in eastern Yunnan and western Sichuan was formed during Late Sinian transgression following Lower Sinian rifting. Formations of the 'subsided belt' located east of the Kam-Yunnan axis (Yang et al., 1986) correspond to this passive margin. A deep depression with thick accumulation of salt was formed in south-western Sichuan. Facies change of formations of the passive margin complex into rocks of the Yangtze platform is observable eastwards.

Carbonate-terrigenous formations with phosphorites of the Shapa-Kamdyong region of Laokay also pertain to formations of this ancient passive margin. They are accompanied by rocks of the Dougshantuo suite of China and are regarded as 'shelf formations' of South-east Asia. The suite is represented by shales and sandstones up to 4000 m in thickness (Fan Chyong Tkhi, 1981).

Within the Caledonian Cathaysian mobile belt siliceous and flysch deposits occur south-east and east of the 'black shales' of the Jiannan zone in the Guangdong-Zhejiang area. Some contemporary researchers deny the existence of a Caledonian eugeosynclinal region south-east of the Yangtze Craton (Ren Jishun et al., 1986). It has been suggested that the Cathaysian Caledonides developed on the continental crust between the Yangtze Craton in the north-west and the Cathaysian Craton (South China Sea) in the south-east. This suggestion is not without basis if we take into consideration the oldest ages of rocks from cores of boreholes drilled in the islands and shelf of the South China Sea. At the same time, this does not exclude the existence of an oceanic basin in the Sinian period in the Cathaysian territory (Simpson et al., 1987). From this viewpoint Vendian turbidite formations may be regarded as sediments of an open or marginal oceanic basin. They are typically developed in central and southern Guangxi in Yunsin region. The lower Shenshan group is made up of rhythmically alternating phyllites and siltstones (5000 m), the middle Shanxi group (5000 m) of phyllites and sandstones, while the upper Syafan group is represented by tuff-sandstones, sandstones and carbonate phyllites.

Formations of northern Guangdong and central and southern Guangxi which experienced Qingkai (Salairian, Early Caledonian) orogeny, have been regarded as belonging to island arcs in the palaeotectonic maps of China (*Atlas of Palaeogeography of China,* 1984). Near Husyan, at the Guangdong-Guangxi border, Upper Precambrian formations are partly correlated with the Banxi group and partly with the Sinian. They are represented by mafic, intermediate and felsic volcanics, shales and sandstones. It is possible that similar complexes are present in south-western Fujian in the form of volcanoterrigenous rocks of the Nanyang and Huanglian formations resting above the Dinguling formation pertaining to the pre-Sinian. As shown on the palaeotectonic maps of China, elevated islands arising at the site of Riphean volcanic arcs existed in the Sinian period in this part of Cathaysia. These island arcs served as zones of supply of terrigenous material into the adjoining sedimentation basins.

Evidently a new stage of ophiolite formation commenced in the second half of the Riphean in South-east Asia and continued in the Vendian and Cambrian. It reflected destruction of the continental crust that had formed just before.

This rifting led to formation of basins with oceanic crust whose relics are represented by Shong Ma, Tamki Thanmi and Shongtai Banka ophiolites. A system of rift troughs originated there and gave rise to the formation of geosynclines in the Cathaysian region (Le Zui Bath, 1985). Formation of the geosynclinal system of the northern Pamirs-Kuen Lun-Qinling evidently also took place in this period. Thus the Vendian-Cambrian interval represented an important epoch of extension and ophiolite (oceanic crust) formation over a significant expanse of South-east Asia.

8.1.14 South American Craton

The Vendian represented an epoch of culmination and completion of the Brasiliano cycle of tectonics and active diastrophism in the interval 680–450 m.y.

8.1.14.1 *Amazon and Sao Francisco cratons.* These events were manifest in the Amazon Craton by intrusion in its eastern and north-eastern margins of dykes and sills of mafic rocks aged 600 m.y. and formation of local grabens (Takutu).

Deposition of upper carbonate-terrigenous formations of the Bambuy group occurred on the Sao Francisco Craton. These formations were later replaced by the molassoid Tres Marias formation. The upper age limit of the Bambuy group falls at 600 ± 50 m.y. according to Rb-Sr and Pb isotope determinations of bulk samples of rocks from this group.

8.1.14.2 *Brazilides.* The main mobile zones arising in the Late Riphean continued tectonic development in the Vendian. About the Middle Vendian intense folding, metamorphism and intrusion of granites occurred in the intracratonic 'Brazilian' troughs. Strata were crumpled into linear folds with a general strike parallel to the margins of the adjoining cratons. Together with general vergence of folding towards the latter, substantial lateral movement of rock masses has been noticed in this direction. Thus along the margin of the Guapore Craton, from the side of the Paraguay-Araguaia fold belt, thrusting occurred along a series of faults, each over 100 km long. Major thrusts and nappes that collided while moving in an easterly direction towards the Sao Francisco Craton are discernible in the Brazilian fold belt.

The Brazilian rock complex experienced five phases of deformation which can be divided into two stages: syntectonic and post-tectonic (Costa Campos, 1984). In the first two phases, composing the first stage, cylindrical folds with a north-easterly strike and cleavage were formed. This stage concluded with major thrusting of the Canastra group over the Paranoa group. In the post-tectonic stage (three phases) less intense folds with a north-easterly strike, without cleavage, were formed. Strike-slip faults with a north-westerly strike are associated with this same stage.

Rocks underwent metamorphism to greenschist facies and also intrusion of syntectonic stocks. Geochronological dating has placed the main phase of metamorphism and granitisation at the level of 650 or 620 m.y. (de Almeida et al., 1976). Transcurrent displacements along sublatitudinal faults intersecting the present-day Atlantic coast play a significant role.

The concluding movements of the Brasiliano cycle after or during which typical red molasse of the type Jaibaras were formed, occurred at the end of the Vendian and beginning of the Cambrian. Numerous radiometric age determinations recorded by various methods in the Brazilian fold belts fall in the interval 650–600 m.y. and pertain to post-tectonic granites. Further, an extensively developed magmatic episode is noticed around 540 m.y.

Similar events with formation of fold systems occurred in the Kariri mobile region in the Ribeira belt; in addition to folding and metamorphism of Riphean formations in geosynclinal troughs, intense tectonothermal reworking of the basement occurred in the horst-like salients ('median massifs') separating these troughs. Movements along faults formed already at the beginning of the Riphean were renewed and often had the character of transgression. Significant dextral transcurrent displacements are observed along the Pernambuco and Paraiba faults in the Kariri belt. Dextral displacement of 300 km occurred in the Ribeira belt, along the Sao Paolo thrust zone.

Diastrophism of the Brasiliano cycle also led to reworking of the old massifs separating the geosynclinal zones of north-eastern and south-eastern Brazil, as manifest in the corresponding age determinations of granites as well as rocks of the substratum.

Granite formation is an essential feature of the Brazilian fold belts. Granites are widely developed in the north-eastern (Kariri) folded region where they have been grouped into four plutonic provinces. A close spatial relationship is noticed between the major granite batholiths and the faults separating this region into zones filled with Upper Precambrian formations of the Sierra group and into reworked ancient basement massifs dividing these zones. Granitoid magmatism manifested in marginal faults occurring among Upper Precambrian complexes as well as within the adjoining 'median massifs', i.e., salients of the reworked ancient basement.

In north-eastern Brazil syntectonic granitoids are most extensively (90% of total volume) developed in the form of major batholiths aged 650±30 m.y. Among them a lower heterogeneous complex represented by diorites and normal granites is distinguishable from a more felsic intrusive series represented by tonalites and normal granites. The synorogenic granitoid plutons of north-eastern Brazil were formed in the lower parts of Late Precambrian troughs by anatexis of volcanosedimentary formations filling them while late and post-tectonic granitoid complexes represent products of partial melting of the lower crust.

A similar development of Late Precambrian granitoid magmatism has been established by Wernick and associates (1988) for the Ribeira belt in south-eastern Brazil. Here too granite intrusions are controlled by the main faults separating the ancient massifs from the folded systems. Polyphase intrusions have been noticed, as in the north-east, but in the Ribeira belt late tectonic intrusions are prevalent. These intrusions are mainly confined to trough complexes and syntectonic bodies to the ancient massifs.

Some late and post-tectonic granitoids of the Ribeira belt are accompanied by migmatites. There was continuous formation of polyphase intrusive complexes without significant intervals from syntectonic to post-tectonic granites.

The number of intrusions in the various zones and massifs of the Ribeira belt varies. A total of 52 have been described, covering about 60% of the area of the belt. The formation period of syntectonic granites has been put at 650 m.y., of late tectonic at 610 m.y. and of post-tectonic at 550 m.y.

Thus E. Wernick emphasises the polyphase nature of the granitoid intrusions of north-eastern and south-western Brazil, their formation by repeated anatexis at different crust levels and the role of faults in the structural control of the distribution of granitoids. He evaluated the period of intrusion of syntectonic diorite-porphyries and tonalites as 650 ± 30 m.y., ago, late tectonic 540 ± 25 m.y. and post-tectonic 510 ± 60 m.y.

Molasse was quite extensively developed in the Vendian of Brazil, occurring in foredeeps and intermontane basin-type structures. Thus in the 900-km long Alto Paraguay trough, along the southern margin of the Amazon Craton adjoining the Paraguay-Araguaia fold belt, the red terrigenous Alto Paraguay series of up to 5 km in thickness was formed. Red terrigenous formations (Pirapora group) also accumulated on the eastern margin of the Brazilian belt. Within the Ribeira belt, molasse formed in the intermontane basins of Bom-Jardin, Camarinja, Marica and Itagi; analogous molasse formed in the Sergipe and Kariri belts: Estancia, Giuca and Correros groups. A distinctive feature of these molasse formations is the extensive development of felsic volcanics in them. Molasse accumulation commenced about 550 ± 30 m.y. ago (Goodwin, 1991).

Formation of the Tocantins-Araguaia marginal geosutures at the junction of the Amazon (Guapore) Craton and Paraguay-Araguaia fold belt was evidently associated with this Brasiliano orogeny. These sutures were marked by bodies (protrusions) of ultramafic and mafic composition extending up to Isla de Marajo in the north for a distance of 1200 km (de Almeida et al., 1976). They are concentrated in rocks of the Tocantins group but are also encountered more westward in the Araxa group and basement (Xingu complex).

8.1.14.3 *Andean belt.* Available data suggests that the passive marginal regime prevailing in the Proto-Andean system changed in the Vendian to an environment that is close to the present-day active margin of the Andean type. This is supported by radiometric age determinations obtained on granites in the Precambrian formations of the Andes. Thus on the basis of Rb-Sr isochron, granitoids of Sierra Pampeanas in Argentina were dated 600–500 m.y. and granites of Sierra Australes 575 ± 10 m.y. Metamorphic rocks of the Eastern Cordillera in Peru are aged 600 m.y. (de Almeida et al., 1976) and 660 ± 30 m.y. in Venezuela. Information is available on the accumulation of Late Precambrian molasse in the Proto-Andean belt, in particular for the Catamarca and Cordoba provinces of Argentina (Ambato group). These molasse top the geosynclinal evolution of the Andean system in the Late Precambrian-Cambrian. The intrusion of granitoids and accumulation of molasse was accompanied at places by felsic volcanism. Thus rhyolites aged 672 ± 35 m.y. are present in the Sierra Australes (Argentina). At the same time, some Lower Cambrian limestones deposited on Precambrian formations have not been affected by metamorphism. Unconformable deposition of some Upper Cambrian rocks on Late and Early Precambrian granitoids has also been described.

Proto-Andean fold belt made up of rocks metamorphosed to greenschist and at places to amphibolite facies intruded by granitoids arose as a result of these orogenic 'Pan-American' processes at the end of the Precambrian within the western continental margin of South America.

8.1.15 African Craton

The Vendian was an epoch of tectonic activity involving all the Riphean mobile zones of Africa. Accumulation of platform cover continued in a quiescent tectonic environment only in those cratons that had appeared in the Late Riphean.

8.1.15.1 *West African, Congo and Kalahari cratons.* In the Taoudenni and Congo syneclises, tillites and supratillite strata deposited with an unconformity on underlying formations correspond to the Vendian. In the Mauritanian Adrar sequence, suite II commences with Jbelia tillites aged 630–595 m.y. It is overlain by the carbonate-terrigenous Teniaguari group (Bthat-Ergil) aged 595 ± 43 m.y., which is replaced upwards along the sequence by sandstones and red shales of Falaish d'Atar with Cambrian-Ordovician fauna (Clauer, 1973; Deynoux et al., 1978; Boullier et al., 1991).

A striking persistence of the upper part of the Precambrian sequence is noticed everywhere in the supratillite complexes of Western Africa. Two lithostratigraphic units are distinctly delineated. These are the lower one usually commencing with tillites overlain by a horizon of limestones and dolomites passing into a suite of greenish-grey shales and siltstones in flyschoid alternation and the upper molassoid formation made up of variegated sandstones, shales and conglomerates. In Mauritania and north-western Mali the Oua-Oua and Aguanet, Kiffa and Tagant-Assaba series correspond to the above formations; in Senegal the Faleme and Bundu series; in Guinea the Mali series; in Burkina-Faso the Kayes-Nara; and in Ghana the Oti and Obosum suites of the Tamale series. N.A. Bozhko detected fragments of Upper Riphean-Vendian stromatolites in carbonate breccia that replaced tillites in the Oti suite and complexes of Vendian microfossils in the supratillite part. Glauconite from formations directly overlying tillites in the Volta basin was dated 620 m.y. (Bozhko, 1984). Thus tillites and supratillites of Western Africa may be placed in the Vendian. The stratigraphic position of the molassoid upper formation is not wholly clear. Possibly, its upper part passes into the Cambrian.

Correlation of the above strata with Gourma trough formations is not entirely clear due to the absence of tillites in the latter sequence. Data published on the Vendian age of the Sarnieri formation from the basement of formation 3 (West African Orogens . . ., 1991) helps compare formations 3 to 5 (Ombori group) with suite II of the Taoudenni syneclise and Tamale series of Ghana. Platform Riphean-Vendian formations of the Gourma and Volta troughs increase in thickness eastwards and form a pericratonic subsidence zone (miogeosyncline). In the concluding stage of evolution these structures played the role of foredeeps filled with red molasse (Bozhko, 1984).

Formations topping the sequence of the Congo syneclise pertain to the Vendian. In the Linda supergroup they are represented by Banamsa arkoses. These arkoses correspond to the interval 700–590 m.y. and pass under the Phanerozoic cover into upper Inkisi red beds of the Western Congolides system falling within the age frame of 734–625 m.y. (Goodwin, 1991).

The platform cover began forming in Namibia directly south of the Orange River. Here it is represented by the Namama group consisting of three subgroups: Kuibis (400 m) arkose-carbonate; Schwarzrand (up to 1700 m) shale-limestone; and Fish River red terrigenous formations. The lower half of the sequence (Kuibis-Schwarzrand) contains Vendian Ediacaran fauna and the upper half Lower Cambrian fauna (Semikhatov, 1974). Palaeomagnetic data also indicates Vendian (650 m.y.) age for the lower part of the Namama group and Early Cambrian for the Fish River subgroup (Kröner and Clauer, 1979). The Namama group is intruded by Bremen syenites aged 521 m.y.

8.1.15.2 *Trans-Sahara belt.* In eastern Hoggar carbonate-terrigenous formations of the Tiririne belt experienced metamorphism from greenschist to mid-amphibolite facies and intrusions of gabbro, diorites, granodiorites aged 660 ± 5 m.y. (U-Pb), adamellites 604 ± 13 m.y. and granites 585 ± 14 m.y. (Boullier et al., 1991).

In the polycyclic Central Hoggar belt, the Eburnean basement experienced active tectonothermal reworking manifest in the thrusting and intrusion of syntectonic granites aged 615 m.y. and post-tectonic magmatism at the level of 580 m.y.

In the Pharusian belt active formation of calc-alkaline volcanics and greywackes took place in the Vendian under conditions of subduction of the Série Verte in the west and Pharusian II (Amded and Irrelouchem series) in the east. In western and central Iforas the magmatic post-Kidal complex of ultramafics, gabbro, diorites and mafic lavas was formed in an environment of back-arc spreading (Boullier et al., 1991).

The Série Verte unconformably overlies quartz diorites aged 696 m.y. and is intruded by a rhyolite dyke aged 634 m.y. Volcanics of the Irrelouchem series have been dated 680 ± 36 m.y. (Rb-Sr) and comagmatic intrusions are aged about 650 m.y. Collision of the West Pharusian belt with the West African Craton occurred in the interval 620–580 m.y. and, correspondingly, intrusion of the calc-alkaline late tectonic Central Iforas batholith aged

595 ± 25 to 581 ± 15 m.y. (Rb-Sr), Immezaren pluton aged 583 ± 7 m.y. and accumulation of the Série Pourprée (intruded by granite aged 581 ± 7 m.y.) and Nigritien molasse (Boullier et al., 1991).

Thus development of the western part of Hoggar corresponded to a Wilson cycle and ended in collision and formation of a suture between the West African Craton and mobile belt. The last phase of diastrophism (about 600 m.y.) was manifest in considerable compression of the belt manifest in turn in formation of folds with a submeridional strike and a gigantic system of dextral strike-slip faults.

In the Atakora system the supratillite Buem series has a similar sequence but ends in volcanics of significantly mafic composition aged 620 ± 50 m.y. (Bozhko, 1984).

In the Tiririne trough of the Central Hoggar tectonic events were some-what delayed compared to the Western Hoggar where the age of synoro-genic granites was close to 585 m.y.

In the southern segment of the Trans-Sahara belt, within Nigeria, closure of ensialic troughs of the Anka type occurred with intrusion of synorogenic granites aged 585 m.y. and ubiquitous Pan-African reworking at the level of 500 ± 100 m.y. during which the so-called 'ancient granites' intruded. These granites are coarsely crystalline, biotitic and schistose with porphyroblasts of microcline and associated with diorites and syenites characterised by a com-paratively low $^{87}Sr/^{86}Sr$ of 0.705. In the much later, weakly metamorphosed pegmatites this ratio rises to 0.718.

In the Anka trough (Holt et al., 1978), polymictic conglomerates and sandstones regarded as Pan-African molasse overlie granites aged 600 ± 70 m.y.

In the Atakora zone the Buem supratillite series has a sequence similar to that of group II of the Taoudenni syneclise but is topped by basalts aged 620 ± 50 m.y. (Bozhko, 1984).

According to Ajibade and Wright (1989), the basement of Nigeria and adjoining regions (Togo-Benin-Nigerian Shield) represents an ensemble of allochthonous terranes of different origin amalgamated in the Pan-African epoch. This explains the extensive width of the Pan-African belt in this region, scatter of Pan-African age determinations (750–450 m.y.), presence of Lower Precambrian and Kibara age rocks and sharp difference in the com-position of volcanosedimentary formations of different supracrustal zones.

Closure of a small oceanic basin occurred in the Mauritanides (Lecorche et al., 1991) in the interval 680–640 m.y., when compression and deformation corresponding to Pan-African folding I took place. This led to partial melting of the crust and formation of the western ensialic volcanic arc and calc-alkaline granites of Niocolo Koba and others aged 680–685 m.y. (Pb). In the interval 640–550 m.y. uplift and cooling of the orogen thus formed took place as well as formation of supratilloid flysch series on the foreland. Pan-African deformation II manifested between 550 and 525 m.y.

In the immediately more southern segment of the Mauritania-Senegalese belt, i.e., the Bassarides, an oceanic rift of small width that was formed in the interval 700–675 m.y., began to close due to subduction under its western margin. This caused formation of the calc-alkaline complexes of the Niocolo Koba and Kuluntou groups aged 680–650 m.y., collision in the Kuluntou zone between two rift margins and overthrusting on the margin of the West African Craton. The interval 650–555 m.y. corresponds to the accumulation of molasse in small basins, Late Precambrian glaciation and formation of the Faleme trough.

Pan-African folding II occurred in the interval 555–550 m.y. Thus in the Bassarides two phases of Pan-African folding, 650 and 550 m.y., manifested distinctly and are clearly seen in the corresponding angular unconformities. Ritz and Robineau (1986) suggested that the Central Mauritanides and Bassarides separated by Bisseukidira differ in geodynamic history. While the former developed on thinned or limited oceanic crust, the latter arose in an intracratonic ensialic trough (Villeneuve et al., 1991).

The Rokellides belt (Culver et al., 1991), extending for 600 km from western Guinea through Sierra Leone into Liberia, falls on the southern continuation of the Bassarides. In Sierra Leone the belt comprises the Rokell River group. Its lower Tabe formation, deposited on the basement, is represented by sandstones with tillites at the base, which are correlated with the tillite horizon of the other regions of West Africa (630–595 m.y.). On top lie turbidites, shales and arkoses of the Teis and Mabole formations. The sequence is topped by tuffs and lavas of andesite and dacite composition with spilites of the Kaleve Hills formation. The terrigenous Kolente group and volcanosedimentary Bani group in Guinea and the Gibi Mountains formations in Liberia are correlated with the Rokell River group.

A single stage of metamorphism and folding occurred in the Rokell belt in the interval 500–550 m.y., corresponding to Pan-African folding II.

The geodynamic evolution of the Rokellides is discussed on the basis of two hypotheses. According to the first, the belt is regarded as a plate-tectonics orogen with the Marampa cryptosuture manifest in a chain of isolated ultramafic bodies in the Marampa group of rocks. This hypothesis is presently denied on the basis of the absence of signs of a real cryptosuture in the old, Lower Precambrian Marampa rocks and manifestation of only one phase of diastrophism at 550 m.y. The second hypothesis assumes development of the Rokellides at the site of a limited rift trough enlarging northwards and possibly with an oceanic crust. This model accords by and large with the above-mentioned variant of Ritz and Robineau for the evolution of the Bassarides (Culver et al., 1991).

The region situated south and south-east of the Trans-Sahara belt and Eastern Saharan Craton including northern Cameroon, the Central African Republic and southern Sudan remains poorly studied. There is no doubt, however, that it was affected by Pan-African diastrophism. According to recent investigations in northern Cameroon (Toteu et al., 1987), an extensive fold belt filled with metamorphosed volcanoclastics, greywackes and bimodal volcanics of the 'intermediate group' was established. The period of formation of this belt has been determined by the age of rhyolites from the Poli shale formation at 830 ± 12 m.y. (U-Pb). These rocks experienced two phases of deformation and metamorphism. Syntectonic granitoids have age values predominantly of 600 m.y. while post-tectonic Gode granites are 546 ± 15 m.y. old. The proposed tectonic model of the belt (Toteu et al., 1987) is based on the development of an intracratonic trough associated with the evolution of an ocean along the eastern margin of the West African Craton. Such an interpretation appears wholly logical taking into consideration the overall structural plan of Late Riphean Gondwana (see Fig. 69) characterised by a network of intracratonic geosynclines with triple junctions in which ensialic zones form apophyses branching from ensimatic zones.

Diachronism of Late Precambrian diastrophism becomes even more evident when compared with the data for geosynclinal systems of Central Africa. As mentioned earlier, the main phase of diastrophism here occurred already at the end of the Late Riphean and accumulation of molasse in the Early Vendian while continental molasse in West Africa is mainly Cambrian in age.

8.1.15.3 *Western Congolides, Katangides and Damarides.* In the Western Congolides, red sandstones, conglomerates and shales of the Inkisi series (1 km) fill the foredeep of the fold edifice. The Inkisi series is characterised by an age of 613 ± 20 m.y.

Two epochs of folding, pre- and post-Inkisi, have been established in the evolution of the Western Congolides. The first has been fixed by Noqui granites aged 734 ± 10 m.y. (Rb-Sr). The second is placed in the interval 734 ± 10 to 625 or 600 m.y. Tectonic models of the Western Congolides vary from an intracratonic ensialic geosyncline (Bozhko, 1984) to a marginal basin deformed as a result of collision between the South American and Congo cratons with an assumed suture in the submerged margin of Africa or in Brazil (Goodwin, 1991).

Deposition of the Upper Kundelungu series that commenced roughly 850 m.y. ago continued in the Katanga fold belt and ceased at around 602 m.y. Its upper subdivision (Kundelungu plateau) made up of arkoses and red shales up to 1500 m thick evidently corresponds to the Vendian. Lufilian orogeny, as a result of which the folded Lufilian arc was formed, occurred in the interval 656–602 m.y. (Cahen et al., 1984).

Evidently the main tectonic movements manifested in the Damaran system somewhat later. The age of diastrophism preceding accumulation of the Mulden group molasse and causing formation of an edifice with an east-north-easterly strike is 650 (665) m.y. (Kroner and Clauer, 1979). Sedimentation of the red Mulden group occurred in the interval 610–560 m.y. (Hedberg, 1979) or 660–570 m.y.

At the same time, an important phase of regional metamorphism, granitisation and block deformation has been established at the level of 550 m.y. (Kröner, 1980), marking a transition from predominantly horizontal movements to vertical.

Geodynamic models of the Damara belt vary from aulacogens (Martin and Porada, 1977), limited spreading (Kasch, 1983), delamination of lithosphere (Kröner, 1980), intracratonic geosyncline (Bozhko, 1984) to subduction at the site of an open ocean (Dewey and Burke, 1973). The most recent model of Kukla and Stanistreet (1991) regards the Damara belt as an immense accretionary prism formed in the process of convergence of the Congo and Kalahari cratons. Matchless amphibolites and pelagic sediments are regarded as the upper part of the oceanic crust sequence (pillow lavas and deepwater deposits).

The age of diastrophism of the Gariep system is 598 ± 25 m.y. (Kröner, 1974) and the intrusion of post-tectonic Kubus granites occurred 525 ± 60 m.y. ago.

8.1.15.4 *Arabia-Nubian region.* The African continental margin of Proto-Tethys was active everywhere in the Vendian. In the Arabia-Nubian region development of marginal eastern parts of the accretionary complex ceased, the modern structural plan (Fig. 72) was formed and suites of small thickness accumulated.

In the south-western part of the Arabian peninsula the Murdama and Shammar groups and their equivalent Fatima belong to the Vendian. These groups are represented by andesites, rhyolites, greywackes and conglomerates resting unconformably on the Upper Riphean Halaban group and intruded by quartz monzonites aged 570–550 m.y. The age of the Shammar and Murdama groups has been determined in the interval 678–650 m.y. and of the Fatima group at about 700 m.y. In Egypt the Hammamat group resting on Dokhan volcanics and intruded by post-tectonic granites aged 620–590 m.y. corresponds to the above level; their equivalents in Sudan and Yemen are the Awat and Gabar groups respectively. In these groups terrigenous, including red molassoid, formations play a significant role. They reflect shallow-marine and continental conditions of sedimentation on the new crust formed in this part of the belt already in the Late Riphean as a result of the accretion of island arcs (see Fig. 63).

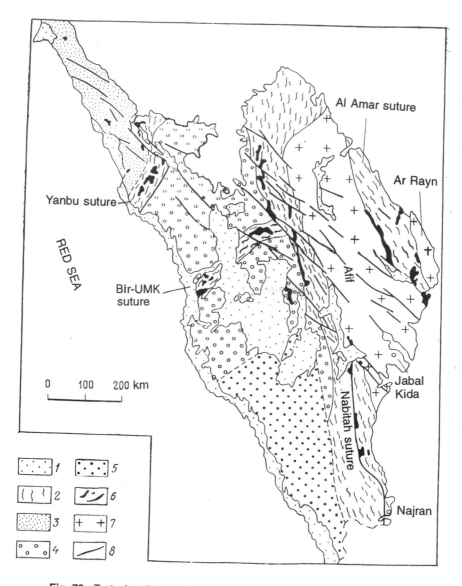

Fig. 72. Tectonic scheme of Arabian Shield (after Stoser and Camp, 1985).

1—Phanerozoic cover; 2—orogenic belt; 3–5—island-arc complexes (3—700–600 m.y.; 4—800–700 m.y.; 5—over 900–800 m.y.); 6—ophiolites; 7—continental blocks; 8—Najd fault system.

8.1.15.5 *Anti-Atlas.* In the Anti-Atlas zone, rifting and oceanic stages of basin development and obduction were replaced by subduction of the oceanic crust under the northern margin of the West African Craton (Leblanc and Lancelot, 1977).

That part of the oceanic crust surviving obduction was subducted under the margin of the craton, which caused formation of greywacke flysch, felsic and mafic volcanics of the Tiddilin series with a maximum age of 623 ± 10 m.y. (Goodwin, 1991), the Bleida granodiorites aged 615 ± 12 m.y. as well as the Ouarzazate calc-alkaline volcanic complex (578 ± 15 m.y.). The concluding collision movements are fixed by subvolcanic granitoids aged 563 ± 20 m.y. (Leblanc and Lancelot, 1977).

8.1.15.6 *TTR basement zones.* The maximum manifestation of TTR basements over extensive expanses of Africa pertains to the Vendian-Early Palaeozoic (500 ± 100 m.y.). Its largest region is the Mozambique belt of Eastern Africa and the Libya-Nigerian belt of Sahara and Nigeria.

In the Kenyan segment of the Mozambique belt the Baragonan phase of reworking caused by compression east-north-east, resulted in formation of upright folds and displacement along transcurrent faults with a north-north-westerly strike. This phase was accompanied by metamorphism from amphibolite to greenschist facies and intrusion of synkinematic granites at the level of 620 m.y. The next, Barsaloian phase (570 m.y.), led to the formation of distinct echeloned meridional strike-slip faults, intrusion of migmatites and granites, and massive rejuvenation of K-Ar and Rb-Sr datings (Key et al., 1989).

Madagascar constituting then a single whole with Africa was also involved in Pan-African reworking in the interval 650–485 m.y., manifested in the formation of superposed meridional structures, metamorphism and granite intrusions.

The ancient Precambrian formations of the Peri-Atlantic part of Central Africa, Zambezi belt etc. were also involved in Pan-African reworking. Tectonothermal reworking developed synchronously with events in the adjoining intracratonic geosynclines and was manifest in the early stages mainly under conditions of extension (diffuse rifting) of Early Precambrian crust, causing augmentation of its permeability, changing into general compression in the concluding stages. At this time transcurrent movements, formation of mylonite zones, intrusion of pegmatites and granites, metasomatism and isotopic 'rejuvenation' were intensely manifested. Tectonic deformation and magmatism in the zones of reworking had mostly concluded in the Vendian but thermal processes continued even in the Early Palaeozoic (up to 450 m.y.), as fixed by the 'rejuvenation' of datings of metamorphic rocks. The nature of this phenomenon will be considered at the end of this chapter.

8.1.16 Indian Craton

Vendian formations have a limited development in India. Among them is evidently part of the Bander group (1 km) topping the sequence of the

Vindhyan basin. It is made up of red sandstones and shales with layers of limestones, gypsum and conglomerates. Moralev (1977) places the sandstones of the Jodhpur-Sind suite of the Punjab basin also in the Vendian. These sandstones are underlain by Upper Riphean Malani rhyolites. TTR zones in Rajasthan and the Eastern Ghats continued to remain active in the Vendian as suggested by radiometric ages of pegmatites (700–580 m.y.). Intense movements along faults occurred in Singhbhum.

In Sri Lanka the Vendian time interval (610-550 m.y.) corresponds to the manifestation of high-grade regional metamorphism in Highland and Wanni (former Western Vijayan) complexes (Curray, 1994).

8.1.17 Australian Craton

Formation of platform cover continued in the Vendian in the basins of northwestern and central Australia. Many carbonate rocks are present in them.

In Kimberley basin (Labbok trough) the Kuniandi group is overlain with an unconformity by the Louisa Downs group (about 4 km). In its base are deposited Egan tillites, with shales, sandstones and dolomites of the Urabi, MacAem, Tin and Labbok formations on top. The age of the Egan tillites is 680–660 m.y. (Chumakov, 1978). The MacAem shales have an Rb-Sr dating of 665 m.y. The sequence is overlain by Cambrian Antrim basalts.

On the western margin of Sturt plateau, the Albert-Edwards group is equated to the Louisa Downs group, tillites being absent in the sequence of the former. Rb-Sr datings of clay minerals from this group are 666 and 653 m.y.

In central Australia tillites of the Olympic formation in the Amadeus basin aged 680–660 m.y. are equated to the Egan tillites of the Kimberley basin. Thus the upper part of the Precambrian sequence of this basin including the Olympia-Pertataka formations and Arambera I sandstones was formed in the Vendian. The upper part of the Arambera formation contains Archaeocyatha and pertains to the Cambrian. The reference horizons of the Olympic formation are traced to the Ngalia and Georgina basins marking the lower part of the Vendian sequences of these structures.

8.1.18 Adelaide System and Tasmania Island

Elatina tillites in the roof of the Umberatana group of the Adelaide system are equated to the Egan tillites of Western Australia. The overlying Wilpena group (up to 6 km) is made up of shales, siltstones, dolomites, mudstones and limestones. The sequence is topped by Pound quartzite from which M. Glaessner described the famous Ediacaran invertebrate fauna. The Pound quartzite is overlain with traces of erosion by Cambrian formation. Isochrons of 750 ± 53 m.y. and 724 ± 40 m.y. were obtained from the Umberatana group of formations and 670 ± 240 m.y. from the base of the Wilpena group (Goodwin, 1991). Thus the upper part of the Adelaide sequence, from Elatina tillites to Pound quartzite inclusive was formed in the Vendian.

Basal formations of the Palaeozoic sequences of Dundas, Dian range, Smitton and Adamsfield troughs studied by Williams (1978) in Tasmania belong to the Vendian. Dismembered ophiolites, possibly of Late Riphean age, are associated with these troughs. It is generally thought that siltstones and sandstones of the Success Creek (970 m) group in Dundas trough were deposited on top of the Una formation, Rocky Cape region, but a tectonic contact is in fact noticed between them. The overlying Crimson Creek (2.5 km) formation is represented by turbidites to which protrusions of ultramafics and gabbro are confined.

Thus palaeotectonic analysis reveals two distinctive features:

1) 'Pan-African' reworking (Vendian-Early Palaeozoic) was negligible in Australia. This is associated to some extent with the absence, at that time, of adjoining active continental margins. Thus the presence of reworking of the ancient basement of eastern Australia has been noticed in the Devonian when the tectonic regime of passive continental margin in the adjoining Tasman belt was replaced by active processes of interaction of plates in an environment of island arcs.

2) Folding was weakly manifest in Australia at the end of the Precambrian. This period was marked in all the mainlands of Gondwana by the inception of such intracratonic fold structures as the Brazilides, Katangides, Western Congolides, Damarides and others. None of these events were observed in Australia, however, except for local manifestation of 'Peterman Range orogeny' in the Amadeus aulacogen, which did not constitute the concluding stage of the latter's development, nor of the western part of Tasmania which later experienced Penguin orogeny. Formations of the Adelaide trough experienced Delamerian folding together with Cambrian formations of the adjoining Kanmantoo trough while formations of Amadeus, Ngalia and Officer aulacogens were deformed in the Early Carboniferous.

8.1.19 East Antarctic Craton

The structural plan laid in the Late Precambrian continued to prevail in the Vendian as well. The eastern margin of the craton from Enderby Land to Victoria Land was involved in tectonothermal reworking, as fixed by numerous Rb-Sr and K-Ar datings of magmatic and metamorphic formations reflecting Vendian-Early Palaeozoic rejuvenation in the corresponding diaphthorite zones. Aulacogens that arose in the Late Riphean continued development. Thus accumulation of the Friendship rock series continued in MacRobertson Land aulacogen, as supported by the finds of Vendian acritarchs and invertebrate bivalve tests.

Evidently deposition of the upper half of formations in the Penck depression of the western part of Dronning Maud Land took place in the Vendian, a fact supported by the results of determining the period of micatisation of shales by the argon method (590–515 m.y.). At the end of the Vendian to commencement of the Palaeozoic, aulacogen formations underwent metamorphism and tectonic deformation.

8.1.20 West Antarctic Continental Margin

At the end of the Late Precambrian (650 ± 50 m.y.), intense tectonic movements manifested in the region of the Transantarctic range. These movements corresponded to the main phase of folding and formation of structural unconformity at the base of the Early Cambrian complex (Grikurov et al., 1976, 1978; Elliot, 1975). In English literature this phase of movements is termed the 'Beardmore orogeny' but would be better named the Early Ross tectonic epoch, emphasising the incompleteness of the Ross tectonic cycle at this time. The most important endogenic events of Early Ross orogeny in the Transantarctic range were intense synkinematic mobilisation of the pre-Ross basement in the form of rheomorphic diorite-granodiorite batholith-like intrusions in supracrustal formations of the type Carlion granodiorite with an Rb-Sr age of 568 ± 10 m.y., etc. as well as the subsequent growth of infrastructure as a result of contact metamorphism of these formations under conditions of amphibolite facies; farther away from these intrusions metamorphism did not exceed greenschist facies. The formation of batholiths proceeded simultaneous with the deposition of felsic and intermediate volcanics on shale-greywacke formations, usually overlain by Cambrian ones. The Wayatt formation, represented by felsic metavolcanics, developed in the Wayatt and Gardiner mountains in the upper reaches of Scott glacier. The age of these rocks determined by the Rb-Sr method is 633 ± 13 m.y. and of granites intruding them 627 ± 22 m.y. (Rudyachenok, 1974). Metavolcanics (Fairweather formations) similar in composition and formation conditions have been detected in the Duncan mountains and in the region of Axel-Heiberg and Liz glaciers. Reworking of the ancient basement of the Transantarctic range is associated with Beardmore orogeny.

Complete closure of the Ross geosyncline occurred in the Late Ross epoch (ϵ_3 − 0) (Elliot, 1975). The conditions in this epoch greatly resembled those of an active margin of the Andean type. Its origin was evidently associated with partial closure of this small oceanic basin west of the East Antarctic Craton and formation of the Benioff zone dipping eastwards under the East Antarctic Craton (Bozhko, 1984).

The final Precambrian phase of diastrophism and corresponding magmatism in the Riphean fold zones of the north-eastern extremity of Victoria Land and western extremity of Mary Byrd Land has not been established. This points to the continued existence of a marine basin until the end of the Vendian, this basin having separated a microcontinent from the East Antarctic Craton. Alpine-type folding and intrusion of granitoids in the Upper

Precambrian formations of this region has been associated with Devonian Borchgrevink tectonics (Elliot, 1975).

Within the microcontinent of Mary Byrd Land, sedimentation evidently continued in the Vendian. In the region of Sondre coast, the Upper Riphean West series is overlain by the Passel series made up of homogeneous chlorite-sericite and quartz-chlorite-sericite microschists (metashales) of 1 km total thickness. Vendian-Cambrian acritarchs and microphytolites have been detected in the rocks of the Passel series.

Thus the palaeotectonic environment in the region of the Transantarctic range underwent a sharp change in the Vendian from conditions of a passive continental margin to an active margin of the Andean type.

Nevertheless it must be emphasised that most of the interpretations of the palaeotectonic environment of the western Antarctic region and south-eastern Australia in the Riphean-Vendian given here and in Chapter 7 are rather hypothetical. What can be asserted with greater certitude is that an oceanic basin did exist between the Robertson Bay and Wilson blocks, i.e., along the line of Bowers zone with the Late Precambrian-Cambrian Sledgers volcanics. The age of Robertson Bay turbidites has been given in recent works as post-Cambrian—Ordovician. Yet granites of Sedge Island in the north-eastern part of the block are dated about 600 m.y. (Borg et al., 1990). With such an age evaluation, the Robertson Bay turbidites cannot represent a complex of the western passive margin of the aforementioned ocean and would constitute an independent terrane. Furthermore, a new variant of the correlation of the formation of Severnaya Zemlya, Queen Victoria Land and south-eastern Australia, without Tasmania, has been proposed: Tasmania was displaced eastwards along the zone of the Main Tasmanian strike-slip fault. Bowers terrane corresponds in this correlation to the Grampian-Stavely terrane of Australia and Robertson Bay to the Stowell terrane (Borg and Paolo, 1991).

8.2 TYPES OF STRUCTURES AND TECTONIC REGIME

8.2.1 Cratonic Basins (Syneclises)

Formation of extensive platform sedimentary cover on large areas that had commenced in Gondwana in the Late Riphean extended in the Vendian to the East European, Siberian and Sino-Korean cratons, while formation of structures of the type of gentle troughs, basins and syneclises characteristic of the platform stage was initiated in this epoch (Fig. 73).

Differences are thus perceptible in the evolution of the platform cover of Gondwana and Laurasia. The platform cover of the large syneclises of Laurasia began to form later, in the Vendian, and was preceded by a prolonged aulacogen (rifting) stage of evolution, corresponding to almost the entire Riphean.

368

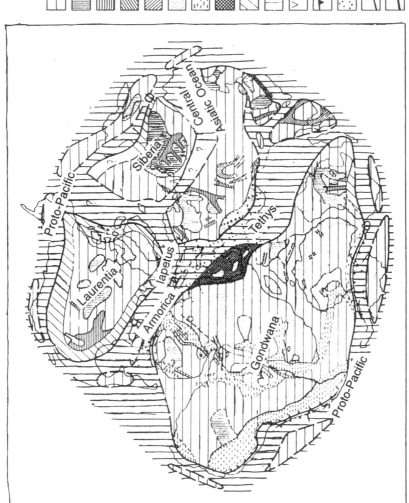

Fig. 73 Global Vendian palaeotectonic scheme (after Bozhko, 1984):

1—continental Early Precambrian crust; 2—platform cover; 3—aulacogens; 4—ensialic intracratonic geosynclines; 5—ensimatic intracratonic geosynclines; 6—zones of non-geosynclinal tectonothermal reworking of the basement; 7—newly formed fold systems; 8—accretionary complexes; 9—passive margins; 10—oceanic crust; 11—volcanoplutonic belts; 12—subduction zones; 13—Cadomian folding in the region of Proto-Tethys (640–600 m.y.); 14—faults; 15—boundaries of continental crust.

A rift regime and accumulation of sparagmite type formations intensified in the Urals (Kurbatskaya, 1985), the indicators of these processes being tillites and tillite-like conglomerates. Their analogues are also traceable in the north-eastern and north-north-western framing of the East European Craton. Closure of aulacogens, expansion of sedimentation area and formation of shales, sandstones and siltstones directly on the basement occurred towards the end of the Early Vendian. The platform cover of the Valday series, characterised by a persistent sequence of sandstones, siltstones and shales with layers of tuffs and tuffites, was formed. Sedimentation proceeded in a shallow, desalted, intracontinental marine basin.

In the Siberian Craton the palaeotectonic environment of the Early Vendian led to an extremely close connection of processes of sedimentation with outbursts of volcanism (Shpunt, 1984). In the Late Vendian the craton was involved in transgression with predominance of carbonate accumulation.

In syneclises of South America and Africa a reverse tendency is discernible in the Vendian, namely some reduction in areas of sedimentation and their concentration in the central parts of basins. Tillites topped by red molassoid deposits were formed. A sharp reduction of zones of accumulation of platform cover has been noticed in India. Evaporites are present here among carbonate-terrigenous beds. A homogeneous carbonate-terrigenous cover continued to accumulate in Australia almost in those same regions as in the preceding stage. On the whole, formation of Vendian sedimentary cover occurred everywhere in a typical platform regime. Its transition into synchronous formations of foredeeps (Western Congolides, Mauritanides and Atakorides), miogeosynclines (Adelaide) and major aulacogens (Proto-Urals western slope) has been noted.

8.2.2 Aulacogens

A sharp reduction in number of aulacogens not only in Gondwana, where they were generally less typical, but also in Eurasia occurred in the Vendian. Most active were the aulacogens of East Gondwana: Australia (Georgina, Amadeus, Officer and Ngalia) and Antarctica (Dronning Maud, MacRobertson and Adele).

8.2.3 Intracratonic Geosynclines

Data for the Vendian enables analysis of the fold systems that arose as a result of the closure of intracratonic geosynclines as well as the formation or early stages of development of the third generation of Late Precambrian structures of this type.

As a result of 'Pan-African' diastrophism (650 m.y.), the geosynclinal systems of South America and Africa were closed (see Fig. 73). This process proceeded unevenly. As mentioned earlier, the mobile systems of Central Africa (Western Congolides and Katangides) had experienced folding

already by the end of the Late Riphean and been transformed into fold mountain systems, as supported by the origin of foredeeps in their frontal parts.
These foredeeps were filled with typical continental molasse. This reveals
yet one more similarity to the Phanerozoic style of geosynclinal development. One such basin, the Inkisi, is noticed on the margin of the Western
Congolides; in the Katangides the red-bed Plateau series was formed in a
corresponding basin in front of the Lufilian arc. Molasse of the Mulden series
accumulated in the Damarides in a similar basin on the northern periphery
of the system.

Folding manifested later, at about 650 m.y., in the geosynclines of
West Africa and Brazil but led to the formation of similar structures. The
ensialic troughs of Tiririne, Nigeria and north-eastern Brazil and the Ribeira
belt experienced deformation, intrusion of granitoids and metamorphism to
greenschist facies. In the later appearing foredeeps and intermontane basins
volcanosedimentary Vendian-Cambrian molasse began to form (Série
Pourprée, Nigritien, Obosum, Tambaras, Kamarinya etc. series). These
systems experienced complex deformation: linear folding and overthrusts,
even nappes of the type Naukluft at places on the southern margin of
the Damarides. Granitisation was manifest extensively in most of these
geosynclinal troughs. As in structures of the preceding generation, the
intrusion of granites was polyphase. Syntectonic and post-tectonic granitoids
are distinguishable with separation into several phases at places (i.e.,
SE Brazil). Such intense granitisation is not manifest in all the troughs,
however. It was far more moderate in the Katangides and Damarides and
altogether absent in the Western Congolides. Regional metamorphism of
ensialic systems was not uniform, ranging from almost nil in the Katanga
belt to zonal metamorphism and migmatisation in troughs of the Ribeira
belt. Metamorphism to greenschist facies is most typical.

Ensimatic troughs of this type in Africa and South America closed after
'Pan-African' diastrophism simultaneous with the ensialic ones. Syntectonic
granites aged 650 m.y. in Western and Central Ahaggar (Hoggar) point to
significant processes of crust melting during the peak period of the early
phase of Pan-African orogeny. This orogeny was associated with uneven
Hoggar uplift as a result of the development of a system of strike-slip faults
and later, to a lesser extent, block tectonics, which led to a complex structural and metamorphic pattern. The degree of metamorphism and deformation in the belt depends on the nature of the crust. Very weak deformation
and metamorphism have been noticed in Central Hoggar resting on granulite crust while amphibolite facies metamorphism, anatexis and syntectonic
granitisation are localised westwards. Polyphase deformation of cover formations is confined to the deeper structural levels of Western Hoggar while
only folding with meridional and vertical axial planes is noticed in Central

and Eastern Hoggar. In the Atakorides folding weakens westerly and has a western vergence towards the foredeep.

Development of the Mauritania-Brazilian zone, already formed in the Early Riphean, was also completed in the Vendian. The concluding stage of its evolution is associated with the birth of extended foredeeps: Faleme on the eastern margin of the Mauritania-Senegalese belt and Alto Paraguay on the western margin of the Paraguay-Araguaia system. The thickness of red molasse in the latter trough reaches 5 km. Basins filled with the Nemchanka suite of the Yenisei range occupy a similar structural position.

A complex structural plan resulted from destruction, which had commenced already at the end of the Late Riphean in Central Asia. This plan may be delineated in a very general form characterised by the main trunk of the Central Asiatic ocean (Zonenshain, 1974) on both sides of which were situated segments of continental crust: Kazakhstan-Tien Shan in the west and Altay-Sayan-Transbaikalia in the east. In these segments a series of median massifs—detached blocks that survived destruction—were separated by complex branches of predominantly ensimatic intracratonic geosynclines, many of which (Dzhida zone and others) represented apophyses of the main ocean trunk. Intracratonic troughs are often arcuate, conforming to the oval plan (central Kazakhstan). But recent palaeomagnetic investigations have shown that this arcuate pattern is secondary.

Zaitsev (1984) studied in detail the development of destruction during formation of the Late Precambrian-Palaeozoic geosynclines of the Kazakhstan-Tien Shan region and palaeotectonic conditions of the embryonic geosynclinal stage. According to his model, relatively narrow but extended troughs arose in the outer part of the territory of the Kazakhstan Palaeozoides. An example of such a zone is the system of troughs up to 2500 km long and 60–150 km wide covering the KalmykKul and Baykonur synclinoria and synclinoria of Greater Karatau and Chatkal-Naryn zones. In these troughs, volcanoterrigenous chert-diabase formations accumulated and may have been replaced along the strike by molasse deposits. Variation in the 'degree of geosynclinal condition' along the troughs noticed in the intracratonic troughs of Kazakhstan represents the most characteristic feature of structures of this type. In his model Zaitsev adopts the view that the Vendian was absent in the eugeosynclinal complex of the Caledonian Kazakhstan-Northern Tien Shan region and distinguished on this ground a relatively stable central core; eugeosynclines with ophiolites arose at the site of this core only in the Cambrian. Nevertheless, Zaitsev himself acknowledged that the problem of the Vendian in the eugeosynclinal system of Kazakhstan has not yet been resolved. If its presence is accepted, the palaeotectonic environment is supplemented with ophiolite troughs of the eastern eugeosynclinal zone. In this case the ensimatic character of intracratonic troughs would diminish from east to west, i.e., from the inner zone adjoining the axial Zaisan system of the Central Asiatic palaeo-ocean to the Kokchetav-Karatau outer zone falling on the periphery of a vast region of destruction.

A similar environment has been established in a very general form in the symmetrically disposed eastern flank of the main oceanic basin in the eastern Altay-Sayan region: Kuznetsk Alatau, Gornaya Shoria, Eastern Sayan and Tuva. At the end of the Late Riphean-Vendian-Cambrian, the structural plan here too was determined by a 'complex mosaic combination of narrow, linear volcanic zones oriented variously and isometrical terranes with carbonate and carbonate-terrigenous sedimentation' (Mossakovsky and Dergunov, 1983). Small bodies of ultramafics, gabbroids and serpentinite melange present in the aforesaid linear zones indicate that destruction during their formation led to spreading of the Red Sea stage. The arcuate character of the troughs, similar to that of the Kazakhstan type, and the 'opening' of some troughs (Dzhida etc.) into the main oceanic basin merits attention (see above). In the Transbaikalia and Baikal-Olekma region, synchronous Vendian-Cambrian basins, placed by Bulgatov (1983) in the category of deuterogenic troughs, were filled with molassoid and carbonate-terrigenous formations and represent mainly ensialic structures. This is in consonance with their more peripheral position relative to the main zone of destruction (palaeo-ocean) compared to the eastern part of the Altay-Sayan region.

N.A. Bozhko considers these basins post-accretionary. Their origin followed accretion of terranes in the middle of the Late Riphean and they developed synchronous with the western Transbaikalian Caledonian geosyncline.

8.2.4 Belts of Tectonothermal Reworking (TTR)

Reworking of continental crust over vast areas of Africa and South America manifested most intensely in the Vendian and led to the formation of the largest belts of reworking, viz., Libya-Nigerian, Mozambique and Brazilian (Atlantic). Reworking renewed in Rajasthan, the Eastern Ghats and Satpura belts of India as well as in the Antarctic region.

The substratum in the 'Pan-African' belts of reworking is heterogeneous. It includes Lower Precambrian formations as well as the Middle Riphean systems Alexod, Lurio and Namama. In the Libya-Nigerian and Brazilian belts, reworked blocks of the basement alternate with intracratonic Late Riphean-Vendian troughs. In the Mozambique belt the presence of rocks of this latter age has not been established but the belt forms a triple junction with the Katangides while structural 'wedges' diverging from the Arabia-Nubian orthogeosynclines pass directly into the Mozambique belt structure in the region of Sudan and Ethiopia. Thus a close spatial and evolutionary relation of TTR zones with synchronous geosynclinal systems is noticed. All the 'Pan-African' zones of reworking experienced total isotope rejuvenation, as reflected in numerous Rb-Sr and K-Ar datings falling predominantly in the interval 650–500 m.y.

Within the Mozambique belt, it is difficult to distinguish granitoids synchronous with 'Pan-African' reworking from much older ones, especially on taking into consideration the increasingly revealed role of Grenville diastrophism in the belt formation. Among the Pan-African granites are evidently the small granite massifs of the type Cape McLyre and Chapalapata in Malawi and Zambia; numerous pegmatites, often bearing uranium, are observed throughout their territory.

During 'Pan-African' reworking of the Nigerian basement so-called 'old granites' intruded, i.e., coarsely crystalline, biotitic, schistose granites with microcline porphyroblasts associated with diorites and syenites characterised by a comparatively low $^{87}Sr/^{86}Sr$ ratio of 0.705. In much later weakly metamorphosed pegmatites, this ratio rises to 0.718.

Granitoids are extensively developed in north-eastern Brazil and in the Ribeira belt. They intrude intracratonic trough deposits but are mainly found in the extensive salients of the ancient basement. These are essentially syntectonic granites (described above) aged about 650 m.y.; post-tectonic granites aged 560–510 m.y. also developed similarly but to a lesser extent. Continuous formation of polyphase intrusive complexes without significant interruptions from syntectonic to post-tectonic granites is observable.

Granitoids of the TTR zones of India, as demonstrated by V.M. Moralev, are represented mainly by adamellites. Compared to basement granites, these are more felsic, the role of plagioclase greater in them, total alkali content less, the thorium to uranium ratio low and the uranium to potassium ratio high.

It is not easy to explain the Vendian-Early Palaeozoic metamorphism of these zones given their polycyclic character. There is no doubt about the development of diaphthoresis along the zones of superposed deformation and amphibolisation of granulites. Judging from the greenschist metamorphism of rocks that are close in age to reworking, a very similar metamorphism was superposed also on the surrounding substratum. At the same time, the view that the granulites in the zones of reworking are Early Archaean in age is no longer undisputable since reliable Riphean datings have been obtained for granulite complexes in Australia and India. In the Mozambique belt granulite metamorphism manifested in the Grenville epoch judging from its development in the Lurio belt of Mozambique.

Similarly, a close relation is detected between the degree of deformation and intensity of metasomatic recrystallisation and in particular potassium metasomatism, leading to microclinisation of felsic and intermediate rocks. TTR zones are characterised by their richness in metamorphogenic pegmatites. Moralev (1975) demonstrated that in India the formation of these rocks proceeded in zones of high permeability that experienced horizontal displacements. Isotope age determinations of minerals from pegmatites and host rocks point to the simultaneity of different phases of pegmatite formation and phases of superposed metamorphism. This aspect was studied in depth on the example of the reworking of the Precambrian basement of western Canada (Burwash and Krupicka, 1969). It was demonstrated that the supply of potassium manifest here in microclinisation is closely associated with the destruction of plagioclase during cataclasis and its replacement by potash feldspar. In this process the upward migration of intergranular fluid enriched with potassium, rubidium and other ions during the rise of mantle diapirs was of vital importance. The process of cataclasis and subsequent meta-somatism led to an increase in rubidium and potassium and reduction in strontium, which makes radiometric age determination of rocks subjected to tectonothermal reworking extremely difficult. Radiometric rejuvenation of the ancient substratum invariably arises during its repeated heating. During reorganisation of the crystalline lattice of minerals, leakage of the products of radioactive disintegration occurs and hence most of the K-Ar and Rb-Sr age determinations reflect the age of the most recent metamorphism and heating. The mechanism of rejuvenation is a complex one requiring special studies. It may only be pointed out that although this is a distinctly geochem-ical problem, the true importance of radiometric age values, rejuvenated as well as relict, can only be reliably established by taking into due consideration the geological data.

During TTR rocks of the upper structural layer are subjected to progres-sive metamorphism and the phases of metamorphism are distinctly manifest.

The main part of the TTR zones experienced repeated deformation dur-ing reworking. This deformation is often associated with the formation of granite-gneiss mantled domes, as pointed out for example by the evolution of deformation in the Pan-African (500 m.y.) TTR Zambezi zone and western part of the Mozambique belt on the margin of the Zimbabwe Craton.

The complex evolution of superposed folding during reworking of the substratum of the Mozambique belt has been reconstructed in northern Malawi, north-eastern Hoggar, Tanzania, Kenya and other places. The preva-lence of an extension environment noticed in the initial stages of reworking was only later replaced by compression.

This view is supported by the presence of preserved graben-shaped riftogenic troughs in some TTR zones. Subsequently, in the period of com-pression these troughs filled with terrigenous and volcanoterrigenous com-plexes experienced folding and granitisation and were transformed into fold belts. Examples of such instances are the southern segment of the TTR Ribeira zone and Nigerian region in the TTR Libya-Nigerian belt.

The superposed folding and cleavage arising in the process of tectonic reworking complicate the already-formed basement structure but do not completely mask the primary folding, which is morphologically inferior. On the whole, the (reworked) substratum is relatively brittle and broken up into a network of linear faults. Movements along them represent the main manifestation of tectonic reworking. They led in particular to the development of tectonite zones of considerable thickness within which rocks experienced fracturing and shearing and acquired secondary folding. They are characterised by comparatively simple internal structure but, in principle, represent zones of maximum deformation in which the primary structure disappeared due to total reworking. The extent of such linear zones of diaphthorites reaches hundreds and thousands of kilometres. They arose in the earliest stages of reworking and exerted considerable influence on its future course, ultimately dividing the crust into independent blocks which later experienced differential displacement of different types. The result of such reworking of rocks by shearing has been described, for example, by D. Hepworth in north-eastern Tanzania, and D. Allen in central Australia. The primary gneiss-amphibolite complex here was reworked into flag gneisses and dipped, together with newly formed fold hinges, eastwards across the main strike of the belt. 'Islands' of unreworked gneisses are preserved among such rocks.

Many transcurrent faults trending across the main structural strike of TTR zones are associated with the Precambrian TTR process of Gondwana. Among them are the Aswa zone in Uganda intersecting the Mozambique belt west-north-west for a distance exceeding 1000 km and abutting the marginal zone of mylonites limiting the Mozambique belt from the west; Mwembeshi zone in Zambia, also terminating at the margin of the belt; Taksakuara and Sao Paolo faults in Brazil; fault zones of Angola; and transverse zones of reactivation of Riphean age in India, Australia, the Tibesti region of Africa and elsewhere.

Movements along faults in TTR zones were not exclusively transcurrent or vertical. The significant role of overthrusts and nappes formed in the process of tectonic reworking has recently been recognised. Overthrusing in the Pan-African epoch was very widespread and played an important role in particular in the evolution of TTR Ribeira belts on the Brazilian continental margin. The crystalline basement highly metamorphosed to granulite facies was remobilised and overlay here the lower grade metamorphic rocks. Superposed Late Precambrian deformation, strike-slip faults and tectonic nappes of reworked basement are also known in south-western Africa, along the eastern and western margins of the West African Craton, in central Australia and other places. The immense overthrust structure of central Australia, known as the Peterman nappe, lies directly south of the Riphean Amadeus basin. Its area exceeds 30,000 km^2. Formation of this nappe is associated with the same-named folding that affected sedimentary Riphean formations and led to remobilisation of the ancient basement and its displacement along overthrusts in a northerly direction.

Thus factual data on 'Pan-African' and 'Brazilian' reworking of continental crust demonstrates the close relationship of this process with the development of intracratonic troughs falling directly within the tectonothermally reworked zones or adjoining them. The probable development of the zones of reworking, using the southern part of West Africa as an example, is schematically depicted in Fig. 74. It must be pointed out that this model of the TTR process prevailing at the end of the Precambrian-commencement of the Cambrian in the Gondwana continents represents a striking but last manifestation of this form of tectonics on such a scale.

8.2.5 Marginal-continental Geosynclines

Much of the Pacific mobile belt continued to develop in a rift regime. Passive margins existed in the Cordillera, on the margin of the Siberian Craton and in Australia. In South-east Asia, after temporary stabilisation, a new stage of destruction set in, leading to formation of basins with an oceanic crust whose relics are represented by ophiolites of Shong Ma, Tamki-Thanmi and Shong Tai-Bakha and signify formation of eugeosynclinal Phanerozoic belts. Within the major Precambrian blocks (Yangtze, Indosinian and Sinoburman), formation of sedimentary cover continued and the passive margin environment of the Cathaysian ocean was renewed on the south-eastern margin.

At the same time, within some segments of the Pacific 'rim' there existed already in the Vendian a palaeotectonic environment of active margin of the Andean type. This pertains to the Andes proper and the region of the Transantarctic range. As pointed out above, the regime of passive margin in the Proto-Andean belt was replaced in the Vendian by intrusion of granitoids, formation of molasse and felsic volcanics. In the Transantarctic range calc-alkaline mafites formed and Beardmore orogeny manifested at this time. Thus the Pacific margin of West Gondwana was first involved in active marginal-continental processes after formation of a global Proto-Pacific miogeosynclinal belt in the Late Riphean.

The Vendian-Cambrian period was remarkable for the opening of Iapetus—the first ocean basin of the North Atlantic—arising at the site of an extended Late Riphean rift system. The Iapetus together with its Innuitian branch is represented in the global plan as a strait connecting the Proto-Tethys with the Proto-Pacific. Its opening led to isolation of the Laurentia palaeocontinent. In the region of articulation between the Iapetus and the Proto-Tethys complex geodynamic processes occurred in the Vendian, which were associated with formation of the so-called Cadomian terranes.

Evolution of the northern Armorican massif was similar to that of other Cadomian terranes, viz., Avalonian and Dalradian. Brioverian turbidites, calc-alkaline volcanics and granites comagmatic with them reflect an environment of ancient island arc and marginal sea against a background of subduction

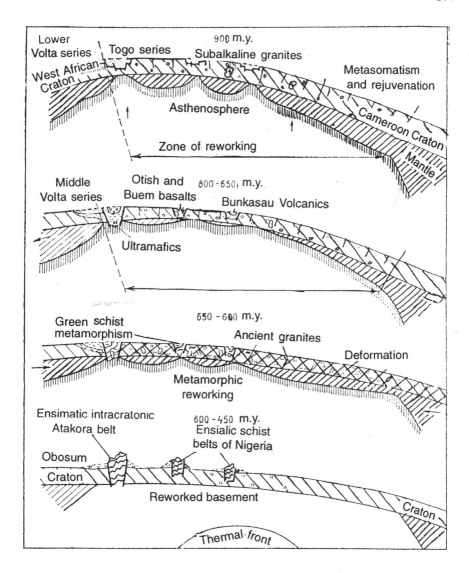

Fig. 74. Schematic depiction of a possible model of tectonothermal reworking of the basement, using the southern part of Western Africa as an example (after Bozhko, 1984).

in the interval 700–540 m.y. Subduction in the Avalonian arc occurred in the interval 675–550 m.y. and in southern Britain 700–660 m.y. (Nance et al., 1991). The diachronous subduction process in these terranes is explained by the interaction of transform faults with a subduction zone. According to palaeomagnetic data, Cadomian terranes constituting the Avalonian-Cadomian belt at the end of the Precambrian to the Early Palaeozoic were attached to the northern margin of Africa, a constituent of the Gondwana supercontinent. This reconstruction conforms to geological data and, in particular, bears similarity to Pan-African accretionary complexes. The origin of subduction zones in the Avalonian-Cadomian belt may thus be associated with its accretion to the supercontinent and subsequent displacement of subduction zones from interarc basins to the periphery of Gondwana. Cessation of subduction in the Avalonian-Cadomian system and transition to a shallow-water platform condition resulted from formation of the supercontinent and reorganisation of the cratonic ensemble (Nance et al., 1991). The foregoing highlight the problem that the ocean surrounding the Avalonian-Cadomian belt may have been older than the Iapetus with its passive margin in the eastern part of the American continent or represented a segment of the Iapetus that opened long before the segment adjoining the aforesaid passive margin.

Another, somewhat more complex model was proposed by V. Van der Voo et al. (Rast and Skehan, 1983). According to the works of V. Van der Voo, N. Rast and others, the succession of events was as follows. Small microcontinents and blocks of sialic crust remaining after the separation of Africa from Laurentia and Europe, grouped and formed the Avalonian microcontinental island arc of Japanese type. Continental fragments were represented in it by the Midland Platform (now part of the British Isles), Avalonian zone of Newfoundland, New Brunswick block and so forth. This arc surrounded on one side by the Cadomian (Celtic) ocean and on the other by the Mona marginal sea was built up with volcanoplutonic material in the process of closure of the Cadomian basin. Ultimately, roughly at the level of 650–640 m.y., a collision took place between the Avalonian arc and the northern part of the African Craton, which had already developed in a regime of active margin in the Anti-Atlas region where the Ouarzazate calc-alkaline volcanoplutonic belt was formed. This collision provoked an extensive Cadomian orogeny while the less significant Mona orogeny manifested as a result of collision of the Avalonian microcontinental arc with Laurentia.

Thus the bulk of the crust of Western Europe was formed during Cadomian orogeny. The rest of the basin evidently continued to exist in the Carpathian-Balkan region where folding manifested only in the Cambrian (Salairian).

Cadomian diastrophism coincided in time with the Pan-African, especially in the Hoggar-Atakora (Pharusian) belt. It coincided also with completion of formation of an accretionary complex and closure of the basin at the rear of the Arabia-Nubian island arcs, which caused deformation and metamorphism of Anatolia and Iran strata. In any case, Vendian formations in these territories bear the characteristics of a platform cover.

A comparatively short period after Cadomian-Mona events, in the Early Cambrian, a new impulse of rifting restored the marine environment almost to its original form but there was already a very large microcontinent, Armorica (or Avalonia), between the Tethys and the Iapetus. This microcontinent comprised not only the former Avalonian arc, but also part of the newly formed crust arising during the Cadomian events.

Opening of the Central Asiatic ocean basin in the Vendian-Cambrian has been assumed on the basis of development of spilite-diabase and deep-water chert series in association with ultramafics and serpentinite melange in the Lake zone of Western Mongolia. Similar formations are recognised in the eastern part of central Kazakhstan, Western Sayan, Gorny (High) Altay and up to Tom. The Kazakhstan-Siberian oceanic basin, according to Mossakovsky and Dergunov's (1983) reconstructions, was finally formed at the end of the Vendian and Early Cambrian as a distinct deepwater basin with a minimum dimension in cross-section of 1600–2000 km. In the Vendian this basin was bound from the west by a passive continental margin and in the east by an active margin, with a chain of marginal volcanic arcs. The northern continuation of the basin is not clearly established. According to one reconstruction, it tapered out under the West Siberian Platform in a meridional direction. The Peive and Savel'ev (1982) version that the early Caledonides continued in the region of the Polar Urals, is not wholly unjustifiable.

Concluding this review of Vendian palaeotectonics, let us recall the predominant role of plate-tectonics processes and oceanic formation in the complex global structural plan. During the Vendian all types of tectonic regimes developed, new continents were isolated and centres of orogenic activity arose. Further, distribution of the latter was not universal but, rather, subordinate to the geodynamics of the various segments of the Earth. In some regions extensions prevailed and, in others, compression. The contrasting geodynamic environments are distinctly seen in Gondwana and Laurasia. The interior parts of Gondwana were consolidated under the influence of 'Pan-African' orogeny while Laurasia disintegrated due to the formation of new oceans and rift structures. It may be said that the transitional stage to plate tectonics concluded in the Vendian and this style of geodynamics became thereafter the dominant form of tectonic activity.

9

General Features of the Structural Development of the Earth's Crust in the Precambrian

The data and ideas discussed in this book shall now be summarised and an overall view of the structural evolution of the Earth's crust in the Precambrian presented. Bearing in mind the dramatic differences in the tectonic regimes of the Early and Late Precambrian, however, the two are separately reviewed. In presenting each of these major subdivisions, we shall commence with the different stages of development and the main structural units, briefly summarise the findings, and then consider the possible geodynamic models.

9.1 STAGES OF DEVELOPMENT AND MAIN STRUCTURAL UNITS OF THE EARLY PRECAMBRIAN

The very f i r s t   s t a g e in the history of the Earth is that of its formation as a planet or the stage of its accretion. As shown in Chapter 1, this accretion may have been homogeneous or heterogeneous. We prefer an intermediate version, which assumes primordial accretion of the inner core of the Earth with formation later of an outer core through differentiation of the mantle material.[1] The assumption of a general primary homogeneity of the latter does not, however, exclude the possibility that the composition of this matter may have undergone a gradual change within certain limits during the planet's growth. As a consequence the composition of the lower part of the mantle may have been enriched with iron and other siderophile elements while the upper part, as suggested by Galimov and colleagues (1982), was already close to carbonaceous chondrites. The total duration of accretion was probably quite brief, about a hundred million years.

In the s e c o n d (far more prolonged) 'p r e g e o l o g i c a l'   s t a g e, differentiation of the material of the Earth should have proceeded more intensely. In this stage the outer core of the Earth formed, the source of its main magnetic

---

[1] Some recent publications have proposed another course of events—a much later separation of the inner core from a primordial undifferentiated core body.

field, and this should have been accompanied by the liberation of a considerable amount of heat. In fact, direct palaeomagnetic data confirms the existence of a magnetic field commencing only 3.5 b.y. ago but nothing else could be expected due to the deep thermal reworking of all the much older rocks.

In this stage the primary crust of the Earth should have begun to form, comprising basalt, komatiite-basalt or anorthosite. Once again we have no reliable relics of this primary crust (in spite of serious attempts at detection), barring xenoliths in grey gneisses or inadequately confirmed radiometric assumptions about the antiquity of some mafic rock salients (for example, the Sutam and Kurul'ta series of the Aldan Shield). Indirect data indicates the existence of a primary, evidently mafic crust, however. Marine basalts older than 4.2 b.y. were detected on the Moon. The possibility cannot be ignored that rocks as old will be found on the Earth in future also, although the probability of their preservation unaffected by reworking is rather remote.

The melting process of primary crust from the mantle should have been stimulated on the one hand by high thermal flux generated by differentiation of the mantle material and its natural radioactivity, which was still very significant, and on the other by intense meteorite bombardment. The most probable picture of the surface of the Earth at this time was that of extensive oval-round basins similar to the Lunar Maria filled with tholeiite-basalts and separated by low plateaus, most likely made up of alkaline basalts and products of their differentiation. On being heated to a few hundred degrees, this surface should have been enveloped in a secondary dense atmosphere, still totally devoid of oxygen. In a word, the picture was far more similar to that of present-day Venus than of the Moon, Mars or Mercury.

Formation of the protocontinental grey gneiss crust is associated with the t h i r d, E a r l y A r c h a e a n (4.0–3.5 b.y.) *stage* of development of the Earth. The origin of this volcanoplutonic association (with the participation of sediments) can only be conjectured. All the more so since melanocratic rocks of that age point to the possible division of the Earth's surface in that remote past into protocontinental and proto-oceanic areas[2] far more distinctly than in the preceding stage. Grey gneisses per se, as pointed out in Chapter 2, probably represent a product of the remelting of the primary, more basic crust.

How compact the cover of grey gneisses on the Earth's surface was remains an unanswered question. The existence of melanocratic equivalents of grey gneisses shows that this cover did not form a single shell although some grey gneiss uplifts may have been hundreds, even a thousand kilometres in cross-section. The thickness of the grey gneiss crust may

---

[2] Having nothing in common with the present-day continents and oceans, however, the concept of Rudnik and Sobotovich (1984) notwithstanding.

have reached 30 km, judging from the period of the earliest manifestation of granulite metamorphism (3.7 b.y., southern Greenland).

Already by commencement of this third stage the temperature of the Earth's surface should have fallen sufficiently to permit the existence of liquid water, i.e., formation of the hydrosphere. The atmosphere maintained its oxygen-free composition, however, albeit its density evidently decreased.

Greenstone belts represent the most characteristic structures of the f o u r t h, M i d d l e t o L a t e A r c h a e a n (3.5–2.5 b.y.) *stage* of the evolution of the Earth's crust. Initially they would have been more extensively spread than is observed in contemporary shields; in regions of their total absence the Archaean was metamorphosed to granulite facies, suggesting a very deep erosion of the crust. Greenstone belts are most typical of the Archaean, in which three generations with the following age ranges are known: 1) older than 3.0 b.y., 2) 3.0–2.8 b.y. and 3) 2.8–2.6 b.y. They are less common or typical (containing less komatiites and more intermediate and felsic volcanics and clastic rocks) in the Early Proterozoic. Protoaulacogens of the Pechenga type, in our opinion, are often erroneously placed in this category.

Greenstone belts are superposed on a protocontinental grey gneiss crust in an environment of its extension forming multiple rift systems. This extension was more plastic than brittle, however. In many cases (first type of belts) it did not cause total disruption of the compact protocontinental crust but produced concentrations of scattered ruptures with a general increase in permeability of the crust for magmatic melts of mantle origin. In other cases (second type belts), however, such extension resulted in the formation of new oceanic crust and its subsequent subduction.

Belts of the first type continued to develop mainly in an environment of extension or experienced no significant compression at all in the concluding phase, such as belts of eastern Finland (Martin et al., 1984), or experienced it to a minor extent. The magmatism of these belts was bimodal; manifestation of felsic lavas in the middle stage of their development was evidently caused by melting of the protocontinental crust. Granitoid plutonism in the concluding stage can be explained by remobilisation of the same crust and diapirism under the influence of inversion of density: the heavy upper crust made up of basalts and komatiites with the much lighter lower grey gneiss crust at its base. This has been termed (Martin et al., 1984) sagduction (to distinguish it from subduction).

The suggested process of development of greenstone belts of the first type actually regards this development as a product of the action of mantle plumes i.e., manifestation of plume tectonics, as described by Australian geologists R. Hill and I. Campbell. According to Kröner (1991) and some other researchers, belts of the first type, and hence tectonics of this type, were characteristic of Middle Archaean belts (Barberton and Pilbara).

Plate tectonics was already manifest, albeit in embryonic form, in the development of belts of the second type. This is supported by the continuously differentiated series of volcanics, so characteristic of this type of belts. In view of the high density and less depleted asthenosphere in the Archaean, the lithosphere rapidly lost its buoyancy and underwent total subduction. This feature explains the relatively brief duration of development of such greenstone belts, usually running into tens, not hundreds, of millions of years.

In the concluding phase of their evolution, belts of the second type underwent extremely intense deformation right up to the formation of nappes, pointing to a significant horizontal compression. It is quite possible that the moderate compression of some belts of the first type resulted from the transmission of these compressive stresses from belts of the second type through intermediate hard blocks. Inclusion of the lower part of the protocontinental crust in melting above the subduction zones led to replacement of soda granitoids by potash granitoids at the end of development of greenstone belts of the second type. A very similar phenomenon has been noticed in belts of the first type but the mechanism here differed.

By the Late Archaean transition from plume tectonics to plate tectonics had concluded, with the establishment of predominance of greenstone belts of the second type.

Granite-gneiss fields separating greenstone belts in granite-greenstone terranes are made up of material originating from two sources. The first is the grey gneiss protocrust formed already in the Early Archaean and outcropping at places in cores of granite-gneiss domes (for example, in the Aldan Shield; Drugova et al., 1984). The second are the Archaean granitoids, soda and potash-soda, formed most probably by remobilisation of this protocrust and preserving its characteristic strontium isotope ratio. In the absence of radiometric age determinations, these soda granitoids are sometimes erroneously classified as grey gneisses. The lower part of the crust of granite-gneiss fields, judging from outcropping sections (for example the Wheat belt in Yilgarn block of Western Australia and Kapuskasing zone in Lake Superior province of the Canadian Shield), had reached the granulite stage of metamorphism already by the Archaean.

Commencing with the borderline period between the Early and Late Archaean (3.0 b.y. ago), significant areas within the present-day shields (and evidently the platforms as a whole) entered the stage of cratonisation (South Africa, east Antarctic region and others). Towards the end of the Archaean, it covered almost the whole area of contemporary ancient cratons. Everywhere in this area, Early Archaean and Middle Archaean and, at places, even Late Archaean formations experienced intense deformation, right up to the origin of interleaving of tectonic nappes and regional metamorphism only in the central part of greenstone belts downgrading to the greenschist stage. Thus the protocontinental crust as well as the products of Archaean accretion added to it, experienced general piling up, granitisation and metamorphism, soda and later potash metasomatism and had transformed into a mature continental crust already towards the end of the Archaean. This crust most probably formed a single massif or supercontinent, which could be designated as Pangaea O. Although Borukaev (1985) places the origin of Pangaea I already at the end of the Early Archaean, there is no adequate proof as yet to support this view.

If Pangaea O manifested as a result of the contraction of sial in one hemisphere with a corresponding thickening of the crust, Panthalassa may have formed in the other hemisphere at this time as a result of sial removal. Unlike rifting and spreading, which gave birth to the much younger oceans, this process should have proceeded, so to speak, in the form of a more plastic centrifugal flow (relative to Panthalassa) of material compensated by the uplift of newer portions of basalt melt from deep within the asthenosphere.

However, there is no direct evidence of the existence of Panthalassa from the end of the Archaean. Such evidence begins to appear only from the end of the Early Proterozoic, if not in the Late Proterozoic. At the same time, it is quite possible that the requisite conditions for separation of the Earth's crust and the entire lithosphere into sialic and simatic hemispheres may have arisen already in the Early Archaean or even earlier. The unequal impacts on the different sides of the Earth may have been responsible for the unequal melting rates of the crust while the concentration of protocontinental cores in one hemisphere could be the reason for its subsequent growth in this hemisphere.

One way or the other, the Earth entered a new stage of development, i.e., the f i f t h, E a r l y P r o t e r o z o i c s t a g e (2.5–1.7 b.y.) with much of its surface already covered by a wholly mature continental crust that barely differed from the present-day crust in composition and rheological properties (viscosity). An excellent proof of this, as pointed out in Chapter 4, is the formation of gigantic dyke swarms of basaltoid composition, including the mafic-ultramafic Great Dyke of Zimbabwe and later the formation of large fields of continental tholeiites (flood basalts). The cratonisation process had commenced already in the Late Archaean, immediately following consolidation of different sections of ancient cratons, but its total completion occurred at the beginning of the Proterozoic. At this time, under the influence of rotational stresses, the regmatic network of the Earth was formed and then continuously renewed.

In this same epoch the continental lithosphere of Pangaea O underwent significant break-up, evidently as a result of its cooling. It was separated into isometric circular-polygonal protocratonic blocks with a cross-section running into several hundred, sometimes exceeding a thousand kilometres; the linear protogeosynclinal belts separating these blocks have a cross-section of up to 200–300 km and are several hundred kilometres long. Archaean granite-greenstone terranes generally serve as the basement of the former while the latter are of ensialic riftogenic origin. Two parameters are of particularly great importance, namely the magnitude of initial extension and the magnitude of final compression. Two situations indicate that extension should have occurred on a relatively limited scale: firstly, the rarity of ophiolites (Transhudsonian, Svecofennian, northern Singhbhum and Aravalli protogeosynclines) and secondly palaeomagnetic data, which reveals no significant relative displacements of protoplatforms divided by mobile belts (for example, Damara belt in Namibia). The latter condition indicates that the scale of extension should mostly correspond to the accuracy of palaeomagnetic measurement, which is about 500 km.

Thus the Early Proterozoic protogeosynclines may either be partly ensimatic with an opening of the Red Sea-Aden type, or have developed on continental crust and crust of the transitional type or be wholly ensialic. The second type was probably more widespread. This conclusion needs to be reviewed, however, relative to at least some Early Proterozoic mobile belts, such as the Transhudsonian and Svecofennian. Not only do these belts reveal fairly typical ophiolites, but palaeomagnetic data obtained for the former points to the opening of an ocean expanse about 5000 km wide.

A characteristic feature of the fairly fully developed protogeosynclines is their zonal structure, distinctly manifest for the first time in the history of the Earth. Protogeosynclines of the Canadian Shield, Wopmay, Circum-Ungava and Southern are the most indicative in this regard. Zones of shelf, continental slope and rise and basin floor are distinctly manifest in them. The character of the facies interrelation and the presence of turbidites here and there suggest the relative depth of the axial parts of the protogeosynclines. It may be assumed that depths here could have reached 2–3 km (in a transitional type crust) or 3–4 km (crust of oceanic type).

Closure of the first generation of Eoproterozoic protogeosynclines was evidently accompanied by formation of the second generation, which ceased development towards the end of this aeon. The scale of compression in them was generally proportional to the scale of the initial extension and was fairly significant in zonal protogeosynclines with a general detachment of sediments of the continental margin from their basement and their thrusting on the protoplatform foreland. Manifestation of subduction in protogeosynclines of the first two types has been documented by the formation of huge volcanoplutonic belts on the north-western Canadian Shield, in Scandinavia, Siberia, Australia and other regions. These belts more readily resemble those arising at the end of the Cretaceous-Early Palaeogene north of the Himalayan ophiolite suture than marginal belts of the Andean type, thereby suggesting that they probably belong to collision-type belts.

Granulite (granulite-gneiss) belts or, more accurately, tectonothermally reworked belts represent a special and very important type of linear mobile belts separating protoplatforms. These were manifest already in the Late Archaean (Limpopo in South Africa) and flourished in the Eoproterozoic but most continued to develop in the Late Proterozoic while some 'survived' until the beginning of the Palaeozoic (Mozambique in Africa and Atlantic in Brazil). The internal structure of these belts was extremely complex and often consisted of nappes; their overthrusts on granite-greenstone regions, large thickness of crust, high degree of metamorphism and deep denudation—all these factors suggest an extremely high degree of compression, probably exceeding that noticed during the closing of protogeosynclinal systems. Thus granulite belts (TTR belts) undoubtedly represent collision, even hypercollision structures arising on the boundaries of Early Precambrian microplates—granite-greenstone terranes—in the course of one plate thrusting under another and doubling of crustal thickness. Against this background, in the concluding stages of development of the belts extension with formation of riftogenic structures was manifest in their central parts, having been reactivated in many cases (East Africa, for example) in the Phanerozoic.

The conditions of formation of granulite belts are not clear, however. Evidently they should be characterised by extension but to what degree was this significant, to what extent was it responsible for destruction of Archaean continental crust? Several studies (Dolginov, 1982; and others) indicate the melanocratic composition of the basal rocks of granulite belts: Lapland, Stanovoy, Mozambique, Eastern Ghats, etc. Recent works have shown that ophiolites are present in the Mozambique belt. This shows that granulite belts arose at the site of hiatuses with an oceanic type crust. Yet it must be pointed out that the outer zones of TTR belts generally lie unconformably on the adjoining granite-greenstone terranes and were formed along their substratum.

Given the concept of granulite belts as equivalent to that of TTR belts, two circumstances should be noted. Firstly, not all the salients on the surface of granulite complexes should be interpreted as granulite belts. These could be simply strongly uplifted and very deeply eroded sections of granite-greenstone terranes (Bug-Podol' block of the Ukrainian Shield, Wheat belt of Western Australia and some regions of the Canadian Shield). Secondly, rocks of these belts need not necessarily be metamorphosed everywhere to granulite facies. At places, their metamorphism may not exceed the amphibolite stage; moreover, diaphthoresis in this facies is widespread.

The most protracted stage (megastage) in the history of the Earth, Protogaeicum (H. Stille) concluded at the end of the Early Proterozoic. This was no brief cataclysm but quite a prolonged transitional epoch involving not only the preceding one-third of the Early Proterozoic, but also the beginning of the Middle Proterozoic-Early Riphean (for data on the characteristics of this epoch, see Sec. 4.3). Its main content led to restoration of the entity of Pangaea (already Pangaea I by now) by closing of the protogeosynclines and piling up of the crust within the granulite belts. It is in this epoch that numerous volcanoplutonic belts and fields originated. Belts generally occupy a peripheral position relative to the present-day ancient cratons. If they are regarded as collision features, then the break-up of Pangaea I and its subsequent disintegration into individual continents and cratons can be interpreted as having occurred along these sutures.

Thus, in this interval polygonal contours of cratons were formed in which polygons of a higher order were, so to speak, inscribed with dome-like cores of granite-greenstone terranes and framing formed by TTR belts and protogeosynclines that experienced compression, folding and orogeny. This should have been the characteristic pattern of the basement of ancient cratons, now best seen in aerial photographs and in geophysical fields, especially in the magnetic field. Volcanoplutonic belts and fields occupying already an intracratonic (relative to present-day cratons) position are confined to structures of the second order within them.

Formation of Pangaea I should have had as its inevitable corollary the final formation of Panthalassa in the opposite hemisphere. There is some proof that this was actually so. Volcanoplutonic belts on the periphery of the Pacific Ocean provide such proof: Great Bear Lake on the north-western Canadian Shield, York Peninsula in Australia as well as the Early Riphean ophiolites of south-eastern China. Later, the regime of active margins of the Andean type was replaced everywhere within the frame of the Pacific Ocean by a regime of passive margins due to the break-up of microcontinents from Pangaea I; these microcontinents played the role of borderlands (see Sec. 5.2).

The question arises: from whence came the water that filled the primordial ocean? Evidently it came from the emergence and desiccation of Pangaea. It has been demonstrated that a fairly large number of relatively deepwater protogeosynclinal basins existed on its surface. The first partial desiccation should have occurred at the end of the Archaean.

9.2 NATURE OF TECTONICS AND ITS EVOLUTION IN EARLY PRECAMBRIAN

A turning point for our discussion on the mechanism of tectonic activity in the Early Precambrian is the fact that the Earth, already in the process of accretion and soon thereafter, underwent intense heating during its initial differentiation (separation of the outer core and commencement of crust formation) and that the heat flow in the Early Archaean, according to the computations of Monin (1983), could have exceeded the present-day level by 5 or 6 times. Since there was no complete melting of the planet in this process, it must be assumed that even in the pregeological stage, the mantle was affected by convective motion and that this convection proceeded far more energetically than in the much later period. Based on the assumption that the temperature of the outer mantle exceeded the present-day level by 200°C, convection should have been chaotic and the zone of partial melting as well as the asthenosphere, possessing uniform viscosity, should have enveloped the entire planet (Campbell and Jarvis, 1984). Under these conditions localisation of magmatic extrusions on the surface could have depended only on the sites where meteorites and asteroids fell.

Transition from chaotic to a more orderly, multicellular convection should have occurred in the Early Archaean. The origin of the magma which produced the grey gneisses is not yet fully understood. Some researchers (Borukaev, 1985; Monin, 1983; Campbell and Jarvis, 1984) believe that these soda granitoids already represented a product of subduction of primary basalt or komatiite crust. Yet some South African geologists, such as de Wit and colleagues (1992) and earlier, in a more general form, Sorokhtin (1974), have assumed that grey gneisses were formed as a result of the piling up of obducted plates of proto-oceanic crust. Obduction is explained by the fact that this crust underwent serpentinisation, which diminished its density compared to that of a normal oceanic crust. But in our opinion neither assumption is appropriate because granitoids of similar composition (trondhjemites and plaglogranites) formed in an environment of midoceanic ridges and entered the composition of ophiolite complexes. Therefore, an alternative to the subduction model for the Archaean is that of mantle currents generating centres of primary accretion of protocontinental crust by remelting of a pre-existing crust of mafic-ultramafic composition.

In any case, by commencement of the Archaean, the crust was already differentiated into protocontinental cores and intermediate proto-oceanic sections while convection assumed an orderly fashion with currents flowing from minioceans to minicontinents. The latter experienced uplift and secondary extension with the formation of plastic rifts, i.e., greenstone belts. This should have led to an increase in their area at the expense of minioceans and to subduction of the crust of the latter. Have not granulite belts formed subsequently at the site of these primordial minioceans?

Rifting in some parts of Archaean greenstone belts (belts of the second type, ensimatic) passed into spreading, naturally accompanied by subduction along their periphery. This embryonic form of plate tectonics differed significantly from the evolved Phanerozoic form not only in scale of plates and convective cells, but also in the high rate of spreading and subduction caused in the latter case, as mentioned before, by the high density of crust and low density of asthenosphere.

The high density of crust and lithosphere as a whole could not have promoted their obduction on protocontinents and this perhaps would explain the absence of true Archaean ophiolites on the surface even though information about them has been published from time to time (Harper, 1984). The young lithosphere underwent subduction, which led to its submergence under the protocontinent at low angles. It has been opined that the submergence of oceanic plates to relatively shallow depths could also explain the absence of low-temperature but high-pressure metamorphites in the Archaean and Early Proterozoic. Their equivalents could be the eclogites and kyanite schists formed at much lower pressures but at higher temperatures.

It is possible that this form of plate tectonics is similar to that presently suggested for Venus on which extension structures (rifts) and compression structures have been detected and on which major lithospheric plates are absent.

On conclusion of the development of convective cells, the direction of flow of mantle currents should have changed from their granite-greenstone cores towards granulite belts, which consequently experienced intense compression, piling up, uplift and later secondary extension and rifting (Fig. 75).

It is quite likely that the Archaean granite-greenstone terranes inherited the grey gneiss Early Archaean protocontinents and the latter the impact,

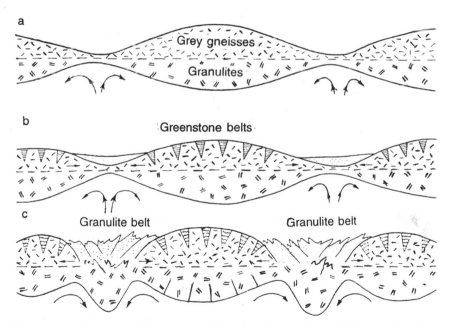

Fig. 75. Possible model of development of the Earth's crust in the Archaean (after Khain, 1984):

a—end of Early Archaean; b—Middle-Late Archaean; c—end of Archaean.

390

ring-shaped structures of the pregeological stage of the evolution of the Earth. If this view is correct, grey gneiss crust should be regarded as a product of transformation, replacement and accretion of primary mafic-ultramafic crust. It was overlain by products of Archaean magmatism, i.e., remobilisates of grey gneisses, mantle extrusives (material of greenstone belts) and intrusive magmatites. In turn, pre-Proterozoic protoplatforms (eocratons) arose on the base of the Archaean granite-greenstone terranes. Finally, at the end of the Early Proterozoic the internal structural plan of the areas of Early Precambrian cratonisation, which constitute the basement of present-day ancient platforms (cratons proper), consisted of dome-ring structures of granite-greenstone terranes inscribed within polygons formed by a lattice of Late Archaean-Early Proterozoic granulite belts and protogeosynclinal fold systems. This type of structural pattern (Fig. 76) was recently detected independently by one of the authors of this book (Khain, 1984) and by an Australian geologist Katz (1985). Katz illustrated this concept by an example of the structure of part of the African Craton.

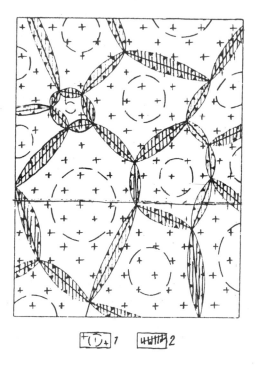

Fig. 76. Sketch showing relationship between granulite belts and granite-greenstone terranes (after Khain, 1984):

1—granite-greenstone terranes; 2—granulite belts.

As pointed out by M.A. Goncharov, such a structural pattern corresponds well to the pattern of Rayleigh-Benard convection cells and, in all probability, could serve as an objective proof of the correctness of the concept regarding manifestation of multicellular mantle convection in the Early Precambrian.

A wider spaced cellular plan of the structure of the continental crust and lithosphere marked by volcanoplutonic belts along the perimeter of future ancient cratons began to form at the very end of the Early Proterozoic and the beginning of the Riphean. In principle, this plan was similar to the multicellular plan of their inner structure and can be regarded as striking proof of the growing scale of convective cells. This accretion ultimately resulted in the formation of supercontinent Pangaea I and its antipode Panthalassa. The present author (V.E. Khain) shares the view of Monin and Sorokhtin (1982) who explained this structural plan by the manifestation already of unicellular convection. Unicellular convection could have been established in a relatively short period at the end of the Archaean if our assumption about the formation of Pangaea O in this epoch is correct. Let us recall that according to Borukaev (1985), the same could have occurred right at the end of the Early Archaean.

Development of the structural plan established in the Early Precambrian persisted even in the later evolution of the Earth. In the Riphean and Early Palaeozoic the break-up of Pangaea occurred along major ancient sutures with formation of palaeo-oceans and geosynclinal belts on their active margins delimiting the large cells of cratonic scale marked by granulite and volcanoplutonic belts. Aulacogens formed along sutures of the second order at the boundary of much smaller cells corresponding to protoplatforms (eocratons and minicratons). Rift systems which gave birth to young Meso-Cenozoic oceans and also the much younger aulacogens and continental rifts, right up to the Late Cenozoic, largely utilised the very same weakened zones of continental lithosphere. Thus this pattern of succession has been traced throughout the evolution of the Earth.

But one must guard against carte blanche acceptance of this pattern. Firstly, it should be remembered that the internal structure of elementary cells that mark the basic division of the continental crust experienced significant changes over time, commencing from the Archaean and especially in the Early Precambrian, on transition from impact craters to grey gneiss cores and later to granite-greenstone terranes. Therefore, direct identification of ring structures observed in present-day space imagery with these oldest astroblems is not entirely correct. Secondly, repeated recombination of these elementary cells and their increasingly new regrouping occurred throughout the evolution of the Earth's crust. This regrouping made for significant renovation of the structural plan of the surface of the Earth.

Attention is drawn to the fact that at the commencement of each new stage of evolution of the Earth's crust, i.e., Archaean, Early Proterozoic, Middle and Late Proterozoic, extension processes, at least of continental crust, were widely manifest. Is this not proof of partial increase in the radius of the Earth and some expansion of it which, towards the end of the stages, may have been replaced by compression giving rise to a general pulsating pattern of planetary evolution? Expansion could also serve as an additional, exclusively so, but not the major factor of ocean formation as suggested, for example, by Glikson (1980) and Milanovsky (1982).

9.3. STAGES OF DEVELOPMENT AND MAIN STRUCTURAL UNITS OF LATE PRECAMBRIAN

The Late Precambrian represents an era of fundamental change in the pattern of global tectonics, a change which was fundamental for the evolution of the Earth's crust. All types of structural units noticed in the Late Precambrian existed already in the Early Riphean but the role of different types at different stages varied sharply. Hence it would be useful to emphasise first and foremost the outstanding aspects of the various stages, i.e., to offer a general review of evolution, and then describe the main structural units based on this presentation.

9.3.1 Stages of Development

The main events of the first, *Early Riphean stage,* were renewal of destruction and formation of a new generation of intracontinental structures—aulacogens, intracontinental geosynclines, zones of non-geosynclinal tectonothermal reworking as well as one of the first reliable manifestations of marginal-continental processes in a limited region of South-east Asia. Significantly ensialic structures predominated among intracratonic ones. The existence of ensimatic intrageosynclines can only be conjectured in principle since the Early Riphean age of strata constituting these zones has nowhere been strictly proved (Araxaides and Siberian systems). Right at the end of the stage, some systems were involved in the main diastrophism (Kibarides). As pointed out in Chapter 5, a group of microcontinents appeared in the region of South-east Asia and active margins, island-arc complexes and ophiolite sutures formed around these microcontinents. Thus the stages of the Wilson cycle were restored from new basins with oceanic crust formation to their closure. The reason for the formation of this microcontinental swarm is not wholly clear: either it resulted from the break-up of this segment of Pangaea or fragments separated as a result of Panthalassa wedges entering Pangaea. It is significant that Stille (1958) in his analysis of Assyntian tectonics assigned much importance to the South-east Asia-Australian region and regarded it as an angular projection of the ocean which facilitated penetration there of a deep orthogeosynclinal breach into Megagaea during the Algonkian.

Secondary tectonic phenomenon of this stage of evolution were the continuing formation of volcanoplutonic complexes in association with molassoids of the Akitkan type, formation of platform cover in deeply subsiding basins of comparatively small size as well as manifestation of non-geosynclinal tectonothermal reworking of older substratum of Elsonian and Paraguaçan types in zones located close to the regions of felsic volcanism, plutonism and molassoid formation.

On the whole, a distinct predominance of intracontinental structures and intraplate tectonic regime is a characteristic feature of the Early Riphean.

The most notable events of the *Middle Riphean stage* are the birth of active island arcs in south-eastern China on the margin of the Yangtze Craton, development and inversion of ensimatic intracratonic geosynclines and birth and evolution of the Grenville belt, a classic structure of non-geosynclinal tectonothermal reworking of older sialic substratum. Calc-alkaline volcanic complexes and batholiths aged about 1 b.y. in Saudi Arabia and Egypt and ophiolite melanges known among them point to the migration of ocean formation from the region of South-east Asia westward and the origin of perioceanic eugeosynclines in this region in which the oldest manifestation of the Wilson cycle in the evolution of the Earth has been described most reliably and in detail. The folding of intracratonic geosynclines of western Africa, central Brazil and southern Siberia is compensated by the formation of elongated, significantly ensialic, marginal troughs of the type Paraguay-Araguaia and Peri-Baikal parallel to the above systems. Formation and development of the Grenville-Dalslandian tectonothermally reworked belt proceeded simultaneous with the formation and development of the Midcontinent Rift System.

In the Middle Riphean the evolution of some felsic volcanoplutonic and molassoid belts was completed while new areas (Madagascar, Namaqua-land and Transantarctic range), in addition to the Grenvillides, were involved in tectonothermal reworking processes. Platform covers and aulacogens developed in a regime similar to that of the Early Riphean. Post-tectonic magmatism in the Kibarides and in other edifices of the same age continued for a fairly long period, almost throughout the Middle Riphean (with intervals). On the whole, intraplate tectonics continued to dominate in spite of intensification of marginal peri-oceanic processes.

The *Late Riphean stage* was of the greatest importance in the Late Pre-cambrian evolution. It was marked by break-up and splitting of Pangaea with formation of oceanic areas between northern Africa and northern America, the western margin of Proto-Tethys and separation of two supercontinents, Gondwana and Laurasia. The process of dismembering of Pangaea from east to west that commenced in the Early Riphean was now completed. Formation of the Circum-Pacific mobile belt in the form of a planetary ring of passive margins and chainlets of oceanic basins separated by borderlands from Panthalassa was associated with this stage. A new generation of intracontinental structures, ensimatic and ensialic geosynclinal troughs, appeared in the Late Riphean. These troughs had an even more complex internal structure compared to that of the first Late Precambrian generation. At places (Central Kazakhstan, Altay-Sayan region and Western Ahaggar), platform conditions prevailed in the brief interval between completion of development of the first Middle Riphean generation and birth of the second. In the Late Riphean the embryonic structure of the Proto-North Atlantic ocean (Proto-Iapetus) evolved in the form of a system of continental rifts extending from Newfoundland to Spitsbergen and connected with the Urals. Zones of long polycyclic development, i.e., aulacogens (Proto-Urals, East European and Siberian cratons) and intracratonic geosynclines (Yenisei range, Mauritania-Senegalese and central Brazilian), continued to grow during this period.

This stage was of vital importance in the history of formation of platform covers. Huge blankets of sedimentary formations in syneclises began replacing deep, but comparatively narrow troughs and basins. Absence of a Late Riphean cover on the North American Craton is noteworthy.

The period around 850 m.y. played a special role as the formation of miogeosynclines of the Cordillera at the site of much older rifts, folding in the Carpathian-Balkan region, the Yenisei range, the Baikal-Vitim belt, folding in the Yangtze region in China and Lufilian folding in Africa were confined to this time slot. In principle, diastrophism in the Grenville belt experienced total extinction at this level. It would possibly be more correct to draw the upper boundary of this stage in the middle of the Late Riphean.

Global evolution in the Late Riphean was characterised by an almost equal manifestation of intra- and marginal-plate processes.

The *Vendian-Early Cambrian stage* is noteworthy for many events, primarily commencement of the disintegration of Laurasia consequent to opening of the Iapetus and Central Asian oceans.

The third generation of intracratonic geosynclines formed simultaneously on the flanks of the Central Asian ocean within Central Kazakhstan Altay-Sayan region and Transbaikalia and rift processes regenerated in South-east Asia.

Some sectors of the Circum-Pacific ring (Andes and Transantarctic range) were involved in active marginal-continental processes. In the Vendian formation of the accretionary complex of the Arabia-Nubian region was completed and Cadomian folding of Newfoundland, Western Europe, Anatolia and Pan-African diastrophism covering a significant part of Gondwana occurred. Baikalian folding was less distinct in its tectonotype region and moreover, this region, as shown earlier, experienced rifting in the Vendian-Cambrian. Vendian conditions clearly illustrate characteristics opposite to the geodynamics of the Gondwana and Laurasia mobile zones. Everywhere, now already in the East European and Siberian cratons, continuous sheets of platform cover were formed over vast territories. Platform sedimentation continued to be absent within North America.

The above palaeotectonic analysis reveals the progressive complexity of the global structural plan from the Early Riphean to the Vendian, gradual replacement of the predominant intraplate regime by marginal-plate and transition from mosaic arrangement of mobile zones to large belts, at some places occurring already in the middle of the Late Riphean. Basic features of the modern structural plan of continents were laid towards the end of the Precambrian. Gondwana was involved in Pan-African orogeny including diastrophism in the intracratonic zones and culmination of tectonothermal reworking of older substrata. With completion of these processes, the inner parts of Gondwana experienced consolidation and were transformed into a stable craton.

9.3.2 Main Structural Units

9.3.2.1 *Structures of platform cover of the Late Precambrian* may be divided into basins or troughs that developed in the Early-Middle Riphean and were filled with extremely thick and comparatively deformed strata (of the type Bangemoll and MacArthur basins in Australia) and syneclises (of the type Taoudenni and Congo in Africa) which are characteristic of the Late Riphean-Vendian. Eurasian syneclises were formed on an ancient substratum cut across by Riphean aulacogens while their formation in Gondwana occurred directly on the basement. An absence of anteclises in the Riphean and Vendian should also be noted. Some degree of diachronism has been established in the development of syneclises. They appear on Gondwana continents in the Late Riphean and on Eurasian continents in the Vendian. As shown earlier, accumulation in small grabens and small basins of irregular contours is characteristic of quasi-platform Lower-Middle Riphean volcanosedimentary formations.

9.3.2.2 *Aulacogens-Palaeorifts* were extensively developed in Laurasia and Australia in the Early-Middle Riphean. They have been preserved in the Late Riphean of Eurasia and Australia and were present in Central Africa. These structures were mainly 'sealed' in the Vendian by the platform cover and developed almost exclusively in Australia.

Two main types of aulacogens are recognisable: inverted (Amadeus, Belt, Mackenzie and others) and uninverted (Pachelma etc.). Ensimatic aulacogens of Keweenaw (Midcontinent) type and 'aulacogeosynclines'[3] of the West Congolese type fall in a special category.

9.3.2.3 *Intracontinental (intracratonic) geosynclines.* This type of linear structures was widely developed in the Late Precambrian. With a distinct intracontinental character, intracratonic geosynclines experienced subsidence in the early stage of their development and uplift accompanied by fold-overthrust deformation, metamorphism and granitisation in the concluding stage. Thus, they fully come under the common definition of the term 'geosyncline'. A tendency towards increase in number of ensimatic intracratonic geosynclines has been noticed over time.

The most characteristic features of ensialic intracontinental geosynclines are: 1) presence of unaltered ancient sialic basement; 2) characteristic pattern in the plan resembling the structure of continental rift systems; 3) terrigenous-carbonate, shallow-water, 'miogeosynclinal' character of strata with uneven development of bimodal or mafic volcanics; 4) negligible transverse zoning; and 5) development of linear folding, thrust deformations, granitisation and metamorphism in the concluding stage of development.

Features characteristic of ensimatic intracontinental geosynclines, usually more extended than ensialic ones, include: 1) presence of ultramafics, often together with other rocks of an ophiolite complex but usually not forming a complete sequence of ophiolite association; 2) distinctly manifest transverse structural sedimentation zoning; 3) significant role of volcanics in the filling of troughs, with localised calc-alkaline series in some together with greywacke flysch; 4) intense fold-thrust-nappe deformation, zonal regional metamorphism and formation of granitoid batholiths in the concluding stage of development; 5) sharp variation in space manifest in the transition from ensimatic to ensialic segments, especially during wedging out of these systems in the body of the craton; and 6) formation of 'triple junctions' with some systems of ensialic aulacogens and perioceanic belts. In the latter case ensimatic intracratonic geosynclines represent apophyses of these belts.

The close spatial association between linear intracontinental mobile zones and the mutual transition of one type into another along a particular strike is a consequence of the different scale of rifting during their formation. An evolutionary series of riftogenic intracontinental structures ranging from aulacogens to ensimatic zones can be outlined (see Table).

Differentiation of intracontinental structures on the basis of duration of their development can be seen but the type of mobile zone has no special

---

[3] This term was introduced by G.P. Leonov and extensively used by E.E. Milanovsky. Nearly all ensialic intracratonic geosynclines have been included in this class. In our opinion, it should be applied only to those very rare structures which have experienced folding and inversion accompanied by formation of foredeeps but in which granitisation and metamorphism are *totally absent*. The use of this term is inappropriate in a wider context.

Table: Evolutionary Series of Late Precambrian Linear Structures

No.	Type of structure	Characteristics of types and subtypes	Examples
1	Aulacogens (palaeorifts)	Absence of granitisation and metamorphism	Amadeus, Georgina and Pachelma
2	Intracratonic ensialic geosynclinal systems	1) Absence of granitisation with insignificant mafic and felsic volcanism and metamorphism; moderate folding	Western Congolides, Espinhaco and Proto-Urals*
		2) Features of subtype 1 with granitisation	Kibarides, Katangides, Paraguay-Araguaia and systems of north-eastern Brazil, Pensacola, Alexod, Apiai, Mayombe and Damarides
		3) Significant manifestation of mafic and felsic volcanism, metamorphism to amphibolite facies and granitisation	
3	Intracratonic ensimatic geosynclinal systems	1) Mafic volcanism, metamorphism, granitisation, single small bodies of alpine-type ultramafics and gabbro	Northern part of Atakorides and Sergipe
		2) Intrusions and protrusions of ultramafics and gabbro	Araxaides, Yenisei range and Atakorides
		3) Intensely manifested features of subtype 2 and also calc-alkaline magmatism	Hoggar
		4) Presence of dismembered ophiolite association and serpentinite melange without calc-alkaline magmatism	Gariep, Malmsberry and southern Araxaides
		5) Dismembered ophiolite association and calc-alkaline magmatism	Mauritanides and systems of Central Kazakhstan
4	Marginal-cratonic geosynclinal belts	Presence of fragments of a complete ophiolite association and continuous calc-alkaline series. Characteristics of active or passive continental margins and development of complex geosynclinal belts	Anti-Atlas, Arabia-Nubian region. Proto-Andean belt, eastern Australia, Rossides, Central Asian belt including Baikal-Vitim zone, Proto-Iapetus and South-east Asia

* This applies only to the western slope of the modern Urals.

significance in this respect. The ensialic Proto-Urals developed throughout the Late Precambrian like an aulacogen simultaneous with the ensimatic Yenisei range or Mauritanides. The ensimatic western Hoggar zone developed only from the middle of the Late Riphean to the Cambrian. The factors responsible for mono-, di- or polycyclic evolution of intracontinental troughs are not wholly clear.

9.3.2.4 *Zones of non-geosynclinal tectonothermally reworked older substrata* represent structures that developed exclusively in the Precambrian and the beginning of the Palaeozoic. They are characterised by superposed deformation mainly in the form of extended strike-slip shear zones along which thick diaphthorites and blastomylonites are noticed. Tectonic reworking may be reflected in the form of mantle domes and interference folding. A typical feature of these zones is the presence of charnockites, associated with other rocks of granulite facies. Further, granulite metamorphism may also arise directly in the period of tectonothermal reworking as happened in Australia and Eastern Africa. In the process of Early-Middle Riphean reworking of the basement (Grenville and Namaqualand belts) major anorthosite bodies arose. Anorthosites are no longer characteristic of Pan-African zones of tectonothermal reworking and magmatism is manifest here mainly by granites and pegmatites. Although the question of formation of supracrustal strata similar in age to tectonothermal reworking has not yet been solved conclusively, available data indicates that single rift troughs which experienced deformation, metamorphism and magmatism simultaneous with very similar but already superposed processes in the surrounding basement are usually present in reworked zones. Among such troughs are the Central Metasedimentary Belt of the Grenville belt, Koras in the Namaqua zone and Lurio in the Mozambique belt.

Zones of alternation of reworked basement blocks with narrow rift troughs represent a special structural type. They appear in the Late Riphean. Examples are the Nigeria, Kariri and Ribeira zones. Reworked zones may be connected with synchronous intracratonic structures in the form of a triple junction, for example Irumide-Mozambique belt and Zambezi-Katangides-Mozambique belt.

Spatial and temporal association of TTR zones with intracratonic geosynclines is an important and persistent feature of the zones of non-geosynclinal reworking.

One of the most distinctive features of TTR zones whereby they are recognised, is the total isotope rejuvenation of the substratum, as established mainly by Rb-Sr and K-Ar age determinations.

Increasing evidence points to repeated reworking of the same regions and its polycyclic character. Thus in the Mozambique belt 'Grenville' (about 1000 m.y.) stages of TTR and 'Pan-African' (650–500 m.y.) reworking have been established quite reliably. Apart from other evidence, this is further supported by 'triple junctions' with the Middle Riphean Irumide fold system and Late Riphean-Vendian Katangides system.

At the same time, there are also monocyclic zones of non-geosynclinal tectonothermal reworking. Evidently the Namaqua and Nimrod zones belong to this category.

9.3.2.5 *Marginal-continental (perioceanic) structures.* The entire succession of palaeotectonic environments described in the Wilson cycle can be reconstructed in the Late Precambrian: rifts of continental and the Red Sea type, passive margins, active island-arc margins and continental active margins of the Andean type. In the modern structure accretionary complexes (Arabia-Nubian) and marginal volcanoplutonic belt systems (Anti-Atlas, Andes and Antarctic Andes) arose as a result of plate-marginal processes. The internal structure of accretionary complexes is characterised by fold-thrust complexes wholly similar to those of corresponding Phanerozoic belts. As pointed out earlier, they contain allochthonous nappes with fragments of complete ophiolite association, extensive zones of serpentinite melange, greywacke flysch and calc-alkaline volcanic series intruded by batholiths. A volcanoplutonic belt with mighty calc-alkaline volcanism (Ouarzazate) developed at the site of an active margin of Andean type in the region of Anti-Atlas. Large bodies of granite batholiths (Transantarctic belt) together with molassoid troughs are more typical for similar belts of the Pacific margin.

9.4 NATURE OF TECTONIC ACTIVITY AND ITS EVOLUTION IN LATE PRECAMBRIAN

Tectonic activity of the Late Precambrian reflects primarily the process of destruction and transformation of the ancient continental crust of Pangaea right up to its partial replacement by oceanic crust. Newly formed oceanic crust began to be increasingly distinctly manifest only at the end of the Late Precambrian. Rifting lies at the base of this global process. Data for the Late Precambrian best enables a study of the evolution of different rift structures and deciphering the possible course of development and transformation of rifts. The mechanism of suspended rifting and abortive spreading predominated during much of the Middle and Late Precambrian, leading to development of various types of mobile zones, ranging from aulacogens to ensimatic intracratonic geosynclines. The structural plan of the Proto-Tethys and Proto-Pacific evolved during this period and a branched network of intracratonic troughs largely diverging from them was formed. These troughs were of different extent and often interconnected, forming triple junctions. In the second half of the Late Precambrian the number of marginal-continental orthogeosynclines increased. This distinct tendency of gradual appearance and subsequent growth of the role of ensimatic structures over time right up to the manifestation of true oceans constituted the leitmotif of Proterozoic tectonics. The data presented earlier enables broaching a solution to the main problem of the importance of the plate-tectonics mechanism in Precambrian tectonics: from which period can this style of tectonic activity be reliably established? The available factual data shows that in the present-day form, i.e., against the background of continental margins spread all over the planet and scale of oceans corresponding to them, plate tectonics began to be manifest in the present form from the Late Riphean. At the same time, during the Early-Middle Riphean in South-east Asia and possibly some other regions, development proceeded according to the Wilson cycle, including the opening of oceanic basins, formation of passive margins, accretionary island arcs and volcanoplutonic complexes. These processes attained a very significant scale compared to the Early Proterozoic but nevertheless proceeded simultaneous with a predominance of intraplate tectonics in the rest of Pangaea.

The mechanism of tectonic development of intracratonic geosynclines is not entirely clear. A possible model is depicted in Fig. 77. Development of these structures was closely associated with the evolution of non-geosynclinal zones of tectonothermal reworking of the substratum.

The TTR process is based on diffuse or areal rifting and subsequent collision. It may proceed under conditions of a still quite plastic, not cooled crust. This process evidently commenced already in the prerift stage associated with the uplift of an asthenosphere diapir. At this time heating of a large zone of lithosphere occurred and centres of crust melting appeared; melting of rocks with alkaline tendency and areal extension and increase in permeability also took place. Here and there, i.e., in the domal parts of thermal anticlines, the crust experienced arched uplift accompanied by the formation of rifts. With cooling of the diapir, extension was replaced by compression, i.e., contraction, leading to inversion in the rifts. Beyond the limits of the rifts proper, diaphthorites formed in the zones of reworking of the basement along linear zones of deformation and intense metasomatism, isotope rejuvenation of the substratum and formation of pegmatites and granites occurred. Thus reworking is an inseparable part of the whole process associated with reactivation of the mantle, leading to reworking of the crust by general heating as well as by linear rifting. The process of reworking commences somewhat earlier and ends later than rifting itself. Its ending coincides with total cooling of mantle diapir and results in thickening of the crust.

Development of the zone of reworking is schematically depicted in Fig. 74 on the example of the southern part of Western Africa. Reworking in the Precambrian has not been noticed around aulacogens. Evidently a fairly large and intensely heated thermal anticline lying close to the crust and capable of involving the rift in the geosynclinal process, simultaneously causing significant heating of the substratum around it, is necessary for

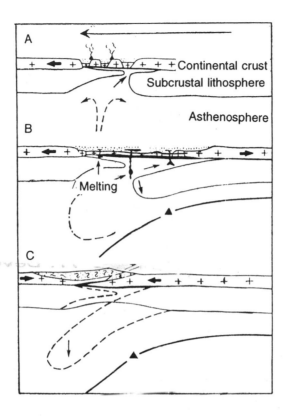

Fig. 77. Sketch showing three stages of evolution of an 'ensialic' fold belt (after Kröner, 1991):

A—rift formation; B—extension of crust and delamination of cover; C—piling up of crust; or A—subduction causing horizontal compression and orogeny.

manifestation of this process. A typical Precambrian reworking is not seen in the Phanerozoic as though it were replaced by continental, epicratonic orogeny.

Conclusions

The main problem of Precambrian tectonics is elucidating the period of effective onset of the mechanism of plate tectonics since the importance of this mechanism for the Phanerozoic is irrefutable, at least in the kinematic sense. With this problem is directly associated the methodological problem of the range of applicability of the principle of uniformity and the method of actualism when analysing the Precambrian history of the Earth. There is no doubt that the principle of uniformity, with minor corrections, has held good from the Late Precambrian onwards, i.e., has been effective for the preceding one billion years of the Earth's evolution. Corrections, however, arise from the fact that this history is cyclic. Hence the character of contemporary tectonic processes corresponds more completely to processes which occurred in the epoch of high tectonic activity. Contrarily, in the epoch of diminished tectonic activity the nature of movements and deformation of the Earth's crust could have differed significantly from that perceived today. Thus the tectonic regimes of some remote geological periods may be closer to modern ones than to regimes of periods which are closer in time (for example, Early Palaeogene and Late Cretaceous).

Commencing from the Middle and especially the Early Proterozoic, it is increasingly difficult to speak of uniformity of tectonic processes, more so of those pertaining to the Archaean. This should not deter use of the method of actualism, however, which ought to be broader based than relying exclusively on the principle of uniformity. Considering that our concern extends to the very earliest stages of evolution of the Earth's crust, we should first look for analogies with the present stage, taking into consideration the similarity between earlier types of rocks and types of dislocations and those that exist today. Only after identifying the characteristics of movements, deformation, structures and rocks which have not persisted to the present, should we turn to ascertaining the differences in those processes under which they were formed and the fluid-thermodynamic conditions of their formation. Only then will we be able to understand objectively the tendencies and stages in the tectonic history of the Earth as well as its geological history as a whole.

Proceeding thus and concentrating our attention on the periods of manifestation of the 'plate' form of tectonics, we can distinguish the following four
stages in the Precambrian tectonic history of the Earth:

1) preplate-tectonics stage, roughly up to 3.5 b.y.; corresponds to the
   Eogaean of V.I. Shuldiner;

2) embryonic multiplate stage; comprises the Archaean and Early Proterozoic and consists of two substages (AR and PR$_1$); corresponds to the
   Protogaean of H. Stille;

3) transitional stage covering the Early and Middle Riphean (Middle Proterozoic); corresponds to the Deuterogaean of H. Stille; and

4) present stage, comprising two substages (Late Proterozoic (Riphean)-
   Palaeozoic and Mesozoic-Cenozoic); this is H. Stille's Neogaeicum.

Replacement of one stage by another did not occur by simple progression. Each stage, except the first perhaps, and even each substage concluded with the formation of a single massif of continental crust, a single
supercontinent, Pangaea, i.e., transition to unicellular convection according
to A.S. Monin and O.G. Sorokhtin. Each successive stage or substage commenced with disintegration of the preceding Pangaea and establishment of
the convection regime characteristic of that stage. It is quite possible that
disintegration of Pangaea was stimulated by some relatively brief and minor
increase in the radius of the Earth (by a few per cents) although there was
gradual reduction of it during the next stage.

The p r e p l a t e - t e c t o n i c s *stage* was characterised by high,
perhaps 5 or 6 times greater than at present, heat flow and correspondingly
shallow disposition, large thickness, low viscosity (due to high content of
molten matter) of the asthenosphere, small thickness and low viscosity of the
lithosphere and high content of fluids in the mantle. Mantle convection (probably whole-mantle) was chaotic; meteorite bombardment exerted considerable influence on the lithosphere, promoting melting of basalts (komatiite-
basalts) from the asthenosphere and formation of a primary melanocratic
crust. It is possible that the impact of a major asteroid led to formation of
the embryonic basin of the Pacific Ocean while 'contraction' of the newly
formed sialic crust in the opposite hemisphere of the Earth laid the foundation for its fundamental asymmetry. Ring-shaped, caldera-like and domal
structures with protosialic (grey gneiss) cores represented the predominant
form of tectonic units at this stage.

As mentioned before, the s t a g e   o f   e m b r y o n i c,   m u l t i -
p l a t e   p l a t e   t e c t o n i c s comprised two substages. In the early,
Archaean substage, heat flow was still 3 or 4 times that of the present
level; protosialic crust and the entire protocontinental lithosphere were characterised by low viscosity and high plasticity. However, mantle convection

gradually acquired a more orderly form. 'Plastic and possibly diffuse rifting' occurred above its ascending branches and greenstone belts with thinning or total break-up of protosialic crust arose. Their development occurred relatively rapidly (in tens of millions of years) and was accompanied by absorption of proto-oceanic or transitional crust characterised by high density[1] and island-arc calc-alkaline magmatism. It concluded with compression and deformation of greenstone belts and intrusion of diapiric plutons of granitoids, initially soda and later potash type. Evidently rearrangement of the system of convective cells and migration of mantle diapirs to a new place occurred with repetition of the process. Further, a general tendency has been observed for successive accretion of volcanic arcs and the crust of back-arc basins (greenstones proper) of the belts towards protosialic islands formed at the end of the Early Archaean with formation of granite-greenstone terranes so typical of the Middle-Late Archaean. As a result, towards the end of the Archaean significant accretion of sialic crust had taken place, which could readily be regarded as continental crust. At depth, it underwent metamorphic differentiation with formation of a granulite-basite layer in the lower part and melting of granite material that accumulated in the midportion of the crust and formed diapiric intrusions in its upper part. Late Archaean crust already possessed a fairly high viscosity and brittleness for the formation of a global regmatic network under the influence of rotational forces (change in rate of the Earth's revolution). Intrusion of dykes of basaltic magma commenced along the fissures and faults of this network. The system of minor convective cells (hundreds of kilometres in diameter) in the asthenosphere at the end of the Archaean acquired definite orderliness; it is reflected in the formation of a polygonal (predominantly hexagonal) lattice of granulite belts with in-built domes of granite-greenstone terranes. These domes could have been inherited to some extent from protosialic cores of the Early Archaean arising at the site of meteorite craters of the preplate-tectonics stage. The final formation of this structure falls towards the last substage, i.e., the E a r l y P r o t e r o z o i c.

In this substage which had already commenced at places at the end of the Archaean, large areas within present-day continents, ancient cratons and median massifs (microcontinents) in the Phanerozoic mobile belts were already quite stabilised, as convincingly supported by the accumulation of weakly deformed, weakly or even wholly unmetamorphosed cover, often with the participation of sheets of continental tholeiites on top of Archaean crust. These areas are regarded as protoplatforms, eocratons or, in a different terminology, microplates (in this case together with their margins).

---

[1] A contrary point of view was recently expressed (Cloud and Glaesser, 1982; De Wit et al., 1992) that rapid serpentinisation of oceanic crust in the Archaean increased its buoyancy, prevented subduction and promoted obduction with subsequent melting due to loading of obducted plates and formation of protocontinental crust of tonalite-trondhjemite-granodiorite composition.

Mobile zones separating them (protogeosynclines) already possessed many features of similarity with genuine intercontinental Neogaean geosynclines; outer zones are delineated in them on Archaean continental crust that did not experience total destruction. These outer zones represent analogues of passive margins of present-day oceans and miogeosynclines (miogeoclines) of the geological past. Inner zones are delineated on a crust of transitional (suboceanic) or even oceanic type and represent analogues of meso- or eugeosynclines corresponding to the deepwater parts of these basins. But in terms of scale and depth these were, for the most part, interior seas of the type Black Sea or Mediterranean basins rather than true oceans; only the Palaeo-Pacific ocean, less deep then than at present, may have been a true ocean. Destruction of continental crust in protogeosynclines only relatively rarely reached the level of formation of new oceanic crust, judging from the rarity of its finds.[2] Nevertheless such examples are known: Cape Smith, Svecofennides, Dalma and Aravalli. In characteristics, however, these ophiolites are more the Red Sea type. It is wholly possible that we have underestimated here the scale of opening of the Early Proterozoic ocean basins; at least the width of basins such as the Transhudsonian in North America and the Svecofennian in Europe is undoubtedly significant (thousands of kilometres). In this context let us mention the interesting attempt of the British geologist A. Park (1991) to apply the model of terrane tectonics to the Svecofennides.

Protogeosynclines most probably arose as a result of break-up of granulite belts, the unaffected part of which remained as though 'frozen' in the basement of protocratons. Some greenstone belts (Guyanan, Leon-Liberian and Canadian shields) still continued to develop in the first half of the Early Proterozoic. Subduction of oceanic or transitional crust in protogeosynclines occurred on a limited scale, as supported by limited manifestation of island-arc calc-alkaline magmatism, which is associated primarily with the small opening of protogeosynclines. Only at the end of the stage did collision of the protocratons (microcontinents) lead to manifestation of mighty felsic volcanoplutonism as a result of formation of melting zones within the thickened continental crust.

Towards commencement of the t r a n s i t i o n a l   s t a g e of development of the Earth's crust (Middle Proterozoic or Early-Middle Riphean), a huge massif of wholly mature continental crust, Pangaea I, was formed and had Panthalassa as its counterpart. The margins of Pangaea in the Early-Middle Riphean were almost exclusively passive except for south-western Scandinavia and some parts of modern south-eastern China (Cathaysia). Thus, in this stage intraplate tectonics prevailed, replacing the miniplate tectonics of the Early Proterozoic.

---

[2] This rarity of Early Proterozoic ophiolites could only be partly a consequence of their rapid absorption in subduction zones.

Thus intraplate tectonics was manifest in the development of palaeorifts: aulacogens and intracratonic geosynclines with different degrees of extension and destruction of continental crust in the formation stage and compression and resultant fold-overthrust deformation in the concluding stage. On the whole, development of these structures covered the first few hundreds of millions of years within one, rarely two Riphean subdivisions. It was accompanied by mantle-crust magmatism, initially mafic and predominantly extrusive and later significantly felsic and intrusive. Manifestation of a magmatic underplating of crust could be assumed at depth. This would be reflected in the larger thickness of the Lower Proterozoic crust compared to the Archaean and in the development of a high-velocity layer at its base (Durrheim and Mooney, 1991).

Many important details of the evolution of the continental hemisphere in this stage are not yet clearly understood. It is quite possible that a complete Wilson cycle manifested during the Middle Proterozoic, i.e., disintegration of Pangaea I was followed by its restoration towards the end of the aeon, in the Grenville stage, according to A.M. Nikishin. In this case there should be two Pangaeas: Pangaea I^1 and Pangaea I^2.

The fourth stage of development of the Earth's crust, commencing in the Late Proterozoic, was characterised by the disintegration of Pangaea I (or Pangaea I^2) with formation of mobile (geosynclinal) belts of Neogaeicum at the site of the largest continental and intercontinental rift systems of the Middle Proterozoic and isolation of independent continental blocks, ancient platform-cratons, within the northern hemisphere. In the southern hemisphere a reverse process of 'closing' of destructive Riphean structures took place, with formation of the Gondwana supercontinent separated from northern continents by the newly formed Proto-Tethys. Its southern active margin is now traced from Maghreb to Arabia and north-western India, eastern Australia and Antarctica. A similar picture is noticed along the western periphery of South American and northern, western and southern peripheries of the Siberian Craton. Thus occurred the transition from predominantly intraplate tectonics of the preceding stage, partly manifest in the northern cratons up to the Early Vendian inclusive, to marginal-plate tectonics. Intercontinental and marginal-continental mobile belts of the Late Riphean-Vendian do not differ at all in principle from corresponding Phanerozoic belts, either in the sequences of ophiolite associations or in the structure of accretionary complexes with their extensive allochthonous ophiolite sheets, melange, thick greywacke flysch, calc-alkaline volcanics, major granitoid batholiths etc.

A definite correlation is noticed between the onset of continental margins of Gondwana and northern continents in the active stage of their development and disappearance of intracontinental destructive Riphean structures with their folding and partly with metamorphism and granitisation. Formation of tectonothermally reworked zones, partly inherited from the Early Protero-zoic, also falls in this period. This entire intracontinental diastrophism can be explained by the fact that continents were as though squeezed in a 'vice' of active margins with their major zones of subduction.

On the whole, evolution of the Earth's crust during the Precambrian reveals a distinctly manifest tendency, with systematic reduction of heat and fluid flow serving as a general prerequisite.[3] There is a corresponding increase in thickness, viscosity and brittleness of the lithosphere and reduc-tion of thickness and plasticity of the asthenosphere. The crust accreted at the expense of the mantle and the bulk of the volume of present-day continental crust was already formed by the Middle Proterozoic. Continental crust 'matured' simultaneously, with an increase in potassium content and increasingly distinct differentiation into layers differing in composition and degree of metamorphism, especially with formation of a sedimentary cover commencing from the Early Proterozoic, especially the Riphean.

Magmatism underwent a wholly evident evolution. At the end of the Archaean potash granites appeared and, very rarely, plutons of peralka-line rocks (nepheline syenites on the Canadian Shield) and also carbon-atites (Palabora in the Republic of South Africa) and kimberlites. Komatiites vanished and intense intrusion of dyke swarms of mafic magma occurred in the Early Proterozoic. By the end of the Early Proterozoic, formation of large granite batholiths had commenced and later already the first plutons of gabbro-anorthosites and rapakivi granites. The latter continued to form in the Early Riphean. Massive manifestation of ignimbrite sheets falls in this same epoch. In so far as sedimentary rocks are concerned, quartzose and arkose sandstones and turbidites had already formed in the Archaean, pointing to differentiation of tectonic conditions, but this phenomenon becomes quite distinctly manifest only in the Early Proterozoic. The affinity of quartzose sandstones to platform covers, turbidites and greywackes to the more inte-rior zones and upper parts of protogeosyncline sequences, arkoses to oro-gens and outer zones of protogeosynclines and manifestation of tillites are noticed. Carbonate rocks, predominantly dolomites, begin to accumulate in large quantity for the first time in the Early Proterozoic. Ferruginous-siliceous rocks, i.e., jaspilites etc., appeared in the Archaean, reached the peak of development in the Early Proterozoic and disappeared in the Riphean.

A definite cyclicity is distinctly perceived in the background of this evo-lution. It consists of the alternation of epochs of accretion and contraction of continental crust with formation of supercontinents: Pangaea O at the end of

---

[3] Some researchers assume even a reduction of lithostatic pressure gradient to explain the formation of granulites in the Archaean when the thickness of the continental crust was admittedly small. Such a reduction of pressure could only be a consequence of the overall significant expansion of the Earth, the probability of which is rather dubious.

the Archaean, Pangaea I at the end of the Early Proterozoic, Gondwana at the end of the Late Proterozoic and epochs of destruction of this crust and its break-up into individual fragments. Further, a comparison of the outcome of the Early Proterozoic and Late Proterozoic (Riphean) destruction shows that with an increase in thickness and brittleness of the lithosphere, greater fracturing into increasingly larger fragments took place than was possible on transition from miniplate tectonics of the Early Proterozoic to large-scale plate tectonics at the end of the Proterozoic and Phanerozoic.

It would be tempting to associate this cyclic alternation of destructive and constructive phases with alternate expansion and contraction of the Earth, with pulsation of its volume, but on a limited scale (a few per cents), and with the general tendency of contraction and not expansion, as assumed by some researchers. A confirmation of this pulsating nature of evolution of the Earth is provided by the repeated opening and closing of ocean basins almost at the same places. Examples are regions of the North Atlantic that opened in the Late Riphean-Vendian and again in the Jurassic-Cretaceous, South Atlantic in the Late Riphean and the Cretaceous, western part of the Indian Ocean in the Late Riphean (Mozambique belt) and Late Jurassic, Tethys in the Late Riphean and Jurassic-Cretaceous, Urals in the Riphean and Ordovician etc. This pattern accords with the palaeomagnetic data of several authors on the permanency and similarity of geocentric angles between Africa, Australia and North America as emerging from their identical location relative to each other in the Precambrian and at present; with our conclusions regarding predisposition to rifting; with data on the permanency of a regmatic network of continents commencing from 2.3 b.y.; and with the reconstructions of Le Pichon and Huchon (1984) showing the amalgamation of continents in Pangaea into a single hemisphere of the Earth. All the aforesaid contradicts the concept of random motion of the fragments of continental crust during the opening and closing of ancient oceans and suggests the existence of a stable structural plan within the framework of which plate-tectonics processes occurred. An important conclusion emerging from this is that the situation of continents in the Late Precambrian Pangaea could not have differed much in spatial orientation from groupings best known to us in the composition of Pangaea II, i.e., Pangaea I should have been similar to Pangaea II. However, some researchers, E. Moores, I. Dalziel and P.F. Hoffman, have recently arrived at significantly different reconstructions for the Late Proterozoic. According to them, the future East Gondwana at this time adjoined North America from the west and only towards the beginning of the Cambrian occupied a position closer to the present one, having traversed a long course around West Gondwana. The main argument in favour of this hypothesis is the amazing similarity of the Riphean and Vendian sequences of Australia and Antarctica on the one hand and North America on the other, and the distribution of mobile belts of Grenville age.

But as they say in French—*comparaison n'est pas raison* (comparison is no proof). Nevertheless, new data (palaeomagnetic and other) confirm this hypothesis.

Another problem concerns the degree of synchronism in the manifestation of epochs of diastrophism in the Precambrian. All regional, continental and global schemes of subdivision and correlation of Precambrian formations are essentially based on the assumption of such a synchronism and the statistics of radiometric age of granitoid magmatism and regional metamorphism (K. Stockwell, L.I. Salop, L. Cahen, N.L. Snelling, and others). But to what extent is this assumption justified? We have made no special study in this direction by plotting appropriate histograms, but such a study was recently made by Yu.D. Pushkarev (1990), who identified the following megacycles in the history of the Earth: 3.6, 2.6, 1.65, 1.1 and 0.4 b.y.

It must be said that objective identification of defnite boundaries in the evolution of the Earth's crust up to the boundary Archaean/Proterozoic is quite difficult due to inadequacy and unreliability of factual data. An exception is the period 4.0 b.y.; it is no mere accident that the ages of the oldest rocks on various continents converge towards this period. The tentativeness of the boundary Archaean/Proterozoic at about 2600 ± 100 m.y. was discussed earlier. The same applies to the boundaries of Early/Middle Proterozoic at 1700 ± 50 m.y. although the great importance of this boundary is undeniable. Levels of about 2300, 2100 and 1900 m.y. within the Early Proterozoic are of relatively secondary importance. The period 1000 ± 50 m.y., used to draw the boundary between the Middle and Late Proterozoic, is extremely important. The boundary between the Early and Middle Riphean at 1350 ± 50 m.y. is of definite significance within the Middle Proterozoic. Levels of 850 m.y. within the Late Riphean and the boundary between the Riphean and Vendian (680 m.y.) are quite significant. From the viewpoint of development of organic life and partly tectonic history, the 620 m.y. level within the Vendian and also the Vendian/Cambrian boundary at 550 m.y. are likewise of interest. From both viewpoints it is more logical to include the Vendian in the Palaeozoic and thus in the Phanerozoic and not the Proterozoic.

In a recent publication by Goodwin (1991) the following were adopted as the boundaries of the main stages in the evolution of the Earth: 3.8, 2.5, 1.8 and 1.0 b.y. The first stage (up to 3.8 b.y.) was called formational, the second (3.8–2.5 b.y.) nuclear-cratonic, the third (2.5–1.8 b.y.) protoplatformal, the fourth (1.8–1.0 b.y.) taphrogenic-anorogenic and the fifth (1.0–0.0 b.y.) platformal.

In concluding this study it must be stated that quite a large number of problems remain unresolved in Precambrian geology. The global ones are: 1) mode of formation of our planet—was it a homogeneous or a heterogeneous accretion?; 2) period of formation of the Earth's core and the temperature of the primordial Earth; 3) composition and age of original crust;

4) origin of grey gneisses (tonalite-trondhjemite-granodiorite association); 5) conditions of formation of earliest granulite complexes; 6) model of initiation and development of greenstone belts; 7) model of initiation and development of tectonothermally reworked (gneiss-granulite) belts; 8) period of formation of Pangaea I—end of Early Archaean or end of Archaean itself?; 9) age and origin of the World Ocean and the Earth's dissymmetry; and 10) possible change of the radius of the Earth in the Precambrian.

Literature Cited

Abbott, D.H. and Hoffman, S.E. (1984). Archaean plate tectonics revisited. 1. Heat flow, spreading rate and age of subducting oceanic lithosphere and their effects on the origin and evolution of continents. *Tectonics*, vol. 3, pp. 429–448.

Abbott, D.H. and Menke, W. (1990). Length of the global plate boundary at 2.4 Ga. *Geology*, vol. 18, pp. 58–61.

Abstracts of International Symposium on Precambrian Crustal Evolution. Beijing, China (1983).

Ahmad, F. (1971). Geology of the Vindhyan system in the eastern part of the Son valley in Mirzapur district, U.P. *Records Geol. Survey India*, 96 (2): 2–41.

Ajibade, A.C. and Wright, I.B. (1989). The Togo-Benin-Nigeria Shield: Evidence of crustal aggregation in the Pan-African belt. *Tectonophysics*, vol. 165, pp. 126–129.

*Akhmetov, R.N., Biryul'kin, G.V., Knyazhev, A.S. and Kudryavtsev, V.A. (1983). Problems of stratigraphy of the Upper Archaean of Aldan Shield. In: *Stratigraphy of the Precambrian of Central Siberia*, pp. 6–16. Nauka, Leningrad.

*Aksenov, E.M., Keller, B.M. and Sokolov, B.S. (1978). General scheme of the stratigraphy of the Upper Precambrian of the Russian Craton. *Izv. AN SSSR, Ser. Geol.*, no. 12, pp. 17–34.

*Aksenov, E.M., Baranov, V.V., Kaveev, I. Kh. and Solontsov, L.F. (1984). New data on the Upper Precambrian of the eastern part of the Russian Platform. *Izv. AN SSSR, Ser. Geol.*, no. 7, pp. 144–148.

Al-Shanti, A.M. and Gass, I.G. (1983). The Upper Proterozoic ophiolite melange zones of the easternmost Arabian Shield. *J. Geol. Soc. London*, 140 (6): 867–876.

*Altukhov, E.N. (1986). *Tectonics and Metallogeny of Southern Siberia*. Nedra, Moscow.

Ancient Structure of the Earth's Crust in Eastern Siberia. Nauka, Novosibirsk (1975).

Andersen, J.S. and Unrug, R. (1984). Geodynamic evolution of the Bangweulu block, Northern Zambia. *Precambr. Res.*, 25 (1–3): 187–212.

Anderson, I.H., Bickford, M.E., Odom, A.L. and Berry, A.W. (1969). Some age relations and structural features of the Precambrian volcanic terrane, St. François Mountains, southeastern Missouri. *Geol. Soc. Amer. Bull*, vol. 80, pp. 1815–1818.

Anderton, R. (1982). Dalradian deposition and the Late Precambrian-Cambrian history of the North Atlantic region: A review of the Iapetus Ocean. *J. Geol. Soc. Lond.*, 139 (4): 421–431.

Anderton, R. and Bowes, D.R. (1983). Precambrian and Paleozoic rocks of the Inner Hebrides. *Proc. Roy. Soc., Edinburgh*, vol. 838, pp. 31–45.

Anhaeusser, C.R. (1973). The evolution of the Precambrian crust of Southern Africa. *Phil. Trans. Roy. Soc. Lond.*, ser. 4, no. A273, pp. 359–388.

Archaean of Central India. Geol. Survey of India, Spec. Publ. no. 3 (1980).

Armstrong, R.L. (1991). The persistent myth of crustal growth. *Austral. J. Earth Sci.*, vol. 38, pp. 613–630.

Arndt, N.T. (1983). Role of a thin, komatiite-rich oceanic crust in the Archaean plate tectonic process. *Geology*, 11 (7): 372–375.

Atlas of the Palaeogeography of China. (1985) Beijing, China.

---

* Asterisked works are in Russian—General Editor.

412

Azizi Samir, M.R., Ferrandini, I. and Tane, I.L. (1990). Tectonique et volcanisme tardif Pan-Africains (580–560 m.a.) dans l'Anti-Atlas Central (Maroc): Interpretation geodynamique a l'echelle du NW de l'Afrique. *J. Afr. Earth Sci.,* 10 (3): 549–563.

Baadsgaard, H., Nutman, A.P., Bridgwater, D., Rosing, M., McGregor, V.R. and Allaart, J.H. (1984). The zircon geochronology of the Akilia association and Isua supracrustal belt, West Greenland. *Earth Planet. Sci. Lett.,* vol. 68, pp. 221–228.

BABEL Working Group. (1990). Evidence for Early Proterozoic plate tectonics from the seismic reflection profiles in the Baltic Shield. *Nature,* vol. 348, pp. 34–38.

Baer, A.J. (1976). The Grenville Province in Helikian times: A possible model of evolution. *Phil. Trans. Roy. Soc. Lond.,* no. A280.

Baer, A.J. (1981). Two orogenies in the Grenville belt? *Nature,* vol. 290, pp. 129–131.

Bakor, A.R., Gass, I.G. and Neary, C.R. (1976). Abel al Wask, north-west Saudi Arabia: An Eocambrian back-arc ophiolite. *Earth Planet. Sci. Lett.,* vol. 30, pp. 1–9.

Banerji, A.K. (1977). On the Precambrain banded iron formations and manganese ores of the Singhbhum region, eastern India. *Econ. Geology,* vol. 72, pp. 90–98.

Barbey, P., Condert, J., Moreau, B., Capdevila, R. and Hameurt, J. (1984). Petrogenesis and evolution of an Early Proterozoic orogenic belt: The granulite belt of Lapland and the Belomorides (Fennoscandia). *Bull. Geol. Soc. Finl.,* 56 (1–2): 161–187.

*Barsukov, V.L. (1981). Comparative planetology and the early history of the Earth. *Geokhimiya,* no. 11, pp. 1603–1615.

*Barsukov, V.L, Bazilevsky, A.T. and Kuz'min, R.O. (1984). Geology of Venus based on the results of analysing the radar images recorded by Automatic Interplanetary Stations Venus-15 and Venus-16 (preliminary results). *Geokhimiya,* no. 12, pp. 1811–1820.

Barton, J.M. (1983). Our understanding of the Limpopo belt—a summary with proposals for future research. In: *The Limpopo Belt* (W.J. van Biljon and J.H. Legg, eds.). Geol. Soc. S. Africa, Spec. Publ. no. 8, pp. 191–203.

*Basharin, A.K., Shcheglov, A.P., Abramov, A.V., Arsent'ev, V.N. and Bognibova, R.T. (1980). In: *Baikalian Megacomplex of Eastern Siberia and Altay-Sayan Region.* Nedra, Moscow, pp. 44–76.

Basu, A.R., Goodwin, A.M. and Tatsumato, M. (1984). Sm-Nd study of Archaean alkalic rock from the Superior Province of the Canadian Shield. *Earth Planet. Sci. Lett.,* vol. 70, pp. 40–46.

*Belichenko, V.G. (1977). *Caledonides of the Baikalian Montane Region.* Nauka, Novosibirsk.

*Belov, A.A. (1981). *Tectonic Development of Alpine Fold Regions in the Palaeozoic.* Nauka, Moscow.

*Belyakov, L.N. and Dembovsky, B. Ya. (1984). Some characteristics of the tectonics of the northern Urals and Pay-Khoy. *Geotektonika,* no. 2, pp. 51–57.

Berak, W.H., Bonavia, F.F., Getachew, T., Schmerold, R. and Tarekegn, T. (1989). The Adola fold and thrust belt, southern Ethiopia: A re-examination with implications for Pan-African evolution. *Geol. Magq.,* 126 (6): 647–657.

Bertrand, J.M.L. and Lassere, M. (1976). Pan-African and pre-Pan-African history of the Hoggar (Algerian Sahara) in the light of new geochronological data from the Aleksod area. *Precambr. Res.,* 3 (4): 342–362.

Bertrand, J.M.L. and Caby, R. (1978). Geodynamic evolution of the Pan-African orogenic belt: A new interpretation of the Hoggar Shield (Algerian Sahara). *Geol. Rdsch.,* vol. 67, pp. 357–388.

Bertrand, J.M.L., Caby, R., Ducrot, J., Lancelot, J., Moussine-Pouchkine, A. and Saadallah, A. (1978). The Late Pan-African intracontinental linear fold belt of the eastern Hoggar (central Sahara, Algeria): geology, structural development, U/Pb geochronology, tectonic implications for the Hoggar Shield. *Precambr. Res.,* 7 (4): 349–376.

*Berzin, N.A. (1981). Structure of the Earth's crust in the Vendian to Early Palaeozoic and its evolution. In: *Main Structural Elements of the Earth's Crust in the Territory of Siberia and Their Evolution in the Precambrian and the Phanerozoic*, pp. 24–34. Nauka, Novosibirsk.

Bessoles, B. (1977). Geologie de l'Afrique. Le craton Ouest Africain. *Mem. BRGM*, no. 88.

Bessoles, B. and Trompette, R. (1980). Geologie de l'Afrique. La chaine panafricaine. Zone mobile de l'Afrique centrale (partie sud) et zone mobile soudanaise. *Mem. BRGM*, no. 92.

*Bibikova, E.V., Gracheva, T.V. and Duk, V.L. (1984). Isotope age of Ungrian magmatic complex. *Dokl. AN SSSR*, 278 (3): 1283–1286.

*Bibikova, E.V., Boiko, V.L., Gracheva, T.V. and Makarov, V.A. (1985). Oldest ultramafics of the Ukrainian Shield. In: *Isotope Age Determination Processes of Volcanism and Sedimentation*, pp. 132–140. Nauka, Moscow.

*Bibikova, E.V., Khil'tova, V. Ya., Gracheva, T.V. and Makarov, V.A. (1982). Age of Peri-Sayan greenstone belts. *Dokl. AN SSSR*, 267 (5): 1171–1174.

Bickle, M.J., Betteney, L.F., Boulter, C.A., Groves, D.I. and Morant, P. (1980). Horizontal tectonic interaction of an Archaean gneiss belt and greenstones, Pilbara block, Western Australia. *Geology*, vol. 8, pp. 525–529.

*Biryul'kin, G.V., Kudryavtsev, V.A. and Nuzhnov, S.V. (1983). Lower Proterozoic structures of the Aldan Shield. *Geol. i Geof.*, no. 2, pp. 16–24.

Black, I.P., Williams, I.S. and Compston, W. (1986). Four zircon ages from one rock: The complex history of a 3930 Ma old granulite from Mount Sines, Enderby Land, Antarctica. *Contrib. Mineral. Petrol.*, vol. 94, pp. 427–437.

Black, R. (1980). Precambrian of West Africa. *Episodes*, no. 4, pp. 3–8.

Bluck, B.J. and Dempster, T.I. (1991). Exotic metamorphic terranes in the Caledonides: Tectonic history of the Dalradian block, Scotland. *Geology*, vol. 19, ppl 1133–1136.

*Boyalikov, O.A., Bogdanova, S.V. and Markov, M.S. (1980). Archaean grey gneisses and magmatism of the early stages of formation of continental crust on the Earth. In: *Precambrian, Internat. Geol. Cong., Moscow*, session 27, pp. 17–24.

*Bogdanova, S.V. (1986). *The Earth's Crust of the Russian Platform in the Early Precambrian*. Nauka, Moscow, 222 pp.

Bonhomme, M.G., Gauthier-Lafaye, F. and Weber, F. (1982). An example of Lower Proterozoic sediments: The Francevillien in Gabon. *Precambr. Res.*, 18 (1/2): 87–102.

Borello, A.V. (1969). Los geosinclinales de la Argentina. *Dir. Anal., Buenos Aires*, no. 24, pp. 17–73.

Borg, S.G. and De Paolo, D.I. (1991). A tectonic model of the Antarctic Gondwana margin with implications for south-eastern Australia: Isotopic and geochemical evidence. *Tectonophys.*, vol. 196, pp. 339–358.

Borg, S.G., De Paolo, D.I. and Smith, B.M. (1990). Isotopic structure and tectonics of the Central Transantarctic Mountains. *J. Geophys. Res.*, 95 (B5): 6647–6667.

Bor-ming, J. and Zong-qing, Zh. (1984). Archaean granulite gneisses from eastern Hebei province, China: Rare earth geochemistry and tectonic implications. *Contr. Min. Petr.*, 85 (3): 224–243.

*Borukaev, Ch. B. (1985). *Precambrian Structures and Plate Tectonics*. Nauka, Novosibirsk.

Bosman, W., Kroonenberg, S.B., Maas, K. and De Roever, W.W.F. (1983). Igneous and metamorphic complexes of the Guiana Shield in Surinam. *Geol. en Mijnb.*, vol. 62, pp. 241–254.

Boullier, A.M. (1991). In: *The West African Orogens and Circum-Atlantic Correlatives*. Springer-Verlag, New York.

Bowes, D.R. (1980). The absolute time-scale and the subdivision of Precambrian rocks in northwestern Britain. In: *Principles and Criteria of Subdivision of the Precambrian into Mobile Zones*, pp. 32–54. Nauka, Leningrad.

Bowles, M., de Bruin, D., de Kock, G., et al. (1992). Some superlatives of geology in southern Africa. Geol. Soc. South Africa.

414

Bowring, S.A., Van Schmus, W.R. and Hoffman, P.F. (1984). U-Pb zircon ages from Athapuscow aulacogen, East Arm of Great Slave Lake, N.W.T., Canada. *Can. J. Earth Sci.,* 21 (11): 1314–1324.

Bowring, S.A., Williams, I.S. and Compston, W. (1989). 3.96 Ga gneisses from the Slave province, North-west Territories, Canada. *Geology,* vol. 17, pp. 971–975.

*Bozhko, N.A. (1988). *Late Precambrian of Gondwana.* Nedra, Moscow.

*Bozhko, N.A. (1994, in press). Riphaean Accretion of Terranes in the Tectonic Evolution of the Baikal Montane Region. Dokl. Akad. Nauk, Moscow.

*Bozhko, N.A. and Demina, L.I. (1974). Geology and conditions of metamorphism of the ancient formations of the central part of western Transbaikalia. *Izv. Vuzov, Geol. i Razved.,* no. 12, pp. 106–121.

Brito Neves, B.B. (1986). Tectonic regimes in the Proterozoic of Brazil. *XII Simposio de Geologia do Nordeste,* pp. 235–251.

Brito Neves, B.B. and Cordani, U.G. (1991). Tectonic evolution of South America during the Late Proterozoic. *Precambr. Res.,* vol. 53, pp. 23–40.

Brito Neves, B.B., de Teixeira, W., Tassinari, C.C.G. and Kawashita, K. (1990). A contribution to the subdivision of the Precambrian in South America. *Rev. Bras. Geocienc.,* 20 (1–14): 267–276.

Buffett, B.A., Huppert, H.F., Lister, J.R. and Woods, A.W. (1991). Analytical model for solidification of the Earth's core. *Nature,* 356 (6367): 329–331.

*Bukharov, A.A. (1987). *Protoreactivation Zones of Old Platforms.* Nauka, Novosibirsk.

*Bukharov, A.A., Glazunov, V.O. and Rybakov, N.M. (1985). Baikal-Vitim Lower Proterozoic greenstone belt. *Geol. i Geof.,* no. 7, pp. 33–40.

*Bulgatov, A.N. (1983). *Baikalides Tectonotype.* Nauka, Novosibirsk.

Burek, P.J., Walter, M.R. and Wells, A.T. (1979). Magnetostratigraphic tests of lithostratigraphic correlations between latest Proterozoic sequences in the Ngalia, Georgina and Amadeus basins, central Australia. *B.M.R. Austr. J. Geol. Geophys.,* no. 4, pp. 47–55.

Burwash, R.A. and Krupicka, J. (1969). Cratonic reactivation in the Precambrian basement of western Canada, I. Deformation and chemistry. *Can. J. Earth Sci.,* 6 (6): 1381–1396.

Caby, R. (1970). La chaine pharusienne dans le nord-ouest de l'Ahaggar (Sahara Central, Algerie): sa place dans l'orogenese du precambrien superieur en Afrique. These, Universite de Montpellier.

Caby, R., Bertrand, J.M.L. and Black, R. (1981). Pan-African ocean closure and collision in the Hoggar-Iforas segment, central Sahara. In: *Precambrian Plate Tectonics* (A. Kröner, ed.). Elsevier, Amsterdam.

Cahen, L., Ledent, D. and Tack, L. (1978). Données sur la geochronologie du Mayumbien (Bas-Zaire). *Bull. Soc. Belge Geol.,* 87 (2): 101–112.

Cahen, L., Snelling, N.J., Delhal, J. and Vail, J.R. (1984). *The Geochronology and Evolution of Africa.* Clarendon Press, Oxford.

Camp, V. (1985). Pan-African microplate accretion of the Arabian Shield. *Geol. Soc. Amer. Bull.,* vol. 16, pp. 817–826.

Campbell, I.H. and Jarvis, G.T. (1984). Mantle convection and early crustal evolution. *Precambr. Res.,* vol. 26, pp. 15–56.

Chase, C.G. and Cilmer, T.H. (1973). Precambrian plate tectonics: The Midcontinent gravity high. *Earth Planet. Sci. Lett.,* vol. 21, pp. 70–78.

Chen Jiangfeng, Foland, K.A., Xing Fengming, Xu Xiang and Zhou Taixi. (1991). Magmatism along the south-eastern margin of the Yangtze block: Precambrian collision of the Yangtze and Cathaysia blocks of China. *Geology,* vol. 19, pp. 815–818.

Cheng Baijin and Sun Dazhong. (1982). The Lower Precambrian of China. *Rev. Brasil. Geocienc.,* 12 (1–3): 65–73.

Chiron, J. (1974). Etude geologique de la chaine des Mauritanides entre le parallele de Moudjeria et le fleuve Senegal (Mauritanie). *Mem. BRGM.*

*Chit, E.I. and Rizun, B.P. (1984). Sedimentogeny and palaeogeography of south-western East European Craton in the Vendian period. In: *Sedimentary Rocks and Ores,* pp. 201–209. Naukova Dumka, Kiev.

Choubert, B. (1969). Les Guyano-Eburneides de l'Amerique du Sud et de l'Afrique Occidentale: essai de comparaison geologique. *Bull. BRGM,* sect. 4, pp. 39–68.

Choubert, G. and Faure-Muret, A. (eds.) (1971). *Tectonique du continent Africain.* Mem. Earth Sci., UNESCO, Paris.

Choudhury, A.K. Gopalan, K. and Sastry, C.A. (1984). Present status of the Precambrian rocks of Rajasthan. *Tectonophys.,* vol. 105, pp. 131–140.

*Chumakov, N.M. (1978). *Precambrian Tillites and Tilloids.* Nauka, Moscow.

Church, W.R. (1979). Granite and metamorphic rocks of the Taif area, western Saudi Arabia: Discussion and reply. *Geol. Soc. Amer. Bull.,* 90 (9): 893–896.

Claoue-Long, J.C., Thirlwall, M.F. and Nesbitt, R.W. (1984). Revised Sm-Nd systematics of Kambalda greenstones, Western Australia. *Nature,* vol. 307, pp. 697–701.

Clauer, N. (1973). Utilisation de la methode rubidium-strontium pour la datation de niveaux sedimentaires du Precambrien superieur de l'Adrar mauritanien (Sahara occidental) et la mise en evidence de transformations precoces des mineraux argileux. *Geochim. Cosmochim. Acta,* 37 (10): 2243–2255.

Cloud, P. (1983). Aspects of Proterozoic biogeology. *Geol. Soc. Amer. Bull.,* vol. 162, pp. 245–251.

Cloud, P. and Glaessner, M.F. (1982). The Ediacaran period and system: Metazoa inherit the Earth. *Science,* vol. 217, pp. 783–792.

Cogne, J. and Wright, A.E. (1980). L'orogene cadomien: vers un essai d'interpretation palaeodynamique unitaire des phenomenes orogeniques fini-precambriens d'Europe moyenne et occidentale. In: *Colloque C 6. Geologie de l'Europe,* pp. 29–55. 26 CGI. Villeneuve d'Ascq.

Compston, R. and Arriens, P.A. (1968). The Precambrian geochronology of Australia. *Can J. Earth Sci.,* 5 (3): 561–583.

Condie, K.C. (1982a). Early and Middle Proterozoic supracrustal successions and their tectonic settings. *Amer. J. Sci.,* 282 (3): 341–357.

Condie, K.C. (1982b). Plate tectonics model for Proterozoic continental accretion in the southwestern United States. *Geology,* 10 (1): 37–42.

Condie, K.C. (1981). *Archaean Greenstone Belts.* Elsevier, Amsterdam.

Condie, K.C. and Shadel, C.A. (1984). An Early Proterozoic volcanic arc succession in southeastern Wyoming. *Can. J. Earth Sci.,* vol. 21, pp. 415–427.

Cooray, P.G. (1994). The Precambrian of Shri Lanka: A historical review. *Precamb. Res.,* vol. 66, pp. 3–18.

Corriveau, L. (1990). Proterozoic subduction and terrane amalgamation in the south-western Grenville province, Canada: evidence from ultrapotassic to shoshonitic plutonism. *Geology,* vol. 15, pp. 614–617.

Costa Campos, Neto Mario Da (1984). Geometria a fases de dobramentos brasilianos superpostos no Oeste de Minas Gerais. *Rev. Bras. Geocienc.,* 14 (1): 60–68.

Craddock, C. and Campbell, K. (eds.). (1980). *Antarctic Geosciences.* Univ. Wisconsin Press, Madison.

Crawford, A.R. (1982). The Pangaean paradox: where is it? *J. Petrol. Geol.,* 5 (2): 149–160.

Cserna, Z. de (1969). Tectonic framework of southern Mexico and its bearing on the problem of continental drift. *Bol. Soc. Geol. Mexicana,* vol. 30, pp. 159–168.

Culver, S.J., Williams, H.R. and Venkatakrishnan, R. (1991). The Rokellide orogen. In: *The West African Orogens and Circum-Atlantic Correlatives.* Springer-Verlag, New York.

Daly, J.S. and McLelland, J.M. (1991). Juvenile Middle Proterozoic crust in the Adirondack Highlands, Grenville province, north-eastern North America. *Geology,* vol. 19, pp. 119–122.

416

Danni, J.C.M., Fuck, R.A. and Leonardos, O.H. (1982). Archaean and Lower Proterozoic units in Central Brazil. *Geol. Rdsch.,* 71 (1): 291–317.

Davison, I. and Santos, R.A. (1989). Tectonic evolution of the Sergipano fold belt, N.E. Brazil during the Brasiliano orogeny. *Precambr. Res.,* vol. 44, pp. 319–342.

de Almeida, F.F.M. (1978). Chronotectonic boundaries for Precambrian time divisions in South America. *An. Acad. Bras. Cienc.,* 50 (4): 527–535.

de Almeida, F.F.M. and Hasui, J. (eds.). (1984). *O Precambriano de Brasil.* Edgard Blucher, Sao Paulo.

de Almeida, F.F.M., Hasui, J. and Brito-Neves, B.B. (1976). The Upper Precambrian of South America. *J. Instituto de Geociencias USP,* vol. 7, pp. 45–80.

de Almeida, F.F.M., Hasui, J., Brito-Neves, B.B. and Fuck, R.A. (1981). Brazilian structural provinces: An introduction. In: *The Geology of Brazil* (I.M. Mabesoone, B.B. Brito-Neves and A.N. Sial, eds.). *Earth Sci. Rev.,* vol. 17, pp. 1–29.

de Carvalho, H. (1983). Notice explicative preliminaire sur la geologie de l'Angola. *Sept. Garcia de Orta, Ser. Geol., Lisboa,* vol. 6, no. 1–2.

De Wit, Roering C., Hart, R.J. et al. (1992). Formation of an Archean continent. *Nature,* vol. 357, pp. 553–562.

Deb, M. and Sarkar, S.C. (1990). Proterozoic tectonic evolution and metallogenesis in the Aravalli-Delhi orogenic complex, north-western India. *Precambr. Res.,* 46 (1–2): 115–137.

Delhal, J. and Ledent, D. (1973). Resultats de quelques mesures d'ages radiometriques par la methode Rb-Sr dans la pegmatite de la haute Luanui, region de Kasai (Zaire). *Rapp. Ann. Mus. R. Afr. Centre Tervuren Belg. Dept. Geol. Min.,* pp. 102–103.

Dewey, J.F. and Burke, K.C. (1973). Tibetan, Variscan and Precambrian basement reactivation: Products of continental collision. *J. Geol.,* vol. 81, pp. 683–692.

Deynous, M., Trompette, M.R., Clauer, N. and Sougy, J. (1978). Upper Precambrian and low-ermost Palaeozoic correlations in west Africa and in the western part of central Africa. Probable diachronism of the Late Precambrian tillite. *Geol. Rdsch.,* 67 (2): 615–630.

Dickin, A.P. and Higgins, M.D. (1992). Sm-Nd evidence for a major 1.5 Ga crust-forming event in the central Grenville province. *Geology,* vol. 20, pp. 137–140.

Dimroth, E., Imreh, L., Goulet, N. and Rocheleau, M. (1982–1983). Evolution of the south-central segment of the Archaean Abitibi belt, Quebec. *Can. J. Earth Sci.,* Part 1 (1982) vol. 19, pp. 1729–1758; Part 2 (1983) vol. 20, pp. 1355–1373; Part 3 (1983) vol. 20, pp. 1374–1388.

*Dobretsov, N.L. (1982). Ophiolites and problems of Baikal-Muya ophiolite belt. In: *Magmatism and Metamorphism of Baikal-Amur Main Railroad Zone and Their Role in the Formation of Economic Mineral Deposits.* Nauka, Novosibirsk.

*Dobrzhinetskaya, L.F., Kapura, I.K. and Dashevskaya, D.M. (1983). Structural position and plastic flow of mafic-ultramafic complexes of Baikal-Vitim greenstone belt. In: *Precambrian Trough Structures of Baikal-Amur Region and Their Metallogeny,* pp. 165–179. Nauka, Novosibirsk.

Dodge, F.C.W., Fleck, R.J., Hadley, D.G. and Millard, H.T. (1978). Geochemistry and $^{87}Sr/^{86}Sr$ ratios of Halaban rocks of the central Arabian Shield. *Precambr. Res.,* 6 (1): A13.

*Dodin, A.L. (1979). *Geology and Mineralogy of Southern Siberia.* Nedra, Moscow.

*Dolginov, E.A. (1982). Precambrian history of continents and global tectogeny. *Itogi Nauki i Tekhniki, Obshch. Geol.,* no. 15. VINITI, Moscow.

*Dolginov, E.A. (1985). Early Precambrian metamorphic complexes in the margins of contem-porary continents. *Obzor VIEMS, Obshch. i Reg. Geol., Geol. Kartirovanie.* Moscow.

*Dol'nik, T.A. and Vorontsova, G.A. (1974). *Biostratigraphy of Upper Precambrian and Lower Cambrian Levels of Northern Baikal and Patom Uplands.* Nauka, Irkutsk.

Don Hermes, O. and Zartman, R.E. (1992). Late Proterozoic and Silurian alkaline plutons within the south-eastern New England Avalon zone. *J. Geol.,* vol. 100, pp. 477–486.

417

*Drugova, G.M., Chukhonina, A.P., Morozova, I.M. and Bogomolov, E.S. (1984). Oldest formations of the Aldan Shield. *Sov. Geol.*, no. 11, pp. 82–89.

Drummond, M.S. and Defant, M.J. (1990). A model for trondhjemite-tonalite-dacite genesis and crustal growth via slab melting: Archaean to modern comparisons. *J. Geophys. Res.*, 95 (B13): 21503–21521.

Drury, S.A. (1983). A regional tectonic study of the Archaean Chitradurga greenstone belt, Karnataka, based on Landsat interpretation. *J. Geol. Soc. India*, vol. 24, pp. 167–184.

Drury, S.A., Holt, A.W., van Clasteren, P.C. and Beckinsale, R.D. (1983). Sm-Nd and Rb-Sr ages for Archaean rocks in western Karnataka, south India. *J. Geol. Soc. India*, 24 (9): 454–459.

Drury, S.A., Harris, N.B.W., Holt, R.W., Reeves-Smith, G.J. and Wightman, R.T. (1984). Precambrian tectonics and crustal evolution in south India. *J. Geol. Soc. India*, vol. 92, pp. 3–20.

Dupret, L. (1988). The Proterozoic of north-eastern Armorican massif. In: *Precambrian in Younger Fold Belts* (V. Zoubek, ed.). Wiley, Chester Publ., New York.

Durrheim, R.J. and Mooney, W.D. (1991). Archaean and Proterozoic crustal evolution. Evidence from crustal seismology. *Geology*, 19 (6): 606–609.

Early History of the Earth (see Windley, B.)

Eastin, R. (1970). Geochronology of the basement rocks of the central Transantarctic mountains, Antarctica. *Inst. Polar Studies Repts.*, no. 5.

Elliot, P.H. (1975). Tectonics of Antarctica: A review. *Amer. J. Sci.*, vol. 275A, pp. 45–106.

Elsasser, W.M. (1963). Early history of the Earth. In: *Earth Science and Meteorites* (J. Geiss and E. Goldberg, eds.). Amsterdam, pp. 1–30.

Elston, D.P. and McKee, E.H. (1982). Age and correlation of the Late Proterozoic Grand Canyon disturbance, Northern Arizona. *Geol. Soc. Amer. Bull.*, vol. 93, pp. 681–699.

Engel, A.E.J., Dixon, T.H. and Stern, R.J. (1980). Late Precambrian evolution of Afro-Arabian crust from ocean arc to craton, *Geol. Soc. Amer. Bull.*, pt. 1, vol. 91, pp. 699–706.

*Esher, A. (1978). Precambrian Shield of western Greenland. Tectonics of Europe and adjoining regions. In: *Tectonics of Europe. Ancient Cratons, Baikalides, Caledonides*, pp. 258–269. Nauka, Moscow.

Etheridge, M.A., Rutland, R.W.R. and Wyborn, I.A. (1987). Orogenesis and tectonic process in the Early to Middle Proterozoic of Northern Australia. In: *Proterozoic Lithospheric Evolution*, pp. vol. 17, pp. 131–147. (A. Kröner, ed.). Amer. Geophys. Union Geodynamic Series.

Evenchik, C.A., Parrish, R.R. and Gabrielse, H. (1984). Precambrian gneiss and Late Proterozoic sedimentation in north-central British Columbia. *Geology*, vol. 12, pp. 233–237.

Evolution of Geological Processes and Metallogeny of Mongolia (N.S. Zaitsev and V.I. Kovalenko, eds.). Nauka, Moscow (1990).

Evolution of Precambrian Magmatism (on the example of Karelia). Nauka, Leningrad (1985).

*Fan Chyong Thi (1981). *Geology of Metamorphic Complexes of South-Eastern Asia*. Ph. D. thesis (abstract), Moscow State University, Moscow.

Faure, G. and Gair, H.S. (1970). Age determinations from northern Victoria Land. *New Zeal. J. Geol., Geoph.*, 13 (4): 1024–1026.

*Fedorovsky, V.S. (1985). Lower Proterozoic of Baikal region. *Trudy GIN AN SSSR*, no. 400.

Ferrara, G., Sacchi, R., Tonarini, S. and Zanettin, B. (1984). *Radiometric Ages of Mozambique Belt*. 27th International Geological Congress (summaries), vol. 2, Sections 04–05, pp. 290–291.

Fleck, R.J., Greenwood, W.R., Hadley, D.G., Anderson, R.E. and Schmidt, D.L. (1980). Rubidium-strontium geochronology and plate tectonic evolution of the southern part of the Arabian Shield. *U.S. Geol. Surv. Prof. Paper* 1131, p. 38.

Fletcher, J.R. and Farquhar, R. (1982). The proto-continental nature and regional variability of the central metasedimentary belt of the Grenville province: lead isotope evidence. *Can. J. Earth Sci.*, vol. 19, pp. 239–253.

Fletcher, J.R., Wilde, S.A., Libby, W.G. and Rosman, K.J.R. (1983). Sm-Nd model ages across the margin of the Archaean Yilgarn block, Western Australia-11; south-west transect into the Proterozoic Albany-Fraser province. *J. Geol. Soc. Austr.*, vol. 30, pp. 333–340.

Friend, C.R.L. (1984). The origin of the Closepet granites and the implications for the crustal evolution of southern Karnataka. *J. Geol. Soc. India*, 25 (2): 73–84.

Fyfe, W.S. (1976). Heat flow and magmatic activity in the Proterozoic. *Phil. Trans. R. Soc. Lond.*, A280, pp. 655–660.

Gaal, G. (1982). Proterozoic tectonic evolution and late Svecokarelian plate deformation of the Central Baltic Shield. *Geol. Rdsch.*, 71 (1): 158–170.

Gaal, G. and Gorbachev, R. (1987). An outline of the Precambrian evolution of the Baltic Shield. *Precambr. Res.*, vol. 35, pp. 15–52.

Gaby, S. and Greiling, R.O. (eds.). (1989). *The Pan-African Belt of North-east Africa and Adjacent Areas.*

*Galimov, E.M., Bannikova, L.A. and Barsukov, V.L. (1982). Matter constituting the upper shell of the Earth. *Geokhimiya*, no. 4, pp. 473–490.

Gansser, A. (1964). *Geology of the Himalayas.* Interscience Publ., London.

Gass, I. (1977). The evolution of the Pan-African crystalline basement in N.E. Africa and Arabia. *J. Geol. Soc. London*, no. 134, pp. 129–138.

* *Geological Structure of the USSR and Pattern of Location of Mineral Deposits*, vol. 8, North-Eastern USSR. Nedra, Leningrad (1988).

Geology of Western Australia. West Austr. Geol. Surv. Mem. no. 2 (1975).

*German, L.L. (1986). Problems of the Early Precambrian of the central Kazakhstan Palaeozoides. *Dokl. AN SSSR*, 288 (1): 1444–1446.

*Ges, M.D. (1988). Geodynamics and magmatism of the Precambrian and Lower Palaeozoic of northern and middle Tien Shan. In: *Magmatism i Geolkarta 50 Srednei Azii*, pp. 152–154. Dushanbe

Gibb, R.A., Thomas, M.D., Lapointe, R.L. and Mukopadhyay, M. (1983). Geophysics of proposed Proterozoic sutures in Canada. *Precambr. Res.*, vol. 19, pp. 349–384.

Gibbs, A.K. and Barron, C.N. (1983). The Guiana Shield reviewed. *Episodes*, no. 2, pp. 7–24.

*Glevassky, E.B. (1989). Early Precambrian palaeogeodynamic reconstructions of the south-eastern part of the Ukrainian Shield. In: *Litosfera Ukrainy.* Naukova Dumka, Kiev, pp. 68–75.

*Gintov, O.B. (1978). *Structure of the Continental Earth's Crust in the Early Stages of Its Evolution.* Naukova Dumka, Kiev, 163 pp.

Glikson, A.Y. (1980). Precambrian sial-sima relations: Evidence for Earth expansion. *Tectonophys.*, vol. 63, p. 193.

Glikson, A.Y. (1983). Geochemical, isotopic and paleomagnetic early sial-sima patterns. The Precambrian crustal enigma revisited. *Geol. Soc. Amer. Mem*, no. 161, pp. 95–117.

*Glukhovsky, M.Z. and Pavlovsky, E.V. (1984). Ring structures of the early stages of the Earth's evolution. *27th Internat. Geol. Cong.*, vol. 19. *Comparative Planetology.* Nauka, Moscow, pp. 65–74.

Goldich, S.S., Nedge, C.E. and Stern, T.W. (1970). Age of the Morton and Montevideo gneisses and related rocks, south-western Minnesota. *Geol. Soc. Amer. Bull.*, vol. 81, pp. 3671–3696.

Goodge, J.W. and Dallmeyer, R.D. (1992). $^{40}Ar/^{39}Ar$ mineral age constraints on the Palaeozoic tectonothermal evolution of high-grade basement rocks within the Ross orogen, central Transantarctic mountains. *J. Geol.*, vol. 100, pp. 91–106.

Goodge, J.W., Borg, S.G., Smith, B.K. and Bennett, V.C. (1991). Tectonic significance of Proterozoic ductile shortening and translation along the Antarctic margin of Gondwana. *Earth Planet Sci. Lett.*, vol. 102, pp. 58–70.

Goodwin, A.M. (1985). Rooted Precambrian ring-shields: growth, alignment and oscillation. *Amer. J. Sci.*, 285 (6): 481–531.

Goodwin, A.M. (1991). *Precambrian Geology*. Acad. Press, London.

Gopalan, K., MacDougall, J.D., Roy, A.B. and Murati, A.V. (1990). Sm-Nd evidence for 3.3 Ga old rocks in Rajasthan, north-western India. *Precambr. Res.*, 48 (3): 287–292.

*Gorbachev, R. (1980). Precambrian of the western part of the Baltic Shield, In: *Principles and Criteria for Classifying the Precambrian into Mobile Zones*, pp. 206–222. Nauka, Moscow.

*Goryainov, P.M. and Fedorov, E.E. (1986). Precambrian geodynamics and continental crust model of the Kola Peninsula. In: *Tektonika i Voprosy Metallogenii Rannego Dokembriya*, pp. 54–72. Nauka, Moscow.

Gower, C.F. (1985). Correlations between the Grenville province and Sveconorwegian orogenic belt—implications for Proterozoic evolution of the southern margins of the Canadian and Baltic shields. In: *The Deep Proterozoic Crust in the North Atlantic Provinces*, pp. 247–257. Nauka, Moscow.

Gower, C.F. (1990). Mid-Proterozoic evolution of the eastern Grenville province, Canada. *Geol. Foren. Stockholm Forhandl.*, 112 (2): 127–139.

Gower, C.F. and Owen, V. (1984). Pre-Grenvillian and Grenvillian lithotectonic regions in eastern Labrador—correlations with the Sveconorwegian orogenic belt in Sweden. *Can. J. Earth Sci.*, vol. 21, pp. 678–693.

*Grachev, A.F. and Fedorovsky, V.S. (1980). Greenstone belts of the Precambrian: rift zones or island arcs? *Geotektonika*, no. 5, pp. 3–22.

Grambling, J. (1981). Pressures and temperatures in Precambrian metamorphic rocks. *Earth Planet. Sci. Lett.*, vol. 53, pp. 63–68.

Greenwood, W., Anderson, R.E., Fleck, R.J. and Schmidt, D.L. (1976). Late Proterozoic cratonisation in south-western Saudi Arabia. *Phil. Trans. R. Soc. Lond.*, vol. A280, pp. 517–527.

Greiling, R. (1982). Precambrian basement complexes in the north-central Scandinavian Caledonides and their pre-Caledonian tectonic evolution. *Geol. Rdsch.*, 71 (1): 85–93.

Grew, E.S. (1984). A review of Antarctic granulite facies rocks. *Tectonophys.*, vol. 105, pp. 177–191.

*Grigaitis, R.K., Il'chenko, L.N. and Kras'kov, L.N. (1989). New palaeontological data on the Precambrian formations of southern Ulytau (central Kazakhstan). *Izv. AN SSSR, Ser. Geol.*, no. 1, pp. 68–79.

*Grikurov, G.E., Znachko-Yavorsky, G.A., Kamenev, E.N. and Ravich, M.G. (1976). *Explanatory Note to the Geological Map of the Antarctic Region. Scale* 1:5,000,000. NIIGA, Leningrad.

*Grikurov, G.E., Znachko-Yavorsky, G.A., Kamenev, E.N. and Kurikin, R.G. (1978). *Explanatory Note to the Tectonic Map of Antarctic Region on Scale 1:10,000,000*. NIIGA, Leningrad.

Grubic, A. (1980). *An Outline of the Geology of Yugoslavia*. Excursions 201A and 202C. 26th Congres Geologique International. Guidebook 15, Paris-Belgrade, pp. 5–49.

*Grudinin, M.I. (1992). Greenstone and ophiolite belts of south eastern Siberia. *Geol. i. Geof.*, no. 12, pp. 15–22.

Guerrot, C., Peucat, J.J., Capdevila, R. and Dasso, L. (1989). Archaean protoliths within Early Proterozoic granulitic crust of the West European Hercynian belt: possible relics of the West African craton. *Geology*, vol. 17, pp. 241–244.

*Gurulev, S.A. (1976). Tectonic structure and development of the northern part of the Baikalian montane region in the Late Precambrian and Early Palaeozoic. In: *Siberian Tectonics*, Vol. 7, pp. 62–68. Nauka, Moscow.

*Gusev, G.S. (1973). Tectonic complexes of Mesozoides of Verkhoyansk-Chukchi region. In: *Main Tectonic Complexes of Siberia*, pp. 130–139. Nauka, Novosibirsk.

*Gusev, G.S., Peskov, A.I. and Sokolov, S.K. (1992). Palaeogeodynamics of Muya segment of Proterozoic Baikal-Vitim belt. Geotektonika, no. 2, pp. 72–86.

Hale, C.J. and Dunlop, D.J. (1984). Evidence for an Early Archaean geomagnetic field: a palaeomagnetic study of the Komati formation, Barberton greenstone belt, South Africa. Geoph. Res. Lett., 11 (2): 97–100.

Harper, G.D. (1984). A dismembered Archaean ophiolite, Wind River Mountains, Wyoming (USA). Ofioliti, no. 9, suppl., p.9.

Hartnady, C.J., Newton, A.R. and Theron, J.N. (1974). Stratigraphy and structure of the Malmesbury Group in the south-western Cape. Precambrian Res. Unit, Univ. Cape Town, Bull. no. 15, pp. 193–213.

Hedberg, R.M. (1979). Stratigraphy of the Ovamboland basin, south-west Africa. Precambr. Res. Unit, Univ. Cape Town, Bull., 24. pp.

Helm, D.G. (1984). The tectonic evolution of Jersey Channel islands. Proc. Geol. Assoc., vol. 95, pt 1, pp. 1–15.

Henger, E., Kröner, A. and Hoffman, A.W. (1984). Age and isotope geochemistry of the Archaean Pongola and Usushwana suites in Swaziland, southern Africa: a case for crustal contamination of mantle-derived magma. Earth Planet. Sci. Lett., vol. 70, pp. 267–279.

Herz, N., Hasui, Y., Costa, J.B.S. and Matta, M.A. da S. (1989). The Araguaia fold belt, Brazil. A reactivated Brasiliano-Pan-African cycle (550 Ma) geosuture. Precambr. Res., vol. 42, pp. 371–386.

Hiroi, I., Shiraishi, K. and Motoyoshi, I. (1991). Late Proterozoic paired metamorphic complexes in east Antarctica, with special reference to the tectonic significance of ultramafic rocks. In: Geological Evolution of Antarctica, pp. 83–87. (M.R.A. Thomas, et al., eds). Cambridge Univ. Press, Cambridge.

Hoffman, P.F. (1989). Precambrian geology and tectonic history of North America. In: The Geology of North America An Overview. Geol. Soc. America.

Hoffman, P.F. and Bowring, S.A. (1984). Short-lived 1.9 Ga continental margin and its destruction, Wopmay orogen, Canada. Geology, 12 (2): 68–72.

Hofmeister, A.M. (1983). Effect of a Hadean magma ocean on crust and mantle evolution. J. Geophys. Res., 88 (B6): pp. 4963–4983.

Holt, R., Egbuniwe, J.G., Atches, W.R. and Wright, J.B. (1978). The relationship between low grade metasedimentary belts, calc-alkaline volcanism and the Pan-African orogeny in NW Nigeria. Geol. Rdsch., 67 (2): 631–646.

Hottin, G. (1976). Presentation et essai d'interpretation du Precambrian de Madagascar. Bull. BRGM, sec. IV, no. 2, pp. 117–153.

Hsü, K.J., Li Jiliang, Chen Haihong, Wang Qingchen, Sun Shu and Şengör, A.M.C. (1990). Tectonics of South China: key to understanding West Pacific geology. Tectonophys., 183 (1–4): 9–39.

*Huan Tszitsin (1984). New data on the tectonics of China. In: Tectonics of Asia, C.05, pp. 11–24. 27th International Geological Congress.

Hutton, J. (1795). Theory of the Earth, Vols. 1 and 2. Cadell A. Davies, London.

Hynes, A. (1982). Stability of oceanic tectonosphere—a model for Early Proterozoic intercratonic orogeny. Earth Planet. Sci. Lett., vol. 61, pp. 333–345.

Hynes, A. and Francis, D.M. (1982). A transect of the Early Proterozoic Cape Smith fold belt, New Quebec. Tectonophys., 88 (1–2) pts. 1–2: 23–60.

*Ilyin, A.V. (1986). Tectonics of southern China. Geotektonika, no. 1, pp. 32–46.

International Symposium on Archaean and Early Proterozoic Geologic Evolution and Metallogenesis. Rev. Brazil. de Geociencias, vol. 12, no. 1–3 (1982).

*Ivanov, S.N. (1979). Baikalian Ural and the nature of metamorphic series in eugeosynclinal framework (preprint). Sredne-Ural 'skoe Knizhnoe Izdatel'stvo, Sverdlovsk, 77. pp.

* Ivanov, Zh., Moskovski, S., Kolcheva, K. and Dimov, D. (1984). Geologic structure of central Rhodope. Lithostratigraphic correlation and characteristics of the sequence of metamorphic rocks in the northern parts of central Rhodope. *Geol. Balcanica*, 14 (1): 3–42.

Jackson, S.L., Sutcliffe, R.M., Ludden, J.N. et al. (1990). Southern Abitibi greenstone belt: Archaean crustal structure from seismic reflection profiles. *Geology*, vol. 18, pp. 1088–1090.

Jantsky, B. (1980). The Precambrian in Hungary: latest results of research. *Anuar. Inst. Geol. Geof.*, vol. LVII, pp. 433–457.

Johnson, P.R., Scheibner, E. and Smith, A. (1987). Basement fragments, accreted tectono-stratigraphic terranes and overlap sequences: elements in the tectonic evolution of the Arabian Shield. In: *Terrane Accretion and Orogenic Belts*. Amer. Geoph. Union Geodynamics Series.

Jones, J.P. (1985). The southern border of the Guapore Shield in western Brazil and Bolivia: An interpretation of its geological evolution. *Precambr. Res.*, vol. 28, pp. 111–135.

Kaiyi, W., Windley, B.F., Sills, J.D. and Yuehua, I. (1990). The Archaean gneiss complex in eastern Hebei province, north China: Geochemistry and evolution. *Precambr. Res.*, vol. 48, pp. 245–265.

Kale, V.S. (1990). Constraints on the evolution of the Purana basins of peninsular India. *J. Geol. Soc. India*, vol. 38, pp. 231–252.

Kalsbeek, F. (1982). The evolution of the Precambrian Shield of Greenland. *Geol. Rdsch.*, 78 (1): 38–60.

Kalsbeek, F., Taylor, P.N. and Henriksen, N. (1984). Age of rocks, structures and metamorphism of the Nagsuktoqidian mobile belt, west Greenland—field and Pb isotope evidence. *Can. J. Earth Sci.*, 21 (10): 1126–1131.

*Kalyayev, G.I. (1973). Tectonics of the Ukrainian Shield and its position in the structure of the Eastern European Craton. In: *Tectonics of the Basement of Ancient Cratons*. Nauka, Moscow, pp. 50–60.

*Kamenev, E.N. (1991). Basic features of geology and evolution of the Antarctic Shield in the Precambrian. Doc. Geol-Min. Sciences, thesis (abstract). Nauka, Leningrad.

*Kapusta, R.A., Sumin, L.V., Shuleshko, I.K., Berezhnaya, N.G. (1985). Zirconometry of volcanic rocks of the Gimola series on isotopes of xenon and plumb. *Geokhimia*, no. 3, pp. 293–299.

*Karapetov, S.S. (1979). Early Precambrian and its position in the structure of the southern part of Central Asia. *Geotektonika*, no. 1, pp. 64–76.

Karlstrom, K.E., Furkey, A.J. and Houston, R.S. (1983). Stratigraphy and depositional setting of the Proterozoic Snowy Pass supergroup, south-eastern Wyoming: Record of an Early Proterozoic Atlantic type cratonic margin. *Geol. Soc. Amer. Bull.*, 94 (11): 1257–1274.

Kasch, K.W. (1983). The structural geology, metamorphic petrology and tectonothermal evolution of the southern Damara belt around Omitara, S.W.A. Namibia, Univ. Cape Town, Dept. Geol. *Precambr. Res. Unit Bull.*, no. 27.

Katz, M.B. (1985). The tectonics of Precambrian craton mobile belts: Progressive deformation to polygonal miniplates. *Precambr. Res.*, 27 (4): 307–319.

Kaufman, A.J., Knoll, A.H. and Awramik, S.M. (1992). Biostratigraphic and chemostratigraphic correlation of Neoproterozoic sedimentary successions: upper Tinder group, north-western Canada, as a test case. *Geology*, vol. 20, pp. 181–185.

Kaz'min, V.G. Shifferaw, A. and Balcha, T. (1978). The Ethiopian basement stratigraphy and possible manner of evolution, *Geol. Rdsch.*, vol. 67, no. 2.

Keller, B.M., Semikhatov, M.A. and Chumakov, N.M. (1984). Type sequences of the upper erathem of the Proterozoic. In: *Precambrian Geology*, pp. 56–76. 27th International Geological Congress, Section C.05, Reports, vol. 5.

Keller, G.R., Lidiak, E.G., Hinze, W.J. and Braile, L.M. (1983). The role of rifting in the tectonic development of the Midcontinent, USA. *Tectonophys.*, 94 (1–4): 391–412.

Key, R.M., Charsley, T.J., Hackman, B.D., Wilkinson, A.F. and Rundle, C.C. (1989). Superimposed Upper Proterozoic collision-controlled orogenies in the Mozambique orogenic belt of Kenya. *Precambr. Res.,* vol. 44, pp. 197–220.

Khain, V.E. (1954). Prominent and most prominent cycles in the history of the Earth. *Nauchn. Doklady Vyssh. Shkoly, Geol. i Geogr. Nauki,* no. 1, pp. 33–58.

*Khain, V.E. (1971). *Regional Geotectonics. North and South America, Antarctic and Africa.* Nedra, Moscow.

*Khain, V.E. (1984a). *Regional Geotectonics. Alpine Mediterranean Belt.* Nedra, Moscow.

*Khain, V.E. (1984b). Origin of ancient cratons. *Vestn. Mosk. Un-ta, Geol.,* no. 2, pp. 32–37.

Khain, V.E. and Rudakov, S.G. (1991). On the present position of the Gondwana original northern boundary in Europe and SW Asia. *Bull. Techn. Univ., Istanbul,* 44 (1–2): 77–96.

*Khomentovsky, V.V. (1976). *Vendian.* Nauka, Novosibirsk (Trudy IGiG SO AN SSSR, no. 243).

*King, Ph. B. (1976). *Precambrian Geology of the USA.* USGS Prof. paper 902, Washington.

*Kiselev, V.V. (1991). *Precambrian of the Palaeozoides of Tien Shan.* Doc. Geol.-Min. Sciences thesis, (abstract) Bishkek.

*Kiselev, V.V. and Korolev, V.G. (1972). *Precambrian Tectonics of Central Asia and Central Kazakhstan.* Ilim, Frunze.

*Kiselev, V.V., Apayarov, F.Kh., Komarevtsev, V.T. et al. (1992a). Geological-geochronological boundaries of the Precambrian history of the Palaeozoides of Tien Shan and Kazakhstan. In: *Precambrian in Phanerozoic Fold Regions.* Nauka, Moscow.

*Kiselev, V.V., Bakirov, A.B., Korolev, V.G. and Maksumova, R.A. (1992b). Precambrian in the structure of the Palaeozoides of Tien Shan. In: *Precambrian in Phanerozoic Fold Regions.* Nauka, Moscow.

*Kitsul, V.I., Petrov, A.F. and Zedintsov, A.N. (1979). Structural and material complexes of Aldan Shield. In: *Major Tectonic Complexes of Siberia,* pp. 16–31. Nauka, Novosibirsk.

*Klitin, K.A., Domnina, E.A. and Rile, G.V. (1975). Structure and age of the ophiolite complex of Baikal-Vitim Uplift. *Byull. MOIP, Otd. Geol.,* 50 (1): 82–94.

Kober, B., Pidgeon, R.T. and Lippolt, H.J. (1989). Single-zircon dating by stepwise Pb-evaporation constraints. The Archaean history of detrital zircons from the Jack Hills, western Australia. *Earth Planet. Sci. Lett.,* 91 (3–4): 286–296.

*Konnikov, E.G. (1991). On the problem of the ophiolites of the Baikal-Muya belt. *Geol. i Geof.,* no. 3, pp. 119–129.

Kontinen, A. (1987). The Jormua mafic-ultramafic complex, north-eastern Finland—an Early Proterozoic ophiolite. *Precambr. Res.,* vol. 35, pp. 313–341.

*Korobeinikov, V.P., Surkov, V.S. and Shcheglov, A.P. (1979). Tectonic complexes of Altay-Sayan region. In: *Major Tectonic Complexes of Siberia,* pp. 91–104. Nauka, Novosibirsk.

*Kosygin, Yu.A. and Parfenov, L.M. (1977). Some problems in the Precambrian tectonics of continents. In: *Problems of the Geology of the Early Precambrian,* pp. 16–29. Nauka, Novosibirsk.

*Kozhukharov, D. (1984). Lithostratigraphy of Precambrian metamorphic rocks of the Rhodopian supergroup in central Rhodope. *Geol. Balcanica,* 14 (1): 43–88.

*Krats, K.O. and Lobach-Zhuchenko, S.B. (1981). What are the grey gneisses? Conjectures and problems. In: *The Oldest Granitoids of the USSR,* pp. 5–13. Nauka, Leningrad.

*Krats, K.O., Khil'tova, V. Ya., Vrevsky, A.B. and Zapol'nov, A.K. (1980). *Stages and Types of Evolution of the Precambrian Crust of Ancient Shields.* Nauka, Leningrad, 164 pp.

Kraütner, H.G. (1984). Lithostratigraphic correlation of the Precambrian in the Romanian Carpathians. *Geol. Balcanica,* 14 (1): 229–296.

*Krestin, E.M. (1980). Precambrian of Kursk Magnetic Anomaly [region] and basic principles of its development. *Izv. Vuzov, Geol. i Razv.,* no. 3, pp. 3–18.

Kröner, A. (1974). The Gariep group Late Precambrian formation in the Western Richtersveld, Northern Cape Province. *Precambr. Res. Unit., Univ. Cape Town, Bull.,* no. 13, 111 pp.

Kröner, A. (1977). The Precambrian geotectonic evolution of Africa: plate destruction. *Precambr. Res.*, 4 (2): 163–213.

Kröner, A. (1980a). Chronologic evolution of the Pan-African Damara belt in Namibia, southern Africa, In: *Mobile Earth. Final Report of the Geodynamic Project* (J. Neumann, ed.). Namibia, Africa.

Kröner, A. (1980b). Chronologic evolution of the Pan-African Damara belt in Namibia, southwest Africa. *Precambr. Res.*, vol. 5, pp. 311–357.

Kröner, A. (ed.) (1981). *Precambrian Plate Tectonics.* Elsevier, Amsterdam.

Kröner, A. (1991). Tectonic evolution in the Archaean and Proterozoic. *Tectonophys.*, vol. 187, pp. 393–410.

Kröner, A. and Clauer, N. (1979). Isotopic dating of low-grade metamorphic shales in northern Namibia (south-west Africa) and implications for the orogenic evolution of the Pan-African Damara belt. *Precambr. Res.*, vol. 10, pp. 59–72.

Kröner, A. and Greiling, R. (eds.) (1984). *Precambrian Tectonics Illustrated.* E. Schweizerbart, Stuttgart.

Kröner, A. and Sengör, A.M.C. (1990). Archaean and Proterozoic ancestry in Late Precambrian to Early Palaeozoic crustal elements of southern Turkey as revealed by single zircon dating. *Geology,* vol. 18, pp. 1186–1190.

Kröner, A. and Layer, P.W. (1992). Crust formation and plate motion in the Early Archaean. *Science,* 256 (5062): 1405–1411.

Kröner, A., Reischmann, T., Wust, H.-J. and Rashwan, A.D. (1989). Is there any Pre-Pan-African (> 950 Ma) basement in the Eastern desert of Egypt? In: *The Pan-African Belt of North-East Africa and Adjacent Areas* (S. Gaby and R.O. Greiling, eds.).

*Kudryavtsev, V.A. and Nuzhnov, S.V. (1981). Upper Archaean structures of the Aldan Shield. *Geol. i Geof.,* no. 6, pp. 28–38.

Kukla, P.A. and Stanistreet, I.G. (1991). Record of the Damaran Khomas Hochland accretionary prism in central Namibia: of a late refutation of an "ensialic" origin of a Late Proterozoic orogenic belt. *Geology,* vol. 19, pp. 473–476.

*Kurbatskaya, F.A. (1985). *Formation and Palaeotectonics of the Urals Margin of the East European Craton in the Late Precambrian.* D.Sc. Thesis (abstract) Moscow State University, Moscow.

*Kuskov, O.L. and Khitarov, N.I. (1982). *Thermodynamics and Geochemistry of the Core and Mantle of the Earth.* Nauka, Moscow.

Kusky, T.M. (1990). Evidence for Archaean ocean opening and closing in the southern Slave province. *Tectonics,* 9 (6): 1533–1563.

Lafliche, M.R., Dupuy, C. and Dostal, J. (1991). Archaean ultrapotassic magmatism: An example from the southern Abitibi greenstone belt. *Precambr. Res.,* 52 (1–2): 71–96.

Lavreau, J. (1982). The Archaean and Lower Proterozoic of central Africa. *Rev. Brasil. Geocienc.,* no. 1–3, pp. 187–192.

*Laz'ko, E.M. (1966). On the Early Precambrian of the southern part of the Indian Platform. *Sov. Geologia,* no. 11, pp. 55–67.

*Laz'ko, E.M., Sivoronov, A.A. and Bobrov, A.B. (1982). Problem of tonalite layer in granite-greenstone regions. *Izv. AN SSSR, Ser. Geol.,* no. 9, pp. 5–15.

Le Cheminant, A.N. and Heaman, L.M. (1989). Mackenzie igneous events, Canada: Middle Proterozoic hotspot magmatism associated with ocean opening. *Earth Planet. Sci. Lett.,* vol. 96, pp. 38–48.

Le Pichon, X. and Huchon, Ph. (1984). Geoid, Pangea and convection. *Earth Planet. Sci. Lett.,* 67 (1): 123–135.

*Le Zui Bath. (1985). Structure of Vietnam and stages of its formation (in the background of South-east Asia as a whole). D.Sc. thesis (abstract). Moscow State University, Moscow.

Leblanc, M. (1976). Proterozoic oceanic crust at Bou Azzer. *Nature,* 261 (555): 34–35.

424

Leblanc, M. and Lancelot, J. (1977). Le domaine pan-african de l'Anti-Atlas (Maroc). In: *Rapport d'activite, Centre Geologique et Geophysique de Montpellier*, pp. 725–749. Montpellier.

Lecorché, J.P., Bronner, G., Dallmeyer, R.O., Rocci, G. and Roussel, J. (1991). Mauritanides. In: *The West African Orogens and Circum-Atlantic Correlatives.* Springer-Verlag, New York.

Ledru, P., N'dong, J.E., Johan, V. et al. (1989). Structural and metamorphic evolution of the Gabon orogenic belt: Collision tectonics in the Lower Proterozoic. *Precambr. Res.,* vol. 44, pp. 227–240.

Lee Dai-sung (ed.). (1987). *Geology of Korea.* Kyohakta, Seoul.

Lemoine, S., Tempier, P., Bassot, J.P. et al. (1990). The Burkinian orogenic cycle, precursor of the Eburnean orogeny in west Africa. *Geol. J.,* vol. 25, pp. 171–188.

Leube, A., Hirdes, W., Mauer, R. and Kesse, G.O. (1990). The Early Proterozoic Birrimian supergroup of Ghana and some aspects of its associated gold mineralisation. *Precambr. Res.,* vol. 46, pp. 139–165.

Light, M.P.R. (1982). The Limpopo mobile belt: A result of continental collision. *Tectonics,* 1 (4): 325–342.

Lindh, A. (1975). Trends in the post-Svecokarelian development of the Baltic Shield. *Geol. Rdsch.,* 71 (1): 130–140.

Liu, D.-Y., Page, R.W., Compston, W. and Wu, J. (1985). U-Pb zircon geochronology of Late Archaean metamorphic rocks in the Tangshan-Wutaishan area, north China. *Precambr. Res.,* vol. 27, pp. 85–109.

Liu, D.-Y., Shen, Q.H., Zhang, Z.Q., Jahn, B.M. and Auvray, B. (1990). Archaean crustal evolution in China: U-Pb geochronology of the Qianxi complex. *Precambr. Res,* vol. 48, pp. 223–244.

*Lobach-Zhuchenko, S.B. et al. (1989). Vodlozero Early Archaean gneiss complex and its structural-metamorphic evolution. In: *Isotope Precambrian Geochronology,* pp. 14–45. Nauka, Leningrad.

Loberg, B.E.H. (1980). A Proterozoic subduction zone in southern Sweden. *Earth Planet. Sci. Lett.,* 46 (2): 257–294.

Lofgren, C. (1979). Do leptites represent Precambrian island arc rocks? *Lithos,* 12 (2): 159–165.

Lucas, S.B., St. Onge, M.R., Parish, R.R. and Dunply, J.M. (1992). Long-lived continent-ocean interaction in the Early Proterozoic Ungava orogen, northern Quebec, Canada. *Geology,* 20 (2): 113–116.

Lumbers, S.B., Tsai-Way Wu, Heaman, L.M., Vertolli, V.M. and Mac Rae, N.D. (1991). Petrology and age of the Atype Mulock granite batholith, northern Grenville province, Ontario. *Precambr. Res.,* vol. 53, pp. 199–231.

*Luts, B.G. (1978). Basalt-andesite-dacite formations of the Early Precambrian and their comparison with contemporary analogues. *Geotektonika,* no. 4, pp. 23–34.

Mabesoone, I.M., Brito-Neves and Sial, A.N. (eds.). (1981). *The Geology of Brazil.*

MacCulloch, M.T. and Black, L.P. (1984). Sm-Nd isotopic systematics of Enderby Land granulites and evidence for the redistribution of Sm and Nd during metamorphism. *Earth Planet. Sci. Lett.,* vol. 71, pp. 46–56.

*Magmatic Formations of the Early Precambrian in the USSR Territory. Nedra, Leningrad (1981).

*Makarychev, G.I. (1992). Primary oceanic nature of the crust of the Ural-Mongolian fold belt. *Geotektonica,* no. 1, pp. 111–124.

*Makarychev, G.I., Ges', M.D. and Palei, I.P. (1983). Basic principles of the formation of continental crust in Precambrian Tien Shan, Kazakhstan and Mongolia. *Byull. MOIP, Otd. Geol.,* 58 (3): 3–18.

*Makarychev, G.I., Ges', M.D. and Pazilova, V.I. (1983). Precambrian ophiolites of Ulytau in the light of stage-wise development of the Earth's crust. *Geotektonika,* no. 4, pp. 60–74.

*Man'kovsky, V.K. and Poroshin, E.E. (1981). Tectonic regime of Altay-Sayan fold region in the Late Precambrian. *Trudy GIN AN SSSR,* vol. 311, pp. 46–55.

*Manuilova, M.M. (1991). Geology and geochronology of the Precambrian Baikal montane region and the Baikalides problem. *Geol. i Geof.*, no. 9, pp. 58–67.

*Markov, M.S. (ed.). (1988). *Archaean of the Anabar Shield and Problems of the Early Evolution of the Earth.* Nauka, Moscow.

Marshak, S., Alkmim, F.F. and Jordt-Evangelista, H. (1992). Proterozoic crustal extension and the generation of dome and keel structure in an Archaean granite-greenstone terrane. *Nature*, 357 (6378): 491–493.

Martin, H. and Porada, H. (1977). The intracratonic branch or the Damara orogen in south-west Africa. *Precambr. Res.*, 5 (4): 311–357.

Martin, H., Auvray, B., Blaise, S., Capdevila, R., Hameurt, J., Jahn, B.M., Piquet, D., Quare, G. and Vidal, Ph. (1984). Origin and geodynamic evolution of the Archaean crust of Eastern Finland. *Bull. Geol. Soc. Finl.*, 56 (1–2): 135–157.

*Martynova, V.P. (1980). Formation of continental crust in Cis-Ladoga [region]. *Geotektonika*, no. 4, pp. 18–28.

Ma Xingyuan and Wu Zhengwen. (1981). Early tectonic evolution of China. *Precambr. Res.*, 14 (3–4): 185–202.

Max, M.P. (1979). Extent and disposition of Grenville tectonism in the Precambrian continental crust adjacent to the North Atlantic. *Geology*, vol. 7, pp. 76–78.

McDougall, J.D., Gopalan, K., Lugmair, G.W. and Roy, A.B. (1983). An ancient depleted mantle source for Archean crust in Rajasthan, India. In: Workshop on a cross-section of Archean crust. Lunar Planet. Inst. Houston Tech. Rept. 83–03, pp. 55–56.

McElhinny, M.W. and Embleton, B.J.J. (1976). Precambrian and Early Palaeozoic palaeomagnetism in Australia. *Phil. Trans. Roy. Soc. Lond.*, A280, pp. 417–431.

McElhinny, M.W. and McWilliams, M.D. (1977). Precambrian geodynamics: A palaeomagnetic view. *Tectonophys.*, vol. 40, pp. 137–159.

McGregor, V.R. (1983). Archaean grey gneisses and the origin of continental crust: Data for the region Gothab, western Greenland. In: *Trondhjemites, Dacites and Associated Rocks*, pp. 131–156. (F. Barker, ed.). Elsevier, Amsterdam-New York.

McLennan, S.M and Taylor, S.R. (1991). Sedimentary rocks and crustal evolution: Tectonic setting and secular trends. *J. Geol.*, vol. 99, pp. 1–22.

*Mel'nikova, T.M. (1992). Palaeozoic granitoid magmatism of Baikal-Patom Upland. Ph. D. thesis (abstract). Irkutsk Univ., Irkutsk.

Metamorphic Complexes of Asia. Nauka, Novosibirsk (1977).

Metamorphism and Tectonics of the Western Zone of the Urals. Sverdlovsk (1984).

*Milanovsky, E.E. (1982). Development and contemporary status of the problem of expansion and pulsation of the Earth. *Izv. Vuzov, Geol. i Razved,* no. 7, pp. 3–29.

*Milanovsky, E.E. (1983). *Rifting in the History of the Earth.* Nedra, Moscow.

*Milov, A.P. (1990). Some new determinations of the isotope age of geological formations of north-eastern USSR and their geodynamic interpretation. In: *Tectonics and Mineralogy of North-Eastern Asia*, pp. 139–140. Magadan.

*Mints, M.V. (1993). Palaeogeodynamic reconstructions for the Early Precambrian of the eastern part of the Baltic Shield. *Geotektonika*, no. 1, pp. 39–56.

*Mitrofanov, F.P. (ed.). (1980). *Principles and Criteria for Delineating the Precambrian in Mobile Zones.* Nauka, Leningrad.

*Mitrofanov, F.P., Kazakov, I.K. and Palei, I.P. (1981). *Precambrian of Western Mongolia and Southern Tuva.* Nauka, Leningrad.

*Mitrofanov, G.L., Nikol'sky, F.V., Taskin, A.P. and Khrenov, P.M. (1984). Upper Precambrian fold belts in the southern part of eastern Siberia. In: *Precambrian Geology*, pp. 119–125. 27th Internat. Geol. Cong., Sec. C 05, Reports, vol. 5.

*Monin, A.S. (1983). Katarchaean. *Geol. i Geof.*, no. 7, pp. 3–15.

*Monin, A.S. and Sorokhtin, O.G. (1982). Evolution of the Earth under conditions of volumetric differentiation of its core. *Dokl. AN SSSR,* vol. 263, pp. 572–575.

426

Moorbath, S. (1977). Ages, isotopes and evolution of Precambrian continental crust. *Chem. Geol.,* vol. 20, no. 2.

Moores, E.M. (1986). The Proterozoic ophiolite problem, continental emergence and the Venus connection. *Science,* 234 (4772): 65–68.

Moores, E.M. (1991). South-west U.S.—east Antarctic (SWEAT) connection: A hypothesis. *Geology,* vol. 19, pp. 425–428.

*Moralev, V.M. (1975). Horizontal movements and tectonic position of metamorphogenic pegmatite provinces (on the example of India). *Geol. Rudn. Mestorozhdenii,* 17 (1): 81–85.

*Moralev, V.M. (1977). The Indian Craton. In: *Ancient Cratons of Eurasia,* pp. 248–272. Nauka, Novosibirsk.

*Moralev, V.M. (1986). *Early Stages of Evolution of the Continental Lithosphere.* Nauka, Moscow, 166 pp.

*Moskovchenko, N.I. (1983). *High-pressure Precambrian complexes in the Phanerozoic fold belts.* Nauka, Leningrad, 160 pp.

*Mossakovsky, A.A. and Dergunov, A.B. (1983). Caledonides of Kazakhstan and Central Asia (tectonic structure, history of development and palaeotectonic setting). *Geotektonika,* no. 2, pp. 16–33.

Muehlberger, W.R. (1980). The shape of North America during the Precambrian. In: *Continental Tectonics: Structure, Kinematics and Dynamics,* pp. 175–183 (M. Friedman and M.N. Toksoz, eds.). *Tectonophys.,* vol. 1, nos. 1–3 (1983).

Muhling, J.R. (1985). The nature of Proterozoic reworking of Early Archaean gneisses, Mukalo Creek area, southern Gascoyne province, Western Australia. *Precambr. Res.,* vols. 40–41, pp. 341–362.

Munasinghe, T. and Dissanayake, C.B. (1982). A plate tectonic model for the geologic evolution of Sri Lanka. *J. Geol. Soc. India,* 23 (8): 369–380.

*Muratov, M.V. (1975). *Origin of Continents and Oceanic Basins.* Nauka, Moscow.

*Musatov, D.I., Fedorov, V.S. and Mezhelovsky, N.V. (1983). Tectonic Regimes and Geodynamics of the Archaean (Regional and Model Aspects). VIEMS, Moscow, 42 pp.

*Musatov, D.I., Fedorov, V.S., Afanas'ev, Yu.T. and Zonenshain, L.P. (1984). Some aspects of the geological structure and history of the development of the USSR territory from the viewpoint of new tectonic theories. In: *Geology of the USSR.* 27th Internat. Geol. Cong., Col. K OI, Reports, vol. 1, Moscow.

Myers, J.E. and Williams, I.R. (1985). Early Precambrian crustal evolution at Mount Narryer, Western Australia. *Precambr. Res.,* vol. 27, pp. 153–163.

Naidoo, D.D., Bloomer, S.H., Saquaque, A. and Hefferan, K. (1991). Geochemistry and significance of metavolcanic rocks from the Bou Azzer-El Graara ophiolite (Morocco). *Precambr. Res.,* vol. 53, pp. 79–97.

*Nalivkina, E.B. (1982). Two types of Precambrian greenstone belts. In: *Greenstone Belts of Ancient Shields,* pp. 47–51. Nauka, Moscow.

Nance, R.D., Murphy, I.B., Strachan, R.A., D'Lemos, R.S. and Taylor, G.K. (1991). Late Proterozoic tectono-stratigraphic evolution of the Avalonian and Cadomian terranes. *Precambr. Res.,* vol. 53, pp. 41–78.

Nance, W.B., Worsley, T.R. and Moody, J.B. (1986). Post-Archaean biogeochemical cycles and long-term episodicity in tectonic processes. *Geology,* vol. 14, pp. 514–518.

Naqvi, S.M. (1982). Early Archean evolution of the Indian Shield with special reference to the Dharwar Craton. *Rev. Brasil. Geosciencias,* 12 (1–3): 223–233.

*Negrutsa, V.Z. (1984). *Early Proterozoic Stages in the Development of the Eastern Part of the Baltic Shield.* Nedra, Leningrad.

Nelson, B.K. and De Paolo, D.J. (1984). 1,700 m.y. greenstone volcanic successions in southwestern North America and isotopic evolution of Proterozoic mantle. *Nature,* vol. 312, pp. 143–146.

Neubauer, F. (1991). Late Proterozoic and Early Palaeozoic tectono-thermal evolution of the Eastern Alps. In: *The West African Orogens and Circum-Atlantic Correlatives*. Springer-Verlag, New York.

Neumann, J. (ed.) (1980). *Mobile Earth. Final Report of the Geodynamic Project*. Namibia, Africa.

Nisbet, E.G. and Fowler, C.M.R. (1983). Model for Archaean plate tectonics. *Geology*, 11 (7): 376–379.

*Nozhkin, A.D. (1983). Early Precambrian gneiss complexes of Yenisei Range and their geochemical characteristics. *Geol. i Geof.*, no. 9, pp. 3–11.

Nutman, A.P., Fryer, B.J. and Bridgwater, D. (1989). The Early Archean Nulliak supracrustal assemblage, northern Labrador. *Can. J. Earth Sci.*, vol. 26, pp. 2159–2168.

Nystrom, J.O. (1982). Post-Svecokarelian andenotype evolution in central Sweden. *Geol. Rdsch.*, 71 (1): 141–157.

*Nystuen, J.P. (1982). The Late Proterozoic basin evolution on the Balto-Scandian craton: the Hedmark group, southern Norway. *Norges Geol. Unders*. vol. 375, pp. 1–74.

Oen, I.S. and Helmers, H. (1982). Ore deposition in a Proterozoic incipient rift zone environment: a tentative model for the Filipstad-Gryttytan-Hjulsjo region, Bergslagen, Sweden. *Geol. Rdsch.*, 78 (1): 182–194.

Oldest Granitoids of the USSR. Nauka, Leningrad (1981).

O Precambriano de Brasil. (F.F.M. de Almeida and J. Hasui, eds.). Edgard Blucher, Sao Paulo (1984).

* *Outlines of Comparative Planetology*. Nauka, Moscow (1981).

Page, R.W. and Williams, J.S. (1988). Age of the Barramundi orogeny in northern Australia by means of ion microprobe and conventional U-Pb zircon studies. *Precambr. Res.*, vol. 40/41, pp. 21–36.

*Page, R.W., MacCulloch, M. and Black, L. (1984). Isotope data on the main episodes in the Precambrian of Australia. In: *Precambrian Geology*, pp. 14–35. 27th Internat. Geol. Cong., Sec. C.05, Reports, vol. 5.

Pallister, J.S., Stacey, J.S., Fisher, L.B. and Premo, W.R. (1988). Precambrian ophiolites of Arabia; geologic settings, U-Pb geochronology, Pb isotope characteristics and implications for continental accretion. *Precambr. Res.*, vol. 38, pp. 1–54.

Park, A.F. (1991). Continental growth by accretion: a tectono-stratigraphic terrane analysis of the evolution of the western and central Baltic Shield, 2.50 to 1.75 Ga. *Geol. Soc. Amer. Bull.*, 103 (4): 522–537.

Park, A.F. and Bowes, D.R. (1983). Basement-cover relationship during polyphase deformation in the Sveco-karelides of the Kaavi district, eastern Finland. *Trans. Ry. Soc. Edinburgh.*, *Earth Sci.*, 74 (2): 95–118.

Park, R.G., Ahall, K.I. and Boland, M.P. (1991). The Sveco-norwegian shear zone network of south-western Sweden in relation to mid-Proterozoic plate movements. *Precambr. Res.*, vol. 49, pp. 245–260.

*Pavlov, A.P. (1922). Attempts to recognise the pre-Archaean era in the history of the Earth and determine its influence on the later history of the geoid. *Byull. MOIP, Otd. Geol.*, vol. 31, pp. 16–22.

*Pavlovsky, E.V. and Glukhovsky, M.Z. (1982). Problem of thermotectogeny. *Geotektonika*, no. 6, pp. 7–25.

*Peive, A.V. and Savel'ev, A.A. (1982). Structures and movements in the lithosphere. *Geotektonika*, no. 2, pp. 5–25.

Percival, J.A. and Williams, H.R. (1989). Late Archaean Quetico accretionary complex, Superior province, Canada, *Geology*, 17 (1): 21–25.

*Perfil'ev, Yu.S. and Moralev, V.M. (1975). Precambrian of the eastern part of the Alpine-Himalayan belt. *Trudy NIL Zarubezhgeologii*, no. 29, pp. 24–43.

428

* *Petrology and Correlation of Crystalline Complexes of the East European Craton.* Naukova Dumka, Kiev (1979).

Pharaoh, T.C. and Pearce, J.A. (1984). Geochemical evidence for the geotectonic setting of the Early Proterozoic metavolcanic sequences in Lapland. *Geol. Rdsch.,* vol. 25, pp. 283–308.

Piper, J.D.A. (1982). The Precambrian palaeomagnetic record: the case for the Proterozoic supercontinent. *Earth Planet. Sci. Lett.,* 55 (1): 61–89.

Piper, J.D.A. (1983). Proterozoic palaeomagnetism and single continent plate tectonics. *Geophys. J.R. Astr. Soc.,* vol. 74, pp. 163–197.

Plumb, K.A. (1979). The tectonic evolution of Australia. *Earth Sci. Rev.,* vol. 14, pp. 205–249.

Plumb, K.A., Derrick, G.M., Needham, R.S. and Shaw, R.D. (1981). The Proterozoic of Northern Australia. In: *Precambrian of the Southern Hemisphere,* pp. 205–307. Elsevier, Amsterdam.

Porada, H. (1979). The Damara-Ribeira orogen of the Pan-African—Brasiliano cycle in Namibia (south-west Africa) and Brazil as interpreted in terms of continental collision. *Tectonophys.,* 57 (2–4): 237–265.

* Postel'nikov, E.S. (1980). Geosynclinal development of the Yenisei range in the Late Precambrian. *Trudy GIN AN SSSR,* No. 141, p. 69.

Powell, D., Brook, M. and Baird, A.W. (1983). Structural dating of a Precambrian pegmatite in Moine rocks of northern Scotland and its bearing on the status of the "Morarian orogeny". *J. Geol. Soc. London,* vol. 140, pp. 813–823.

* *Precambrian Geology.* 27th International Geological Congress, Section C.OH, Reports, Nauka, Moscow (1984).

* *Precambrian in Phanerozoic Fold Belts.* Nauka, Leningrad (1982).

* *Precambrian of Central Asia.* Nauka, Leningrad (1984).

Precambrian Plate Tectonics (A. Kröner, ed.). Elsevier, Amsterdam (1981).

Precambrian Tectonics Illustrated (A. Kroner and R. Greiling, eds.). E. Schweizerbart, Stuttgart (1984).

Priem, H.N.A., Kroonenberg, S.B., Boelijk, N.A.I.M. and Hebede, E.H. (1989). Rb-Sr and K-Ar evidence for the presence of 1.6 Ga basement underlying the 1.2 Ga Garzon-Santa Marta granulite belt in the Colombian Andes. *Precambr. Res.,* vol. 42, pp. 315–324.

* *Principles and Criteria for Delineating Precambrian in Mobile Zones* (F.P. Mitrofanov, ed.). Nauka, Leningrad (1980).

* *Problems of Kazakhstan Tectonics* Nauka, Leningrad (1981).

Proterozoic Basins of Canada. (1981). Geol. Surv. Canada, Pap. 81–10.

* Pushkarev, Yu.D. (1990). *Megacycles in the Development of the Crust-Mantle System.* Nauka, Leningrad.

Qian Xianglin and Chen Yaping (1985). Late Precambrian mafic dyke swarms of the North China Craton. Geol. Assoc. Canada, Special Paper 34, pp. 385–391.

Quesada, C., Bellido, F., Dallmeyer, R.D., Gil-Ibarguchi, I., Oliveira, I.T., Perez-Estaun, A., Ribeiro, A., Robardet, M. and Silva, I.B. (1991). Terranes within the Iberian massif: correlations with west African sequences. In: *The West African Orogens and Circum-Atlantic Correlatives.* Springer-Verlag, New York.

Radhakrishna, B.P. and Naqvi, S.M. (1986). Precambrian continental crust of India and its evolution. *J. Geol.,* vol. 94, pp. 145–166.

Raith, M., Raase, P., Ackerman, D. and Lal, R.K. (1982). The Archaean craton of southern India: metamorphic evolution and P-T conditions. *Geol. Rdsch.,* 71 (1): 280–290.

Raith, M., Raase, P. and Hormann, P.K. (1982). The Precambrian of Finnish Lapland: evolution and regime of metamorphism. *Geol. Rdsch.,* 71 (1): 230–244.

Ramakrishnan, M., Moorbath, S., Taylor, F.N., Anantha Iyer, G.V. and Viswanatha, M.N. (1984). Rb-Sr and Pb-Pb whole rock isochron age of basement gneisses in Karnataka craton. *J. Geol. Soc. India,* 25 (1): 20–33.

Rapp, R.P., Watson, E.B. and Miller, C.F. (1991). Partial melting of amphibolite/eclogite and the origin of Archaean trondhjemites and tonalites. *Precambr. Res.,* 51 (1): 1–25.

Rast, N., Skehan, S.J. and James, W. (1983). The evolution of the Avalonian plate. *Tectonophys.,* 100 (1–3): 257–285.

Ravich, M.G. and Kamenev, E.N. (1975). *Crystalline Basement of the Antarctic Platform.* John Wiley & Sons, Inc., New York.

Read, H.H. and Watson, J. (1975). *Introduction to Geology,* vol. 2. *Earth History,* Part 1. McMillan Press Ltd., London, 221 pp.

Ren Jishun, Chen Tingyu, Liu Zhigang, Niu Baogui and Liu Fengren (1986). Some problems on the tectonics of southern China. *Kexue Tongbao,* 31 (11): 751–754.

Renne, P.R., Onstott, T.C., D'Agrella-Fieho, M.S. and Pacca, I.G. (1990). $^{40}Ar/^{39}Ar$ dating of 1.0–1.1 Ga magnetisations from the Sao Francisco and Kalahari cratons: Tectonic implications for Pan-African and Brasiliano mobile belts. *Earth Planet. Sci. Lett.,* vol. 101, pp. 349–366.

Renne, P.R., Mattison, J.M., Somin, M.L., Onstott, M. and Milan, G. (1989). $^{40}Ar/^{39}Ar$ and U-Pb evidence for Late Proterozoic (Grenville age) continental crust in north-central Cuba and regional tectonic implications. *Precambr. Res.,* vol. 42.

* *Resolutions of the All-Union Stratigraphic Conference on Precambrian, Palaeozoic and Quaternary Systems of Central Siberia.* Nauka, Novosibirsk (1983).

* Rile, G.V. (1991). Tectonics of Precambrian Olokit region (northern Peri-Baikalia). Ph. D. thesis (abstract), Moscow Univ., Moscow.

Ringwood, A.E. (1979). *Origin of the Earth and Moon.* Springer-Verlag, New York.

* *Riphean Stratotypes. Stratigraphy and Geochronology. Trudy GIN AN SSSR,* no. 377, 183 pp. (1983).

Ritz, M. and Robineau, B. (1986). Crustal and upper mantle electrical conductivity of structures in West Africa: Geodynamic implications. *Tectonophys.,* vol. 124, pp. 115–132.

Rivers, T., Martignole, J., Gower, C.F. and Davidson, A. (1989). New tectonic divisions of the Grenville province, south-eastern Canadian Shield. *Tectonics,* vol. 8, pp. 63–84.

Robb, L.J., Davis, D.W. and Kamo, S.I. (1991). Chronological framework for the Witwatersrand basin and environs: towards a time-constrained depositional model. *Terra Nova Abstr.,* vol. 3, p. 25.

Roberts, D. and Gale, G.H. (1978). The Caledonian-Appalachian Iapetus ocean. In: *Evolution of the Earth's Crust* pp. 255–324 (D.H. Tarling, ed.). Academic Press, New York.

Rosen, O.M. (1989). Two geochemically different types of Precambrian crust in the Anabar Shield, northern Siberia. Precambrian Res., vol. 45, pp. 129–142.

Ross, G.M. and Bowring, S.A. (1990). Detrital zircon geochronology of the Windermere supergroup and the tectonic assembly of the southern Canadian Cordillera. *J. Geol.,* vol. 98, pp. 879–893.

* Rudakov, S.G. (1985). Formation and prealpine evolution of ancient complexes of Carpathian-Balkan geosynclinal region. *Vestn. Mosk. Un-ta, Ser. Geol.,* no. 2, pp. 74–85.

* Rudkevich, M.Ya. and Latypova, Z.A. (1979). Pre-Jurassic formations and structural layers of the Western Siberian platform. In: *Major Tectonic Complexes of Siberia,* pp. 67–81. Nauka, Novosibirsk.

* Rudnik, V.A. and Sobotovich, E.V. (1984). *Early History of the Earth.* Nedra, Moscow.

* Rudyachenok, V.M. (1974). Riphean-Early Palaeozoic Fold Complexes of the Antarctic Region. Report of the Interdepartmental Commission for Studying the Antarctic Region, no. 13, pp. 61–84.

Rutland, R.W.R. (1976). Orogenic evolution of Australia. *Earth Science Rev.,* vol. 12, pp. 161–196.

Rytsk, E.Yu. (1991). Factors Controlling Stratiform Lead-Zinc Deposits of Olokit Region (Northern Peri-Baikalia). Ph.D. Thesis (abstract), Moscow Univ., Moscow.

Sacchi, R. (1984). Late Proterozoic evolution of the southernmost Mozambique belt. In: *Geology for Development. The Precambrian of Africa, Newsletters, Bull.* no. 3, pp. 69–73.

*Safronov, V.S. (1969). *Evolution of Preplanetary Cloud and the Formation of the Earth and Planets.* Nauka, Moscow.

Saleeby, J.B. (1983). Accretionary Tectonics of the North American Cordillera. *Ann. Rev. Earth Planet Sci.,* no. 15, pp. 45–73.

*Salop, L.I. (1966). Stratigraphy of the Lower Precambrian of southern India. In: *Problems of Geology.* 22nd Internat. Geol. Cong. Nauka, Moscow, 508 pp.

*Salop, L.I. (1982). *Geological Development of the Earth in the Precambrian.* Nedra, Leningrad.

*Salop, L.I. (1984). Plate tectonics in the light of Precambrian geology. *Byull. MOIP, Otd. Geol.,* no. 4, pp. 15–31.

Saquaque, A., Admou, H., Karson, J. et al. (1989). Precambrian accretionary tectonics in the Bou Azzer-El Graara region, Anti-Atlas, Morocco. *Geology,* vol. 17, pp. 1107–1110.

Sarkar, A.N. (1982). Precambrian tectonic evolution of eastern India: a model of convergent microplates. *Tectonophys.,* vol. 86, pp. 363–397.

Schärer, U. and Allegre, C.J. (1985). Determination of the age of the Australian continent by single-grain zircon analysis of Mt. Narryer metaquartzite. *Nature,* vol. 315, pp. 52–55.

Schmidt, K. and Sollner, F. (1983). Towards a geodynamic concept of the "Caledonian event" in central and SW Europe. *Verhalt. Geol. B.A.,* no. 3, pp. 251–268.

Scott, D.J., Helmstaedt, H. and Bickle, M.J. (1992). Purtuniq ophiolite, Cape Smith belt, northern Quebec, Canada: a reconstructed section of Early Proterozoic oceanic crust. *Geology,* 20 (2): 173–176.

*Semikhatov, M.A. (1974). *Stratigraphy and Geochronology of the Proterozoic.* Trudy GIN AN SSSR, no. 256.

*Semikhatov, M.A., Shurkin, K.A. and Aksenov, E.M. (1991). A new stratigraphic scale of the Precambrian of the USSR. *Izv. AN SSSR, Ser. Geol.,* no. 4.

Sengupta, S., Paul, D.K., de Lacter, J.R., McNaughton, N.J., Bandopadhyay, P.K. and de Smith, J.B. (1991). Mid-Archaean evolution of the eastern Indian Craton: Geochemical and isotopic evidence from the Bonai pluton. *Precambr. Res.,* 49 (1–2): 23–37.

Shaw, R.D., Stewart, A.J. and Black, L.P. (1984). The Arunta inlier: A complex ensialic mobile belt in central Australia. Part 2: Tectonic history. *Austr. J. Earth Sci.,* 31 (4): 457–484.

*Shcherbak, N.P., Bartnitsky, E.N., Bibikova, E.V. et al. (1986). Isotope geochronology of the oldest rocks of the Ukrainian Shield. In: *Methods of Isotope Geology and the Geochronological Scale,* pp. 18–29. Nauka, Moscow.

*Shenfil', V. Yu. (1991). *Late Precambrian of the Siberian Craton.* Nauka, Novosibirsk.

Shen-su-Sun. (1985). Multistage accretion and core formation of the Earth. *Nature,* vol. 313, pp. 628–629.

Sheraton, J.W., Offe, L.A., Tingey, R.L. and Ellis, D.J. (1980). Enderby Land, Antarctica—an unusual Precambrian high-grade metamorphic terrain. *J. Geol. Soc. Austr.,* vol. 27, pp. 1–18.

*Shidlowski, M. (1984). Sedimentary organic matter 3.8 b.y. ago: Life isotopic marks. In: *Comparative Planetology,* pp. 119–125. 27th Internat. Geol. Cong. Nauka, Moscow.

*Shpunt, B.R. (1984). Material composition and conditions of formation of Upper Precambrian strata of the ancient cratons of Eurasia. D.Sc. thesis (abstract). Moscow, Univ., Moscow.

*Shuldiner, V.I. (1982). *Precambrian Basement of the Pacific Belt and Adjoining Cratons.* Nedra, Moscow.

Silver, L.T. (1980). Problems of pre-Mesozoic continental evolution. In: *Continental Tectonics: Structure, Kinematics and Dynamics* (M. Friedman and M.N. Toksoz., eds.).

Silver, L.T., McKinney, R.T., Deutsch, S. and Bolinger, J. (1963). Precambrian age determinations in the western San Gabriel Mountains, California. *J. Geol.,* 71 (2): 196–214.

Simpson, P.R., Gong Yoixun and Gao Bingzang. (1987). Metallogeny, magmatism and structure in Jiangxi province of China: A new interpretation. *Trans. Inst. Mining and Metallogeny*, vol. 96, pp. 77-83.

Sims, P.K., Van Schmus, W.R., Schultz, K.J. and Peterman, Z.E. (1989). Tectono-stratigraphic evolution of the Early Proterozoic Wisconsin magmatic terranes of the Penokean orogen. *Can. J. Earth Sci.*, vol. 26, pp. 2145-2158.

Sinha-Roy, S. (1988). Proterozoic Wilson cycles in Rajasthan. *Geol. Soc. India Mem.*, no. 7, pp. 95-107.

Sisser, W.G. and Dingle, R.V. (1981). Tertiary sea-level movements around Southern Africa. *J. Geol.*, 89 (1): 83-96.

*Sivoronov, A.A. and Malyuk, T.I. (1984). Grey gneisses and the problem of ancient continental crust. *Geol. Zhurnal*, no. 1, pp. 110-117.

Skiold, T. and Cliff, R.A. (1984). Sm-Nd and U-Pb dating of Early Proterozoic mafic-felsic volcanism in northernmost Sweden. *Precambr. Res.*, 26 (1): 1-13.

*Smirnov, A.M. (1976). *Precambrian of North-Western Pacific Mobile Belt*. Nauka, Moscow.

Smith, D.L. (1982). Review of the tectonic history of the Florida basement. *Tectonophys.*, 88 (1-2): 1-22.

*Sobotovich, E.V. (1984). Cosmochemical model of the origin of the Earth. *Geol. Zhurnal*, no. 2, pp. 112-123.

*Sollogub, V.B. and Chekunov, A.V. (1985). Essential characteristics of the structure of Ukrainian lithosphere. *Geofiz. Zhurnal*, no. 6, pp. 43-54.

Solyom, Z., Lindqvist, J.E. and Johansson, I. (1992). The geochemistry, genesis and geotectonic setting of Proterozoic mafic dyke swarms in southern and central Sweden. *Geol. Foren. Stockh. Forhandl.*, vol. 114, pt. 1, pp. 47-65.

*Sorokhtin, O.G. (1974). *Global Evolution of the Earth*. Nauka, Moscow, 184 pp.

Stacey, J.S. and Hedge, C.E. (1984). Geochronologic and isotopic evidence for Early Proterozoic crust in the eastern Arabian Shield. *Geology*, vol. 12, pp. 310-313.

Starmer, I.C. (1991). The Proterozoic evolution of the Bamble sector shear belt, southern Norway: correlations across southern Scandinavia and the Grenvillian controversy. *Precambr. Res.*, vol. 49, pp. 107-139.

Stauffer, M.L. (1984). Manikewan: an Early Proterozoic ocean in central Canada, its igneous history and orogenic closure. *Precambr. Res.*, vol. 25, pp. 257-281.

*Stepkin, E.S. and Samoilovich, Yu. G. (1984). Composition of the Late Precambrian formations in Murmansk coastal shelf, In: *Precambrian Geology of Kola Peninsula*. Apatites, pp. 119-126.

Stern, R.J. and Dawoud, A.S. (1991). Late Precambrian (740 m.y.) charnockite, enderbite and granite from Jebel Moya, Sudan: a link between the Mozambique belt and the Arabian-Nubian Shield? *J. Geol.*, vol. 99, pp. 648-659.

Stevens, B.P.J., Barnes, R.G., Brown, R.E. et al. (1988). The Willyama supergroup in the Broken Hill and Euriowie blocks. *Precambr. Res.*, vol. 40/41, pp. 297-327.

Stewart, A.D. (1982). Late Proterozoic rifting in NW Scotland: the genesis of the "Torridonian". *J. Geol. Soc. London*, vol. 139, pp. 413-420.

Stewart, J.H. (1976). Late Precambrian evolution of North America: plate tectonics implication, Geology, no. 1, pp. 11-15.

Stille, H. (1958). Assyntische tektonik in Geologischen Erdllild. *Beih. Geol. Jb.*, 2 (22).

Stille, P. and Tatsumoto, M. (1985). Precambrian tholeiitic-dacitic rock suites and Cambrian ultramafic rocks in the Pennine nappe system of the Alps—evidence from Sm-Nd isotopes and rare elements. *Contrib. Mineral. Petrol.*, 89 (2): 184-192.

Stöcklin, J. (1968). Structural history and tectonics of Iran: A review. *Amer. Assoc. Petr. Geol. Bull.*, vol. 52, pp. 1229-1258.

Stockwell, C.H. (1982). Proposals for time classification and correlation of Precambrian rocks and events in Canada and adjacent areas of the Canadian Shield, Part 1: A time classification of Precambrian rocks and events. *Geol. Surv. Canada,* Paper 80–19, 135 pp.

Stoser, D. and Camp, V. (1985). Pan-African microplate accretion of the Arabian Shield. *Geol. Soc. Amer. Bull.,* vol. 16, pp. 817–826.

Stowe, C.W., Hartnady, C.J.H. and Joubert, P. (1984). Proterozoic tectonic provinces of southern Africa. *Precambr. Res.,* 25 (1–3): 229–231.

Stratigraphy of the Precambrian Formations of the Ukrainian Shield. Naukova Dumka, Kiev (1983).

Structural and Metamorphic Petrology of the Early Precambrian of the Aldan Shield. Izdatelstvo YaF SO AN SSSR, Yakutsk (1975).

Stump, E. (1976). On the Late Precambrian-Early Palaeozoic Metavolcanic and Metasedimentary Rocks of the Queen Maud Mountains, Antarctica and a Comparison with Rocks of Similar Age from Southern Africa. Inst. of Polar Studies, Ohio State Univ., Report No. 62.

*Stupka, O.S. (1986). *Geodynamic Evolution and Structure of the Earth's Crust in the Southern European Part of the Soviet Union in the Precambrian.* Naukova Dumka, Kiev.

Suk, M. and Weiss, J. (1981). Geological section through the Variscan orogen in the Bohemian massif. *Geologie en Mijnbouw,* vol. 60, pp. 161–168.

Sun, D. and Lu, S. (1985). A subdivision of the Precambrian of China. *Precambr. Res.,* 28 (2): 137–162.

Sun Jiacong (1985). Discovery of the volcanic rock series in Chenginang formations at Luoci area and discussion on the age of the lower limit of the Sinian system, Yunnan. *Sci. Geol. Sinica,* no. 4, pp. 354–363.

*Surkov, V.S. and Zhero, O.G. (1981). *Basement and Development of the Platform Cover of the Western Siberian Platform.* Nedra, Moscow.

Symons, D.T.A. (1991). Palaeomagnetism of the Proterozoic Wathaman batholith and the suturing of the Trans-Hudson orogen in Saskatchewan. *Can. J. Earth Sci.,* vol. 28, pp. 1931–1938.

Tack, L. (1984). Post-Kibaran intrusion in Burundi. In: *UNESCO, Geology for Development, Newsletters,* no. 3, pp. 47–57.

Tankard, A.J., Jackson, M.P.A., Eriksson, K.A., Hobday, D.K., Hunter, D.K. and Minter, W.E.L. (1984). *Crustal Evolution of South Africa: 3-8 b.y. of Earth History.* Springer, New York.

Tarling, D.H. (ed.). (1978). *Evolution of the Earth's Crust.* Academic Press, New York.

Tarling, D.H. (1980). Lithosphere evolution and changing tectonic regimes. *J. Geol. Soc. Lond.,* vol. 137, p. 459.

Taylor, P.H., Kramers, J.D., Moorbath, S., Wilson, J.F., Orpen, J.L. and Martin, A. (1991). Pb/Pb, Sm-Nd and Rb-Sr geochronology in the Archaean craton of Zimbabwe. *Chem. Geol., Isotope Geosc. Sec.,* 87 (3–4): 175–196.

Tectonique du continent Africain (G. Choubert and A. Faure-Muret, eds.). Mem. Earth Sci. UNESCO, Paris (1971).

Teixera, W. and Figueiredo, M.C.H. (1991). An outline of Early Proterozoic crustal evolution in the Sao Francisco craton, Brazil: A review *Precamb. Res.,* 53(1): 1–22.

Teixera, W., Tassinari, C.C.G., Cordani, U.G. and Kawashita, K. (1989). A review of the geochronology of the Amazonian Craton: Tectonic implications. *Precambr. Res.,* vol. 42, pp. 213–227.

Thomas, M.R.A. et al. (eds.). (1991). *Geological Evolution of Antarctica.* Cambridge Univ. Press, Cambridge.

Thorpe, R.S., Beckinsale, R.D., Patchett, P.J., Piper, J.D.A., Davies, G.R. and Evans, J.A. (1984). Crustal growth and Late Precambrian-Early Palaeozoic plate tectonic evolution of England and Wales. *J. Geol. Soc. London,* 141 (3): 521–536.

Thurston, P.C. and Chivers, K.M. (1990). Secular variation in greenstone sequence development emphasising Superior province, Canada. *Precambr. Res.,* vol. 46, pp. 21–58.

*Tikhonova, L.V. (1982). Volcanosedimentary Early Proterozoic complex of Karelia (structure, tectonic nature). Ph.D. thesis, Lomonossov State Univ., Moscow.

Toghill, P. and Chell, K. (1984). Shropshire geology—stratigraphic and tectonic history. *Field Studies,* no. 6, pp. 59–101.

*Tolstikhin, I.N. (1991). Degassing and geodynamics on the 'pregeological' stage of the Earth's evolution and age. In: *The Earth's Crust. Its Composition and Age,* pp. 77–86. Nauka, Moscow.

Toteu, S.F., Michard, A., Bertrand, J.M. and Rocci, G. (1987). U-Pb dating of Precambrian rocks from northern Cameroon: Orogenic evolution and chronology of the Pan-African belt of central Africa. *Precambr. Res.,* vol. 37, pp. 71–87.

Turek, A., Smith, P.E. and Van Schmus, W.R. (1984). U-Pb zircon ages and the evolution of the Michipicoten plutonic-volcanic terrane of the Superior province, Ontario. *Can. J. Earth Sci.,* vol. 21, pp. 457–464.

Umeji, A.C. (1983). Archaean greenstone belts of Sierra Leone with comments on the stratigraphy and metallogeny. *Afr. Earth Sci. J.,* 1 (1): 1–8a.

Upper Precambrian of the Northern European USSR. Syktyvkar, Komi filial Akad. Nauk SSSR, 41 pp. (1986).

Upton, B.G. and Blundell, D.J. (1978). The Gardar igneous province: Evidence for Proterozoic continental rifting. In: *Petrology and Geochemistry of Continental Rifts,* pp. 163–172.

Vachette, M. (1979). Radiochronologie du Precambrien de Madagascar. In: *10-e Colloque Geol. Afric., Resumés,* pp. 20–21. Montpellier.

Van der Velden, W., Baker, J., De Maesschalck, S. and Van Meerten, T. (1982). Bimodal Early Proterozoic volcanism in the Grythytte field and associated volcano-plutonic complexes, Bergslagen, central Sweden. *Geol. Rdsch.,* 71 (1): 171–181.

Van Schmus, V.R. (1980). Chronology of igneous rocks associated with the Penokean orogeny in Wisconsin: In: *Selected Studies of Archaean Gneisses and Lower Proterozoic Rocks, Southern Canadian Shield,* pp. 159–168. (G.B. Morey, ed.). Geol. Soc. Amer. Spec. Pap. no. 182.

Van Schmus, W.R. (1992). Tectonic setting of the Midcontinent Rift System. *Tectonophys.,* 213 (1-2): 1–15.

Vellutini, P., Rocci, G., Vicat, J.P. and Glkan, P. (1983). Mise en evidence de complexes ophiolitiques dans la chaine du Mayombe (Gabon-Angola) et nouvelle interpretation geotectonique. *Precambr. Res.,* 22 (1-2): 1–21.

*Verkhoglyad, V.M. (1985). Plagiogranites—early phase of Kirovograd-Zhitomir complex. *Dokl. AN UkSSR,* 5 (4): 7–9.

Villeneuve, M. (1983). Les sillons tectoniques du Precambrien superieur dans l'est du Zaire: Comparaisons avec les directions du rift est-afrecain. *Bull. Centre Rech. Explor. Prod. Elf-Aquitaine,* 7 (1): 163–174.

Villeneuve, M., Bassot, J.P., Robineau, B., Dallmeyer, R.D. and Ponsard, J.P. (1991). Bassarides. In: *West African Orogens and Circum-Atlantic Correlatives,* Springer-Verlag, New York.

*Vinogradov, A.P. (1962). Origin of the Earth's layering. *Vestnik AN SSSR,* no. 9, pp. 16–29.

Virdi, N.S. (1988). Pre-Tertiary geotectonic events in the Himalayas. *Z. Geol. Wiss.,* 16 (7): 571–585.

*Voitkevich, G.V. (1983). *Origin and Chemical Evolution of the Earth.* Nauka, Moscow.

*Volobuyev, M.I. (1993). Riphean ophiolite belt of the Yenisei Ridge. *Geotektonika,* no. 6, pp. 82–87.

*Volobuyev, M.I., Zykov, S.I., Stupnikova, N.I. and Vorob'ev, I.V. (1980). Lead isotope geochronology of Precambrian metamorphic complexes of the south-western extremity of the Siberian Craton. In: *Geochronology of Eastern Siberia and the Far East,* pp. 14–30. Nauka, Moscow.

Volpe, A.M. and McDougall, J.D. (1990). Geochemistry and isotopic characteristics of mafic (Phulad ophiolite) and related rocks in the Delhi supergroup, Rajasthan, India: Implications for rifting in the Proterozoic. *Precambr. Res.*, vol. 48, pp. 165–191.

*Vorontsova, G.A., Dol'nik, T.A., Sukhanova, N.V. and Fomin, N.I. (1975). Yudomian complex of Sayan-Baikal fold region. In: *Analogues of the Vendian Complex in Siberia*, pp. 169–180. Nauka, Moscow.

*Vrevsky, A.B. and Kolychev, E.A. (1987). Tectonic evolution and metallogeny of the Archaean greenstone belts of the Kola Peninsula. In: *Geologia i Perspektivy Rudonosnosti Fundamenta Drevnikh Platform*, pp. 372–382. Nauka, Leningrad.

Wang, H. and Qiao, X. (1984). Proterozoic stratigraphy and tectonic framework of China. *Geol. Mag.*, 121 (6): 599–614.

Wang, R.-M., He, S., Chen, Z., Li, P. and Dai, F. (1985). Geochemical evolution and metamorphic development of the Early Precambrian in eastern Hebei, China. *Precambr. Res.*, vol. 27, pp. 111–129.

Ward, P. (1987). Early Proterozoic deposition and deformation at the Karelian Craton margin in Southeastern Finland. *Precambrian Res.*, vol. 35, pp. 71–93.

Wardle, R.J. and Bailey, D.G. (1981). Early Proterozoic sequences in Labrador. In: *Proterozoic Basins in Canada*, pp. 331–358. (F.H.A. Campbell, ed.). Geol. Survey Canada, Pap. 81-10.

Webb, A.W. and Horr, G. (1979). The Rb-Sr age and petrology of a flow from the Beda volcanics. *Quart. Geol. Notes, Geol. Surv. S. Austr.*, vol. 66, pp. 10–13.

Werner, K.D. (1985). The Upper Riphean ophiolites of Saxonia (GDR). In: *Riphean-Early Palaeozoic Ophiolites of Northern* Eurasia, pp. 106–119. Nauka, Novosibirsk.

Wernick, E. (1981). The Archaean of Brazil. *Earth Sci. Rev.*, vol. 17, pp. 31–48.

Wernick, E., Hasui, Y. and Brito-Neves, B.B. (1978). As regioes do dobramentos nordeste e sudeste. In: *Anais do XXX Congresso Brasil. Geol., Recife*, Vol. 6, pp. 2493–2511.

West African Orogens and Circum-Atlantic Correlatives. Springer-Verlag, New York.

Williams, E. (1978). Tasman fold belt system in Tasmania. *Tectonophys.*, 48 (3–4): 159–205.

Williams, I.S. and Collins, W.J. (1990). Granite-greenstone terranes in the Pilbara block, Australia, as *coeval* volcano-plutonic evidence from U-Pb zircon dating of the Mount Edgar batholith. *Earth Planet. Sci. Lett.*, vol. 97, pp. 41–59.

Wilson, M.R. (1982). Magma types and the tectonic evolution of the Swedish Proterozoic. *Geol. Rdsch.*, 71 (1): 120–129.

Wilson, M.R., Hamilton, P.J., Fallica, A.E., Aftalion, M. and Michard, A. (1985). Granites and Early Proterozoic crustal evolution in Sweden: evidence from Sm-Nd, U-Pb and O isotope systematics. *Earth Planet. Sci. Lett.*, vol. 72, pp. 376–388.

Windley, B.F. (ed.). 1976. *The Early History of the Earth*. John Wiley and Sons, Inc., London, New York.

Windley, B.F. (1977). *The Evolving Continents*. John Wiley and Sons, London.

Windley, B.F. (1984). The Archaean-Proterozoic boundary. *Tectonophys.*, vol. 105, pp. 43–53.

Windley, B.F. (1989). Anorogenic magmatism and the Grenvillian orogeny. *Can. J. Earth Sci.*, vol. 26, pp. 479–489.

Wit de M., Roering, C., Hart, R. et al. (1992). Formation of an Archaean continent. *Nature*, vol. 357, pp. 553–562.

Worsley, T.R., Nance, D. and Moody, J.B. (1984). Global tectonics and eustacy for the past 2 b.y. *Mar. Geol.*, vol. 58, pp. 373–400.

Wu Geyao (1986). The shear-compression episode of the Jinning cycle and its stress field in western Sichuan and eastern Yunnan. *Geol. Rev.*, 32 (4): 358.

Wyborn, L.A.J. (1988). Petrology, geochemistry and origin of a major Australian 1880–1840 Ma felsic volcanic-plutonic suite: A model for intracontinental felsic magma generation, *Precambr. Res.*, vol. 40/41, pp. 37–60.

Wynne-Edwards, H.R. (1976). Proterozoic ensialic orogenesis: the millipede model of ductile plate tectonics. *Amer. J. Sci.*, 276 (8): 927–953.

Yang, Z., Cheng, Y. and Wang, H. (1986). *The Geology of China.* Clarendon Press, Oxford.

*Yan Zhin-shin, V.A. (1983). *Tectonics of Sette-Daban Horst-Anticlinorium.* Yakutsk.

Yianping Li, Yongan Li, Robert Sharps, R., McWilliams, M. and Zhangiia, Ga.O. (1991). Sinian palaeomagnetic results from the Tarim block, western China. *Precambr. Res.,* vol. 49, pp. 61–71.

Young, G.M. (1979). Correlation of Middle and Upper Proterozoic strata of the northern rim of the North Atlantic Craton. *Trans. R. Soc. Edinb.,* vol. 70, pp. 323–336.

Young, G.M. (1980). Subdivision and correlation of the Precambrian in Canada. In: *Principles and Criteria of Subdivision of Precambrian in Mobile Zones,* pp. 75–96. Nauka, Leningrad.

Young, G.M. (1982). The Late Proterozoic Tinder group, east-central Alaska: Evolution of a continental margin. *Geol. Soc. Amer. Bull.,* vol. 93, pp. 759–783.

Young, G.M. (1984). Proterozoic plate tectonics in Canada with emphasis on evidence for a Late Proterozoic rifting event. *Precambr. Res.,* vol. 26, pp. 233–265.

Young, G.M., Jefferson, C.W., Delaney, G.D. and Yeo, G.M. (1979). Middle and Late Proterozoic evolution of the northern Canadian Cordillera and Shield. *Geology,* vol. 7, pp. 125–128.

*Zabrodin, V. Yu. (1988). Madagascar. Explanatory Note to: Tectonic Map of the World. In: *Tectonics of Continents and Oceans,* pp. 128–129. (V.E. Khain and Yu.G. Leonov, eds.). Nauka, Moscow.

*Zagorodnyi, V.G. and Radchenko, A.T. (1983). *Tectonics of the Early Precambrian of Kola Peninsula.* Nauka, Leningrad.

*Zaitsev, N.S. and Kovalenko, V.I. (eds.). (1990). *Evolution of Geological Processes and Metallogeny of Mongolia.* Nauka, Moscow.

*Zaitsev, Yu. A. (1984). *Evolution of Geosynclines (Oval Concentric Zonal Type).* Nedra, Moscow.

*Zakharov, Yu. I. and Zabiyaka, A.I. (1983). Structural-formational zoning of the Precambrian of Taimyr fold region. In: *Geology and Metamorphogenic Ore Formation of Precambrian of Taimyr,* pp. 26–49. Nauka, Leningrad.

Zhai, M.-G., Yang, R.-Y., Lu, W.-J. and Zhoi, J. (1985). Geochemistry and evolution of the Qinguyan Archaean granite-greenstone terrain. *Precambr. Res.,* vol. 27, pp. 37–62.

Zhang, G.-W., Bai, Y.-B., Sun, I., Guo, A.-L., Zhou, D.-W. and Li, T.H. (1985). Composition and evolution of the Archaean crust in central Henan, China. *Precambr. Res.,* 27 (1–3): 7–35.

Zhang, Z., Liou, J.G. and Coleman, R.G. (1984). An outline of the plate tectonics of China. *Geol. Soc. Amer. Bull.,* vol. 95, pp. 295–312.

*Zhero, O.G., Smirnov, L.V. and Surkov, V.S. (1979). Major tectonic complexes of the pre-Jurassic basement of the Western Siberian Platform. In: *Major Tectonic Complexes of Siberia,* pp. 52–67. Nauka, Novosibirsk.

*Zhuravlev, D.Z., Pukhtel', I.S., Samsonov, A.V. and Simon, A.K. (1987). Sm-Nd age of the relicts of the granite-greenstone basement of central Peri-Dnepr region. *Dokl. AN SSSR,* 201 (5): 1203–1208.

Zolnai, A.I., Price, R.A. and Helmstaedt, H. (1984). Regional cross section of the Southern province adjacent to Lake Huron, Ontario: Implications for the tectonic significance of the Murray fault zone. *Can. J. Earth Sci.,* 21 (4): 447–456.

*Zonenshain, L.P. (1974). Model of development of the geosynclinal process (on the example of the Central Asian fold belt). In: *Tectonics of the Ural-Mongolian Fold Belt,* pp. 11–35. Nauka, Moscow.

*Zonenshain, L.P., Kuzmin, M.I. and Natapov, L.M. (1990). *Geology of the USSR: A Plate-Tectonics Synthesis.* AGU Geodyn. Ser., Monogr. 21.

Zoubek, V. (1984). On the recent state of research of the Precambrian in the European Variscides. *Geol. Balcanica,* 14 (1): 57–99.

Zoubek, V. (ed.). (1988). *Precambrian in Younger Fold Belts.* Wiley, Chester Publ., New York.

INDEX

440

448

Penokean protogeosyncline 92, 101–107,
109, 168, 172–173, 176
Pensacola zone 249, 320–321, 397
Peri-Atlantic protogeosyncline 141,
150–153, 298, 363
Peri-Azov block 127–128
Peri-Baikal trough 131, 177, 275–276,
323–324, 329, 348, 393
Peri-Kolyma massif, rise 175–176, 275–276,
345
Peri-Ladoga aulacogen 323
Peri-Pennine fault 230
Pernambuco-Alagoas block, fault 286, 354
Peterman orogeny 365, 375
Petsapiscau group 187
Pharusian belt, ocean 293–294, 325, 357
Pharusian-Dahomea belt 293, 324
Phennam trough 239, 279
Phu Hoat rise, series 140, 239
Pianco-Alto Brigida zone 286
Pilbara block eocraton 24, 72, 78, 79, 85,
87, 158–159, 161, 165, 210, 247–248,
250, 382
Pine Creek fold system, protogeosyncline 162,
172, 179
Platform basins 213, 250, 322–324
Pokhal series 209
Polousnyi block 276
Polui anticlinorium 233
Pongola supergroup 60, 144, 166, 168
Porongos group 286
Poseidon ocean 227
Pound quarzite 364
Pretoria group 144–145
Preuralides 189–191, 232, 342–343
Proto-Andean belt 234, 287–288, 355–356,
376, 397–399
Proto-Iapetus 261–262, 332, 334, 394, 397
Proto-Pacific 203, 204, 324, 326, 330–331,
350–351, 376, 399
Proto-South Atlantic belt 298, 300, 303–304,
324–326
Proto-Tethys 177, 205, 265–266, 269, 298,
310, 322, 324, 326, 330, 334, 340, 342,
361, 368, 376, 391, 393, 399, 406
Proto-Urals 334, 369, 394, 397
Proto-Yangtze 203–204
Protoaulacogens 113, 170
Protogeosynclines 127–128, 135, 137–138,
140, 143, 154, 155, 161, 170–174, 390
Proto-ophiolites 83, 85–87

Protoplatforms 135, 142, 154, 167–170, 385,
391, 404
Protosyneclises 169
Pudin rise 233
Puncoviscana fold zone 287, 331
Punjab basin, series 342, 364
Purana basins 209
Purcell series 261
Purpol suite 199, 236

Qaidam massif 137–138, 203
Qiantong massif 137
Qilian massif 137
Qingbankou group, system 224, 279–280
Qinling fold belt, ocean 202–203, 281, 350
Qisyan (Yanhan) aulacogen 237
Queen Maud aulacogen 98
Quesh-Quesh group 291
Quruktagh aulacogen 137, 203, 237

Rackland orogeny 222
Rae province 32, 92–93, 97, 102, 112
Rae group 259
Raipur group 312
Rajasthan volcanoplutonic complex 210,
312–314, 372
Rakhmanov suite 338
Rayner complex 24, 75
Redkin suite 338
Reguibat massif 22, 67, 147, 149, 178, 213,
290
Rehobolt fold system
Reindeer belt 98, 112
Rennick group 318
Rewa group 312
Rhodopian massif 231, 267–268, 341
Rhodope supergroup 128, 231
Riacho do Pontal zone 286
Ribeira fold belt 55, 284, 286, 325–327,
354–355, 370, 373–375, 398
Richmond-Belcher aulacogen 100
Riga pluton 177
Rinkides 102–103
Rio das Velhas-Lafayette greenstone
belt 53–55
Rio de la Plata Craton 55, 282, 287
Rio Grande fold belt 206
Rio Negro-Juruena volcanoplutonic
belt 178, 205, 219
Rio Preto fold system 283
Roan group 303, 325, 327
Robertson Bay block, group 318–322, 367

Printed in India